MATERIALS SCIENCE AND TECHNOLOGIES

MODELING AND SIMULATION IN FIBROUS MATERIALS

TECHNIQUES AND APPLICATIONS

MATERIALS SCIENCE AND TECHNOLOGIES

Additional books in this series can be found on Nova's website under the Series tab.

Additional E-books in this series can be found on Nova's website under the E-book tab.

MATERIALS SCIENCE AND TECHNOLOGIES

MODELING AND SIMULATION IN FIBROUS MATERIALS

TECHNIQUES AND APPLICATIONS

ASIS PATANAIK
EDITOR

Nova Science Publishers, Inc.
New York

For permission to use material from this book please contact us:
Telephone 631-231-7269; Fax 631-231-8175
Web Site: http://www.novapublishers.com

Library of Congress Cataloging-in-Publication Data

Modeling and simulation in fibrous materials : techniques and applications / editors, Asis Patanaik and Rajesh D. Anandjiwala.
p. cm.
Includes bibliographical references and index.
ISBN 978-1-62100-116-4 (hardcover)
1. Textile fibers--Simulation methods. 2. Fibrous composites--Simulation methods. 3. Textile fabrics--Simulation methods. I. Patanaik, Asis. II. Anandjiwala, Rajesh D.
TS1540.M64 2011
677--dc23
2011031634

Published by Nova Science Publishers, Inc. † New York

CONTENTS

PREFACE

This book deals with the modeling and simulation techniques and its application in the field of fibrous materials. Different modeling and simulation techniques covered are: finite element analysis, computational fluid dynamics, artificial neural network, fuzzy logic, empirical and statistical modeling. Different fibrous materials dealt with this book are fibers, yarns, woven and nonwoven fabrics, nanofiber based nonwovens, and fiber- reinforced composites. Application of the above modeling and simulation techniques in manufacturing processes, prediction of properties and structure-property interaction are covered for fibers, yarns, fabrics, and composites. The predicted properties are mechanical, thermal, surface, fire, electromagnetic shielding, dielectric, transport, and comfort behavior.

This book is a good reference volume for the undergraduate to graduate level courses covering the background, current trend and applications of modeling in fibrous materials. This book is also a good source of information for a number of inter-disciplinary departments like mathematics, materials science, mechanical, chemical and textile engineering, and computer science.

The editor along with contributors of the chapters acknowledged various sources for granting permissions to reproduce some of the figures and tables used in this book. The editor would like to thank Dr. Rajesh Anandjiwala for going through some of the chapters and making many helpful suggestions.

In: Modeling and Simulation in Fibrous Materials
Editor: Asis Patanaik
ISBN: 978-1-62100-116-4

Chapter 1

INTRODUCTION TO FINITE ELEMENT ANALYSIS AND RECENT DEVELOPMENTS

B. D. Reddy* and A. T. McBride
Centre for Research in Computational and Applied Mechanics
University of Cape Town, 7701 Rondebosch, South Africa

ABSTRACT

Fiber-reinforced composite materials are composed of dispersed fibrous materials (e.g. glass, Kevlar, PET, flax, hemp, sisal, etc.) set within a continuous polymer matrix. The primary benefit of fiber-reinforced composites over traditional engineering materials comes from their impressive strength-to-weight ratio and the ability to design the micro-structure so as to optimize their macro-structural properties. These advantageous properties were first exploited by the space and aerospace industries.

Currently fiber-reinforced composites are an ubiquitous component of modern production and design for a range of products spanning exotic high technology components to more mundane household items. The focus of this chapter is on the second of the aforementioned advantages of fiber reinforced composites over traditional materials: the ability to tailor the macroscopic properties by designing the micro-structural configuration, and the role of the finite element method as a computational tool which makes such multi-scale modeling possible. We emphasize how this bottom-up approach changes the traditional computational modeling perspective where the response of the material is generally formulated upon macroscopic considerations.

The role of the finite element method in micro-macro approaches is described, and the resulting numerical considerations presented and discussed. Also discussed are the statistical techniques needed to interpret the resulting data. Some background to the relevant solid mechanics and the finite element method is presented before discussing the topics of relevance.

*E-mail: daya.reddy@uct.ac.za

1.1. INTRODUCTION

Fiber-reinforced composites have become an integral component of many modern products. The ability to specialize the design of the composite material for its end purpose presents the designer with various challenges such as choosing the optimal fiber volume fraction, fiber orientation and fiber type. Computational modeling allows designers to use virtual product prototyping to assist them in the design process. The objective of this chapter is to provide a clear overview of modern multi-scale computational modeling methodologies for fiber-reinforced composites.

By means of introduction, section 1.1.1. presents a brief overview of several of the key features of fiber-reinforced composites. Thereafter, a discussion of the multi-scale modeling methodology within the context of the finite element method is given. The Introduction concludes with an overview of the use of multi-scale modeling to determine material properties at the macro-scale. These issues will be explored in more depth in subsequent sections.

1.1.1. Features of Fiber-Reinforced Composites

Fiber-reinforced composite materials are composed of dispersed fibrous materials (e.g. glass, kevlar, polyester, flax, hemp, sisal etc.) set within a continuous polymer matrix. The fibrous material is generally of a higher strength than the matrix material. The matrix serves to bond the fibers together, to transfer the stresses due to loading to the fibers, and to protect the fibers against environmental factors. The resulting composite has desirable properties that neither the fiber nor the matrix possess alone.

A fiber-reinforced composite part is generally a laminate composed of layers of stacked fiber-matrix material that are then bonded together. The fibers can either be continuous strands or chopped segments. The orientation of the fibers in each of the layers and the fiber volume ratio can be adjusted to tailor the composite for its final application.

The high degree of flexibility in the design process allows fiber-reinforced composites to be used in a wide range of applications, including aircraft and military components, automotive components, a large variety of sporting goods, construction materials, and in medical and dental applications [1].

Fiber-reinforced composites do however have some potential drawbacks. These include high cost, brittle behavior, susceptibility to deformation under long-term loads, ultra-violet degradation, temperature and moisture effects, and a lack of design codes.

1.1.2. Computational Modeling of Fiber-Reinforced Composites

Computational modeling is now an integral part of the modern design process. It has greatly reduced design times by allowing virtual prototyping to supplement expensive experimental testing. At the heart of any computational model lies a mathematical model predicting the response of the media to applied loading. An understanding of these, often complex, mathematical models and the tools used to solve them numerically is critical for a designer to correctly interpret the results of the model. A computational model is simply that, a model, and its limitations should be understood.

An objective of this chapter is to provide greater insight into one such computational modeling methodology, termed multi-scale modeling. Multi-scale modeling allows one to imbed micro-scale phenomena within a macro-scale model using a process known as homogenization. For an excellent detailed overview of computational micro-mechanics the reader is referred to Zohdi and Wriggers [2]. Multi-scale modeling based upon homogenization has been an active area of research for at least 20 years, but the major works that led to this field becoming well understood both mathematically and computationally have appeared within the last decade (the reader is referred to the recent review article by Geers et al. [3] and references therein).

There has been significant work on the multi-scale modeling of fiber-reinforced composites; examples include the contributions by (Belsky et al. [4], Feyel and Chaboche [5], Sansalone et al.[6]) with specific focus on topics such as fracture and failure (Xia et al. [7], Xia and Curtin [8], González and LLorca [9]), viscoplasticity, (Feyel [10]), biomechanics (Maceri et al. [11]), amongst others. Their heavy computational cost, however, is one of the main reasons prohibiting their inclusion in commercial finite element software currently. This cost must be seen in perspective however; a simulation at the macro-scale that directly includes the detail of the micro-scale without the aid of some sort of homogenization procedure is computationally intractable for all but the simplest of problems.

The multi-scale modeling approach is presented here within the framework of the finite element method (see, for example, Hughes [12] and Zienkiewicz and Taylor [13] which has a chapter dedicated to multi-scale modeling, amongst numerous others, for extensive details). The finite element method is a widely used and mature tool for solving the systems of partial differential equations that typically describe the behavior of solid and fluid continuous media.

A typical finite element simulation proceeds as follows. The domain of the problem is divided into a set of non-overlapping regions termed elements. The solution of the problem is then sought in the approximate form of simple functions such as polynomials over each element. The weak or integral form of the governing equations is used to construct the approximate problem, which is linear if the problem is linear. The contributions from each element are assembled into a global matrix that represents, loosely, the stiffness of the system.

A key step in this procedure is prescribing the material or constitutive model, that is, the relationship between the stress the material experiences and the resulting deformation it undergoes. Conventional macro-scale finite element simulations assume that the material can be described by measurable macroscopic material properties. Typical examples for solids include materials that are modeled as elastic, viscoelastic, plastic, and viscoplastic. The presentation in this chapter will be confined to linear elastic materials.

Multi-scale modeling makes no such assumptions about the underlying constitutive model. Rather, essential features of the micro-scale model are directly linked to the macro-scale model via a homogenization process. The constitutive relationship at the macro-scale is thus allowed to develop from the microscopic behavior. This necessitates the solution of a separate micro-scale model at selected points such as quadrature points, in the macroscopic body.

The objective of the micro-macro approach is to obtain effective material properties that characterize, in an averaged sense and at the macroscopic level, the underlying microscopic details.

The size of the micro-scale model is determined via the concept of a representative volume element (RVE). The RVE should represent the smallest sample at the micro-scale capable of capturing the behavior accurately. If the RVE is too small then a biased and unrepresentative view of the micro-structure is obtained. If the RVE is too large then computational effort is wasted. The procedure to determine the optimal RVE size is based upon physical measurement and numerical tests. Methodologies to determine the optimal RVE size have been presented by various authors (see, for example, Kouznetsova [14] and Zohdi and Wriggers [15]) and will be elaborated on further in this work.

To clarify issues, consider the example of a thermo-mechanically loaded plate presented by Ozdemir et al. [16]. The plate is made of boron fiber reinforced aluminum (see Figure 1.1). The fibers are unidirectionally oriented parallel to the *z*-axis. The plate is clamped on its side surfaces and exposed to a rapidly increasing uniform temperature and mechanical load on the top surface. A plane-strain assumption is used to model the plate. The unidirectional orientation of the uniform fibers in this case makes the determination of the RVE straightforward: it is simply a volume surrounding a fiber cross-section. The ratio of the depth of the plate to the length of the RVE is approximately 182, indicating a clear separation of scale. Figure 1.2 shows the evolution of temperature and plastic strain after 10.0 s of simulation. A key motivation for adopting such a multi-scale model is to be able to capture the highly anisotropic plastic strain distribution shown at point B. One could imagine the extraordinary computational power that would be required to directly account for each fiber directly within the macro-scale model, as would be the case using a conventional macroscopic finite element approach.

The linkage between the scales in the multi-scale framework is based on two key properties. Firstly, the micro-scale features are assumed to be significantly smaller than the macro-scale; that is, we have a separation of scales. Secondly, there is an equivalence between the work done at the micro- and macro-scales. These principles will be elaborated upon further in later sections.

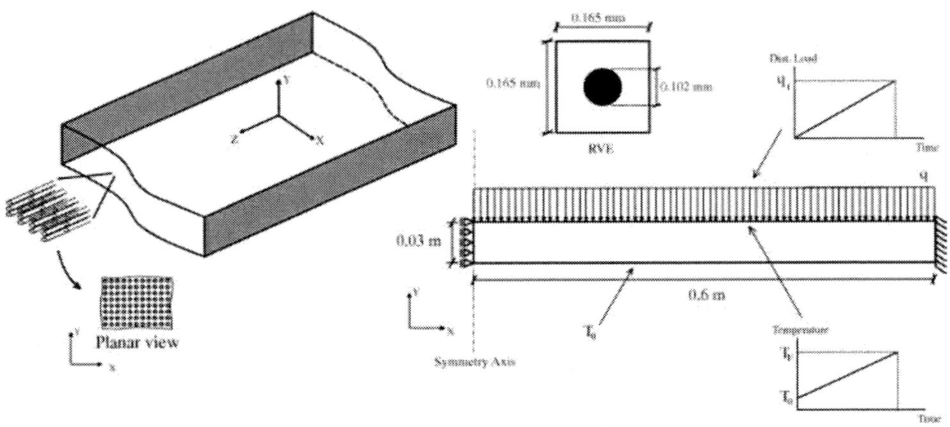

Figure 1.1. Thermo-mechanically loaded plate; geometry, boundary conditions and RVE [Source: Reference [16].

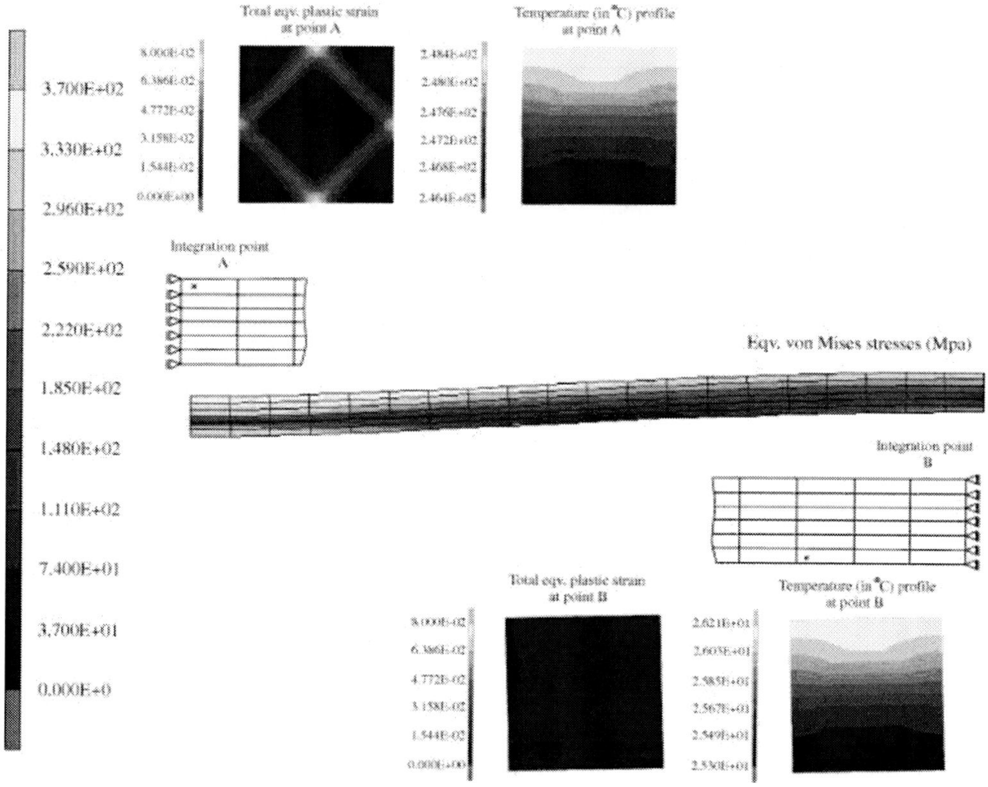

Figure 1.2. Two-scale solution via computational homogenisation at t = 10.0 s [Source: Reference [16].

1.1.3. Determining Material Properties using a Multi-Scale Framework

A key application of the multi-scale modeling formulation discussed previously is to determine numerically the appropriate macro-scale material parameters for use in a macro-scale model, see for example Zohdi and Wriggers [15]. The solution of the macro-scale model can then be performed using mature finite element software in a fraction of the time that it would take to do a full multi-scale simulation. The motivation for adopting this strategy would be to capture as closely as possible the micro-scale material parameters, for use at the macro-scale.

Consider the example of a non-woven needle-punched micro-structure consisting of randomly distributed fibers, as shown in Figure 1.3. The macro-scale response would be isotropic as there is no preferred fiber direction and, if the deformations were sufficiently small, could be approximated as a linear elastic material. Using this methodology, the designer of the fiber-reinforced composite could use a multi-scale methodology to determine, for example, the optimal fiber fraction and fiber type so as to satisfy various criteria. In this approach, a series of micro-scale finite element simulations are performed and the results analyzed using statistical tools to determine the material properties. A rigorous procedure to perform such a series of micro-scale test has been presented by Zohdi and Wriggers [15] and will be elaborated on further in this chapter. The implementation of this procedure within a commercial finite element package and a discussion of the results will also be presented.

Figure 1.3. Scanning electron microscope image of a randomly distributed fibre network [Source: Council for Scientific and Industrial Research (CSIR), Port Elizabeth].

Notation. We will use boldface italic letters to denote vectors and tensors. We adopt the summation convention for repeated indices, unless stated otherwise. Most often, vectors are denoted by lowercase boldface italic letters, and second-order tensors, or 3×3 matrices, by lowercase boldface Greek letters. Fourth-order tensors are usually denoted by uppercase boldface italic letters. We will make use of a Cartesian coordinate system with an orthonormal basis $\{e_1, e_2, e_3\}$. Where it is necessary to show components of a vector or a tensor, these will always be relative to the orthonormal basis $\{e_1, e_2, e_3\}$. Throughout this work we will identify a second-order tensors τ with a 3×3 matrix. We will always use a_i, $1 \le i \le 3$, to denote the components of the vector a, and τ_{ij}, $1 \le i, j \le 3$, the components of the second-order tensor τ. With the basis defined, the action of the second-order tensor τ on the vector a may be represented in the form:

$$\tau a = \tau_{ij} a_j e_i.$$

The scalar products of two vectors a and b, and of two second-order tensors (or matrices) σ and τ, are denoted by a · b and σ: τ and have the component representations:

$$a \cdot b = a_i b_i, \ \sigma : \tau = \sigma_{ij} \tau_{ij}.$$

The magnitudes of a vector a and a second-order tensor τ are defined by:

$$|a| = (a.a)^{\frac{1}{2}}, \ |\tau| = (\tau:\tau)^{\frac{1}{2}}.$$

The tensor product $a \otimes b$ of two vectors a and b is a second-order tensor defined by the relation:

$(a \otimes b)c = (b \cdot c)a \; \forall c.$

Viewed as a matrix, we have the representation:

$a \otimes b = ab^T$.

Since we will be working with a fixed basis, there is little point in making a formal distinction between the tensor τ and the 3×3 matrix of its components, so that unless otherwise stated, τ will represent the tensor *and* the matrix of its components. With this understanding, it is merely necessary to point out that all the usual matrix operations such as addition, transposition, multiplication, inversion, and so on, apply to tensors, and the standard notation is used for these operations. Thus, for example, τ^T and τ^{-1} are, respectively, the transpose and inverse of the tensor (or matrix) τ. Every second-order tensor τ may be additively decomposed into a deviatoric part τ^D and a spherical part τ^S; these are defined by:

$\tau^S = \frac{1}{3}(\mathrm{tr}\tau)I$, $\tau^D = \tau - \frac{1}{3}(\mathrm{tr}\tau)I$, so that
$\tau = \tau^D + \tau^S$.

The invariants of a tensor are defined by:

$I_1 = \mathrm{tr}\tau = \tau_{ii} = \tau_1 + \tau_2 + \tau_3,$
$I_2 = \frac{1}{2}\{(\mathrm{tr}\tau)^2 - \mathrm{tr}\tau^2\} = \frac{1}{2}(\tau_{ii}\,\tau_{jj} - \tau_{ij}\,\tau_{ji}) = \tau_1\,\tau_2 + \tau_2\,\tau_3 + \tau_3\,\tau_1,$
$I_3 = \det \tau = \tau_1\,\tau_2\,\tau_3.$

Here, τ_1, τ_2, and τ_3, are the eigenvalues of τ. The eigenvalues τ_i of a matrix τ are called the principal components of τ.

Scalar, vector, and tensor fields. The gradient of a scalar field $\phi(x)$ is denoted by $\nabla\phi$ and is the vector defined by:

$$\nabla\phi = \frac{\partial\phi}{\partial x_i} e_i$$

The divergence div u and gradient ∇u of a vector field $u(x)$ are respectively a scalar and a second-order tensor field, defined by:

$$div\ u = \frac{\partial u_i}{\partial x_i}$$
$$(\nabla u)_{ij} = \frac{\partial u_i}{\partial x_j}$$

The divergence div τ of a second-order tensor τ is a vector with components:

$$(div\ \tau)_i = \frac{\partial \tau_{ij}}{\partial x_j}$$

1.2. Continuum Mechanics and Linear Elasticity

The continuum approach to the description of mechanical behavior starts with the assumption that a body at the macroscopic level may be regarded as composed of material that is continuously distributed. Such a body occupies a region of three-dimensional space. The region occupied by the body will of course vary with time as the body deforms. The region occupied by the body in the reference configuration at the time $t = 0$ is denoted by Ω, and a material point may be identified by the position vector x. The properties and the behavior of the body can be described in terms of functions of position x in the body and time t. The motion is orientation-preserving; that is, the *Jacobian* J, defined by $j = \det(\partial y_i/\partial x_j)$, must be positive. Hence, every element of nonzero volume in Ω is mapped to an element of nonzero volume in Ω_t (Figure 1.4).

Introduce the *displacement* vector u by:

$$u = y(x,t).$$

The strain tensor, defined by:

$$\eta(u) = \frac{1}{2}[\nabla u + (\nabla u)^T + (\nabla u)^T \nabla u],$$

measures deformation in the body, and is zero if the body undergoes a rigid body motion.

The components η_{ij} may be interpreted as follows: η_{11} equals half the net *change in length* (squared) of a material fiber originally oriented so that it points in the x_1 direction, and the other two diagonal components of the strain are interpreted in a similar way. The off-diagonal component η_{12} gives a measure of the *change in angle* between two fibers originally at right angles to each other and oriented so that they were in the x_1 and x_2 directions, and the remaining off-diagonal components are interpreted in a similar way. The diagonal components are referred to as *direct strains*, while the off-diagonal components are referred to as *shear strains*.

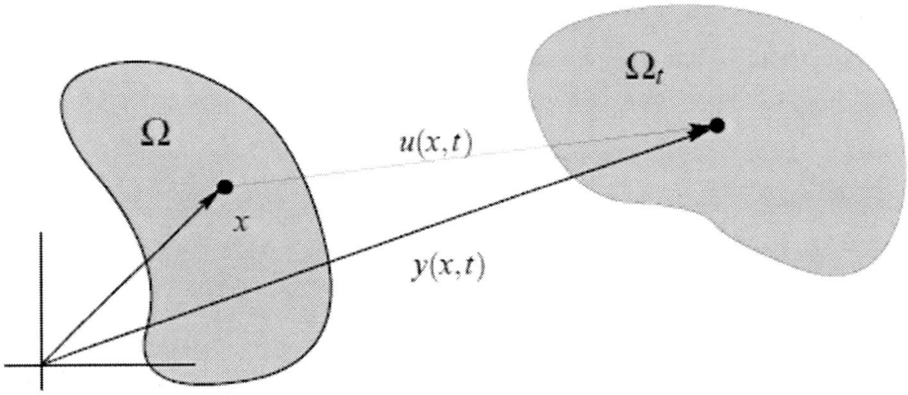

Figure 1.4. Current and undeformed configurations of an arbitrary material body.

Infinitesimal strain. A body is said to undergo infinitesimal deformation if the displacement gradient is sufficiently small so that nonlinear terms can be neglected. When this is the case, we may replace the strain tensor η by the infinitesimal strain ε, which is defined by:

$$\varepsilon = \frac{1}{2}(\nabla u + (\nabla u)^T) \tag{1.1}$$

For infinitesimal deformations the change in volume per unit volume is:

$$\frac{\nabla V}{V} = div\, u. \tag{1.2}$$

An incompressible material is one which is unable to undergo any volume change. For bodies comprising such materials, the displacement must therefore satisfy the condition:

$$div\, u = 0. \tag{1.3}$$

Balance of momentum; stress. Suppose that the body is subjected to a system of forces, which are of two kinds. There is the *body force* $b(x, t)$, which represents the force per unit reference volume exerted on the material point x at time t by agencies external to the body; gravity is a canonical example, the body force in this case being $\rho g e$, where g is the gravitational acceleration, ρ is the mass density, and e is the unit vector pointing in the downward vertical direction. The second kind of force acting on any surface in the body or on its boundary is the *surface traction*, s_n, which denotes the force per unit area acting on a surface with outward unit normal vector n. Cauchy's Theorem states that there exists a second-order tensor or matrix σ with the property that the surface traction on a surface with outward unit normal n is given by:

$$\sigma n = s_n$$

The tensor σ is known as the Cauchy stress.

BALANCE OF LINEAR MOMENTUM. The total force acting on an arbitrary part Ω_t' of the body Ω_t is equal to the rate of change of the linear momentum of Ω_t'; expressed in terms of integrals over the reference configuration,

$$\int_{\Omega'} \rho \ddot{u}\, dx = \int_{\Omega'} b\, dx + \int_{\Gamma'} s_n ds \tag{1.4}$$

An immediate consequence of balance of linear momentum is that the stress satisfies the equation of motion:

$$div\, \sigma + b = \rho \ddot{u}$$

For situations in which all the given data are independent of time, the response of the body will also be independent of time. In this case the equation of motion becomes the *equation of equilibrium:*

$$div\,\sigma + b = 0$$

These equations are valid in the current configuration, but since infinitesimal deformations are assumed throughout, it suffices to solve these equations on the reference domain.

Balance of angular momentum. The total moment acting on Ω_t' is equal to the rate of change of the angular momentum of Ω_t'; expressed in terms of integrals over the reference configuration,

$$\int_{\Omega'} x\wedge \rho\ddot{u}\,dx = \int_{\Omega'} x\wedge b\,dx + \int_{\Gamma'} x\wedge s_n ds$$

An immediate consequence of balance of angular momentum is that the stress tensor is symmetric, i.e.

$$\sigma^T = \sigma \tag{1.5}$$

In summary then, the principles of balance of linear and angular momentum are:

$$div\,\sigma + b = \rho\ddot{u} \tag{1.6}$$

$$\sigma^T = \sigma \tag{1.7}$$

Boundary conditions. In addition to the governing equations, which must be satisfied at every point in the body, it is also necessary to specify a set of boundary conditions. These are of two kinds: a Dirichlet or essential boundary condition, in which the displacement is specified to be equal to a prescribed value on a part Γ_D of the boundary Γ; and a Neumann or natural boundary condition, in which the surface traction is specified on the complementary part Γ_N of the boundary. Thus the boundary conditions are:

$$\begin{aligned} u &= \bar{u} \text{ on } \Gamma_D \\ \sigma n &= \bar{s} \text{ on } \Gamma_N \end{aligned} \tag{1.8}$$

It is possible that no natural boundary condition is specified, in which case Γ_D is the entire boundary Γ. But the converse, that is, of no essential boundary condition, is not considered as such a body could not be in equilibrium, not being fixed at any point on its boundary.

Linearly elastic materials. A body is *linearly elastic* if the stress depends linearly on the infinitesimal strain, that is, if the stress and strain are related to each other through an equation of the form:

$$\sigma = C\varepsilon \tag{1.9}$$

where C, called the *elasticity tensor*. If the density ρ and the elasticity tensor C are independent of position, the body is said to be *homogeneous*. The constitutive equation (1.9) has the component form:

$$\sigma_{ij} = C_{ijkl}\varepsilon_{kl}$$

It is often the case that materials possess preferred directions or symmetries. For example, timber can be regarded as an orthotropic material, in the sense that it possesses particular constitutive properties along the grain and at right angles to the grain of the wood. The greatest degree of symmetry is possessed by a material that has no preferred directions; that is, say, its response to a force is independent of its orientation. This property is known as isotropy, and a material with such a property is called *isotropic*. For an isotropic material the constitutive equation (1.9) can be written in terms of only two material constants. The stress-strain relation in this case is given by:

$$\sigma = \lambda\ (tr\varepsilon)I + 2\ \mu\varepsilon$$

For the purpose of interpreting the moduli, we recall that any second-order tensor τ may be written in the form:

$$\tau = \tau^{D} + \tau^{S}$$

where the deviatoric and spherical parts τ^{D} and τ^{S} of τ are defined, respectively, by:

$$\tau^{D} = \tau - \frac{1}{3}(\mathrm{tr}\tau)\mathrm{I},\ \tau^{S} = \frac{1}{3}(\mathrm{tr}\tau)\mathrm{I} \tag{1.10}$$

The constitutive equation can then be written in the uncoupled form:

$$\sigma^{D} = 2\ \mu\varepsilon^{D} \tag{1.11}$$

$$\sigma^{S} = \lambda\ (tr\varepsilon)I^{s} + 2\ \mu\varepsilon^{s} = 3\ (\lambda + \tfrac{2}{3}\ \mu)\varepsilon^{s} \tag{1.12}$$

The scalar μ is also known as the *shear modulus* (for reasons that are evident from (1.11), while the material coefficient $K \equiv \lambda + \frac{2}{3}\ \mu$ is known as the *bulk modulus* because it measures the ratio between the spherical stress and volume change. Thus an alternative pair of elastic coefficients to the Lamé moduli is $\{\mu, K\}$. Note that the shear modulus is often denoted by G, especially in the engineering literature.

Yet another important alternative pair of material coefficients arises from direct consideration of the behavior of the length of an elastic rod when it is subjected to a uniaxial stress. Suppose that the Cartesian axes are aligned in such a way that an isotropic elastic rod lies parallel to the x_1-axis and is subjected to a uniform stress with $\sigma_{11} \neq 0$ and all other components being zero (Figure 1.5). The effect will be that the rod experiences only direct strains, on account of its isotropy.

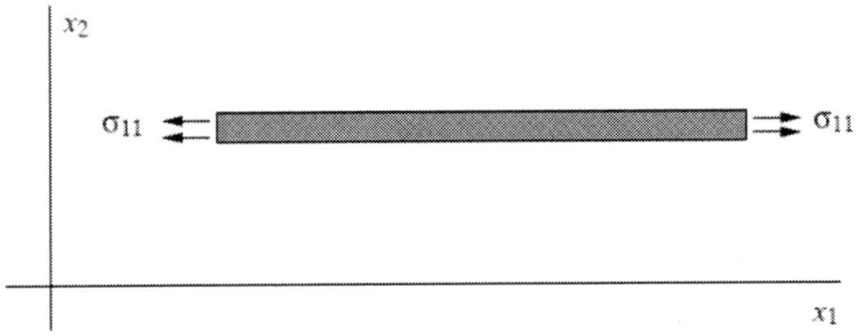

Figure 1.5. A rod in a state of uniaxial stress.

We are interested here first in the ratio $\sigma_{11}/\varepsilon_{11}$ and second in the ratio $\varepsilon_{22}/\varepsilon_{11}$, or, equivalently, $\varepsilon_{33}/\varepsilon_{11}$. The associated material coefficients are known, respectively, as *Young's modulus* and *Poisson's ratio*:

Young's modulus $E = \frac{\sigma_{11}}{\varepsilon_{11}}$

Poisson's ratio $\nu = -\frac{\varepsilon_{22}}{\varepsilon_{11}}$

Thus Young's modulus measures the slope of the stress–strain curve and is analogous to the stiffness of a spring, while Poisson's ratio measures lateral contraction.

The constitutive relation (1.9) can be expressed in terms of E and ν as follows:

$$\varepsilon = E^{-1}\left[(1+\nu)\sigma - \nu(tr\sigma)I\right]$$

Finally, the bulk and shear moduli K and G are given in terms of E and ν by:

$$K = \frac{E}{3(1-2\nu)},\ G = \frac{E}{2(1+\nu)} \tag{1.13}$$

Weak formulation of the problem of elasticity. With a view to using the finite element method to obtain solutions to the problem for elastic bodies, it is necessary to convert the boundary value problem (1.7) to what is known as a weak formulation. To this end, let w be an arbitrary displacement which satisfies the homogeneous essential boundary condition, i.e.

$$w = 0 \text{ on } \Gamma_D \tag{1.14}$$

Now, take the scalar product of the equilibrium equation (1.7) with w and integrate this equation over the domain Ω, this gives:

$$\int_\Omega div\sigma.w\, dV = \int_\Omega b.wdV \tag{1.15}$$

Next, use the divergence theorem to transform the integral on the left hand side as follows:

$$\int_\Omega div\sigma.w\, dV = \int_\Gamma \sigma n.wdA - \int_\Omega \sigma\colon\nabla wdV \tag{1.16}$$

Now, the test function w satisfies $w = 0$ on part of the boundary Γ_D, and on the other part Γ_N the surface traction is given by (1.8). Noting also Cauchy's theorem, it follows that (1.16) can be written as:

$$\int_\Omega div\sigma . w\, dV = \int_{\Gamma_N} \bar{s}. wdA - \int_\Omega \sigma : \nabla w dV \qquad (1.17)$$

Finally, because the stress is symmetric we have $\sigma : \nabla w = \sigma : (\frac{\nabla u + (\nabla u)^T}{2}) = \sigma : \varepsilon(w)$. Putting all of this together, the boundary value problem can now be formulated in weak form as follows: find the displacement u which satisfies $u = \bar{u}$ on Γ_D, and

$$\int_\Omega \sigma : \varepsilon(w) dV = \int_\Omega b. wdV + \int_{\Gamma_N} \bar{s}. wdA \text{ for arbitrary displacement w} \qquad (1.18).$$

It can be shown that under mild conditions, the classical form (1.7) and the weak form (1.18) form are equivalent. The latter will be used to construct finite element approximations.

Voigt notation. It will be convenient when carrying out the finite element formulation to convert all tensorial quantities to Voigt notation. This is simply a way of expressing the components of stress and strain as column vectors, with corresponding modifications to the governing and other equations. Thus the stress and strain are written in Voigt notation as:

$$\sigma = [\sigma_{11}\sigma_{22}\sigma_{33}\sigma_{12}\sigma_{13}\sigma_{23}]^T$$
$$\varepsilon = [\varepsilon_{11}\varepsilon_{22}\varepsilon_{33}2\varepsilon_{12}2\varepsilon_{13}2\varepsilon_{23}]^T \qquad (1.19)$$

Note the factor 2 in the shear strains: this is so that the quantity $\sigma : \varepsilon = \sigma_{ij} : \varepsilon_{ij}$ can be written conveniently in Voigt notation as:

$$\sigma^T \varepsilon \text{ or } \varepsilon^T \sigma$$

Note also that no distinction is made between the original tensor notation and Voigt notation for σ and ε; the context will make clear which version is being used.

Next, consider the expression

$$[div\ \sigma] = \begin{pmatrix} \frac{\partial \sigma_{11}}{\partial x_1} + \frac{\partial \sigma_{12}}{\partial x_2} + \frac{\partial \sigma_{13}}{\partial x_3} \\ \frac{\partial \sigma_{21}}{\partial x_1} + \frac{\partial \sigma_{22}}{\partial x_2} + \frac{\partial \sigma_{23}}{\partial x_3} \\ \frac{\partial \sigma_{31}}{\partial x_1} + \frac{\partial \sigma_{32}}{\partial x_2} + \frac{\partial \sigma_{33}}{\partial x_3} \end{pmatrix}$$

with Voigt notation this is easily written (noting also the symmetry of σ) as:

$$div\ \sigma = \nabla_s \sigma$$

where the 3×3 matrix ∇_s of partial derivatives is defined by:

$$\nabla_s = \begin{pmatrix} \frac{\partial}{\partial x_1} & 0 & 0 & \frac{\partial}{\partial x_2} & \frac{\partial}{\partial x_3} & 0 \\ 0 & \frac{\partial}{\partial x_2} & 0 & \frac{\partial}{\partial x_1} & 0 & \frac{\partial}{\partial x_3} \\ 0 & 0 & \frac{\partial}{\partial x_3} & 0 & \frac{\partial}{\partial x_1} & \frac{\partial}{\partial x_2} \end{pmatrix}$$

The strain can be written in terms of displacements using (1.1) and the same matrix ∇_s, to give:

$$\varepsilon(u) = \nabla_{\mathrm{s}}^{\mathrm{T}}\ u$$

With these additions, the equilibrium equation becomes:

$$\nabla_s \sigma + b = 0$$

Finally, the elastic law for isotropic materials can be written in Voigt form as:

$$\sigma = D\varepsilon = D\nabla_s u \tag{1.20}$$

in which the 6×6 elasticity matrix takes the form:

$$D = \frac{E}{(1+\nu)(1-2\nu)} \tag{1.21}$$

$$\times \begin{pmatrix} 1-\nu & \nu & \nu & 0 & 0 & 0 \\ \nu & 1-\nu & \nu & 0 & 0 & 0 \\ \nu & \nu & 1-\nu & 0 & 0 & 0 \\ 0 & 0 & 0 & (1-2\nu)/2 & 0 & 0 \\ 0 & 0 & 0 & 0 & (1-2\nu)/2 & 0 \\ 0 & 0 & 0 & 0 & 0 & (1-2\nu)/2 \end{pmatrix} \tag{1.22}$$

Thus the weak form (1.18) becomes:

$$\int_\Omega (\nabla_{\mathrm{s}}^{\mathrm{T}}\ w)^T D\nabla_s u\, dV = \int_\Omega w^T b dV + \int_{\Gamma_{\mathrm{N}}} w^T \bar{s}.dA \tag{1.23}$$

1.2.1. The Finite Element Method

In this section we give a brief introduction and overview of those aspects of the finite element method that are relevant to micro-macro modeling. A detailed treatment may be found, for example, in [13].

The point of departure of the finite element method is the weak formulation (1.23) and the Galerkin method, in which an approximate solution of the weak problem is sought. The finite element method is in turn a systematic approach to developing approximate solutions using the Galerkin method. Though the theory to be presented is applicable in three dimensions, for simplicity we will carry out the presentation in two dimensions.

The first step is to write the approximate displacement, which we denote also by u, as a linear combination of R basis or shape functions N_i; i.e.

$$u = N_1 d_1 + N_2 d_2 + \cdots + N_R d_R = Nd \tag{1.24}$$

where the matrix N and row vector d are defined by:

$$N = \begin{pmatrix} N_1 & 0 & N_2 & 0 & \cdots & N_R & 0 \\ 0 & N_1 & 0 & N_2 & \cdots & 0 & N_R \end{pmatrix}, \; d = [d_1 d_2 \ldots d_R]$$

Thus d is a $2R\times1$ vector with entries d_1, d_2, etc., being 2×1 vectors of the unknown coefficients or degrees of freedom, which will need to be solved for. In the same way the arbitrary displacement w can be expressed in the form:

$$w = Nq$$

where q are its degrees of freedom. Next, the strain components are obtained from:

$$\varepsilon(u) = \begin{pmatrix} \frac{\partial u_1}{\partial x} \\ \frac{\partial u_2}{\partial y} \\ \frac{\partial u_2}{\partial x} + \frac{\partial u_1}{\partial y} \end{pmatrix} = \nabla_s^{\mathrm{T}} u = \nabla_s^{\mathrm{T}} N d = Bd$$

where the 3×2R matrix B is given by:

$$B = \nabla_s^{\mathrm{T}} N$$

and whose entries are either 0 or terms of the form $\frac{\partial N_i}{\partial x_j}$. Substituting these quantities into the weak form (1.23), we obtain the equation:

$$c^T \left(\int_\Omega B^T D B \, dV \right) d = c^T \left(\int_\Omega N^T \bar{b} \, dV + \int_{\Gamma_N} N^T \bar{s} . dA \right) \tag{1.25}$$

or, in matrix form,

$$\mathrm{Kd} = \mathrm{F}$$

where the stiffness matrix K and load vector F are defined by:

$$K = \int_\Omega B^T D B \, dV \, , F = \int_\Omega N^T \bar{b} \, dV + \int_{\Gamma_N} N^T \bar{s} . dA \tag{1.26}$$

It can be shown that K is invertible, hence the solution to the approximate problem is found from:

$$d = K^{-1}F$$

and the approximate displacement and stress can now be constructed from

$$u = Nd, \sigma = DBd$$

The finite element method is an approach that allows these steps to be carried out in a way that allows for easy construction of N and refinement of the approximation.

The finite element mesh. We start by partitioning the domain W into a finite number E of subdomains $\Omega_1, \Omega_2, \ldots, \Omega_E$, called *finite elements*. These elements are non-overlapping and cover Ω, in the sense that:

$$\Omega_e \cap \Omega_f = \emptyset \qquad \text{for } e \neq f, \quad \cup_{e=1}^{E} \overline{\Omega_e} = \overline{\Omega}$$

To avoid complicating matters unnecessarily, we assume that the domain Ω is *polygonal* in two dimensions, and polyhedral in three. That is, the boundary Γ of Ω is made up of straight segments in two dimensions and planar surfaces in three. This is illustrated in Figure 1.6 for the two-dimensional case.

It is required that every side (or surface in three dimensions) of the boundary of an element be *either* part of the boundary Γ, *or* a side of another element.

Nodal points. We next identify certain points called *nodes* or *nodal points* in the subdivided domain. Nodes are allocated at least at the vertices of elements, as shown in Figure 1.7(a), but in order to improve the approximation, further nodes may be introduced, for example at the midpoints of the sides of elements as shown in Figure 1.7(b). In any case there is a total of G nodes, say, which are numbered 1, 2, . . . ,G and which have position vectors x_1, x_2, ...,x_G. The set of elements and nodes that make up the domain Ω is called a *finite element mesh*.

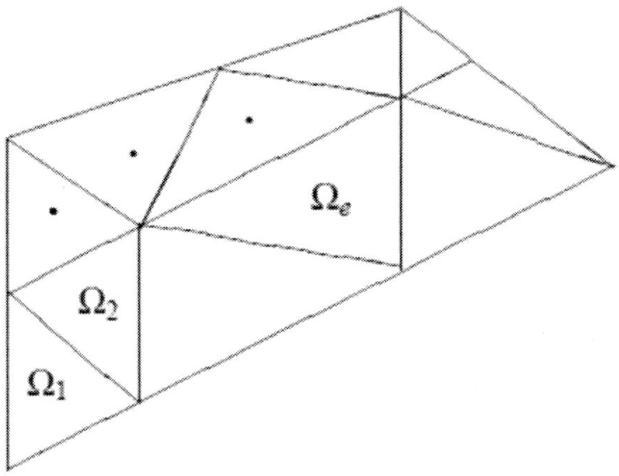

Figure 1.6. A polygonal domain in $\mathbb{R}^2$ and its subdivison into finite elements.

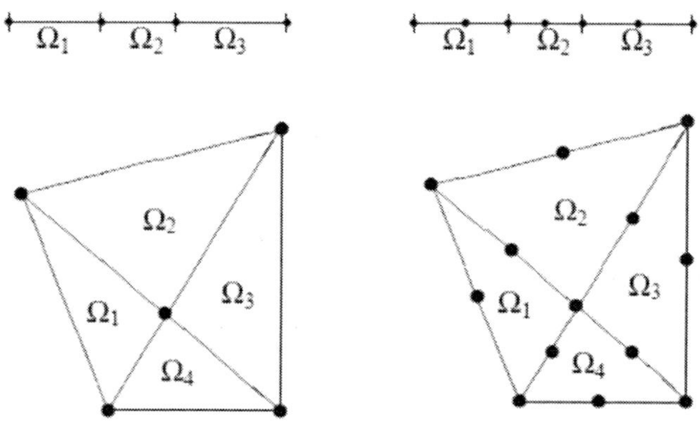

(a) (b)

Figure 1.7. Finite element meshes comprising elements and nodal points.

Basis functions. N_i. We construct the shape functions so that they have the following properties:

I. the functions N_i are *continuous*;
II. there is a total of G basis functions, that is, one for each node, and each function N_i is nonzero only on those elements that are connected to node i:
III. N_i is equal to 1 at node i, and equal to zero at the other nodes:

$$N_i(x_j) = \begin{cases} 1 \text{ if } i = j \\ 0 \; and \text{otherwise;} \end{cases} \tag{1.27}$$

IV. the *restriction* $N_i^{(e)}$ of N_i to an element Ω_e is a polynomial of degree $k \geq 1$.

From (iii) and (iv) it is clear that the function $N_i^{(e)}$ defined on element Ω_e will have the property that

$$N_i^{(e)}(x_j) = \begin{cases} 1 \text{ if } i = j \\ 0 \; and \text{otherwise} \end{cases} \tag{1.28}$$

i and j running over all nodes in Ω_e. We call $N_i^{(e)}$ a *local basis function*. These ideas are illustrated in Figure 1.8.

It should be clear that a typical shape function N_i is built up by patching together the local basis functions $N_i^{(e)}$ associated with node i.

To distinguish the shape functions N_i from the local shape functions $N_i^{(e)}$, we refer to the former as *global shape functions*.

From (1.24) and (1.27),

$$u(x_j) = \sum_{i=1}^{G} d_i \, N_i(x_j) = d_j \tag{1.29}$$

i.e. the coefficient d_j is simply *the value of* u_h *at node* j.

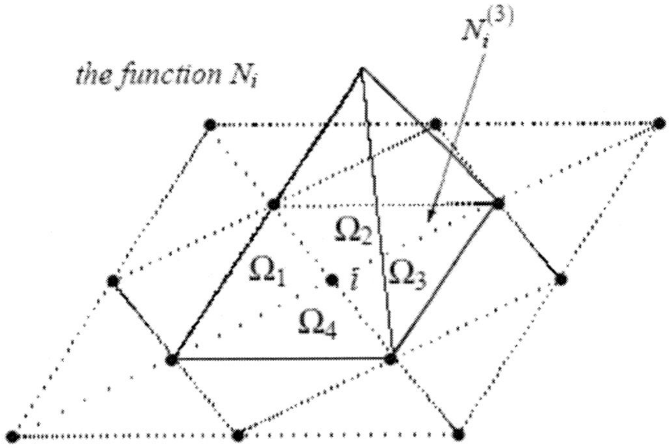

Figure 1.8. Local and global shape functions.

The approximate solution. Recall that the stiffness matrix K is defined by

$$K = \int_\Omega B^T DB\, dV \tag{1.30}$$

Now recognizing that the integral in (1.30) can be carried out element by element, we write:

$$\begin{aligned} K &= \int_\Omega B^T DB\, dV \\ &= \sum_{e=1}^{E} \underbrace{\int_{\Omega_e} B^{eT} DB^e\, dV}_{K^e} \end{aligned}$$

where K^e is the element stiffness matrix, corresponding to the contribution of element e to K.

Likewise, the load vector F can be evaluated at element level, to give;

$$F = \sum_{e=1}^{E} F^e$$

where

$$F^e = \int_{\Omega_e} N^{eT} b\, dV + \int_{\Gamma_N^e} N^{eT}\ s dA$$

Since $N_i = 0$ for all elements which do not have node i as a node, clearly $K_{ij}^{(e)} = 0$ *if nodes i and j do not belong to* Ω_e. It follows that a judicious numbering of nodes will result in the matrix K *having* a *banded* structure in which all nonzero entries are clustered around the main diagonal. From a computational viewpoint this represents a distinct advantage.

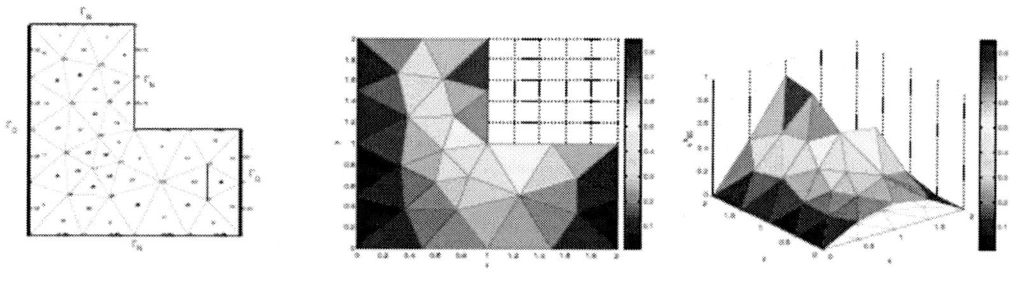

(a) Finite element mesh (b) Finite element solution (c) Finite element solution

Figure 1.9. A finite element mesh and piecewise-linear solution.

As an example, the simplest element is one which yields piecewise-linear approximations. Such an element has nodes only at the vertices, in two and three dimensions. This is illustrated in Figures 1.9, in two dimensions.

This positioning of nodes ensures that if any of the sides of Ω_e is shared with an adjacent element Ω_f, say, the piecewise linear function formed by patching together the functions defined on Ω_e and Ω_f *will be continuous across the interface of these elements*.

It is convenient to adopt a local numbering system when evaluating the element stiffness matrix and load vector, in which the nodes are numbered in a counter clockwise direction, starting with node 1. Once the element stiffness matrix and load vector have been evaluated, the components can then simply be placed in the correct rows and columns of the global matrix and vector by recalling the global node numbers of the element. The process whereby $K^{(e)}$ and $F^{(e)}$ are computed for each element, and then added to the global matrix, is known as *assembly*.

Rather than work directly on the arbitrary element, a reference element $\widehat{\Omega}$ is set up and shape functions defined on it. These are then easily mapped onto the actual element. The reference element is a right-angled isosceles triangle (Figure 1.10), and the transformation:

$$x = x_1^{(e)}\ (1 - \xi - \eta) + x_2^{(e)}\xi + x_3^{(e)}\eta$$

$$y = y^{(e)}(1 - \xi - \eta) + y_2^{(e)}\xi + y_3^{(e)}\eta$$

or

$$x = F_e(\xi) \equiv (1 - \xi - \eta)x_1^{(e)} + \xi x_2^{(e)} + \eta x_3^{(e)}$$

maps the nodal points 1, 2, 3 of $\widehat{\Omega}$ to local nodal points 1, 2, 3 of Ω_e, and indeed maps each point $\xi \in \widehat{\Omega}$ to a point $x \in \Omega_e$.

The local basis functions on $\widehat{\Omega}$ must satisfy (1.28), and are:

$$\widehat{N_1}(\xi) = 1 - \xi - \eta$$
$$\widehat{N_2}(\xi) = \xi$$
$$\widehat{N_3}(\xi) = \eta$$

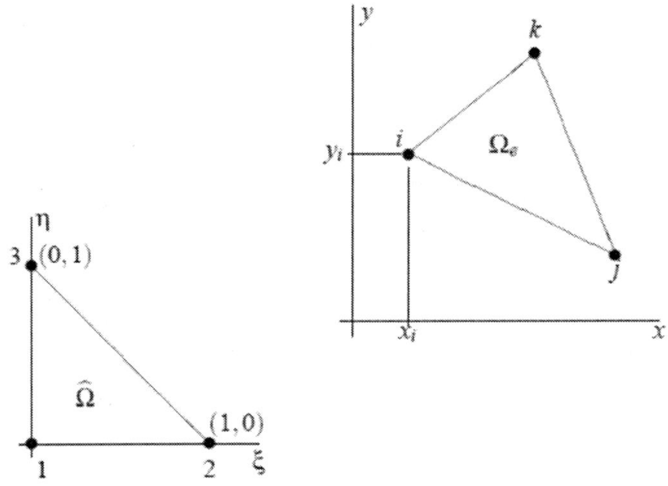

Figure 1.10. A triangular element and the corresponding reference element.

These functions have the property that $\widehat{N_\iota} = 1$ at local node i, and 0 at the other nodes. The basis function N_i formed by patching together all the local functions $N_\iota^{(e)}$ i associated with node i is the two-dimensional counterpart of the "hat" function in one dimension, and is pyramidal in shape. Naturally N_i is piecewise linear, and is nonzero only on those elements that have node i as a node. This is illustrated in Figure 1.11.

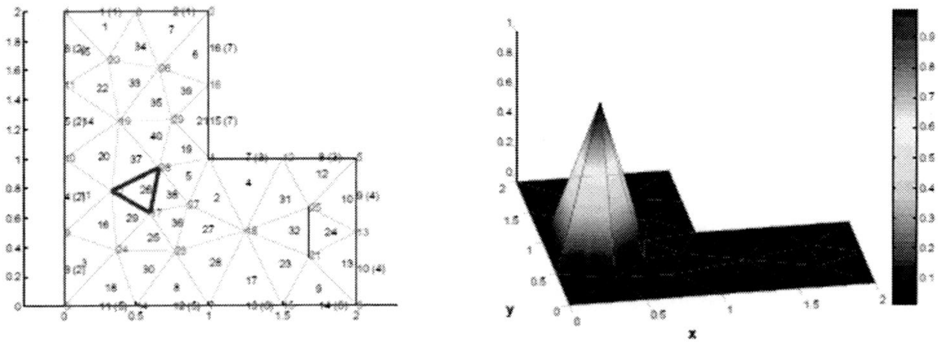

Figure 1.11. A mesh in 2D and a typical global shape function

A key stage in the implementation of the finite element method is the construction of the stiffness matrix and load vector, and these require that a number of terms be integrated, generally over the reference element. The basis of most numerical integration schemes is the identification of selected points, known as sampling points, at which the value of the function is sampled, and the specification of a set of weights, one for each sampling point.

Suppose that integration is to be carried out over one of the reference elements $\widehat{\Omega}$; then if the sampling points are denoted by $\widetilde{\xi_l}$ ($l = 1, \ldots, r$) and the weights by w_l ($l = 1, \ldots, r$), a numerical integration formula *of order r* is defined to be a formula of the kind:

$$\int_{\widehat{\Omega}} f(\xi)d\xi \simeq \sum_{l=1}^{r} w_e f\left(\widetilde{\xi_l}\right) \qquad (1.31)$$

Integration over triangles. An integration rule of order 1 may be defined on a triangle Ω_e by

$$\int_{\Omega_e} f(x,y)dx\,dy \simeq A_e f(\hat{x},\hat{y}) \tag{1.32}$$

in which $(\hat{x},\hat{y})$ are the coordinates of the *centroid* of the triangle (Figure 1.12). Likewise, a rule of order 3 may be defined by:

$$\int_{\Omega_e} f(x,y)dx\,dy \simeq \frac{1}{3} A_e \sum_{l=1}^{3} f(\widetilde{x_l},\widetilde{y_l}) \tag{1.33}$$

where $(\widetilde{x_l},\widetilde{y_l})$ (l = 1, 2, 3) are the coordinates of the midpoints of the sides. The rule of order 1 is exact for polynomials of degree 1, while the rule of order 3 is exact for polynomials of order 2.

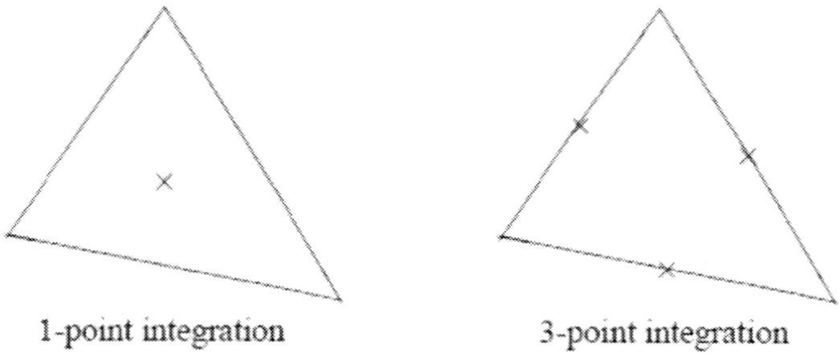

Figure 1.12. Integration rules on the reference triangle.

1.3. Micro-Mechanical Modeling

In this section we go through the details of micro-macro modeling of a random fiber-reinforced composite. The material is assumed to be macroscopically isotropic, but it is micro-structurally anisotropic as a result of the presence of fibers in a random arrangement at that scale.

The objective is to obtain effective material properties that capture, at the macroscopic level and in an averaged sense, the microscopic behavior. These effective moduli are then used in macroscopic finite element analyses.

It is possible, via a homogenization procedure, to find "effective" material properties at the macro-scale by creating and analyzing a model of the micro-structure. In this way the details of microscopic behavior are captured in an average sense, without the need to analyze the entire structure at that scale.

Consider a representative volume element (RVE) of the body. The RVE defines a representative volume of a microscopically heterogeneous material such as a fiber-reinforced composite, a polycrystalline aggregate, or a natural material such as wood. A RVE is associated with each material point in the body (see section 1.2.2), and the idea is to solve a number of boundary value problems for the RVE or micro-structure, in order to obtain effective material moduli. If the material is macroscopically homogeneous, then it suffices to solve a single set of problems for a statistically representative set of RVEs, to obtain effective moduli for the macroscopic body as a whole.

To begin with, we consider the weak boundary value problem as set out in (1.23), for the arbitrary RVE shown in Figure 1.13.

Figure 1.13. Scanning electron microscope image of the entangled fibres [Source: Council for Scientific and Industrial Research (CSIR), Port Elizabeth].

In order to link the macro- and micro-problems, the boundary conditions on the RVE must ensure that the macroscopic stress power is equal to the average microscopic stress power, that is, they must satisfy Hill's condition [17]:

$$\underbrace{\langle \sigma^T \varepsilon \rangle \Omega}_{} = \underbrace{\langle \sigma \rangle_\Omega^T \langle \varepsilon \rangle \Omega}_{} \tag{1.34}$$

macroscopic stress power microscopic stress power

where

$$\langle . \rangle := \frac{1}{|\Omega|} \int_\Omega (.) \, d\Omega \tag{1.35}$$

Hill's condition is satisfied by three types of boundary conditions on the RVE, viz. linear displacement, constant traction, and periodic boundary conditions: the linear displacement and constant traction conditions are specified by:

$$\boldsymbol{u}|\partial\Omega = E.\mathbf{x} \tag{1.36}$$

$$\boldsymbol{t}|\partial\Omega = \mathcal{T}.\mathbf{n} \tag{1.37}$$

where E and $\mathcal{T}$ are constant matrices. Periodic boundary conditions are only relevant in a material with a spatially periodic micro-structure and thus are not considered further in this work.

It can be shown [15] that the linear displacement boundary condition gives an "over-stiff" response, and the constant traction boundary condition an "under-stiff" response. These results thus give upper and lower bounds for the problem.

The analysis on the RVE will give an effective macroscopic linear elasticity tensor, $\mathbb{E}^*$ which relates the averages of the microscopic stress, σ, and the microscopic strain, ε:

$$\langle\sigma\rangle_\Omega = \mathbb{E}^*\langle\varepsilon\rangle_\Omega \tag{1.38}$$

Because the material is assumed to be macroscopically isotropic it is necessary to determine only two material constants. The most convenient constants are the bulk modulus K and shear modulus G, since from (1.13) we have:

$$\langle K\rangle = \frac{tr\langle\sigma\rangle}{tr\langle\varepsilon\rangle}, \langle G\rangle = \frac{tr\langle\sigma^D\rangle}{tr\langle\varepsilon^D\rangle} \tag{1.39}$$

Defining the micro-structure. The fiber-reinforced composite which serves as the model problem consists of a polymer matrix containing thin, randomly orientated, natural fiber reinforcements. Needle-punching is used to entangle the fibers, which are then bonded into the matrix. This process results in a thin, flat composite which is used as the outer layers in a laminate, which sandwich honeycomb filler.

Figure 1.13 shows the entangled fibers, which are of varying length and diameter, of average $62mm$ and $30\mu m$ respectively, with volume fraction approximately 30%. There is a 50% variation in the diameters of the fibers along their length as a result of their natural origin. Due to the nature of the needle-punching process the fibers are orientated randomly within the matrix.

In order to determine the effective elasticity matrix $\mathbb{E}^*$, it is necessary in general to apply six linearly independent loadings defined by linear displacement and constant traction boundary conditions. For a material such as that considered here, which is assumed to be macroscopically isotropic, two constants completely define the macroscopic response, and these can be obtained from a single test. In particular, the two material constants used are the bulk and shear moduli, whose effective values are given by (1.39).

Optimal RVE sizing and sampling procedure. A RVE should capture the nature of the underlying micro-structure of the material in question. For an analysis to give accurate material information the RVE would need to be orders of magnitude larger than the length scale of the heterogeneities (the fibers in this case). Practically, however, this is not possible, therefore analyses of two distinct RVEs will yield different results. It is therefore necessary to carry out a number of tests on a range of samples, and obtain an average from the set of results.

The sizing of the RVEs is important; the larger a sample is, the more accurate the information gained from the test. However, a larger sample is also more computationally expensive, thus it is important to find an RVE size that is statistically relevant as well as computationally feasible. For the example studied here the RVE size was estimated by inspection of a scanning electron microscope image (Figure 1.13) as well as consideration of computational limitations.

The optimal size for the RVE is determined using the following procedure: the geometry of a "large" sample (too large to be computationally feasible) is generated (see Figure 1.14a).

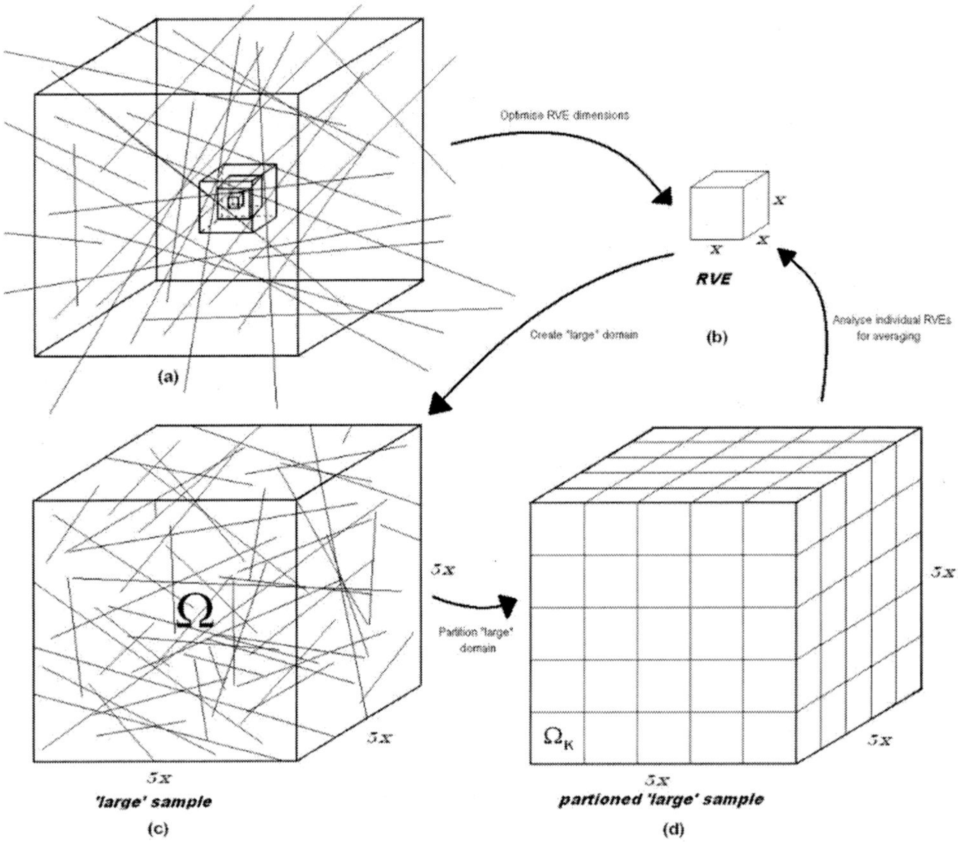

Figure 1.14. Testing methodology to find RVE size and test multiple samples by partitioning a "large" domain.

Thereafter a small subdomain within the sample is defined and analyzed using the prescribed linear displacement loading condition. The subsample is meshed and tested and the values for the effective bulk and shear moduli (K^* and μ^* respectively) calculated. The size of the subsample is then increased within the same geometry and tested again (Figure 1.14a). This process is repeated until the values for K^* and μ^* converge, at which point the subsample dimensions are used to define the RVE (see Figure 1.14b). The size of the first subdomain tested is obtained from an inspection of SEM images (Figure 1.13) and such that it contains the necessary 30% fiber/matrix volume fraction.

Partitioning of a large sample to find average constitutive values. Instead of creating many random sample RVEs, one large sample is created and then partitioned into 125 equal sized domains, each one being treated as an RVE, as shown in Figure 1.14. By solving the boundary value problem on each subdomain Ω_K with the linear displacement boundary condition, the effective moduli for the RVE can be obtained as the weighted average of effective moduli for the sub-RVEs.

The material specimen is a cube. Both the polymer matrix and the fibers are assumed to be linear elastic. The fibers (flax in this case) and matrix have Young's moduli E of 27.6 GPa and 1.35 GPa respectively, and values of Poisson's ratio ν of 0.36 and 0.42 respectively.

The fibers are treated as rectilinear rods of different lengths having an axial stiffness only. The cross sectional area of the fibers is $30\mu m$.

The straight fibers are randomly placed within the matrix. Two points within a domain substantially larger than the RVE size are generated randomly and then joined with a straight line. The generation of the fibers within a domain larger than that of the RVE allows any areas of fiber sparsity around the perimeter of the domain (due to the nature of the randomization algorithm) to be excluded from the RVE that is analyzed. The fibers are added one at a time and the volume fraction within the RVE is monitored until it reaches 30%.

As there is no specific accommodation on the boundary for a fiber entering/exiting the domain, there is no benefit in having fibers terminating within the domain either. Therefore the fibers are assigned an original length equal to twice the diagonal of the domain. A sphere is generated such that its center is somewhere within the large domain. The points of intersection of the sphere with a random vector passing through its center are the endpoints of the fiber. The fiber is then truncated such that the points of intersection with the boundaries of the domain become its new start and endpoints. Thus the midpoint of the fiber is placed randomly within the domain and the fiber is adjusted to start and end at the boundaries of the domain. Fibers are added until the required volume fraction of 30% is reached. The large domain is then partitioned and the intersections of the fibers with the individual RVE boundaries become the new start and end points for the fibers (for each individual RVE). The data then exists for 125 separate boundary value problems. The algorithm to create the fiber geometries is contained in a Matlab code developed for the problem. In developing a finite element analysis of the RVEs, the commercial software, Abaqus is used and the creation of the models is automated using a Python script developed for the problem. The fibers are embedded into the matrix using the embedded element constraint in Abaqus. One of the advantages of this constraint is that it allows the fibers to be meshed separately from the matrix; thus the discretization of the matrix can be regular. This regular mesh is desirable as more control over the element shape is achieved and far fewer degrees of freedom result than from an irregular mesh. The diameter, and therefore the cross-sectional area, of the fibers are very small in comparison to the size of the RVE. Therefore it is important to have a mesh fine enough to capture the effects of the fibers within the matrix but coarse enough to remain computationally viable. The partitioning procedure described previously is achieved by retaining the fiber geometry and shifting the subdomain, Ω_K, throughout the large domain so that the entire large domain is tested. The results for K_K^* , μ_K^* and thus $\mathbb{E}_K^*$ are calculated and averaged to find $\widetilde{\mathbb{E}_K^*}$.

The linear displacement loading condition is applied as an essential boundary condition and the results are interrogated for the effective values κ^* and μ^*. The average values, $\widetilde{\kappa*}$ and $\widetilde{\mu*}$ are then calculated.

Results The methodology is applied to a sample of $1cm^2$, which is divided into 125 RVEs of $4mm^2$ each. The large domain contains 30% fiber volume, with 47,672 fibers randomly dispersed throughout its volume.

One expects effective averaged values to lie somewhere between those of the matrix and the fibers, i.e.:

$$\Phi_m \leq \widetilde{\Phi^*} \leq \Phi_f \tag{1.40}$$

where Φ is the material property in question with subscripts m for the matrix value and f for the fiber value. As before, $\widetilde{\Phi^*}$ represents the effective, averaged value for the composite.

The resulting values are given in Table 1.1. The values show that the inequality in (1.40) is satisfied except for in the case of Poisson's ratio (ν). As truss elements have no transverse strain, effectively the value used for Poisson's ratio is zero. Therefore the fact that the effective averaged value is below the value for the matrix is acceptable.

The histograms in Figure 1.15 show the distribution of the results for the 125 subdomains. These are satisfactory distributions peaking at the average effective values shown in Table 1.1.

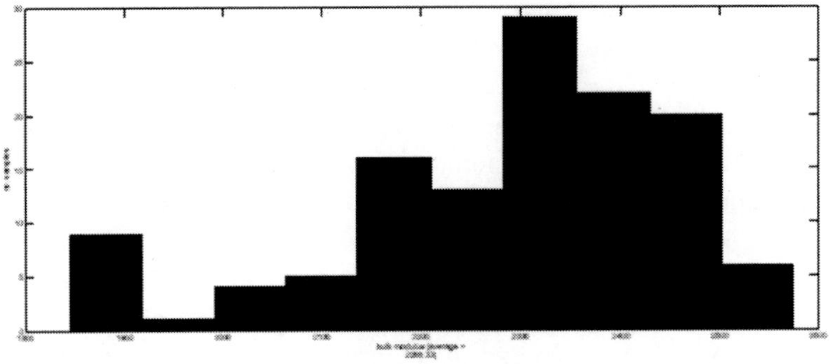

(a) Bulk modulus

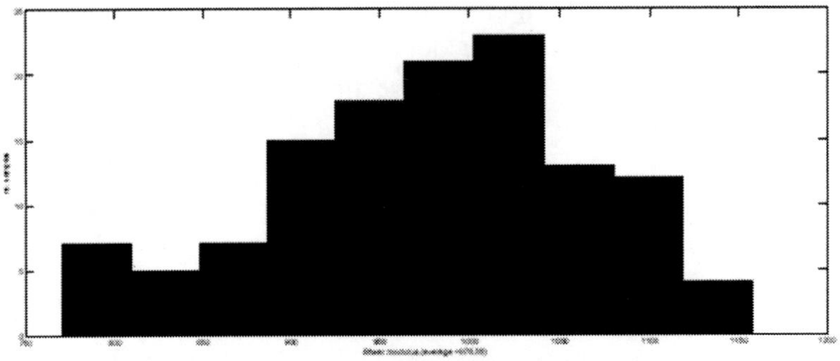

(b) Shear modulus

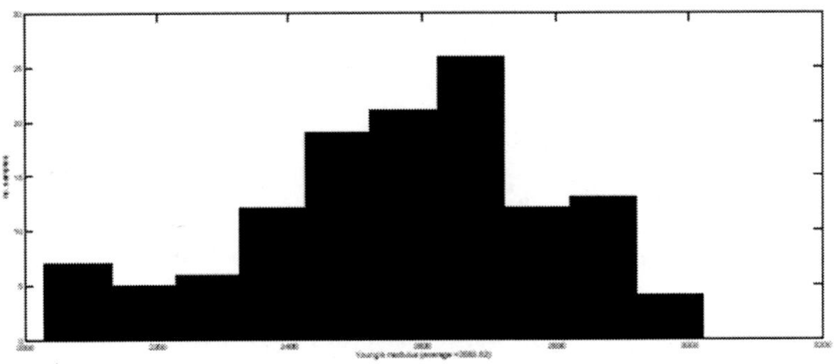

(c) Young's modulus

Figure 1.15. Distribution of data for effective bulk, shear and Young's moduli.

Table 1.1. Material properties

	Φ_m	$\widetilde{\Phi}^*$	Φ_f
K (MPa)	1610	2509.21	172500
μ (MPa)	496	1001.12	9718
$\mathbb{E}$ (MPa)	1350	2649.76	27600
ν	0.36	0.32	0.42

1.4. Open Problems and Future Directions

The objective of this section is to briefly describe some of the key challenges facing multiscale modeling methodologies. Issues related to the modeling of fiber-reinforced composite are emphasized. This section draws heavily on the recent review of Geers et al. [3] which should be consulted, along with the references therein, for further information.

Standard multi-scale methodologies as presented in this review Chapter have developed significantly over the last 15 years. The method is able to describe the large deformation response of inelastic media with complex micro-structure, but certain key limitations exist. One such limitation is the description of shells or plates with complex micro-structure which cannot be captured in a layered-wise composite shell approach. Clearly a large number of fiber-reinforced composite products fall into this category. This limitation exists as conventional first-order approximations can't pass second-order information, such as macroscopic deformation gradients (e.g. in bending), to the RVE boundaries. This information is required in shell theory.

A second-order extension of the framework described in this Chapter provides an effective way to describe shells with micro-structure. The second order methodology also allows for the description of regions of moderate localization that would otherwise render a first-order approach mesh-dependent and incapable of producing reasonable results. Localization is typically associated with material regions experiencing intense gradients in deformation, such as occurs during shear banding.

Open questions still exist concerning the extension of the second-order approaches to a range of shell theories and to more general loading conditions. Another key open issue is multi-scale multi-physics coupling. Progress has been made on thermo-mechanical coupling but many other more complex coupling challenges exist (e.g., electro-mechanical, thermo-electrical, fluid-structure interaction, magneto-electro-elasticity, acoustics, amongst others).

From a design point of view it is important to model not only reinforced composites consisting of random arrangements of fibers, but also structured and periodic arrangements such as composites comprising woven fibers.

In addition, more sophisticated models at the microscopic level would allow a broader range of properties and phenomena to be captured. For example, viscoelastic effects are important in some situations, and depending on the fibrous material, damage as a result of fiber breakage would in certain situations be an important consideration. Furthermore, in the examples treated in this chapter perfect bonding between fiber and matrix has been assumed; it would be important to model, and hence to gain a better understanding of the conditions under which debonding takes place, and its effect on the overall, macroscopic material properties.

ACKNOWLEDGMENTS

A large portion of the work presented in this chapter forms part of a project in collaboration with the Council for Scientific and Industrial Research (CSIR), and forms part of an overarching project Natfibio: Development of a natural fiber/bio-composite cabin interior component, sponsored jointly by Airbus, the Department of Science and Technology, and the Advanced Manufacturing Technology Strategy.

The considerable contributions from Helen Morrissey at the Centre for Research in Computational and Applied Mechanics at the University of Cape Town and Hellmut Bowles from Finite Element Analysis Services are greatly appreciated.

REFERENCES

[1] Mallick, P. K. *Fiber-Reinforced Composites: Materials, Manufacturing, and Design*; ISBN: 0824777964; Marcel Dekker Inc: 2nd Edn, 1993.

[2] Zohdi, T. I.; Wriggers, P. *Introduction to Computational Micromechanics (Lecture Notes in Applied and Computational Mechanics);* ISBN: 3540228209; Springer-Verlag Inc: Secaucus, NJ, 2004.

[3] Geers, M. G. D.; Kouznetsova, V. G.; Brekelmans, W. A. M. *J. Comput. Appl. Math.* 2010, *vol 234(7),* 2175-2182.

[4] Belsky, V.; Beall, M. W.; Fish, J. Shephard, M. S.; Gomaa, S. *Comput. Sys. Eng.* 1995, *vol 6(3),* 213-223.

[5] Feyel, F.; Chaboche, J. –L. *Compu. Method. Appl. Mech. Eng.* 2000, *vol 183,* 309-330.

[6] Sansalone, V.; Trovalusci, P.; Cleri, F. *Acta Materialia* 2006, *vol 54*, 3485-3492.

[7] Xia, Z.; Curtin, W. A.; Peters, P. W. M. *Acta Materialia* 2001, *vol 49,* 273-287.

[8] Xia, Z. H.; Curtin, W. A. *Compos. Sci. Technol.* 2001, *vol 61,* 2247-2257.

[9] González, C.; LLorca, J. *Acta Materialia* 2006, *vol 54,* 4171-4181.

[10] F. Feyel. *Comput. Mater. Sci.* 1999, *vol 16,* 344-353.

[11] Maceri, F.; Marino, M.; Vairo, G. *J. Biomech.* 2010, *vol 43,* 355-363.

[12] Hughes, T. J. R. *The Finite Element Method. Linear Static and Dynamic Finite Element Analysis;* Prentice-Hall: Englewood Cliffs NJ, 1987.

[13] Zienkiewicz, O. C.; Taylor, R. L. *The Finite Element Method for Solid and Structural Mechanics;* ISBN: 9780750663212; Elsevier Butterworth-Heinemann, 2005.

[14] Kouznetsova, V. *Computational Homogenization for the Multi-Scale Analysis of Multi-Phase Materials;* PhD thesis, Eindhoven University of Technology, 2002.

[15] Zohdi, T. I. Wriggers, P. *Int. J. Num. Method. Eng.* 2001, *vol 50,* 2573-2599.

[16] Ozdemir, I.; Brekelmans, W. A. M.; Geers, M. G. D. *Comput. Method. Appl. Mech. Eng.* 2008, *vol 198,* 602-613.

[17] Hill, R. *Proc. Phy. Soc. Lon.* 1952, *vol A65,* 349-354.

In: Modeling and Simulation in Fibrous Materials
Editor: Asis Patanaik
ISBN: 978-1-62100-116-4

Chapter 2

ARTIFICIAL NEURAL NETWORK AND ITS APPLICATIONS IN MODELING

Abhijit Majumdar *

Department of Textile Technology, Indian Institute of Technology Delhi
New Delhi, India

ABSTRACT

Artificial neural network is a very powerful modeling and classification system which can reasonably approximate any kind of functional relationship. In recent years, there has been growing applications of artificial neural network in every field of science, engineering and management. This chapter provides an outline of artificial neural network and its mathematical background. The training algorithm has been also discussed briefly. MATLAB programming has also been introduced to demonstrate a simple modeling problem. Finally, two interesting applications of artificial neural network in engineering and management have been discussed.

2.1. INTRODUCTION

Human brain is one of the most complicated things. The efforts to understand the functioning of human brain have been continuous, which started more than 2000 years ago by Aristotle and Heraclitus. Human brain is having the capabilities like image processing, decision making, pattern recognition and imagination. Researchers have made relentless efforts to mimic the functioning of human brain by devising intelligent machines or computers and thus 'artificial intelligence' has evolved as a fascinating branch of academics and research. Artificial neural network (ANN), fuzzy logic (FL), genetic algorithm (GA), support vector machine (SVM), rough set, particle swarm optimization (PSO), ant colony optimization are some of the major areas of research in artificial intelligence. ANN is mainly used for modeling, prediction and classification. Fuzzy logic is apt to handle imprecision and

---

* E-mail: abhitextile@rediffmail.com

ambiguity and widely used for machine control. Genetic algorithm and particle swarm optimization are used to solve complex optimization problems. This chapter focuses on the theory and applications of ANN.

McCulloch and Pitts laid the foundation stone of ANN research in 1943 by developing first computing model of artificial neuron. The model included all required elements to perform logic operations and thus it was able to function as arithmetic logic computing element. However, the implementation of its compact electronic model was not technologically feasible during the era of bulky vacuum tubes. The next significant development was pursued by Hebb, a Canadian neuro-psychologist, who proposed a learning algorithm (Hebbian learning rule) for updating neuron's connections in 1949 [1]. This rule was subsequently modified by Rosenblatt who proposed the perceptron model in 1958. However, a critical assessment of perceptron model by Minsky (1969) exposed its limitations which led to downfall in the ANN research till 1980s. During 1982 to 1986, several seminal research work were published which significantly enhanced the potential of ANN. The era of renaissance started with the introduction of recurrent ANN architecture for associative memories by Hopfield [2]. His papers formulated computational properties of a fully connected network of units. Unsupervised learning algorithms of ANN were developed for feature mapping by Kohonen [3]. Another revolution of the field was the development of a supervised training algorithm by Rumelhart [4]. The new learning rules and other concepts have removed some of the network training barriers that stalled the mainstream ANN research in the mid 1960s. The researchers are still continuing research to develop more powerful ANN techniques by improving the training algorithm and pruning the superfluous weights.

2.2. THE HUMAN BRAIN

Human brain contains about 100 billion (10^{11}) basic units known as neurons (Greek: nerve cell). Each neuron is connected with about 1000 other neurons. A neuron is a small biological cell composed of a cell body or soma (Greek: body), a nucleus, an axon and dendrites (Figure 2.1). Neurons receive electro-chemical signals through the dendrites (Greek: tree-lings) and in turn respond by transmitting impulses to other neurons through axon. Dendrites are the long and irregularly shaped filaments attached with soma. The dendrites behave as input devices whereas axon serves as an output device. If the cumulative inputs, received by the soma, crosses the internal electric potential of the cell (membrane potential), then the neuron fires by propagating the action potential down the axon to excite or inhibit other neurons. The axon connects with the dendrites of other neurons in a specialized contact known as synapse or synaptic junction. When the signal reaches the end of axon, chemical messengers known as 'neurotransmitters', which are stored in very small spherical 'vesicles', are released. The neurotransmitters are responsible for effective communication between the neurons. The strength of the synapse is very important because it determines the amplification or retardation of a signal. A strong synapse releases more neurotransmitters during the exchange of signals.

As the information about the functioning of human brain was explored, the scientists started to devise systems to mimic the brain's functioning and the quest for an ANN began.

The human brain processes information super-quickly and super-accurately and can recognize pattern even if the input signal is noisy. For example, a person can easily recognize the face of a friend from a distance or distinguish his voice from the noise even inside a stadium laden with thousands of spectators. The signal processing capability of human brain to recognize information and to retrieve it correctly is one of its amazing features. Therefore, an artificial neural system is also highly beneficial for signal processing, pattern recognition, control of robotics, power systems, communication systems and many other fields. A pertinent comparison between human brain and advanced computer is given in Table 2.1.

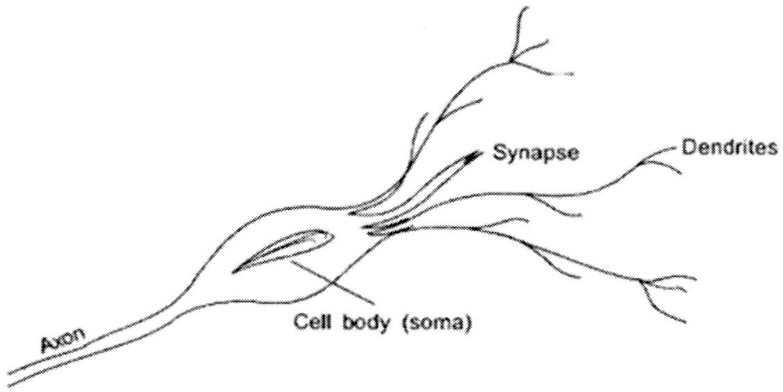

Figure 2.1. Structure of biological neuron [Source: Reference 5].

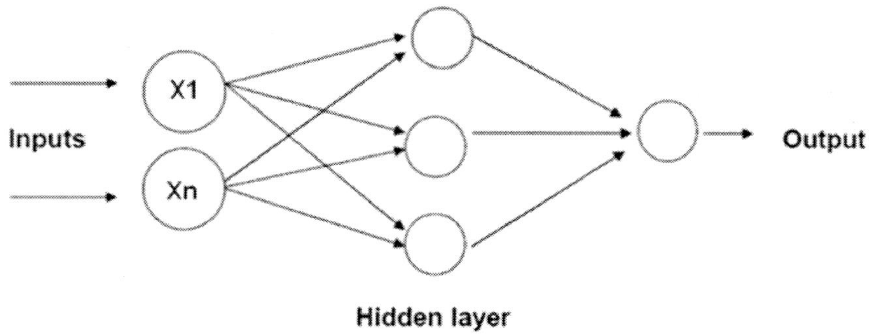

Figure 2.2. Artificial neural network.

Table 2.1. Comparison between human brain and computer [Source: Reference 10]

Parameter	Human brain	Computer
Processing elements	10^{14} synapses	10^{8} transistors
Element size	10^{-6} m	10^{-6} m
Energy used	30 W	30 W (CPU)
Processing speed	100 Hz	$\sim 3 \times 10^{9}$ Hz
Style of computation	Parallel, distributed	Serial, centralized
Fault tolerant	Yes	No
Learning capability	Very high	Very little

2.3. Artificial Neural Network (ANN)

2.3.1. Mathematical Model of ANN

ANN simulates the complex functioning of biological neuron. Every component of ANN bears a direct analogy to the actual constituents of biological neuron. The schematic model of an ANN is shown in Figure 2.2. The ANN is having three or more number of layers each composed of finite number of processing elements or neurons or nodes. The input layer and output layer represent the input parameters and output parameter/parameters. Therefore the number of neurons in these layers is equal with the respective number of parameters. Hidden layers, which are one or more in number, are sandwiched between the input and output layers.

Figure 2.3. shows the simple model of ANN. Here, X_1, X_2,..., X_n are the n input parameters. One neuron j which belongs to the hidden layer is shown in the figure. W_{j1} is the synaptic weight connecting the hidden neuron j with the 1st input neuron. Synaptic weights represent the synapse strength of the biological neuron. Each neuron receives a signal from the neurons of the previous layer and these signals are multiplied by separate synaptic weights. The weighted inputs are then summed up to get the total input I received by artificial neuron. This can be expressed as follows:

$$I = \sum_{i=1}^{n} w_{ji} X_i - \theta \quad (2.1)$$

where w_{ji} is the weight connecting the hidden neuron j and input neuron X_i, n is the total number of input parameters and θ is the bias weight and it represents the membrane potential of the biological neuron. The weighted sum is then passed through an activation function or transfer function, which produces the output Z, as shown below.

$$Z = \phi(I) \quad (2.2)$$

The output of the activation function is then transmitted to the neurons of next layer and similar kinds of computation are done. The final output is produced at the neuron/ neurons of the output layer.

Various types of transfer function can be used depending on the nature of the problem being modeled. If the output has to be scaled between 0 and 1, then the following activation functions can be used.

i. Hard limit or threshold

$$\phi(I) = \begin{cases} 1, & \text{if } I \geq 0 \\ 0, & \text{if } I < 0 \end{cases}$$

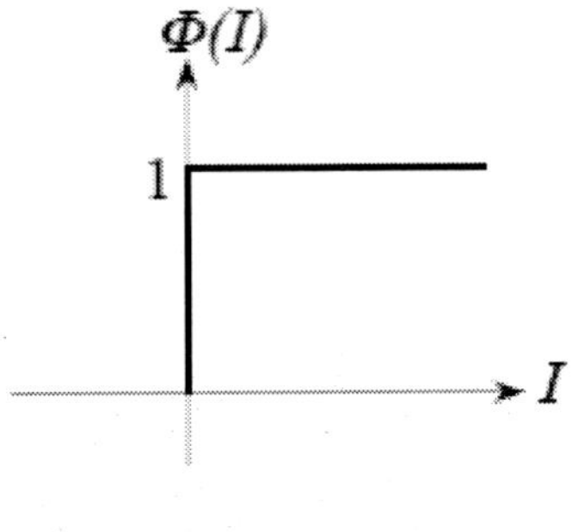

ii. Saturating linear

$$\phi(I)=\begin{cases}1, & \text{if } I>1\\ I, & \text{if } 0\le I\le 1\\ 0, & \text{if } I<0\end{cases}$$

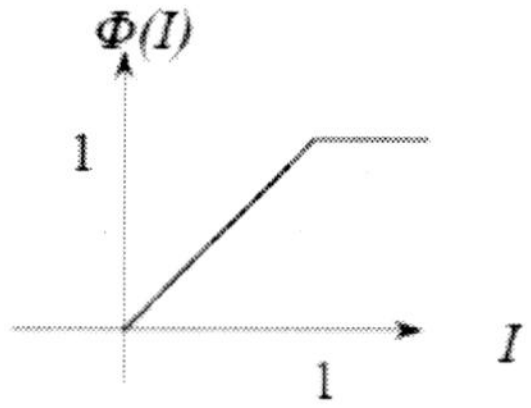

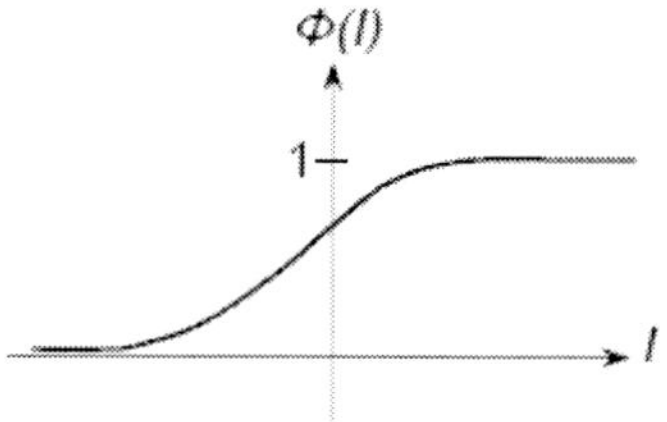

iii. Log-sigmoid

$$\phi(I)=\frac{1}{1+e^{-\alpha I}}$$

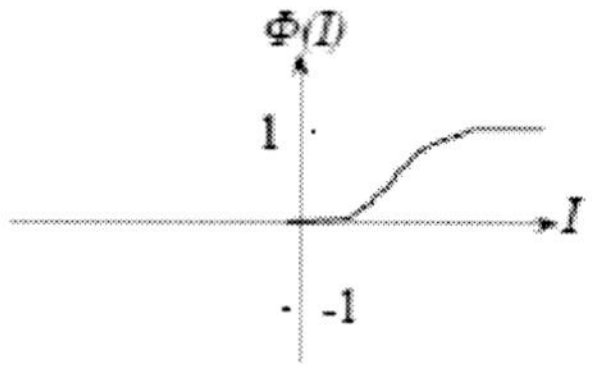

If the output has to be scaled between -1 to 1, then the following activation functions can be used.

i. Symmetrical hard limit

$$\phi(I)=\begin{cases}1, & \text{if } I\ge 0\\ -1, & \text{if } I<0\end{cases}$$

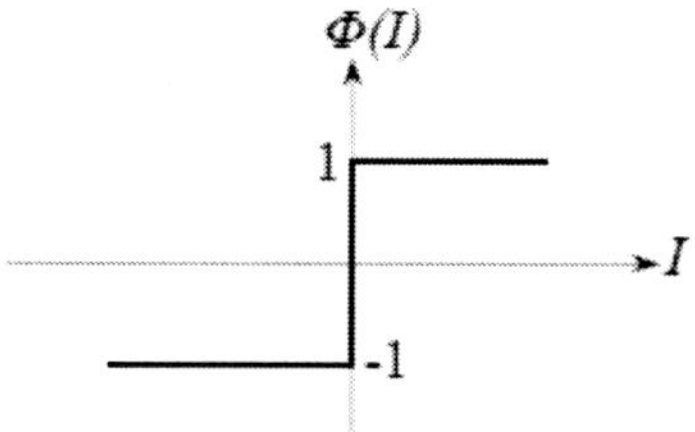

ii. Symmetrical saturating linear

$$\phi(I)=\begin{cases}1, & \text{if } I>1\\ I, & \text{if } -1<I<1\\ -1, & \text{if } I<-1\end{cases}$$

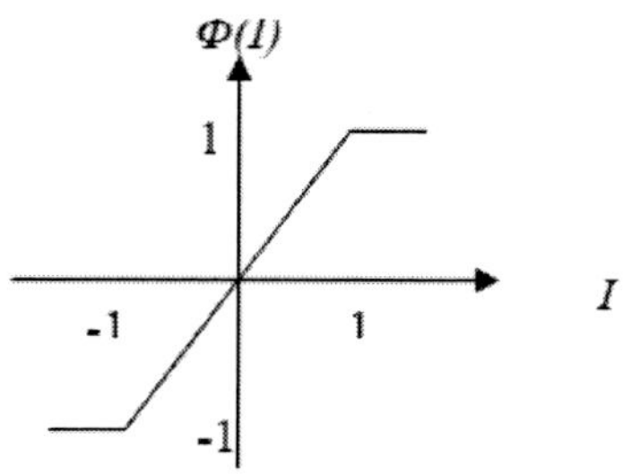

iii. Hyperbolic tangent

$$\phi(I) = tanh\ (I) = \frac{e^{I} - e^{-I}}{e^{I} + e^{-I}}$$

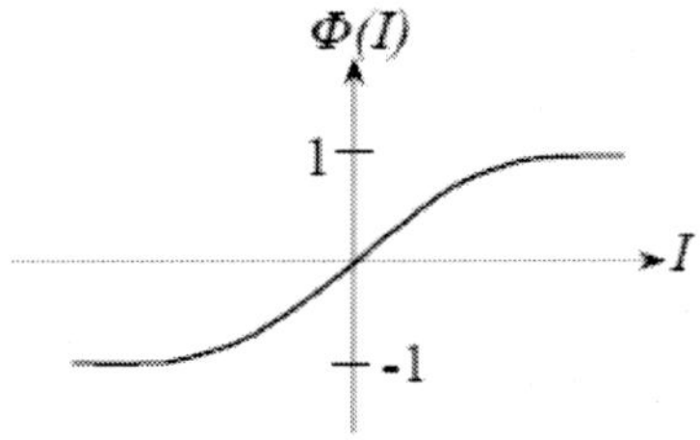

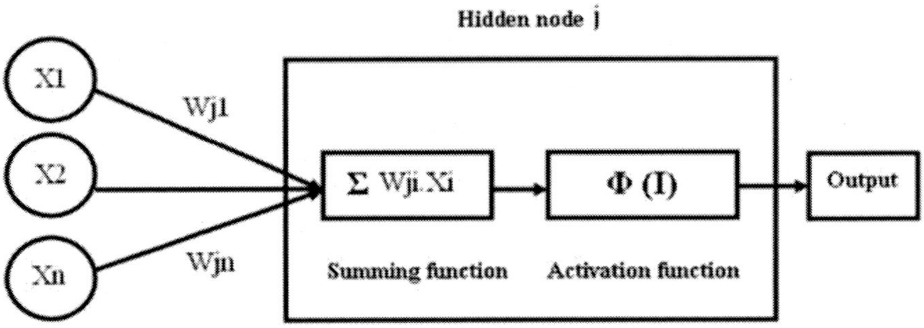

Figure 2.3. Simple model of an ANN.

2.3.2. Learning Methods of Artificial Neural Networks

Learning or training, which is basically the optimization of synaptic weights, of ANN can be broadly classified under three categories as follows.

- Supervised learning
- Unsupervised learning
- Reinforced learning

In supervised learning, the network is presented with a set of input and desired output data. A teacher is assumed to be present during this learning. A comparison is made between the predicted output and the desired output and the error signal is calculated. This error signal is then used to modify the synaptic weights in several steps and thus improved performance is obtained.

In unsupervised learning, no target or desired output is presented to the network. It is analogical to the absence of any teacher. Therefore, the system learns on its own by discovering and adapting to the structural features of the input patterns. Even though unsupervised learning does not require a teacher, it requires guidelines to determine how the groups will be created. Grouping may be done based on color, shape or some other property of the object.

In reinforced learning, although a teacher is present, he does not indicate the target output to the ANN. However, the teacher indicates whether the predicted output is correct or not, just like the ‘pass/ fail’ indication. Thus the error signal generated during training is binary. A reward is given if the predicted output is correct and vice versa. If the teacher's signal is 'fail' then the network readjusts its parameters until the output response is right.

2.3.2.1. Back-Propagation Algorithm

Back-propagation is the most popular among the existing ANN training algorithms. This supervised training algorithm was developed by Rumelhart et al. [4]. The details of this algorithm can be found in many standard textbooks of ANN authored by Rajasekaran and Pai [5], Zurada [6] and Haykin [7]. According to this algorithm, training occurs in two phases, namely forward pass and backward pass. In the forward pass, a set of data is presented to the network as input and its effect is propagated through different layers of the ANN. Finally, a set of outputs is produced. The error vector is calculated according to the following equation:

$$E = \sum_{j=1}^{p} E_j \tag{2.3}$$

where E is the error vector, E_j is the error associated with the j th pattern and p is the total number of training patterns.

The expression of E_j is given in the following equation:

$$E_j = \frac{1}{2} \sum_{k=1}^{s} (T_k - out_k)^2 \tag{2.4}$$

where T_k and out_k, the target output and predicted output, respectively, at output neuron k and s is the total number of output neurons.

In the backward pass, this error signal is propagated backwards to the neural network and the synaptic weights are adjusted in such a manner that the error signal decreases in each iteration process. The corrections necessary in the synaptic weights between output and hidden layer are carried out by a delta rule, which has been pictorially represented in Figure 2.4. At the start of the training, ANN is having random combinations of synaptic weights which results very high error signal.

Therefore, the gradient (slope) of the error function (surface) is determined at the starting point. If it is assumed that the starting point is position 1 in Figure 2.4, then the gradient of the error surface is negative. So, the current weight has to be increased to reach the optimum weight (for error minimization).

On the other hand, if the starting position is position 2, the gradient of the error surface is positive. So, the current weight has to be reduced to move towards the optimum weight. Once the training converges to the point where the error minimization has been attained, the gradient of the error surface becomes zero and no further update of the weights takes place.

Mathematically this process is represented as follows:

$$\Delta W_{kj} = W_{kj(new)} - W_{kj(old)} = -\eta \left[\partial E / \partial W_{kj} \right] = -\eta \left[\frac{\partial E}{\partial out_k} \frac{\partial out_k}{\partial net_k} \frac{\partial net_k}{\partial W_{kj}} \right] \tag{2.5}$$

Now, $$\frac{\partial E}{\partial out_k} = \frac{\partial}{\partial out_k}\left[\frac{1}{2}\sum_k (T_k - out_k)^2\right] = -\ (T_k - out_k) \tag{2.6}$$

$$\frac{\partial out_k}{\partial net_k} = f'_o(net_k) = out_k\ (1 - out_k) \text{ for log sigmoid transfer function} \tag{2.7}$$

$$\frac{\partial net_k}{\partial W_{kj}} = \frac{\partial}{\partial W_{kj}}\sum_k w_{kj} out_j = out_j \tag{2.8}$$

Therefore, $$\Delta W_{kj} = \eta\left[(T_k - Out_k)Out_k(1 - Out_k)\right]out_j = \eta\delta_k out_j \tag{2.9}$$

where W_{kj} is the weight connecting the neurons j of hidden layer and neuron k of the output layer, ΔW_{kj} is the correction applied to W_{kj} at a particular iteration, η is a constant known as learning rate and out_j is the output of neuron j.

The weight change between hidden layer and input layer is calculated as shown below:

$$\Delta W_{ji} = \eta f'_H(net_j)X_i\sum_k \delta_k W_{kj} \tag{2.10}$$

where ΔW_{ji} is the correction applied to the weight connecting input neuron i and hidden neuron j and X_i is the input received by the neuron i.

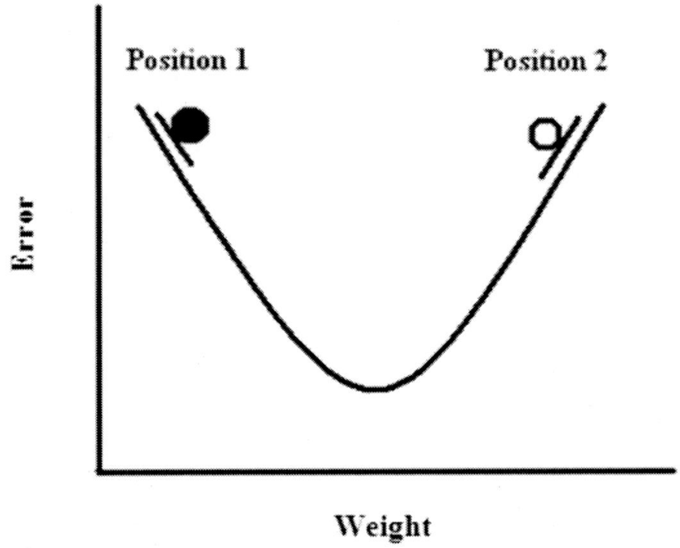

Figure 2.4. Weight change during ANN training.

2.3.3. Learning Rate and Momentum

Learning rate (η) determines the size of weight adjustments made in one iteration and hence it influences the rate of convergence. If the learning rate is very small then the training proceeds in very small steps and training time becomes longer. On the other hand, if the learning rate is too high then the search path will oscillate and 'global minima' cannot be attained. The learning rate has to be chosen as large as possible without inducing oscillations. The effect of very small and very large learning rate is depicted in Figure 2.5.

The error surface, in most of the practical problem, is not as smooth as shown in Figures 2.4 and 2.5. There are several 'local minima' in the error surface as depicted in Figure 2.6 where the back propagation algorithm may be stuck. To overcome those problems, back-propagation algorithm is often modified by adding a momentum term. The momentum is a constant that determines the effect of past weight changes on the current direction of movement in weight space. The modified rule of weight adjustment with momentum term is shown below.

$$\Delta W_{jk}(t+1) = \eta\delta_k out_j + \mu.\Delta W_{jk}(t) \tag{2.11}$$

where $\Delta W_{jk}\ (t+1)$ is the weight change in the $(t+1)^{th}$ iteration, $\Delta W_{jk}\ (t)$ is the weight change in the $t^{\ th}$ iteration and μ is the momentum term and it ranges from 0 to 1.

Momentum term adds a fraction of the previous weight change to the current weight change. The addition of this momentum helps to smoothen the descent path by preventing extreme changes in the gradients due to local minima. If the momentum is zero, the smoothing is least and the weight adjustments are done solely by the newly calculated gradient. In contrast, if the momentum is one then the new adjustment is ignored and previous one is repeated. Between zero and one, the weight adjustment is smoothed by an amount proportional to the momentum factor.

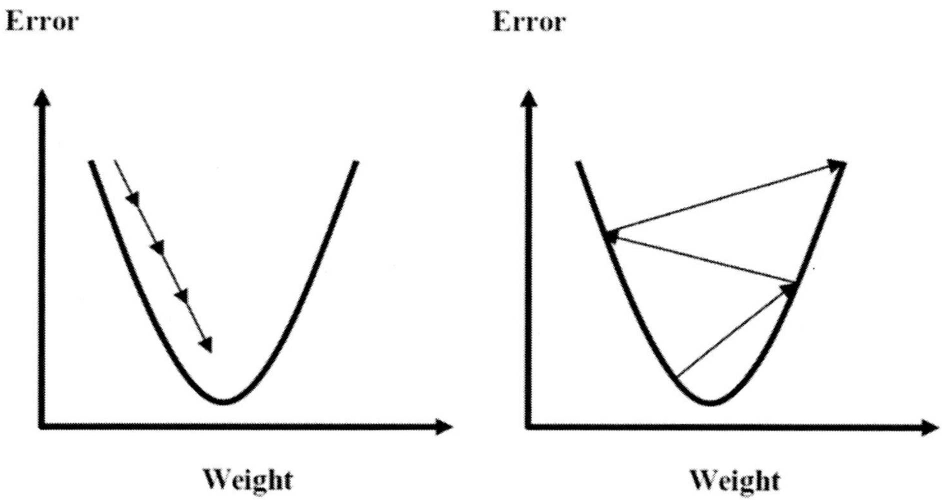

Figure 2.5. Effect of very small and very large learning rate on convergence.

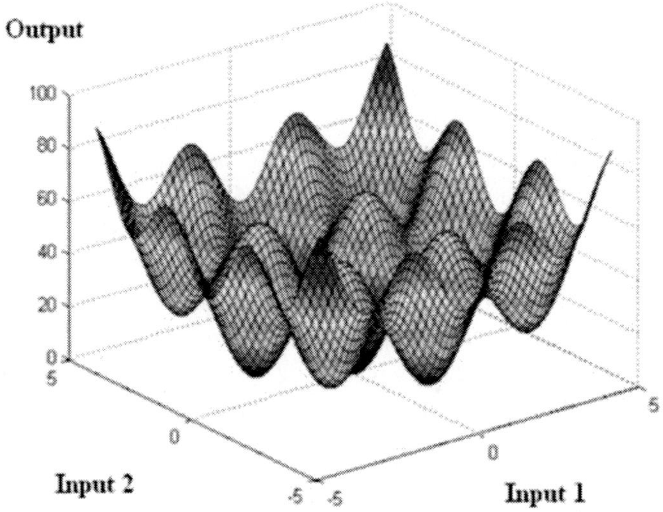

Figure 2.6. Multiple local minima of error surface.

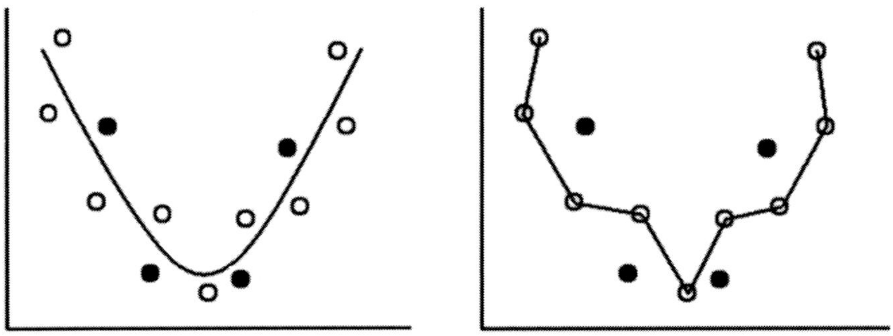

Figure 2.7. Generalization and overtraining.

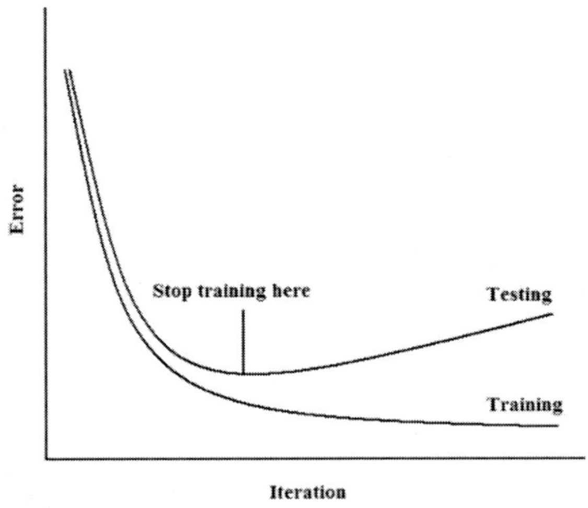

Figure 2.8. Generalization and over-training.

2.3.4. Generalization and Over-Training

The ANN can reasonably approximate any smooth function if optimum neurons are present in the hidden layer. This is known as generalization. However, if too many neurons are present in the hidden layer or if the training is continued for a very long duration then the ANN tends to learn the noise present in the training data and memorize them. Therefore, the network starts to lose its efficacy of prediction or classification in the unseen testing data. This is called over-training. Figure 2.7 schematically represents generalization and overtraining. Out of 13 experimental data 9 (hollow circles) has been used for training and remaining 4 (solid circles) has been used for the testing. The left side of the figure indicates generalization by the ANN as the smooth curve passes over and under the training and testing points. It is expected that at this situation the ANN will show comparable error for the training as well as for the testing data sets. However, the training error continues to reduce with the increased iteration and at extreme case a point will be reached when the function mapped by ANN will pass through all the training points (hollow circles) resulting in almost zero training error. However, the testing points (solid circles) will be far apart from the mapped function as shown in the right hand side of the figure. This will lead to very high error in the testing data set. To prevent the problem of over-training, the training has to be ceased as soon as the error signal in the testing data starts to shoot up, as shown in Figure 2.8.

2.3.5. Performance Evaluation of ANN

2.3.5.1. Statistical Parameters

Once the ANN has been trained, its performance is evaluated separately in the training and testing dada sets. Correlation coefficients (R) between the experimental and predicted values, mean absolute percentage error (MAPE) and mean squared error (MSE) are the statistical parameters with which the performance of ANN is appraised. The expressions of these statistical parameters have been given below:

Correlation coefficient =
$$\frac{\sum_{i=1}^{n} e_i p_i - \frac{\sum_{i=1}^{n} e_i \sum_{i=1}^{n} p_i}{n}}{\sqrt{\sum_{i=1}^{n} e_i^2 - \frac{\left(\sum_{i=1}^{n} e_i\right)^2}{n}}\sqrt{\sum_{i=1}^{n} p_i^2 - \frac{\left(\sum_{i=1}^{n} p_i\right)^2}{n}}} \qquad (2.12)$$

Mean absolute percentage error $= \frac{1}{n}\sum_{i=1}^{n}\left(\frac{|e_i - p_i|}{e_i}\right).100$ (2.13)

Mean squared error $= \frac{1}{n}\sum_{i=1}^{n}(e_i - p_i)^2$ (2.14)

where e_i is the i th experimental value, p_i is the i th predicted value and n is the number of observations.

2.3.5.2. Predictive Importance of Inputs

The analysis of trained ANN model can reveal very useful information about the importance of a particular input parameter and the functional relationship between the input and the output variables. Predictive importance of an input variable is conducted by eliminating one designed input from the optimized ANN model at a time. Training is initiated again and continued up to the optimum level. The difference in prediction performance is measured by the fractional change in mean squared error value in the testing data set. Higher change in the mean squared error (*ΔMSE*) signifies higher predictive importance of the eliminated input and vice versa as expressed below:

$$\Delta MSE = \frac{MSE_n - MSE_{n\text{-}1}}{MSE_n}.100 \qquad (2.15)$$

where MSE_n is the mean squared error of the optimized ANN with all input parameters and MSE_{n-1} is the mean squared error of the ANN with one eliminated input parameter.

2.3.5.3. Trend Analysis

Trend analysis is performed to understand the influence of input parameters on the output parameter. For the trend analysis, all the input parameters are kept constant at their median level and only one parameter is changed, in steps, from its minimum level to the maximum level and corresponding changes in the output is noted. The pictorial representation of output and input reveals the kind of relationship (linear or nonlinear) prevailing between them. Sometimes, the % change in the output parameter is taken as an indicator of the importance of the chosen input parameter as expressed below.

$$\%\ \text{contribution} = \frac{|Y_{x\max} - Y_{x\min}|}{Y_{\max} - Y_{\min}}.100 \qquad (2.16)$$

where Y_{xmax} and Y_{xmin} are the values of the output parameter when the chosen input parameter (x) is at its maximum and minimum level, respectively, and all other input parameters are at the median level, Y_{max} and Y_{min} are the highest and lowest value of the output parameter, respectively, in the entire dataset.

This approach of appraising the % contribution of input parameter is plausible if the relationship between the output and input parameter is monotonically increasing and decreasing.

2.4. ANN and Regression Model

Regression techniques are very popularly used for modeling when experimental data are available. In most of the cases linear form of regression is used due to its simplicity. The linear regression is having the following mathematical form:

$$y_p = \sum_{i=1}^{n} w_i x_i + \theta \quad (2.17)$$

where y_p is the predicted value of the output parameter, x_i is the ith input parameter, w_i is the regression coefficient of the ith input parameter and θ is the regression constant and n is the number of input parameters. The values of w_i and θ are fitted in least square method so that the squared residual (difference between the actual and predicted value of the output parameter) is minimized as shown below:

$$\text{Minimise} \sum e^2 = \text{Minimise} \sum_{1}^{m} (y_a - y_p)^2 \quad (2.18)$$

where y_a is actual value of the output parameter and m is the total number of observation. The limitations of regression model can be summarized as follows:

- A relationship has to be chosen before the actual analysis
- The relationships chosen are mostly linear
- Outliers can distort the computation of mean and variances. So, standard regression is not robust to the influence of outliers.

In case of ANN, nonlinear transfer functions are used in the hidden neurons. The output of the kth hidden neuron having log-sigmoid transfer function can be expressed as follows:

$$out_k = \frac{1}{1 + e^{-\left(\sum w_{kj} Out_j - \theta\right)}} \quad (2.19)$$

So, it is understandable that the ANN can handle linear and nonlinear relationships. There is no need to specify the functional form before modeling. However, the choice of number of hidden layers, number of neurons in each hidden layer, learning rate and momentum are of paramount importance to elicit the best performance from the ANN.

2.5. Applications of ANN

ANN has been successfully used in almost every field of science, engineering and management [8]. It has been used in weather forecasting, speech recognition, medical diagnosis and share market prediction. In civil engineering, ANN has seen successful applications in detection of structural damage, structural system identification, modeling of material behavior, structural optimization, structural control and ground water monitoring. In textile engineering, ANN has been used for fiber identification, prediction of yarn and fabric properties and for fault/defect analysis [9, 10]. Several software packages are available for ANN model development. However, MATLAB provides the most flexible and versatile platform for the ANN model development [11]. A simple MATLAB programme has been explained in the following section of the chapter for the benefit of the readers.

2.5.1. MATLAB Programming

A nonlinear function of the following form has been attempted for modeling using ANN. The range and interval of the input variable (x) is from -10 to 10 and 0.1, respectively.

$$f(x) = \frac{x^2 - 10}{x^2 + 10} \tag{2.20}$$

```
x= -10:.1:10;                                          [input parameter; range and
                                                        interval]
y= (x.^2-10)./(x.^2+ 10);                              [training output data]
net= newff(x, y, 6,{'tansig', 'tansig'}, 'traingd');   [network structure and transfer
                                                        function]
net.trainParam.show= 10;                               [result display interval]
net.trainParam.lr= 0.2;                                [learning rate]
net.trainParam.epochs= 100;                            [training iteration]
net.trainParam.goal= .01;                              [MSE target as stopping criterion]
net.divideParam.testRatio = 0;                         [no division of training data for
                                                        testing]
net.divideParam.valRatio = 0;                          [no division of training data for
                                                        validation]
net =train(net, x, y);                                 [training command]
        tx= -9:.1:9;                                   [testing input parameter;
                                                        range and interval]
ty= (tx.^2-10)./(tx.^2+10);                            [testing output parameter]
py1= sim (net, x);                                     [ANN output of training data]
py2= sim (net, tx);                                    [ANN output of testing data]
A= corr (y', py1')                                     [correlation for the training data]
B= corr (ty', py2')                                    [correlation for the testing data]
plot (x,y);                                            [plotting the actual function]
hold on;
plot (x, py1, 'black');                                [plotting the ANN output
                                                        in training data]
hold on;
plot (tx, py2, 'red');                                 [plotting the ANN output in
                                                        testing data]
```

Line 3 of the above programme builds an ANN structure having only one hidden layer with six neurons. Tan-sigmoid transfer function has been employed in this problem. The training algorithm is back-propagation or gradient descent (*traingd*). Execution of the above MATLAB programme shows the following results. The performance function (MSE) is evaluated after every 10th iteration (epoch). It is observed that after 52nd iteration, the performance goal or targeted MSE (0.01) has been achieved and the training has stopped. The correlation coefficient between the actual output and ANN predicted output in training and testing data sets are 0.9926 and 0.9930, respectively. Figure 2.9 depicts the actual and ANN fitted functions, which are reasonably close to each other.

```
TRAINGD-calcgrad, Epoch 0/100, MSE 1.13663/0.01, Gradient 0.516473/1e-010
TRAINGD-calcgrad, Epoch 10/100, MSE 0.78141/0.01, Gradient 0.45145/1e-010
TRAINGD-calcgrad, Epoch 20/100, MSE 0.0580127/0.01, Gradient 0.163246/1e-010
TRAINGD-calcgrad, Epoch 30/100, MSE 0.0267197/0.01, Gradient 0.0913753/1e-010
TRAINGD-calcgrad, Epoch 40/100, MSE 0.0155128/0.01, Gradient 0.0590749/1e-010
TRAINGD-calcgrad, Epoch 50/100, MSE 0.0104904/0.01, Gradient 0.0412757/1e-010
TRAINGD-calcgrad, Epoch 52/100, MSE 0.00984115/0.01, Gradient 0.0386719/1e-010
TRAINGD, Performance goal met.
A = 0.9926
B = 0.9930
```

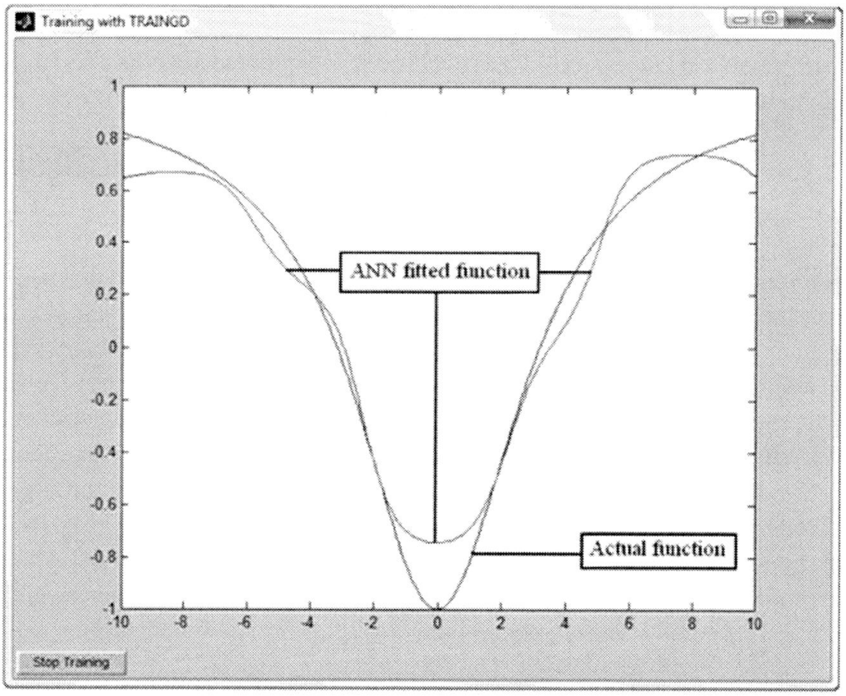

Figure 2.9. Function approximation by ANN.

Table 2.2. Prediction performance of ANN [Source: Reference 12]

Statistical parameters	Training data	Testing data
Coefficient of determination (R^2)	0.999	0.997
Mean absolute percentage error	3.332	5.067

2.5.2. Prediction of Compressive Strength of Concrete by ANN

In a recent work, Topcu and Sardemir [12] have used ANN for the prediction of compressive strength of fly ash filled concrete. The number of days, Portland cement, water, sand, crashed stone I, crashed stone II, high range water reducing agent, fly ash and CaO were the nine inputs to the ANN model. The compressive strength of concrete after 7, 28 and 90 days was the output. At a time only one output parameter was kept in the output layer. Out of 180 specimens, 120 were used for the training of the ANN using back-propagation algorithm. Remaining 60 specimens were used for the testing of the ANN model. The optimized ANN was found to have following parameters:

- Number of hidden layer = 1
- Number of neurons in the hidden layer = 11
- Momentum = 0.9
- Learning rate = 0.75
- Error after learning = 0.000163

The prediction performance of the ANN is shown in the Table 2.2. It is observed that the prediction performance is very good in training as well as in the testing data set implying very good generalization by the ANN model.

2.5.3. ANN Application in Modeling Organizational Behavior

Statistical methods have been used for years to model organizational behavior. However, in complex organizational system, the presence of outliers and non-linearity often make the standard statistical models inapt. Moderate levels of individual self-management may increase the inherent motivation and performance in organizational teams. However, high levels of individual self management may result in some independent decision-making that interfere with the interdependent behavior necessary for effective team functioning. Thus, there may be a nonlinear relationship between individual self-management and organizational team effectiveness. Grznar at al. [13] have reported the application of ANN to model organizational behavior. The data set was collected through structured interviews. A sample of 102 teams was interviewed using a questionnaire. The teams were from several different organizations and included a variety of ongoing work teams. The organizations varied in size from only 7 to over 10,000 employees. The team size ranged from 2 to 30 with its members having an average age of 38.7 years. The 41% of the participants were female. The average length of tenure at the present employer was 8.6 years, with 2.3 years of average tenure with the team. The independent parameters were organizational context, intra-team processes, inter-team processes and team size and the only output parameter was team effectiveness. Two third of the samples were used for the model development or training and remaining one third samples were used for the testing. Standard regression, robust regression based on median and ANN models were developed. The coefficient of determination (R^2) values for the regression, robust regression and ANN models were 0.247, 0.313 and 0.414, respectively in the testing data set. Trend analysis for the input parameter team size on the team's effectiveness showed very interesting nonlinear relationship. The poorest team performance was observed when the team size is around 10. Thus a manager allocating teams within an organization should ensure that the team size is much smaller or much bigger than 10. Team sizes much bigger than 10 were found to be effective, perhaps because larger teams often prefer to break up into smaller sub-groups of four to six members to do their work efficiently.

2.5.4. ANN Application in Modeling Yarn Tenacity

Tenacity is considered to be the most important property for spun yarns. As the properties of cotton fibers have lot of variability, prediction of yarn tenacity from the cotton fiber properties becomes a difficult task. Majumdar [10] predicted the single yarn tenacity of ring spun yarns by using ANN model. Cotton fiber properties measured by HVI (upper half mean length, uniformity index, bundle tenacity, elongation, micronaire, reflectance degree and yellowness) and yarn count were used as the inputs to ANN model. The number of neurons in the input and output layers was eight and one, respectively. The number of neurons in the hidden layer was optimized by trial and error method. Training and testing of ANN was done with 72 and 15 data sets, respectively. Log-sigmoid transfer function was used in the hidden and output layers. The MAPE of prediction was around 5% in the testing samples. Out of 15 testing data, 8 showed error of prediction higher than 5%. However, all the individual prediction error was within the limit of 10%. In order to quantify the relative importance of each input, predictive importance test was conducted by eliminating one designed input parameter from the ANN model at a time. Percentage increase in MSE value was considered as the indicator of importance of the eliminated input parameter. The detailed results of predictive importance test are shown in Tables 2.3. The elimination of fiber bundle tenacity resulted in the maximum increase in MSE, in the testing samples, followed by yarn count and fiber elongation. Length uniformity index was found to be the fourth important input parameter influencing ring yarn tenacity. However, upper half mean length (UHML) was placed in seventh position in the hierarchy which is not in agreement with the perception of spinning technology. This may be due to the lower range of UHML value in the used data sets.

Table 2.3 Relative importance of input parameters for ring yarn tenacity [Source: Reference 10]

Excluded input parameter	Mean squared Error (MSE)	% Increase in MSE	Rank
Fiber bundle tenacity	1.698	188.78	1
Fiber elongation	0.944	60.54	3
UHML	0.666	13.27	7
Uniformity index	0.761	29.42	4
Micronaire	0.703	19.56	6
Reflectance degree	0.704	19.73	5
Yellowness	0.661	12.42	8
Yarn count	1.132	92.52	2

Conclusion

The theoretical background of ANN and some applications have been presented in this chapter. Although ANN is a powerful modeling tool and has strong mathematical foundation, it requires huge among of data for learning of complex relationships. In recent years, ANN has been hybridized by coupling it with other soft computing techniques to make it more versatile and powerful. Neuro-fuzzy and ANN-GA are some of the hybrid systems which have become popular in practice for modeling and optimization.

References

[1] Hebb, D. O. The Organization of Behaviour, A Neurosychological Theory; ISBN: 0805843000; John Wiley: NY, 1949.

[2] Hopfield, J. J. In the Proceedings of National Academy of Sciences, USA, 79, 2554-2558, 1982.

[3] Kohonen, T. A Simple Paradigm For The Self-Organized Formatiom Of Structured Feature Maps In Competition And Cooperation In Neural Nets (Lecture notes in biomathematics: vol 45, Amari, S.; Arbib, M. A.; Eds.); ISBN: 0387115749; Springer-Verlag: Berlin, 1982.

[4] Rumelhart, D. E.; Hinton, G.; Williams, R. J. Learning Internal Representations by Error propagation in Parallel Distributed Processing: *Explorations in the Microstructure of Cognition*, vol 1: ISBN10: 026268053; MIT Press: Cambridge, 1986.

[5] Rajasekaran, S.; Pai, G. A. V. Neural Networks, Fuzzy Logic and Genetic Algorithms: Synthesis and Applications; ISBN: 8120321863; Prentice-Hall of India Pvt. Ltd.: New Delhi, 2003.

[6] Zurada, J. M. Introduction to Artificial Neural Systems; ISBN: 8172246501; Jaico Publishing House: Mumbai, 2003.

[7] Haykin, S. Neural Networks: A Comprehensive Foundation; ISBN-10: 0132733501; Pearson Education: 2nd Edn.; Singapore, 2004.

[8] Bhadeshia, H. K. D. H. ISIJ Int., 1999, vol 39(10), 966-979.

[9] Chattopadhyay, R.; Guha, *A. Text. Prog.* 2004, vol 35(1), 1-46.

[10] Majumdar, A. Quality Characterization of Cotton Fibers for Yarn Engineering using Artificial Intelligence and Multi-criteria Decision Making Process; PhD Thesis, Jadavpur University, Kolkata, 2005.

[11] Demuth, H.; Beale, M.; Hagan, M. MATLAB Neural Network Toolbox, The MathWorks Inc, 2001.

[12] Topcu, I. B.; Saridemir, M. *Comput. Mater. Sci.* 2008, vol 41, 305-311.

[13] Grznar, J.; Prasad, S.; Tata, *J. Eur. J. Oper. Res.* 2007, vol 181, 939-955.

In: Modeling and Simulation in Fibrous Materials
Editor: Asis Patanaik
ISBN: 978-1-62100-116-4

Chapter 3

INTRODUCTION TO FUZZY LOGIC AND RECENT DEVELOPMENTS

Yordan Kyosev*
Niederrhein University of Applied Sciences, Mönchengladbach, Germany

ABSTRACT

This chapter presents an introduction into the fuzzy logic, adopted for non mathematicians, in order to be simplifying its understanding from the textile related personnel and motivate them to use it. At the beginning, the fuzzy number and fuzzy logic rules will be explained. Before explaining the most used applications of the fuzzy logic – the fuzzy controller is reported in section 3.4, both types of reasoning – forward and backward are explained in the section 3.3. So the users can understand the mathematics and the logic behind the normal reasoning and can get idea about the inverse problems. The definition and application of the inverse problems for fuzzy linear systems is the topic for section 3.5. An extension of the fuzzy logic into Atanassov's intuitionistic logic is explained into section 3.6. Some possible applications are reported in the last section.

3.1. INTRODUCTION

The *Logic*, from the Greek λογικός (logikos) is the study of reasoning [1]. According to one of the simple and very early definitions of the logic [2], it is "the tool for distinguishing between the true and the false". For the classical mathematicians both of these states are simple denoted with "1" and "0" and implemented in the modern computers. But the nature is complex enough and cannot be effectively presented as not only "black" and "white" images. If we see just the colors in the nature there are millions intermediate levels between black and white, or blue and red.

*E-mail: yordan_kyosev@gmx.net

We, the humans, are complex creatures as well, able to use emotions in our valuations. Let us see how the humans evaluate one product; it could be a textile one. First, evaluation criteria have to be defined. Depending on these criteria, the evaluated product could satisfy us completely (satisfaction= 1 or "true") or not. In the most cases the evaluated products are created and produced as compromise between "costs" and "quality", and because of this they cannot satisfy all criteria completely. In this case some adjectives are used in the evaluation and the final note is set it to be "almost good", "near good" etc. which could be set mathematically as "satisfaction = 0.7" or "almost true". The area in the mathematics, which deal with such numbers is named *fuzzy logic* and as its founder is Prof. Zadeh [3].

This chapter presents an introduction in the *fuzzy logic,* but adopted for non mathematicians, in order to be simplifying its understanding from the textile related personnel and motivate them to use it.

At the beginning, the basics of the fuzzy sets and fuzzy logic rules will be explained. After this introduction, different types of mathematical treatments of "If-Then" rules are presented, as well as the use of such in the fuzzy controllers. At the end, fuzzy linear systems and their extension to intuitionistic linear systems as an inference engine are discussed.

3.2. Fuzzy Logic and Fuzzy Sets

Let the cotton fiber bales of a some spinning mill are named B_i and the mean HVI strength of the fibers of each of these bales is x_i, where $1 \leq i \leq N$ is the number of the current bale, and currently there are N bales. Some of these fibers have to be used for spinning of yarn with certain tenacity T. The spinning specialists can calculate the required minimum HVI strength of the fibers for this application and let in this case the minimum limit is 27 g/tex.

According this limit, all the bales can be classified into two sets – enough strong, or just "strong" and "weak" for the current application (see Figure 3.1).

Mathematically, the bale B_1 belongs to the set of "Weak" fibers. Two linguistic terms can be defined according the current application "weak tenacity" (*W*) and "strong tenacity" (*S*).

The membership of each bale to one of the sets or terms can be expressed with the membership degree $\mu(x)$, so that:

$$\mu(x) = \begin{cases} 1 & \text{if } B_i \in W \\ 0 & \text{if } B_i \notin W \end{cases} \tag{3.1}$$

The membership degree $\mu(x)$ is 1, if the bale belongs to the set *W*, it means it fibers are "weak", and opposite – is zero, if the fibers are "strong". So, for every bale, the pair of the bale number and its membership degree to the *W* build a finite set of pairs:

$$W = \{(B_1, \mu(B_1)), B_1, \mu(B_1)), \ldots, B_N, \mu(B_N))\} \tag{3.2}$$

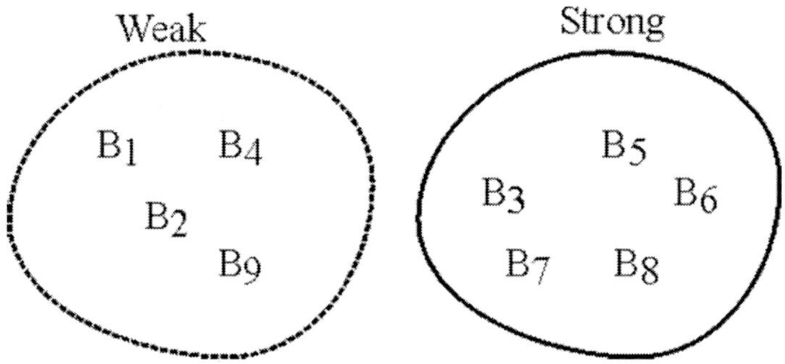

Figure 3.1. Yarn bales, classified into two sets, according their tenacity.

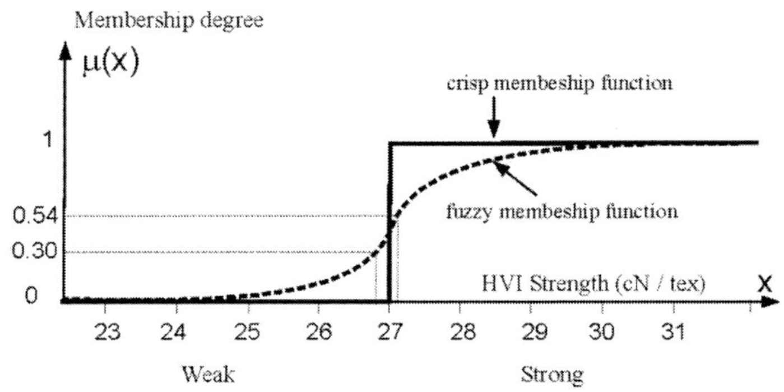

Figure 3.2. Crisp and fuzzy membership functions, connecting the HVI strength of the fibers with the membership into the set of "strong" fibers.

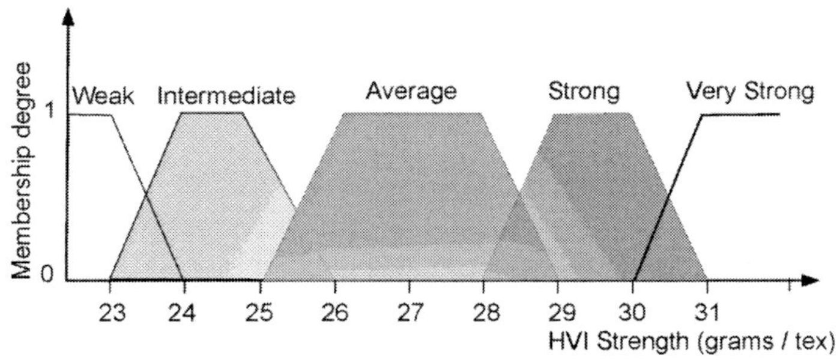

Figure 3.3. Linguistic variables for the classification of cotton depending its tenacity.

This set contains the information only if fibers in certain cotton bale are strong or weak, but classified on the basis on the 27 cN/tex limit. This is the usual way of handling memberships in the classical logic, named as well Boolean logic because of the two possible states (0 or 1) or crisp logic. One statement could be only "true" or "false" or one element could belong to one set or not. Thus, the situations, where the membership to some set is not very clear and sure, cannot be described and treated using the crisp logic. If some bale consists of fibers with tenacity 26.8 cN/tex, it will be still classified as bale with "weak" fibers, even if these could be used for the application exactly so gut, as the "strong" fibers with tenacity 27.1 cN/tex.

This limitation or disadvantage of the crisp logic is eliminated with the allowance for the membership degree to get every value between 0 and 1. In this case every bale can belong to the set of the "weak" fibers with membership degree of 0.3, for instance. This means, that these fibers are weak, but "not very" weak. But if there are "more weak" or "more strong" it could not be seen on the single set *W*. If the bale have membership degree 0.54 to the "strong fibers", then is clear, that these fibers are "more strong, then weak" (Figure 3.2).

According to the current needs, several linguistic terms to the linguistic variable tenacity can be defined and used. Usually the cotton fibers are classified into groups depending on their tenacity as "weak", "intermediate", "average", "strong" and "very strong" fibers (Figure 3.3) ("Cotton Classification").

Using more linguistic terms is possible to represent the human knowledge about this problem more accurately. The fuzzy membership degrees allow more accurate representation in the transition regions between two terms. The accuracy of the representation in the transition regions depends on the form of the membership function. On the following figures are demonstrated graphically some membership functions, which are used in the praxis.

The classical, "crisp" case has a strong limit and is mathematically described as:

$$\mu(x) = \begin{cases} 0 & x \le x_a \\ 1 & x_a < x < x_b \\ 0 & x_b \le x \end{cases} \tag{3.3}$$

and is presented on the Figure 3.4, as well as Figure 3.5. and Figure 3.6. as curve *C*. If instead of a crisp, impulse-like transition, a linear one on the both sides of the interval is used, a trapezoidal form of the membership function can be designed:

$$\mu(x) = \begin{cases} 0 & x \le x_{af} \\ \dfrac{x - x_{af}}{x_{ma} - x_{af}} & x_{af} \le x \le x_{ma} \\ 1 & x_{ma} \le x \le x_{mb} \\ \dfrac{x - x_{bf}}{x_{mb} - x_{bf}} & x_{mb} \le x \le x_{bf} \\ 0 & x \ge x_{bf} \end{cases} \tag{3.4}$$

The trapezoidal membership function is suitable for linguistic terms, which have the maximal degree of membership (1) not in a single point, but in a more wide area, in a range between $x_{ma} \le x \le x_{mb}$. Probably the most common used membership function is a derived from it, if this range is deformed to a single point – it is the triangular fuzzy membership function, presented on the Figure 3.5.

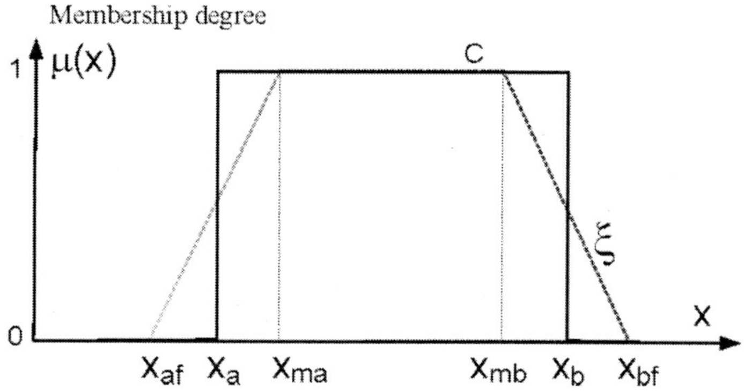

Figure 3.4. Crisp (rectangular) and trapezoidal fuzzy membership functions.

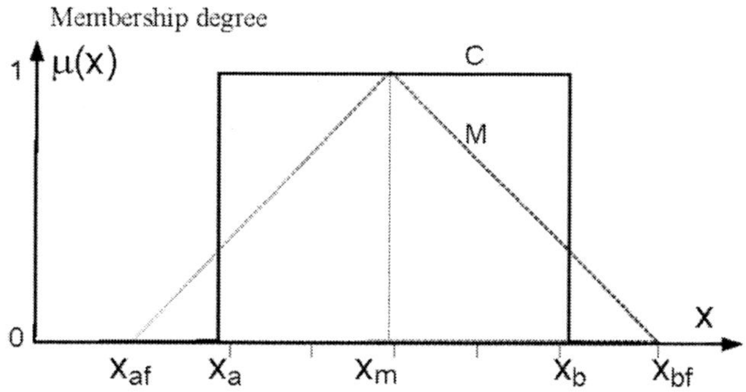

Figure 3.5. Triangular membership function.

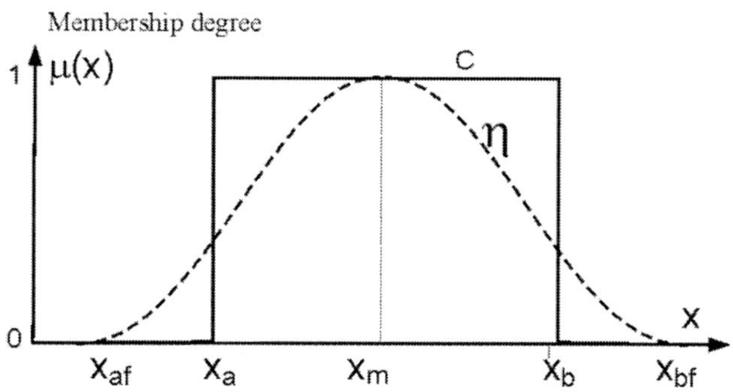

Figure 3.6. Gaussian curve η as a membership function.

$$\mu(x) = \begin{cases} 0 & x \le x_{af} \\ \dfrac{x - x_{af}}{x_m - x_{af}} & x_{af} \le x \le x_m \\ \dfrac{x - x_{bf}}{x_m - x_{bf}} & x_m \le x \le x_{bf} \\ 0 & x \ge x_{bf} \end{cases} \tag{3.5}$$

Both function (3.4) and (3.5) have the advantage, that they are very simple for calculation. More complex, but more naturally way to describe the membership to one value would be the normal (Gaussian) Curve, presented on the Figure 3.6.

$$\mu(x) = e^{-\frac{(x_m - x)^2}{2s^2}} \tag{3.6}$$

where in the case of Figure 3.6 $s = \frac{1}{2}(x_b - x_a)$, but the width of the distribution could be selected in any other way. The most of the measured parameters in the textile industry corresponds to normal distribution and thus using this membership function will have the advantage of clear connection between the membership function and the statistics from the experimental investigation. Of course, if some experimental data does not follow the normal distribution, but some other, suitable equations for membership functions can be selected. Depending on the required accuracy and the computational way and time, from very simple triangle functions up to complex mathematical functions can be used. Some more fuzzy membership functions, not described here could be found in the literature, for instance in [4].

3.3. Logical and Fuzzy Operations

Let two variables A and X is available. The classical logic says, that
If A is "true" and X is "true", then (A and X) is "true".
But if *A* is "almost true" and *X* is "less true", what will be the result of (*A* and *X*)?

Before demonstrate some extension of the classical logic for the fuzzy case, some basic rules about these operations have to be fixed:

- It was mentioned for all examples in the previous subsection, that the fuzzy variables (and membership functions) can have values between 0 and 1.
- If some operations over one variable *X* (unary operations, like "not *X*") or over two variables (binary operations, like "*A* and *B*") have to be calculated, the result have to be as well in the same range – between 0 and 1;
- Join $\wedge$ and meet $\vee$ operations between two sets (or variables) are defined (as intersection and union of the two sets).

In the mathematical language (algebra) these rules define a *bounded chain* L and are written as:

$$\mathsf{L} = (L, \wedge, \vee, 0, 1) \tag{3.7}$$

In case of using max as join $\wedge$ and min as meet $\vee$, this bounded chain in known as *fuzzy algebra:*

$$\mathsf{I} = ([0,1], \max, \min, 0, 1) \tag{3.8}$$

Table 3.1. Comparison between the Boolean and Fuzzy operations [Source: Reference 6]

		Boolean	Fuzzy	Boolean	Fuzzy	Boolean	Fuzzy
A	X	A and X	min(A,X)	A or X	max(A,X)	not A	1- A
0	0	0	0	0	0	1	1
1	0	0	0	1	1	0	0
0	1	0	0	1	1		
1	1	1	1	1	1		

This is the most used, but not only one used composition in the fuzzy logic, where the logical "and" and "or" are represented with "max" and "min" operations, respectively [5]. These operation work as well with the classical values (0 and 1) as with fuzzy numbers, as demonstrated on the Table 3.1 [6]. As it is presented on the Table 3.1, this replacement works as well with the Boolean numbers. This is, because the fuzzy logic is an extension, a superset, of the classical Boolean logic. Except the max-min operations, there are several different other operations, used as extension of the standard logic in for the fuzzy case, as for instance *max* and *product* operations. More about these and some such applications can be found in [7-9].

3.4. If-Then Rules with Look up Table

The important advantage of the fuzzy logic is in the possibility to represent in formal language a set of if-then rules. These rules allow representation of the human knowledge in the form of conditional sentences. Depending on the number and the type of the connection between the single conditions, there are different methods for calculations.

The most simple and popular case is the situation, where two conditions have to be simultaneously satisfied – they are connected with AND. Simple example for such case is the washing procedure (or in case of automated measuring of the clothing dirtiness – washing machine). Let variable X_W is the weight of laundry, which has to be washed and X_D is the dirtiness of the clothing. The required amount of washing time is then *Y*. Logical is the following rule:

If X_W is low and X_D is low, then Y is very low
If X_W is low and X_D is middle, then Y is low
If X_W is high and X_D is high, then Y is very high
If X_W is high and X_D is low, then Y is medium etc.

A simple form a "if-then" rules can looks like:

if X is "low" AND Y is "low" THEN Output is "very low (VL)".

In the case of two variables, using the membership functions an analogue input, like "amount material" and "dirtiness" the so named "look up tables" can be used, as presented on the Figure 3.7.

If the input variables are more than two, the look-up table can still be programmed in the computers as multidimensional array, without significant problems in the computational complexity. Problematic in such cases could be the proper selection of the values of the multidimensional table, if no enough data about the relation between the input variables and the output is available. In such cases, a design of experiments (DOE) can be helpful, in order to keep the possible relations under control with possible small amount of tests.

In some cases the size of the table can be larger and in order the values in the matrix (responses of the system) to be better visualized and easier to be understand, they are presented as a 3D surface (Figure 3.8).

The above example has an advantage – the rules are quite simple - and at least two disadvantages – the results are changing in steps. What to do, if the input variable is "between" two levels? For instance, if the degree of dirty is determined using a measurement of the water dirtiness, one can get a degree of dirtiness, which corresponds with degree of membership 0.8 to the "Low" and 0.2 to "medium". In this case there are two strategies.

First strategy is to fire only those rules, which have higher degree of membership, in our case "dirtiness is "low".

Second strategy is to use both of the rules and to combine the results of these in the answer.

The second strategy is quite wide used in the fuzzy control, because it uses the entire information, available in the fuzzy membership functions and variables. Contrary, the first strategy uses the membership functions only to classify the degree of appearance of the output.

There is no possibility to control the washing time more accurately during selection of sometime between the steps in the look-up table. In order to get continuous result, a "defuzzification" of the end result is required, which is explained in the next section.

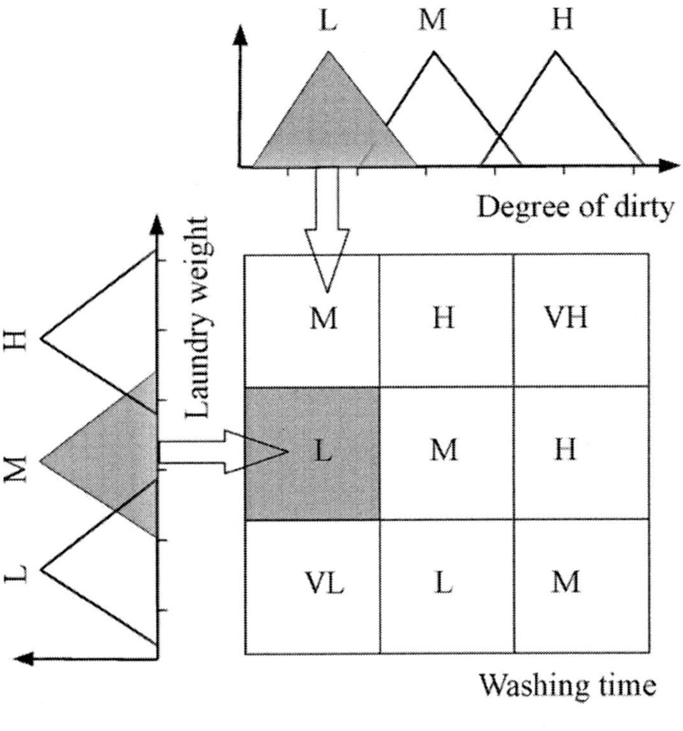

Figure 3.7. Look-up table for implementation of two if-then rules, connected with AND operator. VL- very low, L- Low, M-Middle, H- High, VH – very high.

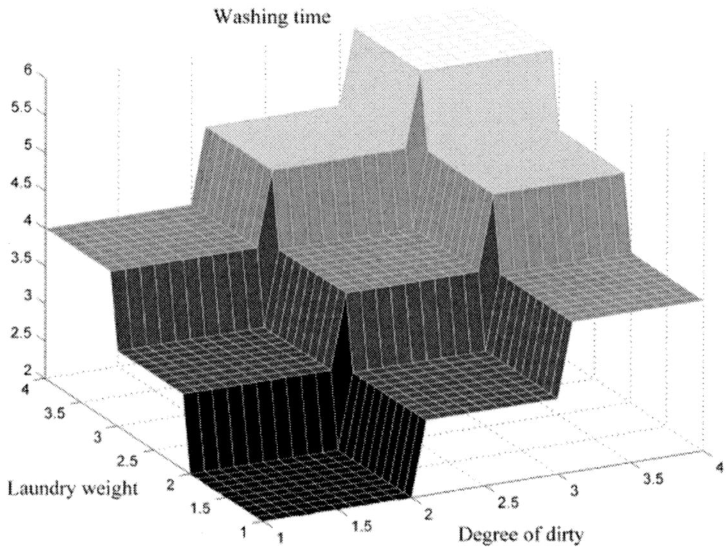

Figure 3.8. Surface representation of the relation between the input variables "Laundry weight" and "Degree of dirty" and the output variable "Washing time" for the look-up table in the Figure 3.7 All scales are in relative dimension.

3.5. If-Then Rules With Defuzzification of the Output

Let several If-Then rules between the two variables "degree and dirty" and "laundry weight" are given. In each rule, both these variables are connected with "AND", because the response of the machine is valid only if both terms are valid. From other side, one can select between the single rules, and because of this, the rules are connected with OR operator.

The entire procedure consists of four steps:

Step 1: Reading the physical variables and finding the membership degrees of the linguistic variables

Step 2: Calculating the AND operation of each rule, using the min operator

Step 3: Aggregation: Applying the OR operation over the results of each active rule (max operator)

Step 4: Defuzzification - calculate the answer according some rule – in the case of Figure 3.9 – this is the exact washing time.

Before the Step 4 – a set of membership functions is received as a solution. There are different possibilities for the calculation of one number as a final result from this set, such like using the centroid formula, using the mean of maxima, left-right maxima or other. In the most cases in the literature is used the centroid method according the following equation:

$$x_{solution} = \frac{\int \mu(x) \cdot x \, dx}{\int \mu(x) \, dx} \tag{3.9}$$

where $\mu(x)$ is the aggregated membership function of the output and $x_{solution}$ is the defuzzified output.

This way of dealing allows consideration of several rules, connected with OR, if every one of them influence one and the same output variable.

There are several applications of this way of thinking, where one good textile application in the textile is for instance reported in [10-11] as a method for prediction of the yarn strength, based on the mechanical properties of the fibers. The most popular other applications are is in the area of the fuzzy control, described in the next section.

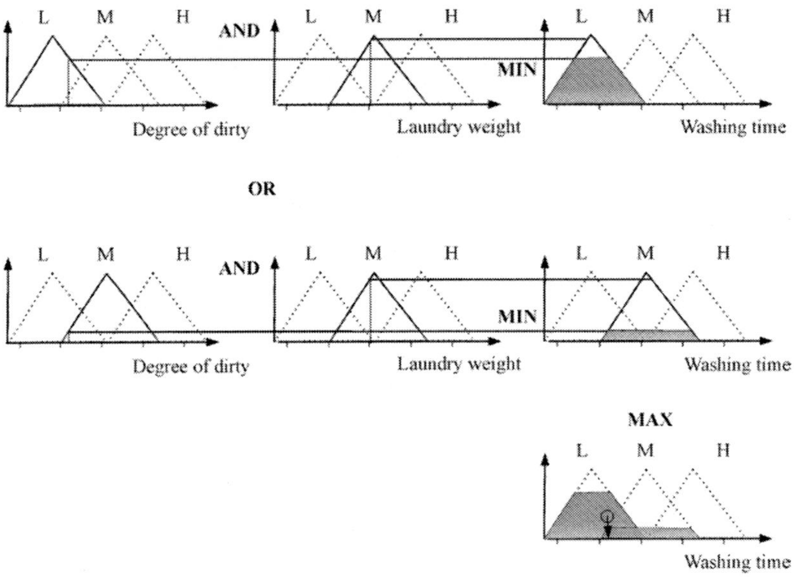

Figure 3.9. Schematically presentation of two "AND" rules, connected with OR.

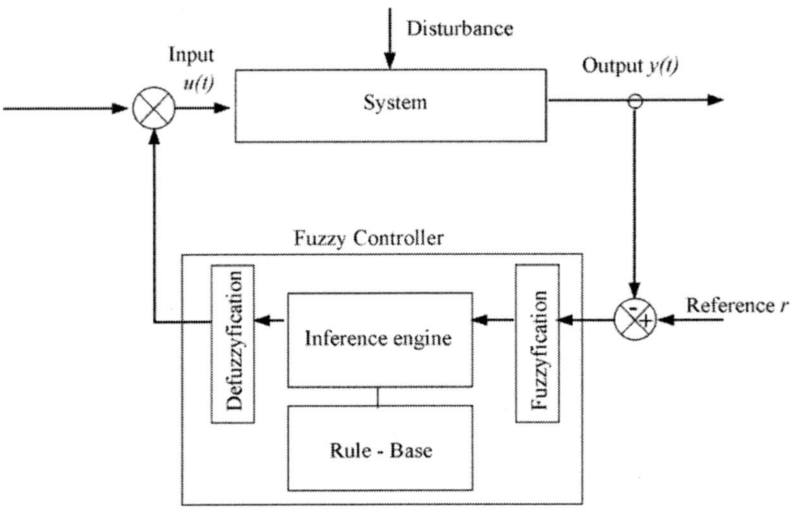

Figure 3.10. System with controller. Here the controller is implemented as a fuzzy controller.

3.6. Fuzzy Controller

Under controller here is mentioned a device, which monitors a given dynamical system and affects the output variables *y(t)* of this system by adjusting certain input variables.

Let the system of the system – "washing machine – laundry" is considered. The degree of dirty of the water in the washing machine is output of the system and required is degree 0 (clear water). If the laundry is not enough clear, "required additional washing time" is needed, and this is the input variable of the system – "washing machine – laundry". The output variable "degree of dirty of the water" is compared with a reference value $\mathrm{r} = 0$ and the difference is given to the if-then rule based fuzzy system, described in the previous section. Based on the rules, an additional time for washing will be calculated and given to the machine until the laundry will be clean. If the system is well designed, the laundry will be washed for the minimal possible time, saving energy, water and time, because for less dirty laundry, less time will be planned.

The output of such systems can be a vector of variables as well and in this case the fuzzy system operates with all of them.

In the similar way can be controlled almost all kind of dying, fixing etc. machines, where some state has to be achieved, but where the exact amount of time, energy, chemicals etc. is unknown and depends on several unknown properties of the system. Such controllers can be used as well as for devices for adjusting of the fabrics at the input of machines, where etc., actually the advantage of the fuzzy control is there, where the input has to be similar to the human language or where several input variables have to be combined.

One has to take into account, that these systems works (until now) with several inputs and only one output. For modeling of information, where several inputs have the influence of certain numbers of outputs, the fuzzy linear systems are proposed by Peeva and Kyosev [7]. The reader has to be careful with starting of these systems, because if the theory is already almost very well developed, the application of such systems is still mainly object of the research projects and there is not enough data from practical applications available.

3.7. Fuzzy Linear Systems

As mentioned in the previous section and as follows from their name, the fuzzy linear systems deal with several variables simultaneously. Let an example about faults and causes for them on a sewing machine be considered. Let m possible faults can detected, and there exists n causes, which could cause these faults separately or in combination. All the causes and faults are presented here using their degrees of membership x_i and b_j , respectively, where x_i is the membership degree of the *i*-th cause and b_j - the membership degree of the *j*-th fault.

Let us introduce the coefficients a_{ij} of influence between cause i and defect j. If $a_{ij} = 0$, there is no dependence between the cause i and the defect j. If $a_{ij} = 1$, there is very strong, clear relation between then, and even if the cause i appear with very small intensity, the defect j will follow with the corresponding intensity, calculated according the composition law. All these coefficients a_{ij} build one relation between the vectors of causes and defects, common named as “expert” matrix A.

The single rules, (relation between faults and defects through the expert matrix A) can be expressed as follow:

IF cause i is presented with degree x_i AND this cause has influence a_{ij} over fault j

OR

IF cause k is presented with degree x_k AND this cause has influence a_{kj} over fault j

THEN the fault j will appear with degree

$$b_j = (x_i \wedge a_{ij}) \vee (x_k \wedge a_{kj}) \tag{3.10}$$

In this case for operations $\wedge$ and $\vee$ are used min and max operations, respectively, as proposed in [6, 9, 12]. If all the causes x_i as set in a vector X and all the effects b_i in a vector B, then all the rules between these vectors build a linear system, which could be written in mathematical form as a system of linear equations:

$$\left|\begin{array}{lllll} (a_{11} \wedge x_1) & \vee \cdots \vee & (a_{1n} \wedge x_n) & = & b_1 \\ \ldots & \ldots & \ldots & \ldots & \ldots \\ (a_{m1} \wedge x_1) & \vee \cdots \vee & (a_{mn} \wedge x_n) & = & b_m \end{array}\right. \Leftrightarrow A \circ X = B \tag{3.11}$$

Here $A = (a_{ij})_{m \times n}$ is the matrix, where the relations between faults $X = (x_i)_{n \times 1}$ and causes $B = (b_j)_{m \times 1}$, $1 \le j \le m$, $1 \le i \le n$ are given. All the elements of these matrixes have to be in interval $[0,1]$. The operation $\circ$ here denotes the “max-min” composition, which has to be calculated according the rule (3.10) – first the minimum between elements and then the maximum of these minimums.

In the mathematical literature there are several possible composition lows cited as for instance “min-max”, “max-prod”, “min-prod” etc. Actually in the practical cases the most used one is the max-min composition.

There are three types of problems, which can be defined and solved, for the system (3.11):

1) Given Expert Matrix A, given faults X, to found the causes B, this corresponds to the forward scheme of reasoning;
2) Given Expert Matrix A, given causes B, to found the faults X, so we have backward scheme of reasoning;
3) Several sets of faults X and causes B are known, to find the expert Matrix A. This is as well in some way backward scheme of reasoning.

3.7.1. Forward Scheme of Reasoning

We suppose the relations matrix A and one vector with faults X are given. We can compose the elements from the left side of (3.11) and to receive the expected faults. In this way we will apply the *forward scheme of reasoning*, or the so called *direct problem*. This way is useful for testing student's knowledge – the computer give the causes X, students write the possible faults B^*, then computer calculate the difference

$$Error = A \circ X - B^* \tag{3.12}$$

and if error is zero vector, then the answer is correct.

This way is useful as well to test the sensitivity of the systems – changing the input variables, to find out which change in the outputs will be achieved. From the computational point of view this problem is the most simple to be solved, because it requires only applying the operations between A and X.

3.7.2. Backward Scheme of Reasoning

This is the most used case by diagnosis and it is connected with the solution of the "inverse problem", finding all solutions of the system (3.11). The vector with current faults B is given, as well as matrix A. Then we have to apply *backward scheme of reasoning*, solving the system (3.11). If the system (3.11) has solution, it has one maximal and one or lot of minimal solutions. The minimal solutions give the easiest and the lightest combinations of causes that could lead to a given defect. The maximal solution gives the most complicated combination, which would give the same defect.

For calculation of maximal (upper point) solution is known easy analytical formula [7], if the system is solvable:

$$X_{Gr} = A^t \alpha B \tag{3.13}$$

where A^t is the transposed matrix of a, and the operation α is defined as follows [13]:

$$a\,\alpha\, b = \begin{cases} 1, & if \quad a \le b \\ 0, & if \quad a > b \end{cases} \tag{3.14}$$

The difficulty is in finding all minimal solutions. There are several papers about the calculation of the minimal solutions of one fuzzy linear system of equations. In [14], it is described as effective algorithm that finds all low solutions, presented as well as in [7] and [15].

This algorithm is realized in a MATLAB$^{(R)}$ Toolbox, with two versions – "standard" or slow working version, available under BSD Lizens as well as compact, "fast solver" which is running several times faster than the standard system as a compiled demo version.

The physical meaning of the maximal solution in the case of diagnosis can be described with the “worst case”, where all possible reasons, which lead to some defect, appear at the same time simultaneously, with the full power. This worst case happens seldom and normally is not difficult to be detected.

The problems, where just only one reason appear and leads to some defect are normally the case. Such simple cases are described with the minimal solutions of the fuzzy linear system (3.11). Such system could have:

- no solutions – this means that the task is not correct defined.
- unique solution – in this case only one solution is available –there is clear reason for the fault.
- unique maximal and a set of minimal solution – where several combinations can lead to the same cause. In the praxis, each of the minimal solutions has to be checked in order to be found the reason for the fault.

Still mathematically open is the problem with the definition of the expert matrix. In the practical case a number of experts are sitting together build such a matrix. If such matrix will give answers in all the cases, and how sensitive is this system – these questions are still subject of the research works.

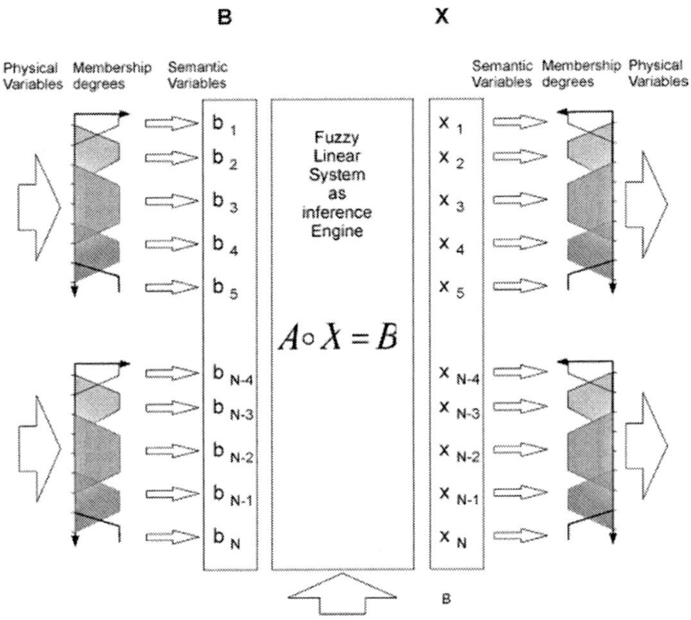

Knowledge (Relations) Matrice A

$$A = \begin{bmatrix} a_{11} & a_{12} & a_{13} & \dots & a_{1N} \\ a_{21} & a_{22} & a_{23} & \dots & a_{2N} \\ a_{31} & a_{32} & a_{33} & \dots & a_{3N} \\ \dots & \dots & \dots & \dots & \dots \\ a_{M1} & a_{M2} & a_{M3} & \dots & a_{MN} \end{bmatrix}$$

Figure 3.11. Block diagram of the expert system.

3.7.3. The Expert System

The symbolical block-diagram of working system, based on the fuzzy linear system is presented on the Figure 3.11. The user could enter the measured parameters of defects as inputs, then the system will translate them using membership function into coded vector of faults variables. It is possible to omit this first stage and the user directly to enter faults as linguistic variables. This way is more intuitive and understandable for users, because the current state is described with near to the human language proposals. After solving fuzzy linear system, each solution vector X could be used directly with linguistic variables, or could be decoded in pure values of parameters using corresponding membership functions.

Two level of access are suitable for such systems:

- User level - at this level the user is allowed only to use diagnosis system for fault detection. After starting, user has to fill the form with currently appearing faults. The normal users, does not have right to change the coefficients in the expert data base.
- Expert level access - experts have right to change the relations matrix and membership functions for each process or machine type. With appropriate shell user is not obligated to now the numbers of faults and causes, because the program shows the text, assigned to each of them.

3.7.4. Practical Example

One of the most important problems in sewing industry is the quality of stitches. There are a lot of factors, which may influence over the stitches quality and not every person can operate within them. If it is possible to ask a computer expert system when troubles appears, it will be easier and effective for a lot of companies and housewives, instead to use the specialist. Here is introduced only a small example of such expert system.

Let us use short list of stitch defects, introducing the variables:

- b1 - no defects found;
- b2 - missing stitches;
- b3 - the interweave of the threads visible on the upper (face) side;
- b4 - the interweave of the threads visible on the bottom (back) side;
- b5 - torn upper thread;
- b6 - torn down thread.

The input "missing stitches" with degree of appearance 0.3 will be coded as

B = [0 0.3 0 0 0 0]

Used causes, presented with relevant linguistic variables are:

- upper thread tension:

x1- low; x2 – normal; x3 - high

- proportion upper / bottom thread tension:

x4 - greater than 1; x5 - near to 1; x6 - less than 1

- upper thread tenacity:

x7 – low; x8 – normal; x9 - high

- bottom thread tenacity:

x10 - low; x11- normal; x12 - high

- hook to needle horizontal distance

x13 - near; x14- normal, x15 - far

- vertical position of the needle

x16 – low; x17- normal; x18 - high.

The relational matrix between given defects and causes for selected type of sewing machine and membership functions for the parameters could be:

$$A = \begin{bmatrix} 0.0 & 1.0 & 0.0 & 0.0 & 1.0 & 0.0 & 0.0 & 1.0 & 0.0 & 0.0 & 1.0 & 0.0 & 0.0 & 1.0 & 0.0 & 0.0 & 1.0 & 0.0 \\ 0.0 & 0.0 & 0.0 & 0.0 & 0.0 & 0.0 & 0.0 & 0.0 & 0.0 & 0.0 & 0.0 & 0.0 & 0.0 & 0.0 & 1.0 & 1.0 & 0.0 & 1.0 \\ 0.0 & 0.0 & 1.0 & 1.0 & 0.0 & 0.0 & 0.0 & 0.0 & 0.0 & 0.0 & 0.0 & 0.0 & 0.0 & 0.0 & 0.0 & 0.0 & 0.0 & 0.0 \\ 1.0 & 0.0 & 0.0 & 0.0 & 0.0 & 1.0 & 0.0 & 0.0 & 0.0 & 0.0 & 0.0 & 0.0 & 0.0 & 0.0 & 0.0 & 0.0 & 0.0 & 0.0 \\ 0.0 & 0.0 & 0.7 & 0.0 & 0.0 & 0.0 & 1.0 & 0.0 & 0.0 & 0.0 & 0.0 & 0.0 & 0.0 & 0.0 & 0.0 & 0.0 & 0.0 & 0.0 \\ 0.0 & 0.0 & 0.0 & 0.0 & 0.0 & 0.7 & 0.0 & 0.0 & 0.0 & 1.0 & 0.0 & 0.0 & 0.0 & 0.0 & 0.0 & 0.0 & 0.0 & 0.0 \end{bmatrix}$$

Let us analyze the result from solving the system (3.11), by the given A and B using the algorithm in [7]. The program output window is given in the Figure 3.12. The systems found 3 minimal solutions and one maximal solution for the case missing stitches. A disadvantage of the fuzzy linear systems is the great number of the variables, which has to be used – for each physical variable at least 3 fuzzy variables (low, middle, high- level) are introduced. This makes the work on such systems more difficult and the finding of the errors require a lot of patience. The next problem is that in several cases as well the low level of some parameter as the high level of this parameter can lead to similar effect. But these to states cannot appear at the same time. Some check of these rules after the solution of the system has to be done in order to avoid results, which could be mathematically correct, but physically wrong. The problem results from the fact, that all variables in the input vectors X and B are assumed to be independent. Actually in the case where 3 variables describe 3 different states of one the same variable, these are already dependent, because only one or two of them could appear at the same time. This problem could be elegantly solved using the intuitionistic variables and the intuitionistic fuzzy logic, described in the following section.

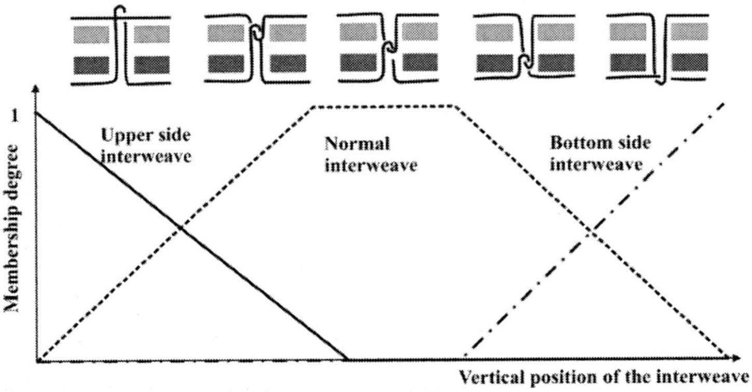

Figure 3.12. Membership functions for the interweave of the sewing threads.

```
Input
B = 0.3 missing stitches

Solution exists

Max solution
X 9 = 1 high upper thread tenacity
X 12 = 1 high botton trhead tenacity
X 13 = 1 near hook to needle - horizontal distance
X 15 = 0.3 far hook to needle - horizontal distance
X 16 = 0.3 low vertical position of the needle
X 18 = 0.3 high vertical position of the needle

Min 1-st solution
X 15 = 0.3 far hook to needle  - horizonal distance

Min 2-th solution
X 16 = 0.3 low vertical position of the needle

Min 3-th solution
X 18 = 0.3 high vertical position of the needle
```

Figure 3.13. Output from the simple diagnosis system to the given symptom “missing stitches”.

3.8. INTUITIONISTIC FUZZY LOGIC

3.8.1. Introducing Intuinitionistic Variables

Let us introduce the intuitionistic variable "high position of the needle". To this intuitionistic variable corresponded three linguistic variables, with relevant “membership” functions:

Membership function $\mu(x_1)$ - “the needle is too high” and

Non-membership function $v(x_1)$ - “the needle is too low”

where $0 \leq \mu(x_1) + v(x_1) \leq 1$ [16]

A third function $\pi(x_1)$ can be introduced - “the needle is in normal state”, so that

$$\mu(x_1) + v(x_1) + \pi(x_1) = 1 \tag{3.15}$$

Actually, the linguistic variable should be named “normal position of the needle”, but then is difficult to make relation between membership and non-membership degrees and the application variables. Because of (3.15) it will be no so possible to appear at the same time “very high position” and “very low position” of the needle. In this example the degree of non-uncertainty $\pi(x_1) = 1 - \mu(x_1) - v(x_1)$ (3.16)

has meaning of the normal state of the variable. In some other cases $\pi(x_1)$ becomes the natural meaning of degree of uncertainty. In both of the cases, independent on this which physical meaning this variable has, the system will operate only with the values of the membership and non-membership functions and the value of the non-uncertainty is calculated as corollary of these degrees.

If the intuitionistic expert matrix is denoted as *A*, the causes as *X* and the effects as *B*, then the intuitionistic fuzzy system will be denoted as:

$$A \otimes X = B \tag{3.17}$$

where

$$A = (a_{ij})_{m\times n}, a_{ij} = (\mu_{ij}^{A}, \nu_{ij}^{A}), B = (b_{i})_{m\times 1}, b_{i} = (\mu_{ij}^{B}, \nu_{ij}^{B})$$

$$X = (x_{j})_{1\times n}, x_{j} = (\mu_{j}^{X}, \nu_{j}^{X})$$

and

$$0 \le \mu_{ij}^{A} + \nu_{ij}^{A} \le 1; 0 \le \mu_{ij}^{B} + \nu_{ij}^{B} \le 1; 0 \le \mu_{ij}^{X} + \nu_{ij}^{X} \le 1$$

and $\otimes$ is the "standard- co-standard composition" (max-min for the membership degrees and min-max for the non-membership degrees) as presented in [7].

3.8.2. Inuitionistic Fuzzy Logic versus Fuzzy Logic

The intuitionistic model logically joins variables that miss in the conventional fuzzy model. In both of the cases, we have to analyze the same number of variables. Because of the dependency (3.16) the intuitionistic model operate with 2/3 from the variables, and the remaining 1/3 are calculated after solving the system. For this reason, the intuitionistic fuzzy logic will be 33.3% faster as computing time in comparison with the conventional fuzzy logic systems.

The next advantage of the intuitionistic model more adequate presentation of the variables – the specialist will orient faster inside the matrices, which dimensions diminish with 1/3 dimensions of the fuzzy model. The problem, which appears in some cases, is the limitation of the intuitionistic variables – there are only two states, stored inside – for instance "low position" and the opposite, the negation of this "NOT low position" as non-membership function.

If the word "thick" is used, then using the intuitionistic membership functions allow the using the membership and non-membership functions and working with "thick" and "not thick".

Contrary, the "normal" fuzzy membership functions are not limited and several variables can be used like "very thick", "thick", "less thick", "medium", "less thin", "thin", "very thin" etc. This wealth of grades between the "thick" and "thin" is not possible with the intuitionistic approach.

APPLICATIONS AREAS

There is no limitation in the use of the fuzzy logic in the praxis and the areas, where it is used are growing. Actually there has to be made difference which part of the fuzzy logic is used in the applications.

The most used "fuzzy" parts are the membership functions. There are several works, where the space of the input variables is defined and then it is *just normalized* in the range between 0 and 1 using some fuzzy membership functions. This is more "normalization" of the input variables as "fuzzy" but is often named "fuzzy" system, because so it is more modern approach, the application seems to be new etc. The possibility to input multiple linguistic variables, describing single physical variable is suitable for the representation of the human way of thinking and classifying the states of some object and is used in the expert systems for storage of knowledge data into computer, so that it is prepared for working with IF-THEN rules.

In several cases this data is manipulated with non-fuzzy methods, and the IF-THEN rules are realized in some other system, like the environment Prolog or similar. This combination of fuzzy input and descriptive rule definition is in some cases easier for realization and more reliable. Contrary, if instance neuronal nets for the data manipulation are used, is not clear if in all the cases the solutions will be enough exact or useful and in some critical application (medical diagnosis) this could be danger.

The If-THEN rules, as described in this section are some simple way of dealing with the data, which their advantages and disadvantages. The advantage of the IF-THEN rules is the simplicity of the methods and the possibility to see and understand the coefficients in the matrix. Once the number of the input and output variables is large, the usual disadvantage appears - working with several variables and large matrices is not convenient for the non-mathematical experts, which has to provide expert data for filling these matrices. The other disadvantage is that only the independent influence of the variables is taken into account. If two variables x_1 and x_2 have some influence on the result, working independently, their combined influence, appearing at the same time could be different. One way to solve this is to input additional, dependent variable $x_3 = x_1 \cdot x_2$ and to build additional rules for this variable. Actually there are no enough investigations which influence will has the input of such variable to the solvability of the systems.

CONCLUSION

This introduction in the fuzzy logic for non-mathematicians is not exhaustive and do not cover all the areas and methods of the fuzzy logic because of two reasons – first – the fuzzy theory and methods are several and the volume of the chapter would goes over the it scope and – second – these methods are in active development and continually new approaches and applications appear. Here are reported the most common used and reported basic techniques like the fuzzy membership functions and the IF-THEN Rules realized as look-up table and as a fuzzy linear system. The short introduction into the relative new intuitionistic fuzzy set theory demonstrate the power of the intuitionistic variables, which condensate more information in one variable (for the membership and non-membership to some set simultaneously). At the same time this power leads to limitation – those only two independent states – instead of several intermediate steps as in the fuzzy logic are used.

References

[1] Hofweber, T. *Logic and Ontology (Stanford Encyclopedia of Philosophy);* Retrieved June 07, 2010, from http://plato.stanford.edu/entries/logic-ontology/.

[2] Al-Ghazali, Averroes. *Islamic Nominalism vs. Islamic Realism, Part II; Societas Christiana;* Retrieved June 07, 2010, from http://tgenloe.wordpress.com/2008/02/23/al-ghazali-and-averroes-islamic-nominalism-vs-islamic-realism-pt-2/.

[3] Zadeh, L. A. *Info. Control* 1965, *vol 8(3),* 338–353.

[4] Ross, T. J. *Fuzzy Logic with Engineering Applications;* ISBN: 0071136371; McGraw-Hill Inc: NY, 1995.

[5] Zimmermann, H. -J. *Fuzzy Set Theory and Its Applications;* ISBN-10: 079239075; Kluwer Academic: Boston, 2nd Edn, 1991.

[6] Klir, G. J.; Yuan, B. *Fuzzy Sets and Fuzzy Logic: Theory and Applications;* ISBN- 10: 0131011715; Prentice Hall PTR: NJ, 1st Edn, 1995.

[7] Peeva, K.; Kyosev, Y. *Fuzzy Relational Calculus: Theory, Applications and Software (with CD-ROM, vol. 22)*, *Advances in Fuzzy Systems, 22;* ISBN: 9789812560766; World Scientific: Singapore, 2005.

[8] Peeva, K. *Soft Computing* 2007, *vol 11(7),* 593-605.

[9] Kyosev, Y.; Peeva, K.; Reinbarch, I.; Gries, T. *In the Proceedings of the International Conference 9th Fuzzy Days in Dortmund,* Germany, 18-20th September, 2006; pp 93-105.

[10] Majumdar, A.; Ghosh, A. *J. Eng. Fiber. Fabrics* 2008, *vol 3(4),* 61–68.

[11] Majumdar, A.; Majumdar, P. K.; Sarkar, B. *J. Text. Inst.* 2005, *vol 96(1),* 55–60.

[12] Sanchez, E. *Info. Control* 1976, *vol 30(1)*, 38-38.

[13] Kyosev, Y. *Paper presented in the 5th International Conference TEXSCI-2003,* Liberec, Czech Republic, 16-18th June, 2003.

[14] Peeva, K. *Methods and Algorithms for Solving Fuzzy Linear Systems – Applications in Artificial Intelligence Areas: DrSc Thesis;* Technical University of Sofia, 2002.

[15] Peeva, K.; Kyosev Y. *Fuzzy Calculus Toolbox,* from www.kyosev.net/fuzzy_calc.

[16] Atanassov, K. *Intuitionistic Fuzzy Sets;* Heidelberg, New York: Physica-Verlag, 1999.

In: Modeling and Simulation in Fibrous Materials
Editor: Asis Patanaik
ISBN: 978-1-62100-116-4

Chapter 4

Application of CFD in Yarn Engineering in Reducing Hairiness during Winding Process

Asis Patanaik*
CSIR Materials Science and Manufacturing, Polymers and Composites Competence Area, Nonwovens and Composites Group, Port Elizabeth 6000, South Africa

Abstract

This chapter deals with the application of computational fluid dynamics (CFD) modeling in reducing yarn hairiness during the winding process. Yarn hairiness increases during the process of winding. This increase in yarn hairiness creates problem in downstream processes like warping, sizing and weaving, where neighboring yarn ends cling to each other causing end breakages. Fabric appearance is also adversely affected with the use of hairy yarns. As increasingly strict requirements are placed on yarns in high performance oriented weaving and knitting, the hairiness of yarns needs to be reduced to avoid entanglement of yarns and end breakages in subsequent processing. This has resulted in the development of new nozzle profiles to address the problem. The airflow pattern in the nozzles plays a significant role in the reduction of hairiness. CFD has been effectively utilized in yarn engineering to understand the actual mechanism of yarn hairiness reduction. The influence of different nozzle parameters such as diameter and axial angle of the nozzle in reducing yarn hairiness are investigated with the help of CFD. Other factors such as air pressure in the nozzles in affecting the tendency of reducing hairiness in different yarns are also investigated. The influence of drag force on reducing yarn hairiness is also discussed. CFD modeling of airflow inside the nozzles provides useful information in understanding the hairiness reduction phenomena.

4.1. Introduction

Yarns spun from staple fibers have protruding ends and loops of fibers projecting out of the yarn surface. These are commonly termed as 'hairs'. The amount of hairs quantified either

*E-mail: patnaik_asis@yahoo.com

in numbers or in total length per unit length of a yarn, is termed as hairiness [1]. Yarn hairiness increases during the process of winding due to rubbing of fiber ends against various machine parts like tensioning device, yarn guide, and the grooved winding drum. This increase in yarn hairiness creates problem in downstream processes viz. warping, sizing and weaving, giving rise to clinging of neighboring yarns resulting in end breakages. Fabric appearance is also affected with the use of hairy yarns.

Therefore, reducing yarn hairiness by using nozzles during winding will be of importance to industry. Placing a nozzle in one winding position is equivalent to placing 40-60 nozzles in ring spinning, as the production rate of one winding unit is higher than that of a ring spindle by that order.

Further any increase in the number of hairs during winding would also be subjected to the action of nozzle. JetWind (or nozzle-winding) is the process of winding the yarn with help of nozzle placed at a suitable position on a winding machine [1].

Some research works are reported on yarn hairiness reduction using nozzles in winding [2-5]. All these reported works were carried out at low winding speed of 400-600 m/min, whereas the commercial winding speeds are in the region of 800-1200 m/min. In the reported work of Wang and Miao [2], ring and rotor spun cotton yarns were wound first, and these wound yarns were subjected to the action of air-nozzle during second winding operation. This additional winding operation (first winding) increases the operational cost. Chellamani et al. [3] studied the influence of air pressure and axial angles of air inlets in JetWind process. Three different pressure levels viz. 0.5, 1.0, and 1.5 bar (gauge) and three different axial angles of air inlets viz. 30°, 45°, and 60° were investigated. A nozzle with axial angle of air inlets 60° and 1.0 bar (gauge) air pressure gave larger reduction in yarn hairiness.

Zeng and Yu [5] studied the influence of air pressure and axial angles of air inlets in JetWind on hairiness of 18 tex combed cotton yarns. They investigated the effects of air pressure levels viz. 1.5, 2, 2.5, and 3 bar (gauge) and axial angles viz. 35°, 45°, and 55°. They concluded that though a higher nozzle pressure and smaller orifice angle cause more hairiness reduction, but it is not helpful to increase nozzle pressure to a very high level or to use too small orifice angle. An air pressure in the range of 1.5-2.5 bar (gauge) and jet orifice angle of 40°-50° gave better results in terms of hairiness reduction.

The airflow pattern in the nozzles plays a significant role in the reduction of yarn hairiness. For better understanding the principles of reducing hairiness by nozzles, a simulation of airflow is essential to obtain velocity profiles, such as the axial, tangential and resultant velocities of air acting on yarn surfaces and hairs. Airflow simulation has been carried out by employing principles of Computational Fluid Dynamics (CFD).

A comprehensive study was undertaken for the reduction of yarn hairiness in nozzle-winding [1, 6-10]. A mechanism of reduction in yarn hairiness was proposed. Optimization of nozzle parameters, fiber linear density and air pressure on reduction in yarn hairiness has been carried out for polyester yarns spun from fibers of different linear densities. All the experiments were carried out in first winding operation i.e. spun yarn is directly subjected to the action of air-nozzle in winding.

Quantification of drag forces acting on hair and yarn surface are of importance in describing the behavior of different nozzles on reduction in yarn hairiness. Drag forces acting on yarn and hairs were also computed.

4.1.1. Nozzles Fabrication and Mounting on the Winding Machine

Different set of nozzles were fabricated by varying the nozzle parameters, axial angle of air inlets and yarn channel diameters. To place the nozzles at appropriate positions during winding, nozzle-housings were fabricated and the nozzles were mounted inside the nozzle-housings [1, 6]. A schematic representation of the nozzle geometry is shown in Figure 4.1. Yarn channel diameter *(D)* and axial angle of air inlets (θ) are shown in the Figure 4.1. Four air inlets were made tangential to the inner walls of the nozzle to create air vortex. For all the nozzles, length and the outer diameters of the nozzles were kept constant. Nozzle was kept at a distance of 10 cm above balloon bracket of a Savio Orion-M winding machine. A small ceramic guide was placed at the bottom of the nozzle to position the yarn at the centre of the nozzle. Compressed air was supplied to the nozzles through pipes with a pressure regulator and an air filter. A schematic representation of the nozzle-winding process is shown in Figure 4.2.

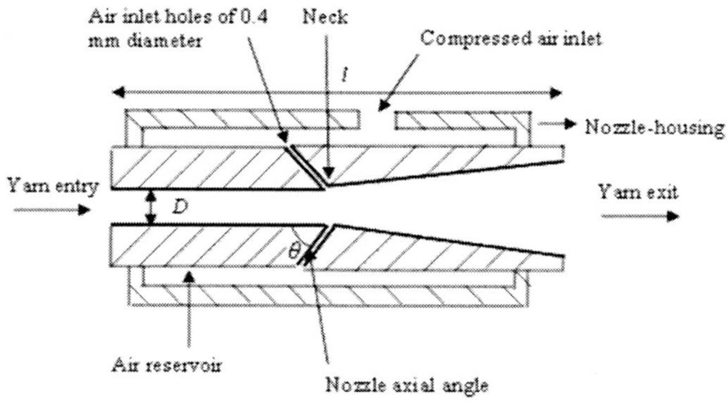

Figure 4.1. Schematic representation of the nozzle [Source: Reference 1].

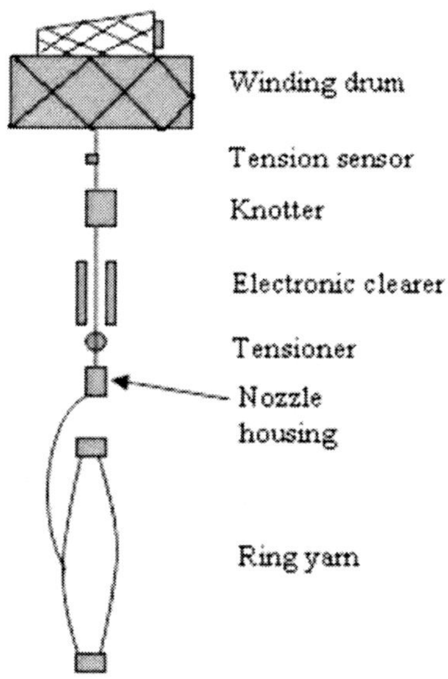

Figure 4.2. Schematic representation of the nozzle-winding process [Source: Reference 1].

4.2. Computational Fluid Dynamics

Simulation of airflow inside the nozzle was carried out by CFD. Using the results of air velocity based on simulation, air drag forces and their directions were calculated.

4.2.1. Airflow Simulation

To analyze the airflow pattern, simulation of airflow was carried out using a fluid flow analysis package, Fluent 6.1 [1, 6-12]. To solve the three-dimensional airflow field inside the nozzles, a CFD model was developed using the above software. Fluid flow and related phenomena can be described by partial differentiation equations, which cannot be solved analytically except in over-simplified cases. To obtain an approximate solution numerically, a discretization method to approximate the differential equations by a system of algebraic equations, which can be then numerically solved on a computer. The approximations were applied to small domains in space and/or time so the numerical solution provides results at discrete locations in space and time. Much of the accuracy depends on the quality of the methodology used, for which CFD is a powerful tool to predict the flow behavior of fluid inside any object. It provides various parameters such as air velocity profiles (axial, tangential, resultant etc.) and path lines trajectory, which are important for subsequent analysis. It was for those reasons that a CFD package, Fluent 6.1, which uses a Finite Volume (FV) method, was employed for airflow simulation.

The Finite Volume (FV) method is the simplest method to understand and to program. The Finite Volume (FV) method uses the integral form of the conservation equations as its starting point. The solution domain was subdivided into a finite number of continuous control volumes (CVs), and the conservation equations were applied to each CV. At the centroid of each CV lies a computational node at which the variables are to be calculated. Interpolation was used to express variable values at the CV surface in terms of nodal (CV-centre) values. Surface and volume integrals were approximated using suitable quadratic formulae. By solving algebraic equations results for each CV were obtained, in which a number of neighbor nodal values appeared. The simplest complete models of turbulence are two-equation models in which the solution of two separate transport equations allows the turbulent velocity and length scales to be independently determined. The standard k-ε model in Fluent falls within this class of turbulence model and has become the workhorse of practical engineering flow calculations. It is widely used for a range of turbulent flows in industrial flow and heat transfer simulations. It is a semi-empirical model and the derivation of the model equations which relies on phenomenological considerations and empiricism. The Fluent package includes the following series of operations during the airflow calculations [11-12]:

- Filters (translators) for import of surface and volume meshes from CAD/CAE packages such as Ansys, Cgns, I-Deas and others.
- Gambit, the preprocessor for geometry modeling and mesh generation.
- TGrid, an additional preprocessor that can generate volume meshes from existing boundary meshes (prePDF, the preprocessor for modeling non-premixed combustion in Fluent).

- Fluent, the solver.

The airflow in the developed nozzles was turbulent and hence the standard k-ε model of turbulence along with standard wall functions was used [1, 6-12]. The density weighted average equation and continuity equations along with energy equation, turbulent kinetic energy equation and ε equation were solved in segregated solver [11-12]. It has been assumed that the flow inside the nozzle affects the yarn but the presence of yarn inside the nozzle has no effect on airflow and hence yarn was not modeled during simulation [6-10]. The cross-sectional area occupied by the yarn inside the nozzle was very small compared to that of yarn channel of the nozzle (1:100). There are issues related to meshing of the zone near the yarn surface, hence it was not directly modeled in the simulation. High pressure and velocities of the air, coupled with considerably low volume of the yarn compared to that of yarn channel of the nozzle also justifies this assumption. A three-dimensional simulation model was used, because of the positioning of air inlets, air velocity is resolved into three components viz., axial (x-direction), tangential or swirling (y-direction) and inward radial velocities (z-direction). A very negligible amount of radial velocity was found because of the fluctuations in Z-axis was very small as compared to fluctuations in X- and Y-axes [1].

In the present configuration, air inlet boundaries are assumed to be "Pressure Inlet" while outflow boundaries are assumed "Pressure Outlet". Pressure inlet boundary conditions were used to define the total pressure and other scalar quantities at flow inlets. Pressure outlet boundary conditions were used to define the static pressure at flow outlets. At the nozzle inlet, the air pressure was varied. At the nozzle outlet, the pressure was supposed to be the external pressure (one atmosphere). At the wall of the nozzle standard wall function boundary condition was applied. Although the high velocity of air stream was a heat source that will increase the temperature in the nozzle, the nozzle length was very short and the process occurs in a very short time. For simplification, it was assumed that the process is adiabatic i.e. no heat transfer occurred through walls. The flow model used was viscous, compressible airflow [1, 6-10]. The following series of equations were used to solve a compressible turbulent flow for airflow simulation [1, 6-12];

Mass conversion equation:

$$\frac{\partial \rho}{\partial t} + \frac{\partial \rho u_i}{\partial x_i} = 0$$

Momentum conversion equation:

$$\frac{\partial}{\partial x_i}\left(\rho u_i u_j\right) = -\frac{\partial p}{\partial x_i} + \frac{\partial}{\partial x_j}\left[\mu\left(\frac{\partial u_i}{\partial x_j} + \frac{\partial u_j}{\partial x_i} - \frac{2}{3}\delta_{ij}\frac{\partial u_l}{\partial x_l}\right)\right] + \frac{\partial}{\partial x_j}\left(-\rho\overline{u'_i u'_j}\right)$$

Turbulent kinetic equation:

$$\frac{\partial(\rho k u_i)}{\partial x_i} = \frac{\partial}{\partial x_j}\left[\left(\mu + \frac{\mu_t}{\sigma_k}\right)\frac{\partial k}{\partial x_j}\right] + G - \rho$$

Rate of dissipation of the turbulent kinetic energy:

$$\frac{\partial(\rho \varepsilon u_i)}{\partial x_i} = \frac{\partial}{\partial x_j}\left[\left(\mu + \frac{\mu_t}{\sigma_\varepsilon}\right)\frac{\partial \varepsilon}{\partial x_j}\right] + c_1 \frac{\varepsilon}{k} G - c_2 \rho \frac{\varepsilon^2}{k}$$

The Nomenclature Used Is Given Below:

ρ = density of air, t = time, u = velocity of air, p = static pressure, u_i = fluctuation in u in x-direction, $\rho\overline{u'_i u'_j}$ = Reynolds stress tensor, δ_{ij} = Kronecker's delta function, G = rate of generation of turbulent kinetic energy, k = turbulent kinetic energy, μ_t = turbulent viscosity, ε = rate of dissipation of turbulent kinetic energy and $c_1, c_2, \sigma_k, \sigma_\varepsilon$ constants of the k-ε model.

The structure of airflow inside the nozzle depends on the following factors: angle at which air enters into the nozzle (or axial angle of air inlets), channel diameter (or yarn channel diameter) and air pressure. From the simulation, airflow pattern, components of air velocity (axial, tangential and resultant) at different normal planes for various nozzles were obtained [1]. To get air velocity profiles on different normal planes of nozzles, various sections of the nozzle were considered at a distance of 1 mm, along the axis of the nozzle, details of which are described in Rengasamy et al. [13].

4.2.2. Computation of Air Velocity on Yarn Surface and Inner Wall of the Nozzle

To calculate air velocity acting on the surfaces of yarn and hairs (inner wall of the nozzle), a 30 tex nozzle-wound polyester yarn was considered. Using a microscope (Leica) and software yarn diameter was measured. The measured diameter of yarn was 0.239 mm. A solid cylinder of 0.239 mm (analogues to yarn) was superimposed on the air velocity profile obtained at various normal planes of the nozzle by using post processing software in such a way that the cylinder was coaxial to the nozzle axis. From this, air velocity acting on yarn surface can be computed [1]. The neck of the nozzle where the divergent portion starts is considered as the 0-normal plane. The values of resultant air velocity at the surface of the cylinder represent the resultant air velocity acting on the yarn surface.

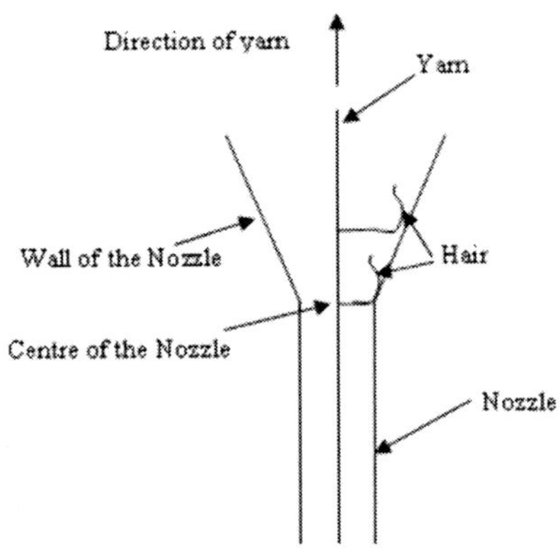

Figure 4.3. Schematic representation of a hair on the yarn in the nozzle [Source: Reference 1].

Hairs are considered as projecting from the cylinder (yarn surface). The values of resultant air velocities acting on each and every point on hairs of different lengths, orientations and lying at different normal planes of the nozzles can be calculated. However, hairs of numerous configurations and orientation are found on a spun yarn, calculation of air velocity acting on each of it is extremely a time consuming job and not so useful also. So, a simplified approach was adopted wherein, a hair is assumed to lie perpendicular to the yarn. Further, the resultant air velocity reduces from inner wall to the axis of the nozzle; a hair is subjected to two extreme air velocities; VRy, a minimum at the base of the hair (i.e. at yarn surface) and VRh, a maximum at the inner wall. Longer hairs may bend inside the nozzle while touching the wall of the nozzle due to space constraint. A schematic representation of this is shown in Figure 4.3.

4.2.3. Calculation of Air Drag Forces on Hair

When a hair that is projecting vertically from the yarn surface is subjected to airflow, two sequential events could be considered. The first, the transverse drag forces acting on the hair trying to bend and fold it over the yarn surface. Secondly, the folded hair being wrapped around the yarn surface by the longitudinal drag forces acting on it.

In fluid dynamics equations have been developed to quantify the drag forces on smooth circular cylinders which are analogous to hair [14-16].

Transverse drag force (F_t) or (TDH) acting on hair is given by:

$$F_t = \frac{1}{2}\rho V_t^2 S_t C_{Dt}$$

The longitudinal drag force (F_l) or (LDH) acting on hair is given by:

$$F_l = \frac{1}{2}\rho V_l^2 S_l C_{Dl}$$

where, ρ = density of the air (kg/m^3)

V_t = relative velocity between air and hair for transverse airflow (m/s)
V_l = relative velocity between air and hair for longitudinal airflow (m/s)
S_t = projected area of hair for transverse airflow (m^2) = $d_h\, l$
S_l = projected area of hair for longitudinal airflow acts (m^2) = $S_l = \pi\, d_h\, l$

C_{Dt} = air-drag coefficient for transverse airflow, related to Reynolds number, $R_e = \dfrac{V_t\, d_h}{\mu / \rho}$

C_{Dl} = the air-drag coefficient for longitudinal airflow, related to Reynolds number, $R_e = \dfrac{V_l\, d_h}{\mu / \rho}$

where, d_h = the diameter of the hair (m)
μ = the viscosity of air (N s/m^2)
l = is the length of the hair (m)

Author considered the bending of hair over the yarn is occurring at an angle to the yarn axis as decided by the direction of resultant air velocity [1]. So the resultant air velocity is used for transverse drag calculation. Because of the reducing gradient of the resultant air velocity from the wall (VRh) to the yarn surface (VRy), the transverse air drag force acting on a vertical hair is due to the average of these resultant air velocities i.e. (VRh + VRy)/2. Subtracting the value of yarn velocity from this value, the relative velocity between air and hair, V_t can be obtained for the transverse airflow.

Now the hair is folded and lying on the yarn surface along the direction of resultant air velocity found at the yarn surface. This resultant air velocity (VRy) is parallel to the fiber axis. The longitudinal drag force acting on a folded hair over the yarn body is due to the presence of resultant air velocity on the yarn surface (VRy).

Under standard testing conditions, ρ = 1.197 kg/m^3 and μ = 18.22×10^{-6} N s/m^2 [15-16]. Hence for a given hair diameter and relative speed between air and hair, *Re* can be calculated for the above equations. Once *Re* is known, C_D can be obtained from the established relationship between the drag coefficient and Reynolds number for circular cylinders [14-16]. The transverse and longitudinal drag forces acting on hairs (1.0 denier, 10 μm diameter) at the divergent section of the nozzle is shown in Table 4.1.

Table 4.1. Transverse and longitudinal drag forces acting on hairs (1.0 denier, 10 μm calculated diameter) of various lengths at the divergent section for the nozzle of 45°, 2.2 mm at 0.5 bar (gauge) [Source: Reference 1]

	Transverse drag force parameters on hair				Longitudinal drag force parameters on hair			
Various planes in the nozzle	Velocity of air on hair in transverse direction (m/s)	Reynolds number (R_e)	Drag coefficient (C_{Dt})	*Drag force (mN)	Velocity of air on hair in longitudinal direction (m/s)	Reynolds number (R_e)	Drag coefficient (C_{Dl})	+Drag force (mN)
y = 0	195	128.109	1.35	0.3379	120	78.836	1.45	0.4316
y = 1	142	93.289	1.45	0.2099	105	68.981	1.45	0.3605
y = 2	97	63.720	1.45	0.1061	60	39.418	1.65	0.1451
y = 3	67	44.017	1.60	0.0601	15	9.854	2.80	0.0165
y = 4	60	39.418	1.60	0.0517	15	9.854	2.80	0.0177
y = 5	52	34.162	1.70	0.0440	15	9.854	2.80	0.0189
y = 6	52	34.162	1.70	0.0467	15	9.854	2.80	0.0201
y = 7	45	29.563	1.85	0.0403	15	9.854	2.80	0.0213
y = 8	45	29.563	1.85	0.0448	15	9.854	2.80	0.0236

4.3. Application of CFD in Yarn Engineering

This sections deals with the application of CFD in conjunction with experimental data in reducing yarn hairiness during winding process. The mechanism of reduction in yarn hairiness is discussed. The influence of different nozzle parameters in reducing hairiness in yarns of different linear densities and optimization of nozzle parameters, fiber linear density and air pressure were also studied. The influences of drag force and impact angle of airflow on tendency to reduce yarn hairiness were also discussed.

4.3.1. Experimental Method

Three different polyester ring yarns (30 tex Z-twisted) were spun each from 1.0, 1.2, and 1.4 denier fibers respectively. The fiber lengths were 38 mm. These yarns were passed through air-nozzles while winding them on a winding machine. Yarns were wound at 800 m/min. The directions of yarn movement and airflow in the nozzle was kept in the same direction i.e. airflow is along the directions of yarn movement.

Using Box and Behnken factorial design, two sets of experiments were carried out. In the first set of experiment (axial angle of air inlet series), all the three nozzles have a constant yarn channel diameter 2.2 mm but air inlets were positioned at different axial angles 40^0, 45^0, and 50^0.

In the second set of experiment (yarn channel diameter series), all the three nozzles have air inlets positioned at a fixed axial angle 40^0 but each having different yarn channel diameters as 1.8, 2.2, and 2.6 mm. In both these experiments air pressures supplied to the nozzles were kept at three levels as 0.5, 0.7, and 0.9 bar (gauge). The length of the yarn channel was 17 mm.

The diameter of air entry orifice was 0.4 mm. The selected variables and their levels are given in Tables 4.2. and 4.3. The hairiness of yarns was tested on Zweigle G 566 hairiness tester and S3 hairiness values were obtained. S3 hairiness values are the summation values of hair lengths equal to or exceeding 3 mm.

Detailed testing procedure has been described elsewhere [13]. Samples were kept in standard testing condition for 24 hours prior to testing.

Table 4.2. Levels of variables (angle series) [Source: Reference 1]

Variables	Coded levels		
	-1	0	+1
Axial angle of air inlets (a), degree	40°	45°	50°
Fibre fineness (b), denier	1.0	1.2	1.4
Air pressure (c), bar	0.5	0.7	0.9

Table 4.3. Levels of variables (diameter series) [Source: Reference 1]

Variables	Coded levels		
	-1	0	+1
Yarn channel diameter (a_1), mm	1.8	2.2	2.6
Fibre fineness (b_1), denier	1.0	1.2	1.4
Air pressure (c_1), bar	0.5	0.7	0.9

4.3.2. Reduction of Yarn Hairiness

One nozzle was used to explain the process of hairiness reduction as the underlying process is similar in other nozzles. The nozzle used was 45° axial angle of air inlets and 2.2 mm yarn channel diameter, operated at 0.5 bar (gauge) pressure. Figure 4.4. shows the air particle path lines obtained from the CFD simulation. Figure 4.5. shows values of resultant of air velocity acting on yarn surface and at the inner wall of the nozzle. The direction of resultant airflow is at an angle to the axis of the nozzle at the divergent section, termed as Z-direction. Distance 0 corresponds to the neck of the nozzle. Distances from 0 to 0.008m represent the divergent section of the nozzle. VRy and VRh represent resultant velocity of air acting on yarn surface and at the wall respectively. Similarly, plots are obtained for other nozzles.

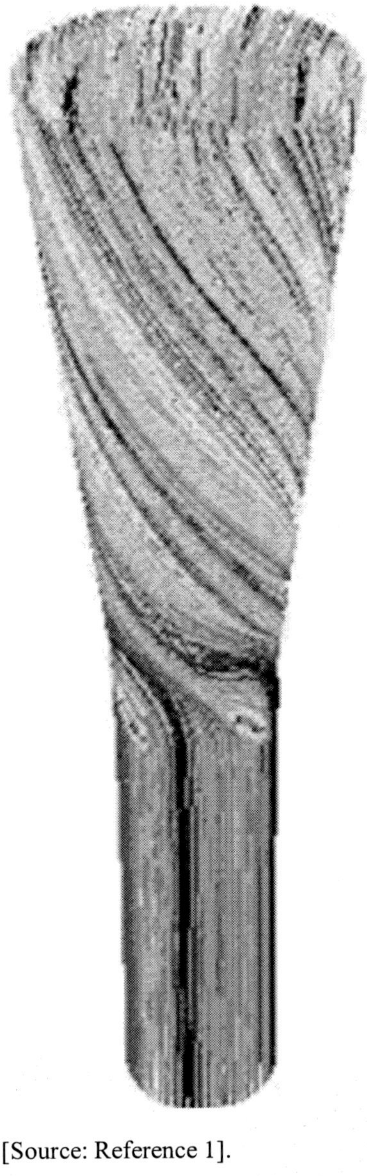

Figure 4.4. Air particle path lines [Source: Reference 1].

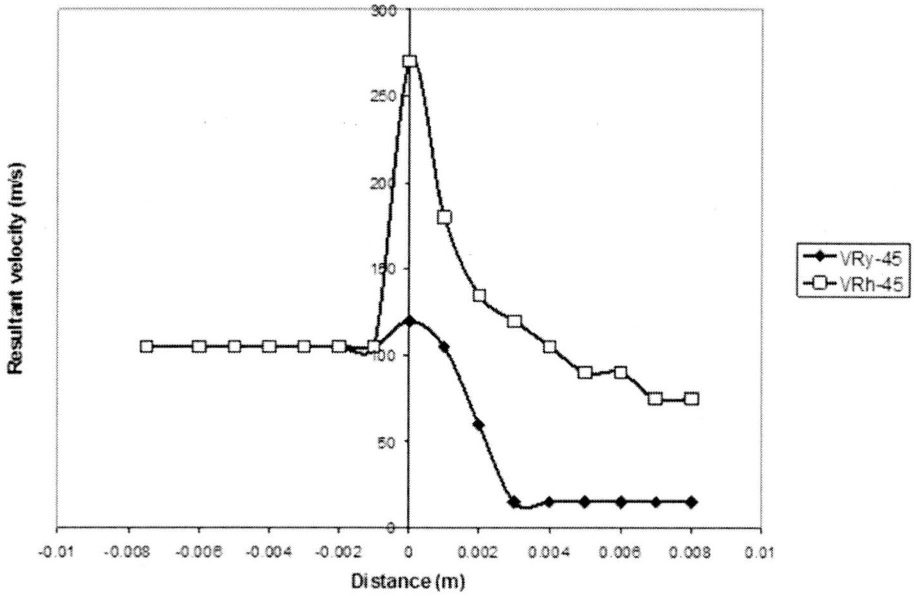

Figure 4.5. Values of resultant of air velocity acting on yarn surface and at the inner wall of the nozzle having 45° axial angle of air inlets and 2.2 mm yarn channel diameter [Source: Reference 1].

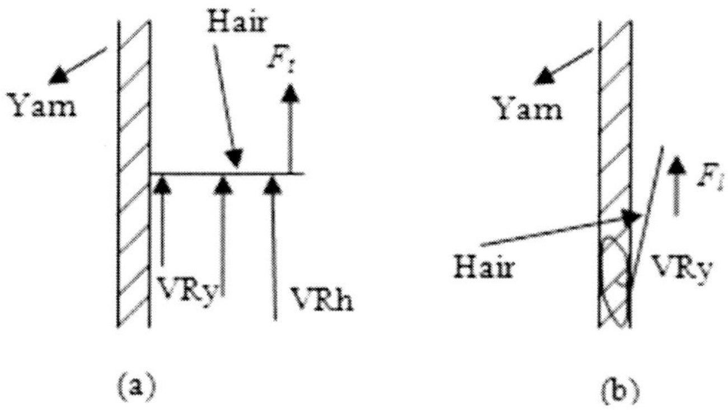

Figure 4.6. Schematic representation of air velocity and drag forces acting on a hair [Source: Reference 1].

Yarn enters from the bottom of the nozzle through the cylindrical portion, where there is no swirling of air (Figure 4.4). From the neck and up through the divergent section of the nozzle the airflow is in S-direction i.e. counter-clockwise facing the nozzle exit. Highest resultant air velocity is found at the neck of the nozzle, where the air streams from the four air-inlets are issued. This implies that yarn and hairs are subjected to maximum drag forces at the neck (y = 0). The drag forces acting on a hair lying at different planes are given in Table 4.1. The drag forces both transverse and longitudinal are found to decrease drastically from the neck to exit of the nozzle. Resultant air velocity inside the nozzle is resolved into three components viz. axial, tangential, and radial. Swirling action is created by the tangential and axial velocity components of air velocity. Since the resultant air velocity is at an angle to the axis of the nozzle, the component of tangential air velocity subjects a Z-twisted yarn to twist

(Z-torque) from yarn entry plane and up to the neck of the nozzle and then followed by untwisting (S-torque) at the divergent section of the nozzle. Due to this false twisting action of the yarn inside the nozzle, yarn retains its original twist. Majority trailing hairs formed during ring spinning become leading hairs (pointing towards the direction of yarn movement) when the yarn goes through the nozzle placed at winding machine during first winding operation. The axial direction of airflow is along the direction of moving yarn. The straight airflow (at cylindrical portion of the nozzle) and especially the swirling airflow (around the neck portion of the nozzle, i.e. y = 0 and 1) create sufficient transverse air drag forces (F_t in Figure 4.6.a) on some of the leading hairs folding them over the yarn (Note that hair is shown as vertical as described earlier in air drag calculation for simplicity). Upon folding them, the presence of longitudinal drag forces (F_l in Figure 4.6.b) wraps them around the yarn body at the divergent portion of the nozzle (i.e. as the yarn leaves the nozzle). Due to the action of air vortex, hairs (especially leading hairs) are wrapped over the yarn body while the yarn is untwisted and in the processes reducing yarn hairiness [1]. So, bending of a hair can be considered as a prerequisite for it to be wrapped over the yarn surface, and hence, reduction in yarn hairiness. Most of the bending and wrapping of hairs believed to take place at the planes y = 0 to y =1 where the drag forces acting on hairs are considerable. S3 hairiness values/100 m length of as-spun yarns spun from 1.0, 1.2, and 1.4 denier fibers are 72, 77, and 189 respectively. For the as-spun yarns, as the fibers become coarser, the hairiness increases. Coarser fibers have higher bending and torsional rigidities than the finer fibers. Rigid fibers have more tendencies to protrude from yarn surface giving higher hairiness [1]. Although the number of fibers in yarn cross-section spun from 1.0 denier fibers is more than that for the yarn spun from 1.2 denier fibers, the bending and torsional rigidities of the fibers are the major influencing factor contributing yarn hairiness during spinning [17-18].

Table 4.4. Percentage reduction in S3 hairiness values from corresponding spun polyester yarns (angle series) [Source: Reference 1]

Run. no.	Level of variables			*% Reduction S3 values
	a	b	c	
1	40°	1.0	0.7	11
2	50°	1.0	0.7	8
3	40°	1.4	0.7	26
4	50°	1.4	0.7	22
5	40°	1.2	0.5	14
6	50°	1.2	0.5	13
7	40°	1.2	0.9	21
8	50°	1.2	0.9	18
9	45°	1.0	0.5	11
10	45°	1.4	0.5	25
11	45°	1.0	0.9	17
12	45°	1.4	0.9	33
13	45°	1.2	0.7	21
14	45°	1.2	0.7	19
15	45°	1.2	0.7	19

*From corresponding as-spun yarns.

Table 4.5. Percentage reduction in S3 hairiness values from corresponding spun polyester yarns (diameter series) [Source: Reference 1]

Run. no.	Level of variables			*% Reduction
	a_1	b_1	c_1	S3 values
1	1.8	1.0	0.7	10
2	2.6	1.0	0.7	7
3	1.8	1.4	0.7	25
4	2.6	1.4	0.7	21
5	1.8	1.2	0.5	14
6	2.6	1.2	0.5	10
7	1.8	1.2	0.9	19
8	2.6	1.2	0.9	17
9	2.2	1.0	0.5	8
10	2.2	1.4	0.5	20
11	2.2	1.0	0.9	14
12	2.2	1.4	0.9	29
13	2.2	1.2	0.7	17
14	2.2	1.2	0.7	18
15	2.2	1.2	0.7	17

*From corresponding as-spun yarns.

4.3.3. Optimization of Axial Angle of Air Inlets and Yarn Channel Diameter on Percentage Reduction in S3 Values

Percentage reduction in S3 hairiness values on the nozzle wound yarns from that of as-spun yarns is given in Tables 4.4. and 4.5 and this reduction in S3 hairiness values is statistically significant. The response surface equation for the same is given in Table 4.6. along with the square of correlation coefficients between the experimental values and calculated values obtained from the response surface equations. Some selective contours using the factorial design are presented in the following sections.

4.3.3.1. Influence of Nozzle Axial Angle, Fiber Fineness, and Air Pressure on Percentage Reduction in S3 Values

Figure 4.7. shows the interactive effect of fiber fineness and air pressure on percentage reduction in S3 hairiness values from that of the corresponding as-spun yarns. With the increase in fiber denier, a more reduction in S3 values is observed. Presence of coarser fibers during yarn formation generates more hairiness. Yarn spun from coarser fibers present more number of hairs to the airflow inside the nozzle, for the air currents to wrap them around the yarn body. Hence, a higher hairiness reduction is observed during nozzle-winding for the yarns spun from coarser fibers.

Transverse drag forces acting on hair constituted by fibers of different deniers is shown in Figure 4.8. The calculated diameters of 1.0, 1.2, and 1.4 denier fibers are 10, 11, and 12 μm. The transverse drag force acting on 1.4 denier hairs is higher in comparison to that on 1.2 and 1.4 denier hairs. Most of the bending of hairs is likely to take place at planes in between 0 to

0.002 m distance along the length of the nozzle as the air velocity is higher in these regions than found in the other regions. The transverse drag forces acting on different denier fibers are nearly the same from 0.004 to 0.008 m planes.

Air drag force F acting on a protruding fiber depends on the projected area of the fiber which linearly varies with diameter of the fiber d. The later varies with square root of the denier D.

$F \propto d \propto \sqrt{D}$

The bending rigidity of a given fiber B varies with fiber diameter as:

$B \propto d^4 \propto D^2$

The ratio of drag forces acting on a fiber to its bending rigidity $\propto 1/D^{1.5}$. This ratio decreases as the fiber becomes coarser indicating the difficulty of bending the fiber. Hence, one would expect a lower percentage reduction in hairiness on yarns spun from coarser fibers provided the transverse air drag forces are insufficient to bend the coarser fibers (i.e. 1.2 and 1.4 deniers). But this is not evident. The statistical effect of presentation of more hairs to the nozzle by the coarser-fiber yarn seems to be playing a dominant role on hairiness reduction.

With the increase in air pressure more reduction in S3 values is observed (Figure 4.7). Table 4.7 indicates the values of resultant velocity of air acting on yarn (VRy) and the maximum resultant velocity of air acting on the hair (VRh) for different air pressures. These values are given for the plane cut just above the neck of the nozzle. It can be seen that with increase in air pressure from 0.5 to 0.9 bar (gauge), resultant air velocity increases, thereby increasing the drag forces, consequently increasing the wrapping of fibers around yarn body and decreasing yarn hairiness.

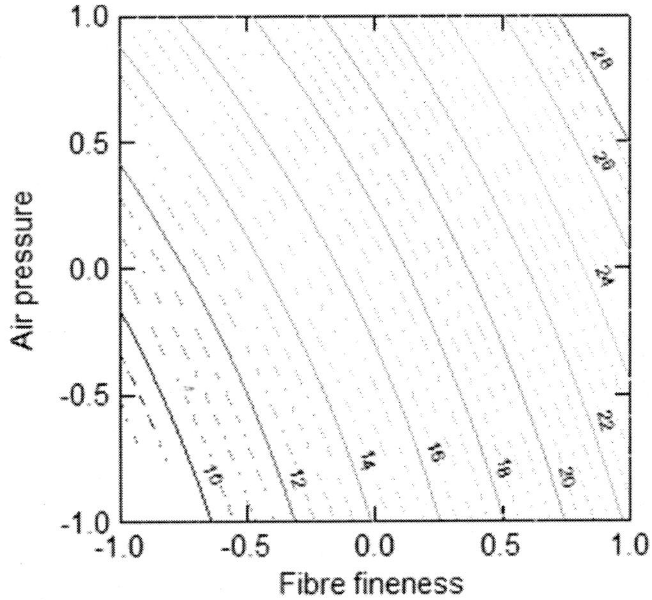

Figure 4.7. Influence of fiber fineness and air pressure on percentage reduction in S3 values from corresponding as-spun yarns during nozzle-winding for the nozzle having 40° axial angle of inlets [Source: Reference 1].

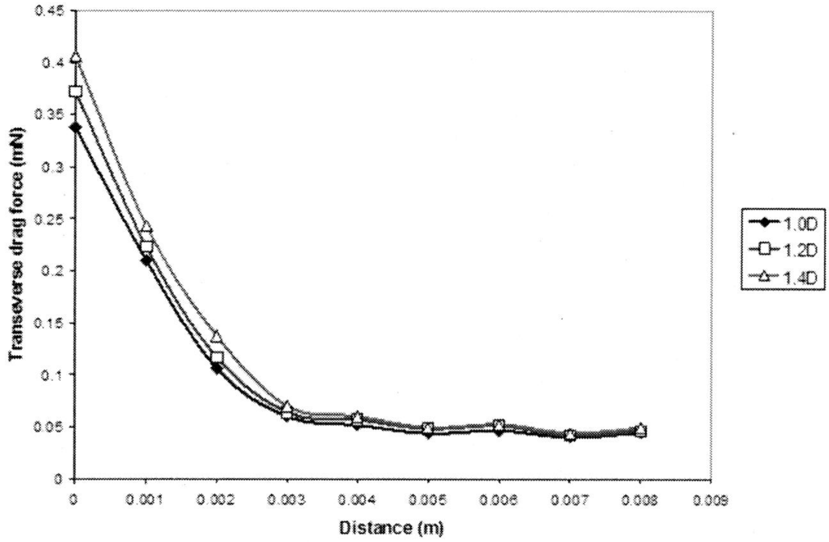

Figure 4.8. Comparison of transverse drag forces (TDH) acting on hairs constituted by fibers of different deniers (Nozzle 45°, 2.2 mm diameter at 0.5 bar, gauge) [Source: Reference 1].

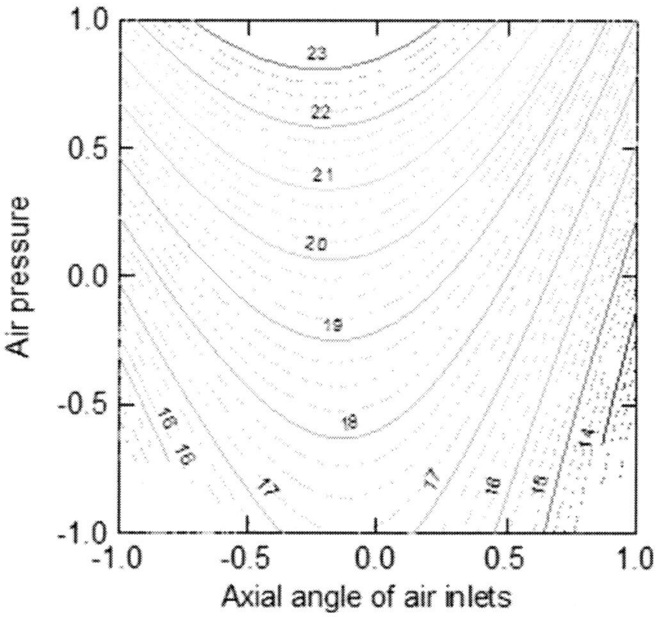

Figure 4.9. Influence of axial angle of air inlets and air pressure on percentage reduction in S3 values from corresponding as-spun yarns during nozzle-winding for yarns spun from 1.2 denier fibers [Source: Reference 1].

The combined effect of fiber fineness and air pressure on percentage reduction in S3 values of nozzle wound yarns from corresponding as-spun yarns indicate that 1.0 level of fiber fineness (1.4 denier) and 1.0 level of air pressure (0.9 bar, gauge) gives the largest percentage reduction in hairiness. The influence of axial angle of air inlets and air pressure on hairiness reduction is shown in Figure 4.9. With an increase in air pressure from -1.0 levels to

1.0 level, more reduction in S3 values is observed. While increasing the axial angle of air inlets from -1.0 to 0 levels, there is not much change in terms of percentage reduction in S3 values, but from 0 to 1.0 level, less reduction in S3 values is observed. From the above contours best result is found for 0.0 level of axial angle (45°); rest other levels i.e. -1.0 (40°) and 1.0 levels (50°) show more or less similar values. CFD modeling has been used to describe the results by comparing resultant velocity of air acting on yarn surface and hair for the nozzles having axial angles of air inlets 40°, 45°, and 50°.

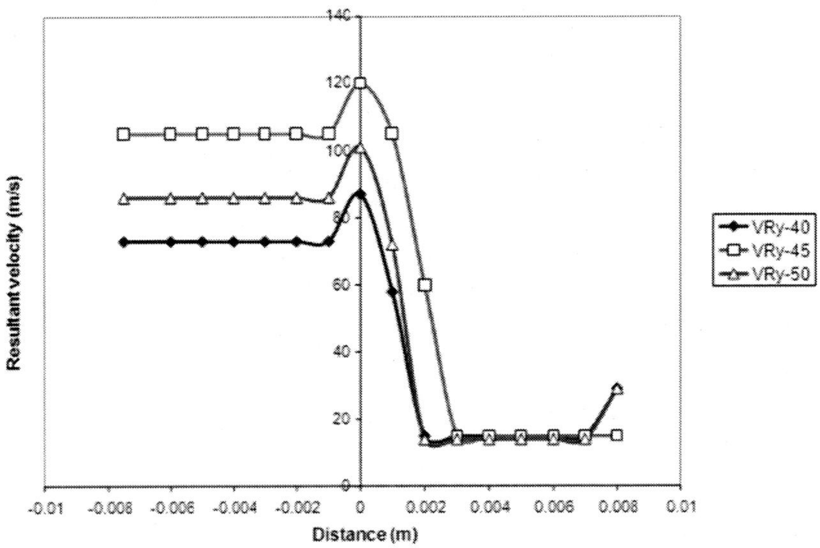

Figure 4.10. Resultant velocity of air acting on yarn surface for the nozzles with air inlets having axial angles 40°, 45°, and 50° at 0.5 bar (gauge) [Source: Reference 1].

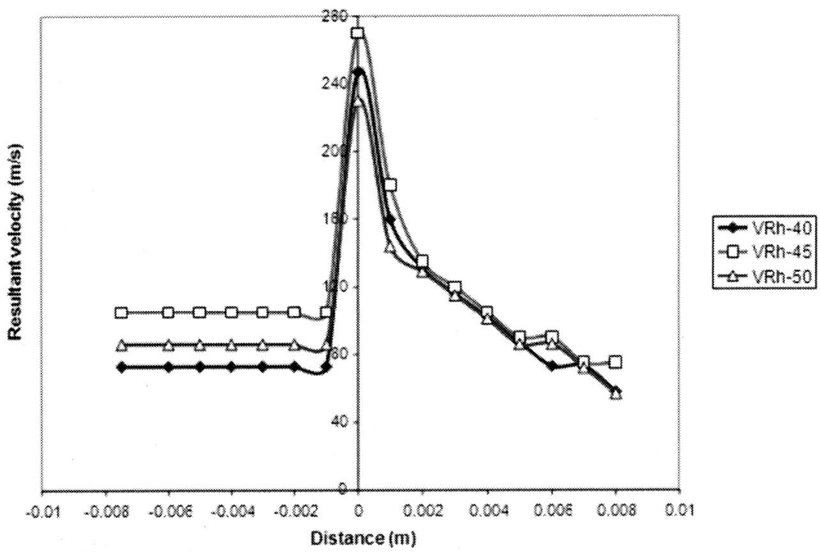

Figure 4.11. Resultant velocity of air acting on hairs for the nozzles with air inlets having axial angles 40°, 45°, and 50° at 0.5 bar (gauge) [Source: Reference 1].

The trend of resultant air velocities acting on the yarn surface and at the wall of the nozzle (i.e. the maximum velocity striking the hair) are similar except that hairs are subjected to a higher air velocity as shown in Figures 4.10 and 4.11. Resultant velocity of air acting on yarn surface and on hairs (near the wall) in the case of the nozzle with 45° axial angles is higher in comparison with the nozzles 40° and 50° axial angles. So swirling intensity created by nozzle with 45° axial angle is stronger than the nozzles with 40° and 50° axial angles, thereby more hairiness reduction by the former. Although yarn is subjected to a slightly lower velocity of air for the nozzle with 40° axial angle in comparison with 50° axial angle, but the resultant velocity of air acting on hairs for 40° axial angle is higher than that of 50° axial angle, that by may be enhancing the efficiency of wrapping of fibers in case of nozzle with 40° axial angle in comparison with 50° axial angle.

The combined effect of air pressure and nozzle axial angle on percentage reduction in S3 values from corresponding as-spun yarns indicate that 0.0 level of nozzle axial angle (45°) and 1.0 level of air pressure (0.9 bar, gauge) gives the optimum zone.

4.3.3.2. Influence of Yarn Channel Diameter, Fiber Denier, and Air Pressure on Percentage Reduction in S3 Values

From Figure 4.12. the combined effect of fiber fineness and air pressure on percentage reduction in S3 values from corresponding as-spun yarns indicate that 1.0 level of fiber fineness (1.4 denier) and 1.0 level of air pressure (0.9 bar, gauge) gives the optimum zone in terms of reducing the hairiness of yarns when the yarn channel diameter is kept 2.2 mm. This result follows the trend observed earlier.

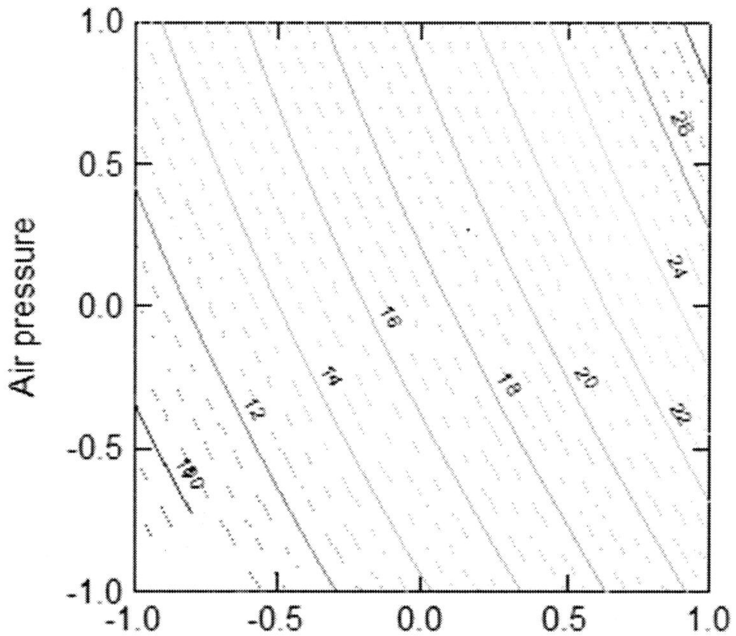

Figure 4.12. Influence of fiber fineness and air pressure on percentage reduction in S3 values from corresponding as-spun yarns during nozzle-winding for the nozzle with channel diameter 2.2 mm [Source: Reference 1].

The influence of yarn channel diameter and air pressure on percentage reduction in S3 values from corresponding as-spun yarns for yarns spun from 1.2 denier fiber is shown in Figure 4.13. With increase in air pressure from -1.0 level to 1.0 level more reduction in yarn hairiness is observed as similar to the earlier observation. With an increase in yarn channel diameter from -1.0 level to 1.0 level, less reduction in yarn hairiness is observed. Using CFD modeling, the nozzles with 1.8, 2.2, and 2.6 mm yarn channel diameters are compared for resultant air velocity acting on yarn surface at an operating pressure of 0.5 bar (gauge).

Higher resultant air velocity is observed in the vicinity of nozzle with 1.8 mm diameter followed by 2.2 and 2.6 mm diameters. Though swirling intensity created by the nozzle with 1.8 mm diameter is stronger than that the nozzle with 2.2 and 2.6 mm diameters, but in terms of hairiness reduction, the nozzle with 2.2 mm diameter is the best one followed by 1.8 and 2.6 mm diameters. The reason may be that with the smaller diameter of nozzle (1.8 mm), the air velocity is very high, yarn may be rubbing at the wall of the nozzle as the diffusing tendency of air velocity is less with the this diameter which may affects the hairiness reduction. Between the nozzles of 2.2 and 2.6 mm diameters, swirling intensity created by nozzle with 2.2 mm diameter is stronger than that of the nozzle with 2.6 mm diameter, thereby more hairiness reduction by the 2.2 diameter nozzle. Swirling intensity depends on the angle of impact of air on yarn or fiber and resultant air velocity. The angle of impact is the tan^{-1} of the ratio of the tangential velocity to the axial velocity components of airflow. A zero angle implies that the direction of resultant air velocity is parallel to the axis of the nozzle, i.e. the air is not swirling. Negative and positive values indicate that resultant directions of airflow are in S- and Z-directions respectively. In other words, the directions of tangential air velocities are different. A particular length of hair of 0.9 mm (this is the distance from centre to wall) is considered at 0-plane and this length is divided into four equal radial increments (i. e. R0, R0.33, R0.67, and R1). From the values of the tangential and axial velocities at each radial positions, angle of impacts are calculated for 0-plane.

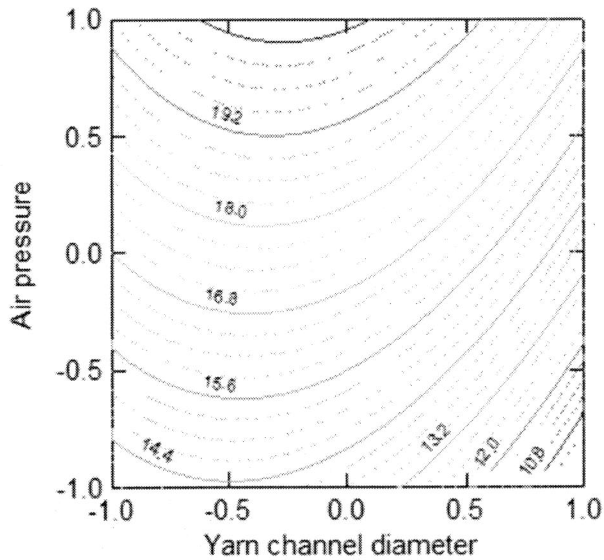

Figure 4.13. Influence of yarn channel diameter and air pressure on percentage reduction in S3 values from corresponding as-spun yarns during nozzle-winding for yarn spun from 1.2 denier fibers [Source: Reference 1].

Table 4.6. Response surface equations for various parameters [Source: Reference 1]

Parameter	Response surface equations	Coefficient of determination
% Reduction in S3 (angle)	$19.667-1.375a+7.375b+3.25c-3.958a^2+1.042b^2+0.792c^2-0.25ab+0.5bc-0.5ca$	0.993
% Reduction in S3 (diameter)	$17.333-1.625a_1+7b_1+3.375c_1-2.167a_1^2+0.583b_1^2-0.167c_1^2-0.25a_1b_1+0.75b_1c_1+0.5c_1a_1$	0.995

Table 4.7. Air Velocities at Different Air Pressure for 45°, 2.2 mm at 0.5 bar (gauge) [Source: Reference 1]

*Air pressure (bar)	VRy (m/s)	VRh (m/s)
0.5	60	135
0.9	123	172

*Gauge.

Table 4.8. Angle of impact of air on hair of 0.9 mm length for nozzles of different axial angle series at 0.5 bar (gauge) [Source: Reference 1]

Yarn channel diameter (mm)	0-plane				1-plane			
	R0	R0.33	R0.67	R1	R0	R0.33	R0.67	R1
1.8	0	0	3	17	10	10	17	30
2.2	0	0	5	20	0	5	5	30
2.6	-3	-3	-3	-3	-3	-3	5	12

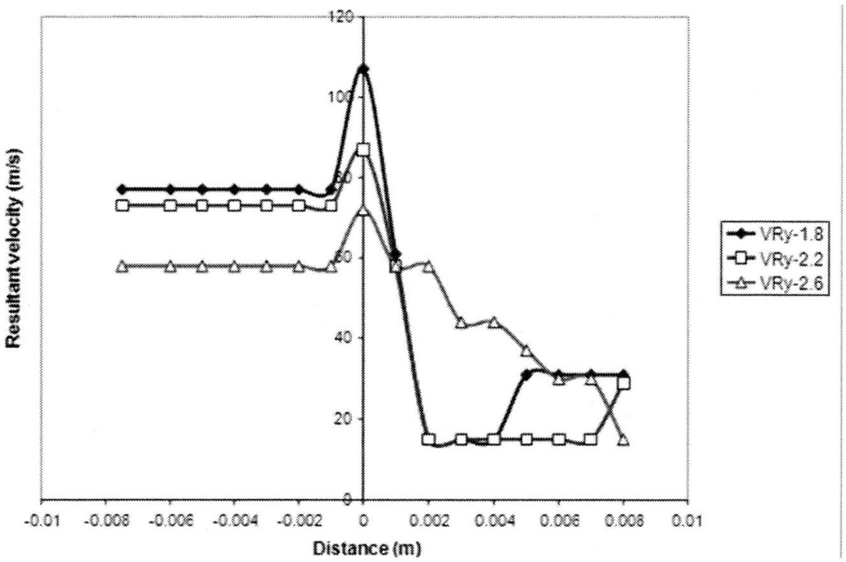

Figure 4.14. Resultant velocity of air acting on yarn surface for the nozzles with yarn channel diameters of 1.8, 2.2, and 2.6 mm at 0.5 bar (gauge) [Source: Reference 1].

Sometimes the impact angle shows negative values as one of the components of the velocity has negative values. Similar calculations are made for 1-plane. The values of the angle of impact of air on hair of 0.9 mm length for the above planes are given in Table 4.8. The angle of impact reduces for all the nozzles from wall (R1) to axis of the nozzle (R0) at 0- and 1-planes indicating a reduction in swirling intensity of air and air velocity.

A close scrutiny of the Table 4.8. and Figure 4.14. reveals a fact that faster the resultant airflow, larger the angle of impacts are observed; in other words higher the swirling intensity of air. There is not much difference in the values of impact angles with the nozzles 1.8 and 2.2 mm diameter especially at the nozzle inner wall. The nozzle having 2.6 mm diameter shows low angles of impact and also negative values towards the axis of the nozzle, indicating reversal of airflow.

However, the magnitudes of negative tangential air velocities are very low (-2 m/sec) and hence its role is limited. The magnitude of resultant air velocity is the predominant factor in reducing the hairiness when comparing the nozzles of different channel diameters as discussed earlier.

The combined effect of nozzle diameter and air pressure on percentage reduction in S3 values from corresponding as-spun yarns indicate that value close to 0 level of yarn channel diameter (2.2 mm) and 1.0 level of air pressure (0.9 bar, gauge) gives the optimum zone.

Conclusion

A CFD model is employed for simulation to get airflow pattern, resultant directions of airflow, velocities of resultant airflow and its components in air nozzles. The transverse and longitudinal drag forces acting on hairs are computed.

Based on simulation and experimental values of hairiness on yarns, the role of axial angle of air inlets and yarn channel diameter, air pressure and fiber denier are explained. Vortex nature of air along with air velocity profile is important phenomenon in reducing yarn hairiness.

The transverse drag forces acting on hairs fold them over the yarn followed by wrapping them around the yarn by the longitudinal drag forces. Increase in air pressure increases swirling intensity, thereby decreasing yarn hairiness. Fiber denier is found to be a major influencing factor in reducing yarn hairiness. An operating air pressure of 0.9 bar (gauge) in combination with 1.4 denier fibers, 45° axial angle and 2.2 mm channel diameter are the optimized values to give maximum reduction in S3 hairiness values.

Future Trend

Application of CFD in yarn engineering is in its early stage. This present chapter tried to address the use of CFD in reducing yarn hairiness during the winding process. There is a vast scope for CFD to be use in the field of yarn manufacturing, as there are number of processes where it can be utilized. The CFD tools should be cost effective and simple to be use by the industry personnel. Also emphasis should be given in developing real-time simulation tools.

REFERENCES

[1] Patnaik, A. Studies on Reduction of Yarn Hairiness Using Air Nozzles during Ring Spinning and Yarn Winding; PhD Thesis, Indian Institute of Technology, New Delhi, 2006; pp 1-208.

[2] Wang, X.; Miao, M. *Text. Res. J.* 1997, vol 67, 481-485.

[3] Chellamani, K. P.; Chattopadhyay, D.; Kumarasamy, K. *Ind. J. Fibre Text. Res.* 2000, vol 25, 289-294.

[4] Doraiswamy, I.; Chellamani, K. P.; Chattopadhyay, D. *Asian Text. J.* 2002, vol 11, 38-43.

[5] Zeng, Y. C.; Yu. C. W. *Text. Res. J.* 2004, vol 74, 222-226.

[6] Patnaik, A.; Rengasamy, R. S.; Kothari, V. K.; Bhatia, S. K. *J. Text. Inst.* 2008, vol 99, 17-27.

[7] Patnaik, A.; Rengasamy, R. S.; Ishtiaque, S. M.; Ghosh, A. *J. Text. Inst.* 2007, vol 98, 243-249.

[8] Rengasamy, R. S.; Kothari, V. K.; Patnaik, A.; Ghosh, A.; Punekar, H. *AUTEX Res. J.* 2005, vol 5, 127-132.

[9] Patnaik, A.; Rengasamy, R. S.; Kothari, V. K. Paper Presented in 34th Textile Research Symposium, Mt. Fuji, Japan, 9-11th August, 2005.

[10] Patnaik, A.; Rengasamy, R. S.; Kothari, V. K.; Ishtiaque, S. M.; Ghosh, A.; Punekar, H. In Proc. of Emerging Trends in Polymers and Textiles, New Delhi, India 7-8th January, 2005.

[11] Fluent 6.1, *User Guide*; 2003, vol 1, pp 1-31.12.

[12] Fluent 6.1, *User Guide*; 2003, vol 2, pp 1-31.12.

[13] Rengasamy, R. S.; Kothari, V. K.; Patnaik, A.; Punekar, *H. J. Text. Inst.* 2006, vol 97, 89-96.

[14] Goldstein, S. Modern Developments in Fluid Dynamics; ISBN-10: 0486613585; Dover Publications Inc: Mineola, NY, 1998, Vol. 2, pp 1-702.

[15] Roberson, J. A.; Crowe, C. T. Engineering Fluid Mechanics; ISBN-10: 0471147354; John Wiley and Sons: NY, 1996, 6th Edn, pp 1-689.

[16] Janna, W. Introduction to Fluid Mechanics; ISBN-10: 0534076327; PWS Engineering: Boston, MA, 1987; 2nd Edn, pp 1-680.

[17] Pillay, K. P. R. *Text. Res. J.* 1964, vol 34, 663-674.

[18] Basu, A.; Chellamani, K. P. *Ind. J. Fibre. Text. Res.* 2000, vol 25, 163-168.

In: Modeling and Simulation in Fibrous Materials
Editor: Asis Patanaik
ISBN: 978-1-62100-116-4

Chapter 5

APPLICATION OF FUZZY LOGIC IN FIBER, YARN, AND FABRIC ENGINEERING

Anindya Ghosh*

Government College of Engineering and Textile Technology,
Berhampore, India

ABSTRACT

This chapter outlines the application of fuzzy logic in various field of textile engineering. The chapter begins by discussing a brief outline of fuzzy logic. It then discusses the applications of fuzzy logic in solving variety of textile problems ranging from trash and color classification of cotton, control of process parameters for spinning, weaving, knitting and nonwovens, to the yarn and fabric property prediction and development of different textile products etc. These are discussed under the heads of fibers, yarns, fabrics, yarn and fabric manufacturing and textile chemical processing. The final part of this chapter gives an overview of the research contribution of the author with regard to the application of fuzzy logic in yarn strength prediction and silk cocoon grading. As the fuzzy logic is tolerant to imprecision and uncertainty its impact in textile engineering will be felt increasingly in the coming years.

5.1. INTRODUCTION

The invention of the von Neumann digital computer initiated an ever growing attempt-how to program them to learn to improve automatically with experience like human beings. The impact has been proven to be dramatic with the successful endeavor of developing learning algorithms. Eventually, it has formed a separate subject area under the head-machine learning. Broadly, the machine learning can be defined as a computer program that improves its performance at some task through experience. It is an established fact that most of the real-world problems are too complex to model mathematically and there exist many physical tasks

* E-mail: anindya.textile@gmail.com

that cannot be solved by classical programming techniques. The machine learning is gaining strategic importance by solving such type of problem where precision is considered to be secondary and we are interested primarily in acceptable solutions. By and large, machine learning is a collection of many methodologies, such as fuzzy logic, artificial neural network (ANN), genetic algorithm (GA) etc. Different members of this family are able to perform various types of tasks. For example, fuzzy logic is a powerful tool for dealing with imprecision and uncertainty, ANN is a potential tool for prediction and classification, genetic algorithm is an important tool for search based optimization. ANN models are superior for prediction and pattern recognition; nevertheless, they are not good for explaining how they reach the decisions. In comparison, fuzzy models are good for explaining their decisions with the aid of linguistic rules; however, they cannot automatically acquire the rules. In combined techniques, either two or more tools are amalgamated to get the advantage of both.

The foundation of fuzzy logic was laid by Prof L. A. Zadeh of the University of California, USA, in 1965 [1]. In general, the real world problems are associated with different type of uncertainties and imprecision. Prior to 1965, the researchers used to consider classical probability theory, which is based on the crisp logic or bivalent logic, as the prime tool for dealing with uncertainties. Prof. Zadeh was the first to argue that there are some uncertainties which cannot be tackled using the probability theory. For a simple example, a little daughter asks her father to bring a red frock for her from the market. There are two uncertainties at least, (i) availability of the baby frock and (ii) a guarantee that the frock is red. There is a probability of obtaining baby frocks (i.e., the frequency of likelihood that an element is in a class), which varies between 0 and 1. According to the crisp set, the frock will be either red (1) or non-red (0). On the contrary, in fuzzy set the color red can be defined using the concept of membership (i.e. similarity of an element to a class) value. If the color is perfectly red, then it may be said red with a membership value of 1, if it is almost red, then it is considered as red with a membership value of 0.8, if it is slightly red, then it is assumed to be red with a membership value of 0.4, even if it is not red, then also it is called red with a membership value of 0, and so on and so forth. Thus, in fuzzy set, an element can be a member of the set with some membership value (i.e. degree of belongingness). In this way, the uncertainty related to color of the frock can be handled using fuzzy logic.

As fuzzy logic is tolerant to imprecision, uncertainty, partial truth, approximation and it is based human reasoning, it has empowered scientists and technologists from diverse engineering disciplines. Fuzzy logic is also an appealing tool for textile product engineering as it can able to handle imprecision that present in the textile data. For an example, a spinner often uses the terms like 'fine' and 'coarse' to assess the fiber and yarn count, although these terms do not constitute a well defined boundary. In textile, the rapidly bourgeoning sway of fuzzy logic is now well recognized in selection of raw material, setting and control of process parameters, classification of patterns and prediction of the properties of various fibrous products.

Although, the basic algorithms of fuzzy logic have been around 45 years; however, the last two decades have been witnessed its more and more application in the variety of textile problems. The fuzzy logic was first applied to the textile problem by Pan et al. in 1988 [2]. However, the publications on fuzzy logic application in textiles have become more pronounced after 2000. This chapter presents an overview of the application of fuzzy logic in various field of textile engineering.

5.2. Brief Outline of Fuzzy Logic

The theoretical aspects of fuzzy logic and fuzzy sets have been explained in many standard textbooks authored by Zimmerman [3], Kartalopoulos [4], Klir and Yuan [5], Berkan and Trubatch [6], Kosko [7], Ross [8], Yen and Langari [9], etc. A brief description of the fuzzy logic theory is discussed as follows.

Classical crisp sets contain objects that satisfy precise properties of membership, contrarily the fuzzy sets contain objects that satisfy imprecise properties of membership, i.e. membership of an object in a fuzzy set is not a matter of affirmation or denial, but rather a matter of a degree of belongingness. Suppose we have an exhaustive collection of individual elements x, which makes up a universe of discourse X. Further, various combinations of the individual elements make up a crisp set, say A. For set A, an element x in the universe X is either a member of A or not. Mathematically, the membership of element x in set A can be expressed by the characteristic function:

$$\mu_A(x) = \begin{cases} 1, & \text{if } x \in A \\ 0, & \text{if } x \notin A \end{cases} \tag{5.1}$$

where $\mu_A(x)$ indicates the unambiguous membership of element x in set A. Suppose set A is the crisp set of yarn count (tex) with $6 \leq x \leq 60$. If two yarns, say x_1 and x_2, having tex values of 30 and 64, respectively, then x_1 has the full membership and x_2 has no membership in set A, or symbolically, $\mu_A(x_1) = 1$ and $\mu_A(x_2) = 0$.

Now, consider a fuzzy set $\tilde{A}$ consisting of coarse yarns which are approximately ranging from 40 to 60 tex. A yarn of 44 tex does not have same belongingness to the set $\tilde{A}$ as much as a yarn of 54 tex. The fuzzy set $\tilde{A}$ covers a range of yarns having different tex, but the belongingness of a yarn to $\tilde{A}$ is represented by its degree of membership with a value between the real continuous interval [0, 1]. Therefore, a key difference between crisp and fuzzy sets is their membership function; a crisp set has a unique membership function, whereas a fuzzy set can have an infinite number of membership functions to represent it. Mathematically, a membership function can be defined as a function which maps its elements onto the interval [0, 1]. Symbolically, the functional mapping is given by:

$$\mu_{\tilde{A}}(x) \in [0,1] \tag{5.2}$$

where $\mu_{\tilde{A}}(x)$ is the degree of membership of element x in fuzzy set $\tilde{A}$. Commonly, fuzzy set $\tilde{A}$ is expressed in terms of ordered pairs as:

$$\tilde{A} = \{x, \mu_{\tilde{A}}(x) | x \in X\} \tag{5.3}$$

For the set $\tilde{D} = \tilde{A} \cup \tilde{B}$, called the union of fuzzy sets $\tilde{A}$ and $\tilde{B}$, the membership function is defined as:

$$\mu_{\tilde{D}}(x) = \mu_{\tilde{A}}(x) \vee \mu_{\tilde{B}}(x) = \max\{\mu_{\tilde{A}}(x), \mu_{\tilde{B}}(x)\}, x \in X \tag{5.4}$$

where the symbol $\vee$ stands for the maximum operator. For the set $\tilde{E} = \tilde{A} \cap \tilde{B}$, called the intersections of fuzzy sets Ã and B̃, the membership function is defined as:

$$\mu_{\tilde{E}}(x) = \mu_{\tilde{A}}(x) \wedge \mu_{\tilde{B}}(x) = \min\{\mu_{\tilde{A}}(x), \mu_{\tilde{B}}(x)\}, x \in X \tag{5.5}$$

where the symbol $\wedge$ represents the minimum operator. The membership function of the complement of a fuzzy set Ã is defined by:

$$\mu_{\tilde{A}^C} = \{1 - \mu_{\tilde{A}}(x)\}, x \in X \tag{5.6}$$

In more general terms, fuzzy intersection is defined by fuzzy AND operator, fuzzy union is defined by fuzzy OR operator, and complement by fuzzy NOT operator. All properties of crisp set are also applicable for fuzzy sets except for the excluded-middle laws. In fuzzy set theory, the union of fuzzy set with its complement does not yield the universe and the intersection of fuzzy set and its complement is not null. This difference is shown below:

For crisp sets, $$A \cup A^C = X \qquad A \cap A^C = \emptyset \tag{5.7}$$

For fuzzy sets, $$\tilde{A} \cup \tilde{A}^C \neq X \qquad \tilde{A} \cap \tilde{A}^C \neq \emptyset \tag{5.8}$$

For each input and output variable of a fuzzy system, the fuzzy sets are created by dividing the universe of discourse into a number of sub-regions, named in linguistic terms (high, medium, low etc.). It accounts for the uncertainty inherent in such a linguistic description by using multi-valued sets. Once the fuzzy sets are chosen, a membership function for each set should be created. The process of assigning membership functions to the sets of data is referred to as fuzzification. In this process the crisp values are converted into fuzzy values to express uncertainties present in the crisp values. Figure 5.1. depicts the degree of membership of various yarn counts (tex) to the fuzzy subsets fine, medium and coarse. Membership function can have various forms, such as triangle, trapezoid, Gaussian etc. which are illustrated in Figure 5.2. Triangular membership function is the simplest one and it is a collection of three points forming a triangle. Dubois and Prade [10] defined triangular membership function as follows:

$$\mu_{\tilde{A}}(x) = \begin{cases} \frac{x-L}{m-L}, & \text{for } L \leq x \leq m \\ \frac{R-x}{R-m} & \text{for } m \leq x \leq R \\ 0, & \text{otherwise} \end{cases} \tag{5.9}$$

where m is the most promising value, L and R are the left and right spread (the smallest and largest value that m can take). The trapezoidal membership curve has a flat top and it is just a truncated triangle curve producing $\mu_{\tilde{A}}(x) = 1$ in large regions of universe of discourse. The trapezoidal curve is a function of a vector x and depends on four scalar parameters a, b, c, and d, as shown below:

$$\mu_{\tilde{A}}(x) = \begin{cases} 0, & \text{for } x \leq a \text{ or } x \geq d \\ \frac{x-a}{b-a} & \text{for } a \leq x \leq b \\ 1, & \text{for } b \leq x \leq c \\ \frac{d-x}{d-c}, & \text{for } c \leq x \leq d \end{cases} \tag{5.10}$$

The Gaussian membership function depends on two parameters, namely standard deviation (σ) and mean (μ) and it is represented as shown below:

$$\mu_{\tilde{A}}(x) = e^{\frac{-(x-\mu)^2}{2\sigma^2}} \tag{5.11}$$

The general bell shaped membership function is defined by three parameters (a, b and c) as shown below:

$$\mu_{\tilde{A}}(x) = \frac{1}{1+\left|\frac{x-c}{a}\right|^{2b}} \tag{5.12}$$

A fuzzy system maps an input space to an output space by means linguistic rules, which is based on human reasoning. The linguistic representation presents an intuitive, natural description of a system allowing for relatively easy algorithm development compared to numerical systems. A fuzzy linguistic rule consists of an IF-THEN statement. A fuzzy rule is evaluated by means of fuzzy operators such as 'fuzzy AND', 'fuzzy OR' etc. For example, in the case of two inputs (I_1 and I_2) and single output (O) fuzzy system, it can be expressed as shown below:

IF I_1 is high AND I_2 is medium THEN O is low

where I_1 and I_2 are two input variables, O is output variable, high, medium and low are the fuzzy sets of I_1, I_2 and O, respectively.

Theoretically there may be r^n number of rules, where n is the number of input variables having r linguistic levels. However, it is not mandatory to construct all these rules. An efficient fuzzy system can be built up with lesser number of rules as well. Moreover, large number of rules invites more complexity to the system. The output of each rule is also a fuzzy set. All rules are evaluated in parallel and output fuzzy sets are then aggregated into a single fuzzy set. This step is known as 'aggregation', which is calculated using 'max' operator. Eventually, the resulting fuzzy set is resolved to a crisp output value by 'defuzzification'. There are several methods of defuzzification like centroid, centre of sums, mean of maxima and left-right maxima etc. The calculation of output in centroid, which is very commonly used, is done as shown below:

$$x^* = \frac{\int \mu_{\tilde{A}}(x)x\,dx}{\int \mu_{\tilde{A}}(x)\,dx} \tag{5.13}$$

where x^* is the defuzzified output and $\mu_{\tilde{A}}(x)$ is the membership value of output fuzzy set after aggregation of individual implication results.

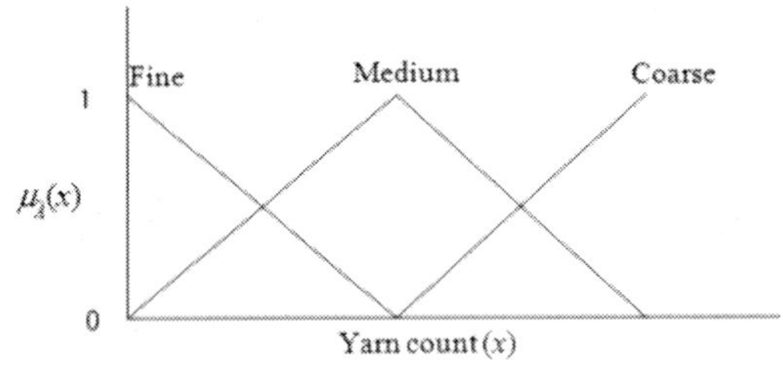

Figure 5.1. Membership function of yarn count (tex).

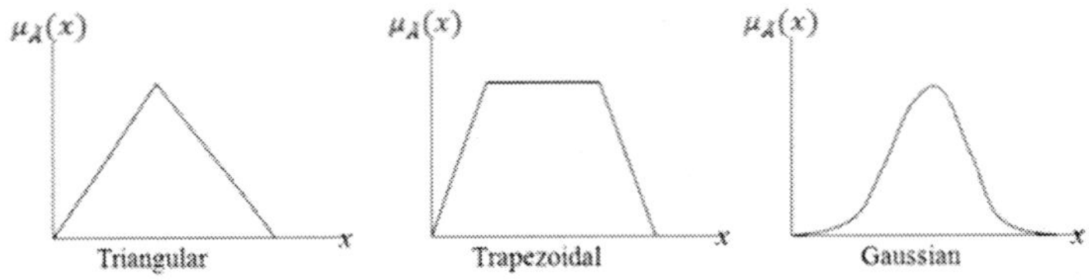

Figure 5.2. Various types of membership functions.

Fuzzy modeling can be categorized by two types, such as subjective modeling and objective modeling [11]. In the subjective modeling approach, it is assumed that a priori knowledge about the system is available and it can be directly solicited from experts. On the other hand, in case of objective modeling, it is assumed that either there is no priori knowledge about the system, or the expert's knowledge is not trustworthy enough. Therefore, instead of any priori interpretation of the system, raw input and output data are used to augment human knowledge or even to generate new knowledge about the system.

In terms of inference process, there are two main types of fuzzy inference system (FIS), namely the Mamdani type and the TSK (Takagi, Sugeno and Kang) type [3]. In case of Mamdani FIS, the consequent membership functions are also fuzzy in nature. On the contrary, the consequent membership functions in a TSK FIS are not fuzzy (either linear or constant). Mamdani FIS more popularly used because it provides reasonably good results with a relatively simple structure and its interpretable nature of the rule base. The interpretability of TSK FIS is lacking. However, the consequent membership functions for TSK FIS can have as many parameters per rule as input variables; this translates into more degrees of freedom in the design as compared to Mamdani FIS, thus providing more flexibility in the design of the system. Mamdani FIS can be used directly for MISO systems (multiple input single output) and MIMO systems (multiple input multiple output), while the TSK FIS can only be used in case of the MISO system.

5.3. Application of Fuzzy Logic in Textiles

5.3.1. Fibers

Xu et al. [12] used fuzzy logic for color grading of cotton fibers to reduce the HVI-classer disagreement in cotton color grading. The authors used R_d and $+b$ as the input variables and five major classes of cotton color as the output variables. They selected five fuzzy sets for the input variable R_d and six fuzzy sets for *+b*. The fuzzy sets for R_d represent the brightness from very low (I), low (II), medium (III), high (IV) to very high (V), and fuzzy sets for *+b* represent yellowness ranging from very low (I) to extremely high (VI). The output variable color grade was split by five fuzzy sets named as white, light-spotted, spotted, tinged and yellow stained. A Gaussian membership function was chosen for the input fuzzy sets and a triangular membership function was used for the output fuzzy sets. In order to make the output clear and unbiased, the symmetric, non-overlapping and equal-size membership functions were used for the output fuzzy sets. The defuzzification was done using controid method. The constructed fuzzy system was tested using U. S. Upland cotton samples in 1996, 1997 and 1998 crop years. For the 'white' class the HVI-classer disagreement was 54.1%, 61.5% and 51.2% for the year 1996, 1997 and 1998, respectively. The proposed fuzzy logic method was able to reduce this disagreement to a level within 6.1% to 6.2%. They concluded that proposed method has great potential for being a more reliable way of grading cotton colors and it could able to eliminate the hard boundary problems associated in cotton color grading.

Xu et al. [13] demonstrated the use of ANN and fuzzy clustering method for the identification of various types of trash based on their color and shape parameters. Three color parameters, namely lightness (*L*), redness (*a*) and yellowness (*+b*) and one shape parameter in the form of roundness were used as inputs. The outputs were four trash categories, namely leaf, bark, hairy seed coat and smooth seed coat. The classification accuracy was higher for ANN clustering than the fuzzy clustering.

Siddaiah et al. [14] investigated the identification of various types of trash in ginned cotton using fuzzy logic. Various types of trash images were acquired using a 3-chip CCD colour camera. The trash objects were separated from the cotton background. A set of twenty five features were measured for each segmented image based on the shape descriptors such as area, shape factor, perimeter, the first order and second order moments, ferret diameters etc. The trash objects were clustered into four categories such as bark1, bark2, leaf and pepper using fuzzy C-means algorithm.

Sarna et al. [15] proposed a method for determining the degree of cotton maturity based on the analysis of fiber fracture SEM image with the help of fuzzy set theory. The application of fuzzy set theory was justified as the categories of images of cotton fracture do not have clearly defined boundaries. Fourteen categories of cotton fracture were described with linguistic variables. The membership function value 1 was assigned for the category of 'completely belonging' to a given maturity degree and 0 was assigned for the category of 'evidently not belonging' to a maturity degree. The membership function values between 0 and 1 evaluated the intermediate state of category, belonging to a given maturity degree.

5.3.1. Yarns

Zaman and Wunsch [16] predicted the CSP of cotton yarns from the fiber properties and spinning parameters using fuzzy ARTMAP. The fiber properties were evaluated by traditional, HVI and AFIS instruments. Seven different models were developed and all of them yielded correlation coefficient higher than 0.96 between the actual and predicted CSP. The model which included the data of traditional and AFIS instruments produces the best prediction accuracy.

Sette et al. [17] used fuzzy efficiency based classifier system (FECS) to automatically generate a rule set for predicting yarn strength. In this system a modified genetic algorithm was used for the optimization of fuzzy rule set. The rule efficiency and reward mechanism were modified to account for the membership degree of each data sample. They considered two, three and five fuzzy strength classes and observed that a three-fuzzy set implementation of yarn strength proves to be a good compromise between accuracy and generalization, resulting in a total prediction accuracy of 92%. They also observed that there exists a danger of over fitting when applying higher fuzzy strength classes. The generated rule set allowed almost 100% acceptable classification of yarn strength in three categories and it provided additional qualitative information about the fiber to yarn process behavior [18].

Kuo et al. [19] applied fuzzy logic for prediction of tensile strength and yarn count of melt-spun fibers. They used the controllable factors affecting the draw ratio of melt spinning as inputs, viz., the extruder screw speed, the gear pump speed and winding speed. The output variables were tensile strength and denier of as-spun fibers. Triangular membership function was used for both input and output fuzzy sets. The defuzzification was done using centroid method. The fuzzy forecast model was verified with fifteen entries of new data set and the result showed that 97.32% and 97.09% of the forecast values lie within one standard difference ($\pm\sigma$), for yarn strength and denier, respectively.

Ceven and Özdemir [20] analyzed the shrinkage behavior of chenille yarns with the fuzzy logic method. They employed chenille yarn count, twist level and pile length as input variables and yarn shrinkage as the output variable. Triangular form of membership function was used for input and output fuzzy sets. They divided input membership functions into two intervals and output membership function into eight intervals. The base widths of each interval for output variable were different as they were determined on the basis of experimental values of shrinkage. They concluded that the coarser and high twisted chenille yarns having low pile length are less prone to boiling water shrinkage property.

5.3.3. Fabrics

Choi et al. [21] reported a fabric defect identifying system using fuzzy inference system. Image pre-processing techniques that enhance raw images were applied before the defect classification. Only a small number of fuzzy inference rules were developed to identify the non-defect, slub (warp direction), slub (weft direction), nep and composite defect. The experimental results with this approach demonstrated that the proposed method could able to identify the fabric defects at an extent comparable to that of a human inspector.

Lin et al. [22] developed an intelligence diagnosis system for fabric inspection by combining a conventional expert system and a diagnostic system based on fuzzy set theory. Diagnosis system was developed using Turbo C language and the expert system was developed in an artificial language PROLOG. Altogether eighteen kinds of fabric defects and forty-four possible causes of these defects were incorporated into the system. The developed system was applied on a water-jet loom during weaving. The proposed system was capable of tracing possible breakdown causes of fabric defects. This system could recognize and evaluate uncertain linguistic description. Using this system, the operator could acquire an immediate response about breakdown caused by the input of the categories of defects appearing on the fabrics. This system could provide the operator a knowledge base for further consultation in fabric inspection. The authors claimed that through the assistance of this intelligent diagnosis system, even a new operator lacking expertise or skill in the weaving engineering field can still easily determine causes of fabric defects in a fraction of a second and then eliminate them. In addition, this system can be regarded as a teaching aid for a new operator to advance the technique of weaving engineering.

Wang and Yao [23] used fuzzy priority similarity comparison method to assess the crease recovery of fabrics. The most similar specified samples were selected from the established set of arguments to the sample on the basis of the fuzzy closeness. The authors inferred that the grade of fabric's crease recovery can be assessed more objectively and correctly by using the fuzzy priority similarity comparison method. In addition, the proposed method of assessing the crease recovery has high consistency with practice.

Kuo et al. [24] studied color and pattern analysis of printed fabrics using an unsupervised clustering method with the help of fuzzy C-means (FCM) clustering algorithm and a specific cluster-validity (SC) criterion. The images of printed fabrics were captured by a color scanner. SC criterion was used to search for a good cluster number and the corresponding color clusters were obtained based on the FCM clustering algorithm. The quality of the fuzzy C-partitions was also measured by SC criterion. The patterns within a cluster were more similar to each other than the patterns belonging to different clusters.

Kuo et al. [25] reported an automatic recognition of fabric weave pattern using a FCM clustering method. Three types of weave such as plain, twill and satin were considered. The images of fabrics were captured using a color scanner and they were converted to gray-level images. After image enhancement, warp and weft crossed areas were located based on the interstices of yarns. A gray-level co-occurrence matrix was used to extract the feature of the warp and weft crossed areas. Unsupervised decision rules for recognizing warp and weft floats were developed using the FCM method.

Pan et al. [2] studied fabric handle sorting using fuzzy cluster analysis, which is a combination of fuzzy set theory and conventional cluster analysis. The fabrics samples were sorted into various groups and the degree of property association was high between the member of same group and low between the members of different group. Marriot's $g^2|W|$ criterion was used to determine the optimum number of groups.

Raheel and Liu [26] developed a fuzzy comprehensive evaluation technique to calculate fabric handle of light weight dress fabrics. The fabric hand was described by thickness, weight, flexural rigidity, wrinkle recovery and 45^0 filling elongation. Five membership functions were chosen for each of these input variables and a decreasing half Cauchy distribution was used to describe the membership degree. The importance of each property selected was ascertained and expressed as a weighted vector on the basis of a survey of judges. The fabric handle was calculated by using the weighted vector and the fuzzy transformation matrix. A similar approach was presented by Park and Huang [27] for predicting total hand value of double weft knitted fabric (knit, tuck and welt stitches) from the selected mechanical properties. Primary hand and total hand values from the KES-FB were analyzed and compared with the fuzzy total hand test result. They claimed that the results from the new method for double weft-knitted fabrics provide experts with meaningful values in comparison with Kawabata's total hand values. Chen et al. [28] also used same approach for grading softness of 100% cotton and 50/50 cotton/polyester blended fabrics. Fabric mechanical properties controlling fabric softness were measured with the KES-FB instruments. Statistical methods of ANOVA and factor analysis were used to determine the weighting-factor vector. The fuzzy conversion matrix was formed on an assumption of linear membership function. Correlation of fuzzy evaluation, subjective evaluation, and the Kawabata primary hand evaluation was estimated statistically and the results was found to be encouraging. Hwang et al. [29] developed an identical approach for hand evaluation of warp knitted fabrics using fuzzy logic. They used eight input parameters, viz., weight (*W*), flexural rigidity (*B*), 45^0 filling elongation (*2HG*), one dimensional elongation and recovery (*EM*, *RT*), one dimensional compression (*WC*, *RC*) and frictional coefficient. These inputs were transformed into meaningful values of scale by fuzzy transformation matrix. 50 warp knitted fabrics were used to establish the fuzzy evaluation method. The difference between the results of Kawabata's THV and the fuzzy method were compared. As per their claim, the fuzzy method was more reliable than that of KES-FB for winter outwear knit fabrics. The fuzzy fabric hand evaluation technique also generated a similar tendency with the subjective assessment.

Lau et al. [30] made an attempt to use a fuzzy expert system with gradient descent optimization for prediction of fabric specimens in fashion product development. They developed a fuzzy fabric advisory system which could provide the most appropriate fabric satisfying individual desires for fabric hand. Firstly, the human knowledge on the selected fourteen fabric hand descriptors for fabric selection was captured through a survey and was formulated into fuzzy IF-THEN rule architecture. In the second step, a self-optimized fuzzy fabric advisory system was implemented which employed the rules created from the survey as preliminary prediction. It was then followed by a gradient descent optimization to minimize the individual predictor error. Finally, a fine tune mechanism was proposed to reduce the structural error on the fuzzy fabric advisory system. They inferred that the proposed system could cope with the subjective fabric hand descriptor to incorporate the psychological effects from individuals into the prediction of fabric specimens. Further, the system could help consumers to select the most appropriate fabric and provide field practitioners appropriate suggestions for effective product development in clothing and fashion industries.

5.3.4. Yarn and Fabric Manufacturing

Huang and Chang [31] presented an experimental study of fuzzy self-organizing control in developing an auto leveling system with a drawing frame. The fuzzy self-organizing controller consists of a basic fuzzy logic and self-organizing scheme. The self-organizing scheme corrects the control action and modifies the fuzzy control rules. They used a laboratory scale drawing frame with two drafting zones and two sliver doubling. The draft ratio was adjusted by regulating the speed of the middle roller using the proposed control system. In fuzzy self-organizing control, they chose the error and the change in error as input parameters and the speed change required for the servo motor as the output. The error was calculated based on the difference between desired sliver linear density and output linear density. The control scheme was closed-loop type because the output linear density was fed back to compare with the desired value. Initially, the control system was generated from the rule base with the basic fuzzy control and then the control action was corrected using the self-organizing scheme based on both inputs. The experimental results showed that the fuzzy self-organizing controller was effective in reducing the *CV*% of output sliver.

Dorrity et al. [9] designed a fuzzy logic controller for controlling drying system of the sizing machine. Two input variables (can pressure and nip pressure) and one output variable (exit yarn moisture content) were monitored. A fuzzy C-means algorithm was applied to the raw data and nine clusters were formed. From these they derived membership functions for each input variable with three linguistic terms. The output variable was assigned five membership functions. The rule base was constructed with nine rules. The proposed system was able to control moisture content in the yarn successfully and its performance was shown to outperform a conventional control law.

Kayacan et al. [33] developed a fuzzy controller to control the weft insertion system in air-jet weaving loom. The effect of yarn linear density (tex) and value of twist coefficient on weft yarn velocity along the tube at a constant air pressure was determined by the fuzzy logic system. They selected five fuzzy sets for both the input variables (yarn tex and twist coefficient), and 16 fuzzy sets for output variable (weft velocity). Triangular form of membership function was chosen for the fuzzy sets and eighteen fuzzy rules were developed based on the expert knowledge. They inferred that coarser yarn with low twist coefficient increases the velocity of weft yarn.

Dayik et al. [34] determined the warp tension in the shedding operation using fuzzy logic. The shed height, weft yarn density and warp yarn elongation were chosen as input parameters. The input membership functions were divided into five equal foot ranges. The determination of the warp tension was based on a combination of human-determined fuzzy rules to solve nonlinear control setting. Experimental studies as well as expert opinions were utilized for rule-base construction. They observed that the experimental results and fuzzy logic results were close to each other. Shed height was found to be the most effective membership function on warp yarn tension, while weft yarn density was the least. The increase in warp yarn tension with the decrease in warp yarn elongation was also observed.

Semnani and Sheikhzadeh [35] developed a method for online control of loop length based on the stitch deformation of knitted fabric using image analysis and fuzzy logic. The stitch deformation was measured using computer vision and analyzing the fabric images based on radon projection intensity matrix. A fuzzy control system was designed to control the setting of cam which controls loop length. Five fuzzy classes were used for stitches deformation. Six fuzzy rules were generated based on the knowledge of skilled operator for changing the cam setting. The fuzzy reasoning was conducted by means of a min-max composition method using triangular membership functions and if-then rules. The developed system was presented for readjusting the cam setting to enhance knitted fabric regularity by controlling the loop length based on the defined stitch deformation index. The authors suggested that it is possible to install the system on high speed knitting machine easily for online control of fabric quality.

Deng et al. [36] presented a fuzzy selection criterion which takes into account both the sensitivity of measured data and human knowledge concerning the relations between process parameters and quality features of nonwoven product. They proposed a fuzzy logic based linguistic selection criterion which can permit to rank the process parameters of the related nonwoven production line and take the most relevant ones as input variables of the model for designing new nonwoven products. The selection criterion was expressed as $F = f(H, S)$, where H represents the human knowledge on nonwoven processes and products and S is the sensitivity of measured data defined according to the two assumptions:

- IF a small variation of an input variable Δx corresponds to a large variation of output variable Δy, THEN S is big.
- IF a large variation of an input variable Δx corresponds to a small variation of output variable Δy, THEN S is small.

These assumptions were transformed into a fuzzy model in which Δx and Δy was taken as two input variables and S as output variable. Three fuzzy sets were considered for each variable. The fuzzy rules were defined according to the experience of experts. Given fixed values of Δx and Δy, the corresponding numerical value of S was obtained from the fuzzy model. They focused on five direct process parameters (take-off apron speed, delivery apron speed, production speed, stroke frequency, penetration depth of needles) and three indirect process parameters (needle loom draft, number of layers, stroke density) related to the cross-lapper and needle-loom. These parameters were preselected by nonwoven experts according to their possible influence on porosity, which is the one of the main critical structural parameter of a nonwoven product. They predefined a selection threshold at 50% of the total amount of inputs in the final selected list. Using the proposed fuzzy selection procedure, they identified four process parameters such as number of layers, penetration depth of needle, needle loom draft and stroke density as the most relevant independent process parameters. The authors concluded that the proposed method is more robust and less sensitive to proximities of measurement and uncertainties, as compared with classical selection criteria.

5.3.5. Textile Chemical Processing

Huang and Yu [37] investigated the use of fuzzy logic controller for controlling concentration, pH and temperature in dyeing process. A fuzzy controller simulates the decision-making activities of experienced experts and employs a reasoning scheme to infer control actions. A small laboratory scale experiment yielded results showing that fuzzy control was able to control dye concentration, pH, and temperature at the desired values.

5.4. Grading of Silk Cocoons Using Fuzzy Logic [38]

In view of the significance of shell ratio (SR) (%), defective cocoon (DC) (%) as well as subjectively evaluable parameter such as cocoon size in governing the quality of cocoons, these three quality parameters were used for cocoon grading. A sample of twenty lots of multibi cocoons were collected from the cocoons market. The SR (%) was estimated as the ratio of the shell weight to the cocoon weight expressed as a percentage. Fifty cocoons were selected from each lot and average SR (%) was calculated. The DC (%) was measured as a percentage of defective cocoons by number from a lot of one kg cocoons. Cocoon size was determined subjectively by asking three experts to evaluate the cocoon size in a five-point scale having the linguistic terms low, below average, average, above average and high based on the characteristics such as perimeter, volume, weight, shape, texture, dryness and hardness etc. As it is required to present the inputs to the fuzzy expert system in the form of crisp data, the subjective data of the cocoon size attribute was converted into appropriate objective data using the method proposed by Chen and Hwang [39]. A five-point scale having the linguistic terms low, below average, average, above average and high was considered for this purpose and their crisp scores were obtained as 0.115, 0.295, 0.495, 0.695 and 0.895 respectively. The process of fuzzification was done by assigning triangular form of membership functions for the crisp quantities of both inputs and output parameters. Three linguistic fuzzy sets namely low, medium and high were selected for each of the input parameters in such a way that they are equally spaced and cover the whole input spaces. Figure 5.3. depicts the triangular membership plots for inputs and output of the fuzzy expert system. Nine output fuzzy sets (level 1 to 9) were considered for cocoon score ranging from 10 to 90, so that the fuzzy expert system can map the small changes in cocoon score with the changes in input variables.

The levels 1 and 9 represent the worst and best cocoon qualities, respectively. Twenty seven fuzzy rules were developed based on the experiences which are depicted in Table 5.1. in the matrix form. For example, the first rule as given in Table 5.1. can be read as: IF (SR% is high) AND (Size is high) AND (DC% is low) THEN (Cocoon score is at Level 9). A schematic representation of fuzzy expert system for cocoon grading is displayed in Figure 5.4. All twenty seven rules were evaluated in parallel. The fuzzy sets that represent the outputs of each rule were combined into a single fuzzy set by the process of aggregation. The input of the aggregation process was the list of truncated output functions evaluated by each rule. The 'max' function was used to aggregate the output of each rule into a single fuzzy set of the output variable. The aggregate fuzzified output was then converted into a single crisp value by the process of defuzzification. The defuzzification was done by the centroid method which returns the center of area under the curve. A MATLAB (version 7.0) based coding was used to execute the proposed fuzzy expert system for cocoon grading. The parameters of twenty lots of cocoons encompassing SR (%), size and DC (%) and corresponding cocoon score generated from fuzzy expert system are given in Table 5.2.

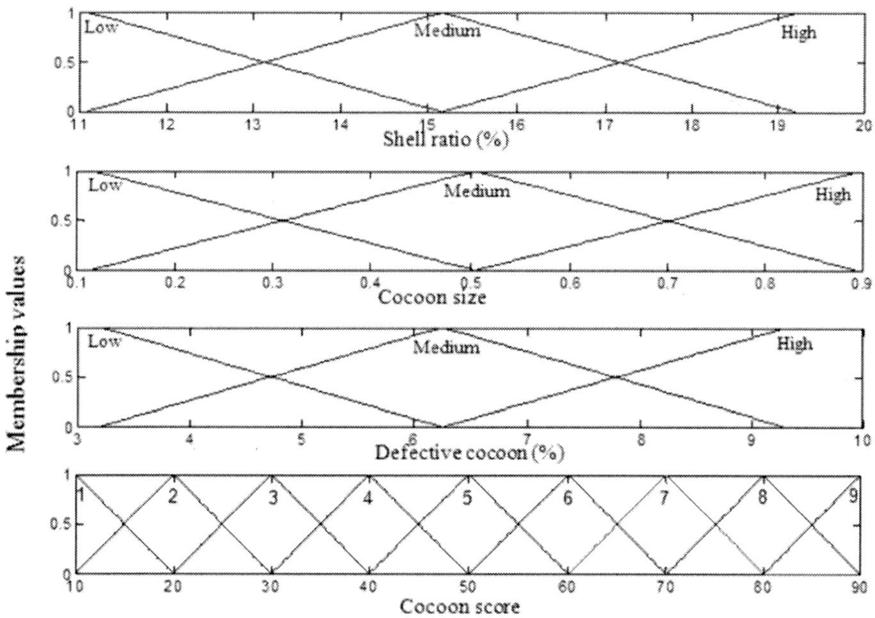

Figure 5.3. Triangular membership function plots of inputs and output.

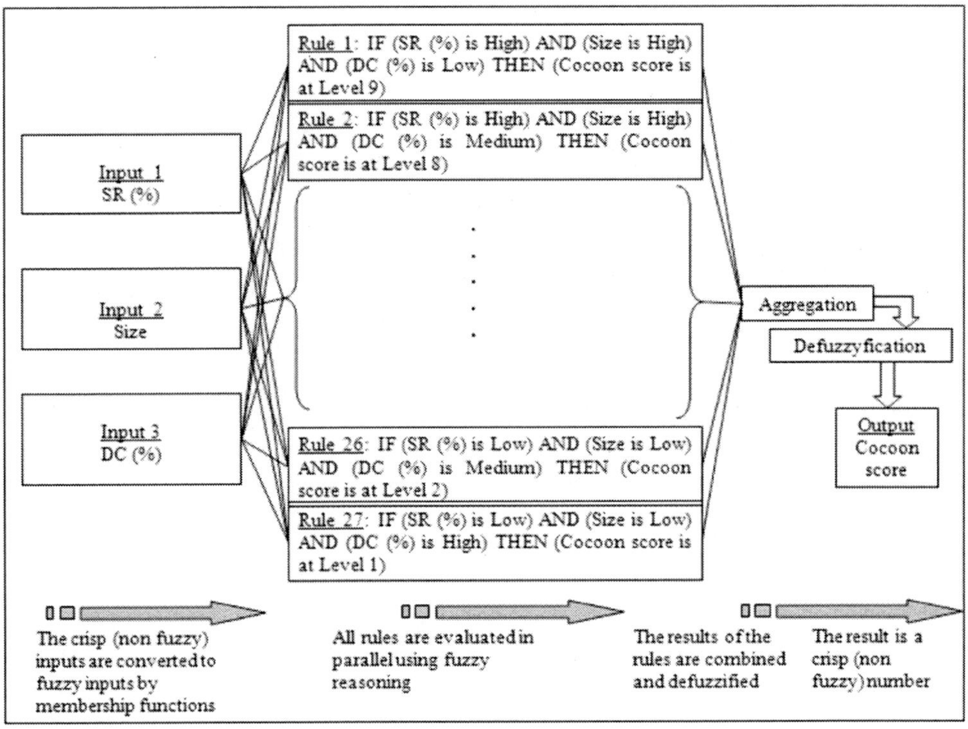

Figure 5.4. Schematic representation of fuzzy expert system for cocoon grading.

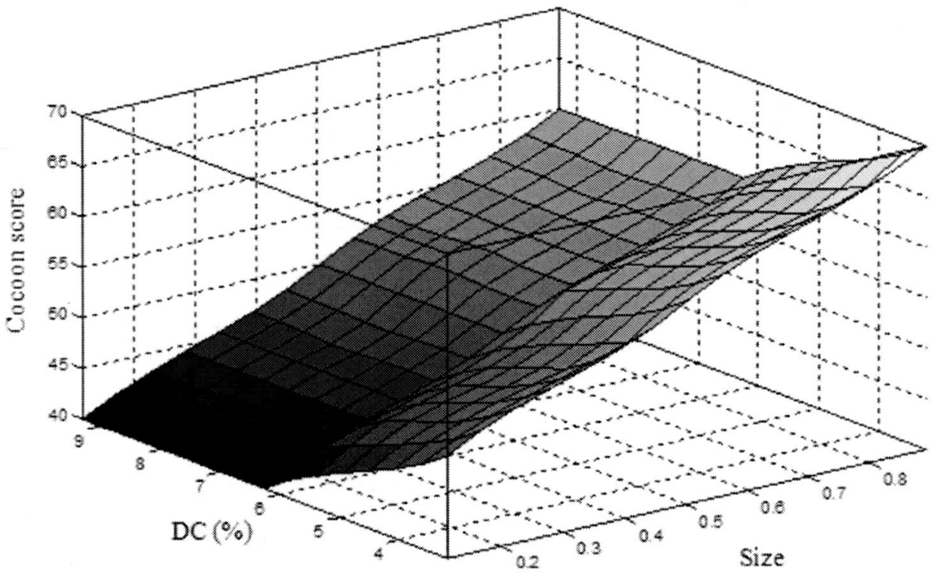

Figure 5.5. Surface plot showing the effect of DC (%) and cocoon size on cocoon score.

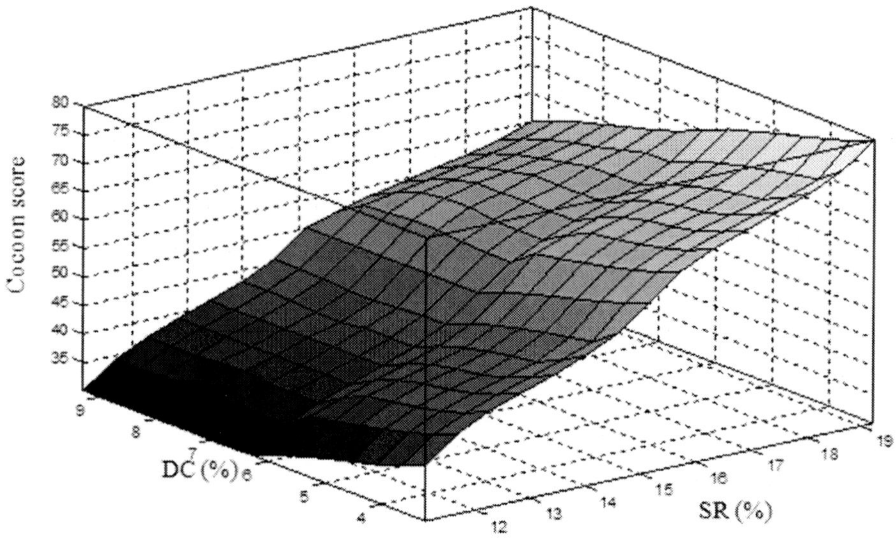

Figure 5.6. Surface plot showing the effect of DC (%) and SR (%) on cocoon score.

Table 5.1. Matrix of fuzzy rules

Rule number	Membership level			
	SR (%)	Size	DC (%)	Cocoon score
1	High	High	Low	9
2	High	High	Medium	8
3	High	High	High	8
4	High	Medium	Low	8
5	High	Medium	Medium	7
6	High	Medium	High	6
7	High	Low	Low	7
8	High	Low	Medium	6
9	High	Low	High	6
10	Medium	High	Low	7
11	Medium	High	Medium	6
12	Medium	High	High	6
13	Medium	Medium	Low	6
14	Medium	Medium	Medium	5
15	Medium	Medium	High	5
16	Medium	Low	Low	5
17	Medium	Low	Medium	4
18	Medium	Low	High	4
19	Low	High	Low	5
20	Low	High	Medium	5
21	Low	High	High	4
22	Low	Medium	Low	4
23	Low	Medium	Medium	3
24	Low	Medium	High	3
25	Low	Low	Low	2
26	Low	Low	Medium	2
27	Low	Low	High	1

Table 5.2. Cocoon parameters and their quality values

Lot No.	SR (%)	Size	DC (%)	Cocoon score
1	16.8	Above average	3.2	70.4
2	15.7	Average	7.6	54.4
3	18.7	High	4.2	77.4
4	19.2	High	6.7	80.0
5	11.6	Low	7.8	22.9
6	15.4	Average	3.2	61.7
7	14.6	Average	9.1	45.2
8	16.4	Average	4.5	62.8
9	13.1	Below average	5.4	38.0
10	14.4	Average	6.9	44.2
11	17.9	Above average	7.2	67.0
12	12.6	Low	4.3	33.2
13	11.1	Low	5.2	20.0
14	15.3	Average	9.3	50.1
15	16.2	Average	6.7	56.2
16	18.1	Above average	8.6	66.3
17	12.2	Below average	5.7	34.9
18	11.9	Below average	6.5	30.5
19	15.7	Average	3.9	60.9
20	13.6	Below average	4.0	41.8

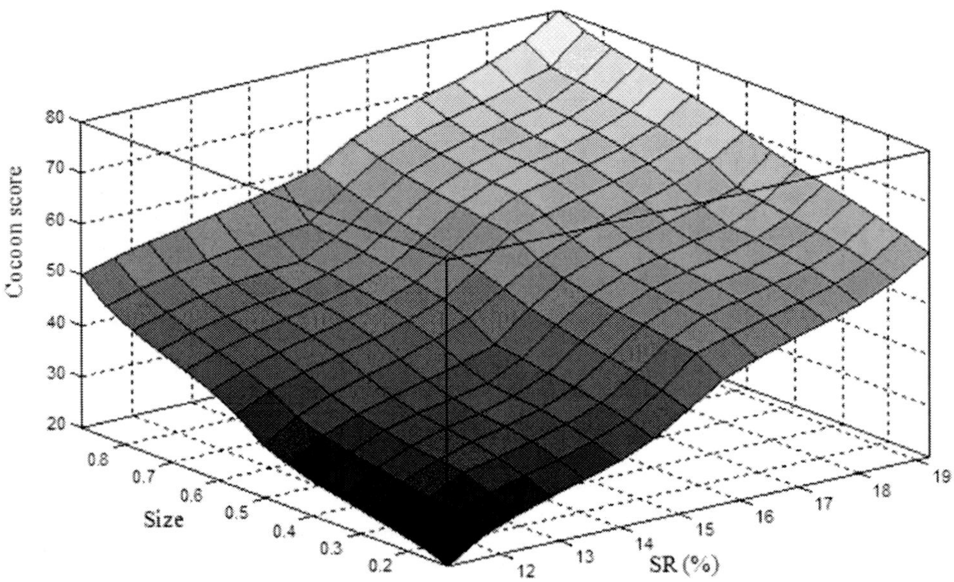

Figure 5.7. Surface plot showing the effect of cocoon size and SR (%) on cocoon score.

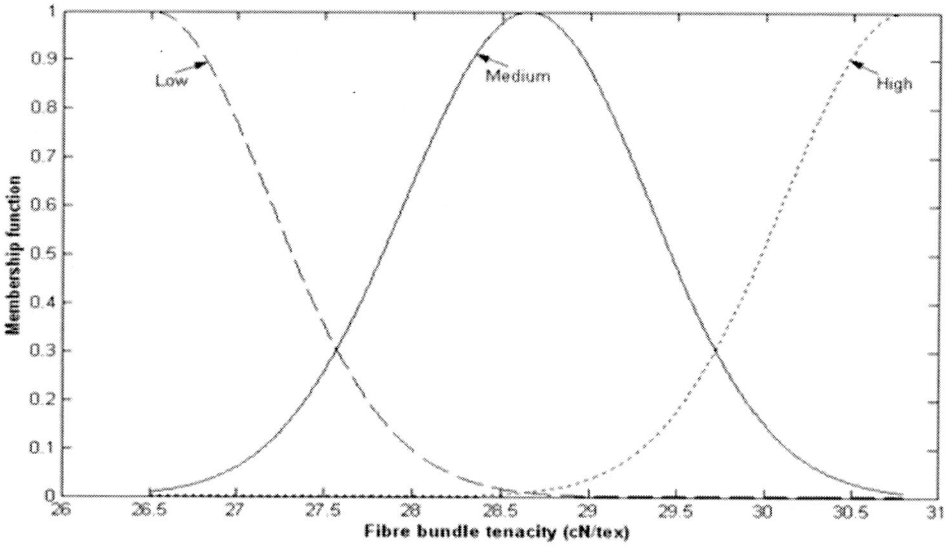

Figure 5.8. Gaussian membership function plots of fiber tenacity .

The surface plots shown in Figures 5.5-5.7 depict the impacts of cocoon parameters on the cocoon score. Figure 5.5 shows higher value of cocoon size and lower proportion of defective cocoons improves the cocoon quality as expected. Figure 5.6 shows that the SR (%) has similar influence on cocoon quality like the cocoon size. The cocoon quality reaches the apex when the SR (%) and cocoon size both reach their respective maximum level (Figure 5.7). The main advantage of the proposed fuzzy system of the cocoon grading is that the subjective aspect of cocoon grading as follows in the cocoon market in India has also been incorporated in the system. It thus provides a good flexibility in reflecting the expectations and visual grading into the results. This method could be used as a reference material for someone who solves cocoon marketing and subjective judgments problems. The present study was concentrated on a single variety of cocoon. The same methodology can be applied by considering more cocoon varieties.

5.5. Yarn Strength Modeling Using Fuzzy Expert System [40]

Four parameters of cotton fibers namely fiber bundle tenacity (cN/tex), HVI mean length (mm), micronaire and AFIS short fiber content (%) were used as the input parameters to the fuzzy expert system. These fiber parameters were exclusively selected since they influence the yarn strength to a greater extent. Three linguistic fuzzy sets namely low, medium and high were chosen for each of the input parameters. Two forms of membership functions (Gaussian and triangular) were tried for inputs as well as for the output. Figures 5.8. and 5.9. depict the Gaussian and triangular membership curves, respectively, for fiber tenacity which is one of the inputs for fuzzy expert system. Nine output fuzzy sets (1 to 9) were considered for yarn tenacity. Figures 5.10. and 5.11 show the Gaussian and triangular membership curves, respectively, for yarn tenacity of 16's Ne yarn. Similar membership curves for yarn tenacity were also developed for 22's Ne and 30's Ne yarns. However, the ranges of yarn tenacity coved by output membership curves were varied by a little, depending upon the yarn count, as coarse yarns show higher yarn tenacity and vice versa when the input variables are at the same level. Theoretically there could be $3^4 = 81$ fuzzy rules, as there are four input variables and each one of them are having three linguistic levels. However, to simplify the expert system only thirty six fuzzy rules were developed as shown in Table 5.3.

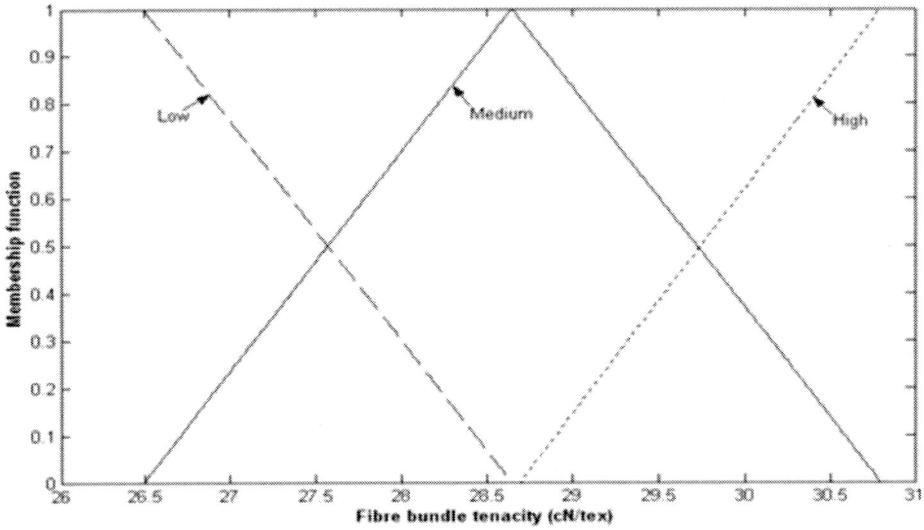

Figure 5.9. Triangular membership function plots of fiber tenacity.

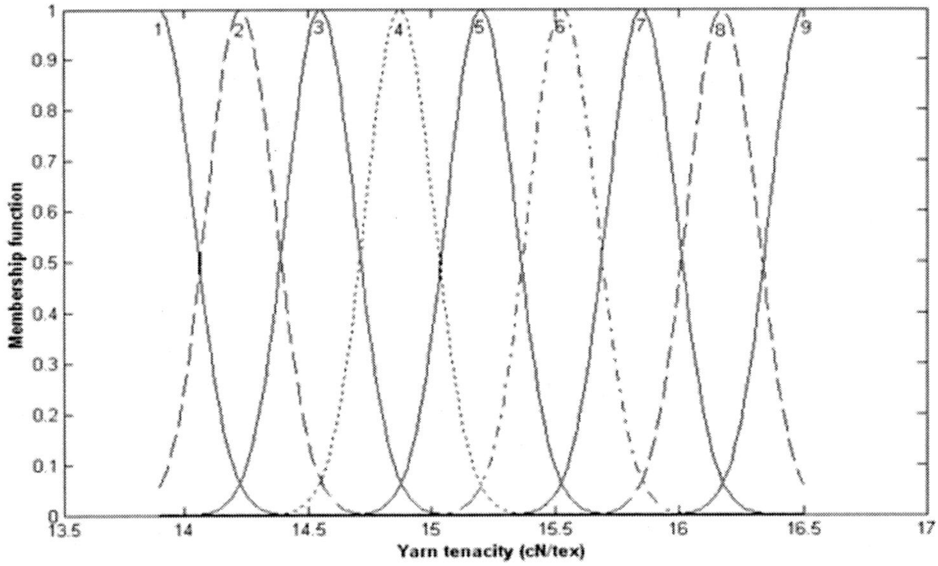

Figure 5.10. Gaussian membership function plots of yarn tenacity for 16 Ne yarn.

Table 5.3. Matrix of fuzzy rules

Rule no	Membership level				
	I_1	I_2	I_3	I_4	*0*
1	L	L	H	H	Level 1
2	H	H	L	L	Level 9
3	H	M	L	L	Level 7
4	H	L	L	L	Level 4
5	M	H	L	L	Level 6
6	L	H	L	L	Level 3
7	H	H	M	L	Level 7
8	H	H	H	L	Level 4
9	H	H	L	M	Level 8
10	H	H	L	H	Level 6
11	M	L	H	H	Level 3
12	H	L	H	H	Level 5
13	L	M	H	H	Level 2
14	L	H	H	H	Level 3
15	L	L	M	H	Level 3
16	L	L	L	H	Level 5
17	L	L	H	M	Level 2
18	L	L	H	L	Level 4
19	M	M	M	M	Level 6
20	L	M	M	M	Level 4
21	H	M	M	M	Level 8
22	M	L	M	M	Level 5
23	M	H	M	M	Level 7
24	M	M	L	M	Level 7
25	M	M	H	M	Level 4
26	M	M	M	L	Level 7
27	M	M	M	H	Level 5
28	H	H	M	M	Level 8
29	H	M	L	M	Level 8
30	H	M	M	L	Level 8
31	L	M	M	H	Level 2
32	L	L	M	M	Level 2
33	L	M	H	M	Level 2
34	M	H	L	M	Level 7
35	M	H	M	L	Level 7
36	M	M	L	L	Level 7

H = high, M = medium, L = low, I_1= fiber tenacity,
I_2= mean length, I_3 = short fiber content,
I_4= fiber micronair, O = yarn tenacity

Table 5.4. Prediction of yarn tenacity with Gaussian and triangular membership functions

Yarn count (Ne)	Actual yarn tenacity (cN/tex)	Gaussian membership function		Triangular membership function	
		Predicted yarn tenacity (cN/tex)	Absolute error (%)	Predicted yarn tenacity (cN/tex)	Absolute error (%)
16 Ne	14.77	15.54	5.23	15.50	4.95
	16.28	15.71	3.53	15.77	3.14
	15.45	15.17	1.83	15.06	2.50
22 Ne	13.15	13.95	6.11	13.93	5.91
	15.21	14.74	3.09	14.75	3.02
	13.24	13.79	4.17	13.83	4.45
30 Ne	13.52	13.76	1.80	13.84	2.35
	12.28	13.28	8.16	13.26	8.02
	14.06	13.99	0.52	14.30	1.71
	14.86	13.97	6.01	13.89	6.55
Mean absolute error (%)			4.04		4.26
R^2			0.75		0.75

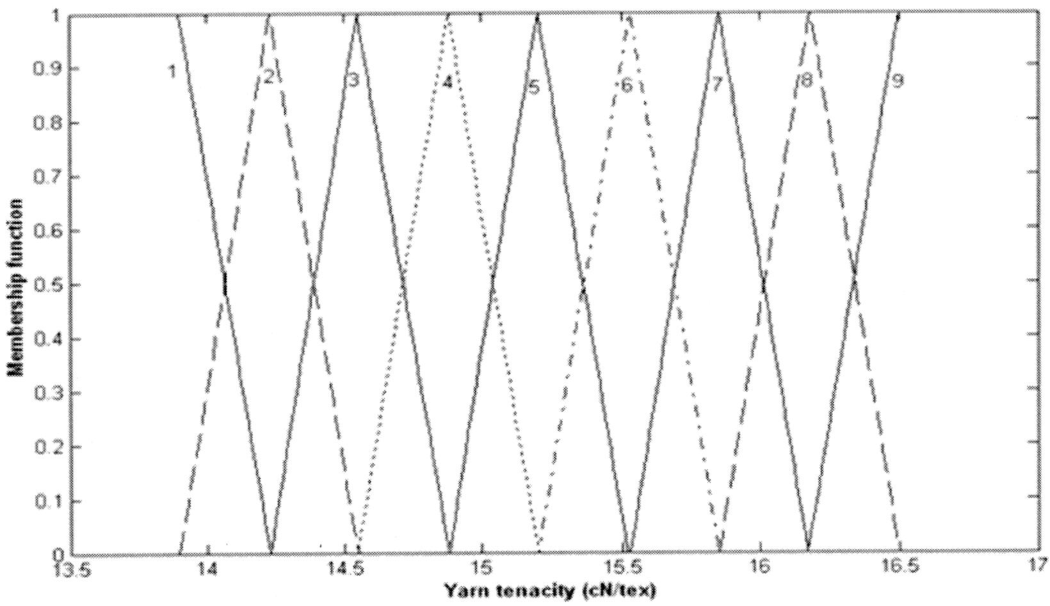

Figure 5.11. Triangular membership function plots of yarn tenacity for 16s Ne yarn.

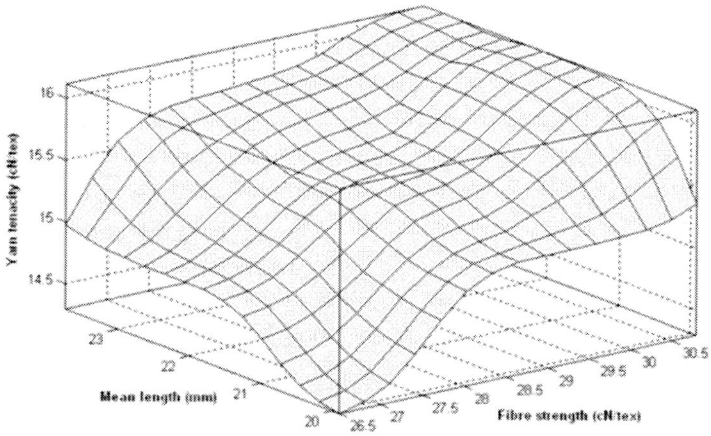

Figure 5.12. Surface plot showing the effect of fiber length and strength on yarn tenacity.

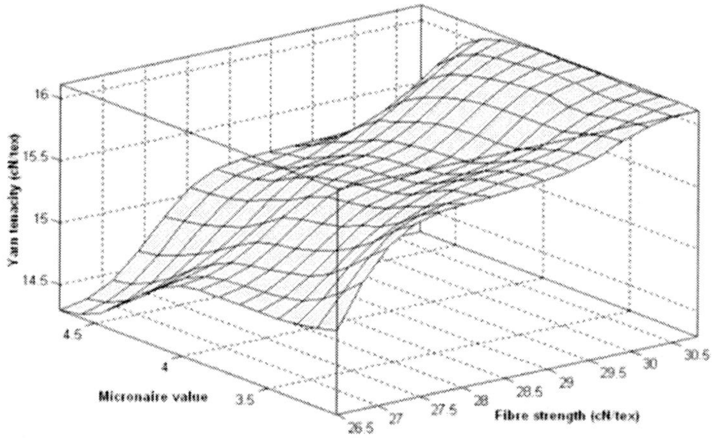

Figure 5.13. Surface plot showing the effect of fiber micronaire and strength on yarn tenacity.

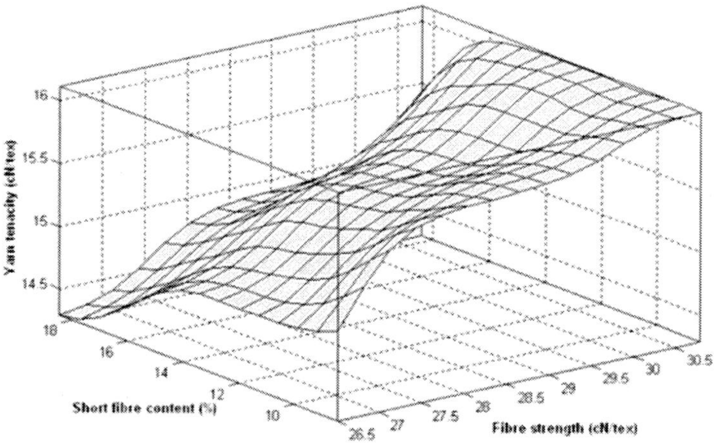

Figure 5.14. Surface plot showing the effect of short fiber content (%) and fiber strength on yarn tenacity.

The surface plots shown in Figures 5.12-5.14 depict the impacts of fiber parameters on the yarn tenacity. Figure 5.12 shows that as fiber tenacity and mean length increases there is concomitant increase in yarn tenacity as expected. Figure 5.13 demonstrates that as the cotton fiber becomes finer the yarn strength increases, although the effect is less prominent at the higher level of fiber tenacity. Figure 5.14 shows that as the short fiber content in cotton increases, especially at lower level of fiber tenacity, yarn tenacity also diminishes. Short fibers do not contribute much towards yarn strength. Besides the short fibers also generate drafting waves during roller drafting operations in draw frame, speed frame and ring frame and deteriorates the evenness of the fiber strand, which in turn reduces the yarn tenacity.

The prediction results of yarn tenacities for three different yarn count (16's, 22's and 30's Ne) are shown in Table 5.4. The coefficient of determination (R^2) was estimated to 0.75 for both the Gaussian and triangular membership functions. Therefore, the proposed fuzzy expert system can explain up to 75% of the total variability of yarn tenacity. However, fuzzy expert system based on Gaussian membership function is showing lower mean error (4.04%) as compared to that of triangular membership function (4.26%). This could probably be attributed to the fact that Gaussian membership function fits better with most of the fiber properties. The expert system for yarn strength prediction was developed by translating the perception of a spinner into fuzzy inference system. The developed fuzzy rules give a very good understanding about the interaction between important fiber parameters and their influence on yarn tenacity. The prediction accuracy of the proposed fuzzy system is reasonably good as the mean error % of prediction was below 5% for Gaussian and triangular form of membership functions. The system is quite easy to develop and it could be modified easily if the spinning technology is changed.

5.6. SUMMARY

This chapter presents an overview of the application of fuzzy logic in various textile problems such as fibers, yarns, fabrics, yarn and fabric manufacturing and textile chemical processing. The results of the two research works reported by the author also included in this chapter. These are silk cocoon grading and yarn strength prediction using fuzzy logic. Although, the basic algorithms of fuzzy logic exist since 1965, it was first applied to the textile problem by Pan et al. [2] in 1988. Since last two decades, fuzzy logic has been applied in the variety of textile problems. In recent years, the publications on fuzzy logic application in textiles have become more frequent. In its textile application, hand evaluation of fabrics shares the most of the publications, but only limited work has been reported in the area of textile chemical processing. Fuzzy logic can provide useful solution to the qualitative parameters which are often used in textile industry. It can be expected that in near future fuzzy logic will emerge out as one of the important modeling tools in textiles.

5.7. FUTURE TREND

Fuzzy logic has become a very popular tool in every field of science, engineering and management. With the advent of very high computational speed and availability of advanced softwares the fuzzy logic has empowered scientists and technologists from diverse fields. Fuzzy logic is an attractive tool for textile product engineering as it can able to handle imprecision that present in the textile data. It has seen successful applications in various stages of fibrous materials such as fibers, yarns, fabrics and their manufacturing process. Before its implementation, domain knowledge is prerequisite for developing the linguistic rules between the input and output parameters. As the fuzzy logic can able to capture the imprecision and uncertainty, its application in textile engineering will be felt more progressively in the years to come. In future, hybridization of fuzzy logic with other soft computing techniques such as ANN and GA is expected to solve various complex problems of modeling and optimization of fibrous materials engineering.

REFERENCES

[1] Zadeh, L. A. *Info. Control* 1965, *vol 8*, 338-353.
[2] Pan, N.; Yen, K. C.; Zhao, S. J.; Yang, S. R. *Text. Res. J.* 1988, *vol 58,* 565-571.
[3] Zimmerman, H. J. *Fuzzy Set Theory and Its Applications;* ISBN: 0792396243; Kluwer Academic Publishers: Boston, 1996, 3rd Edn, pp 1-456.
[4] Kartalopoulos, S. V. *Understanding Neural Networks and Fuzzy Logic: Basic Concepts and Applications;* ISBN: 09017162610; Prentice-Hall of India Pvt. Ltd.: New Delhi, 2000.
[5] Klir, G. J.; Yuan, B. *Fuzzy Sets and Fuzzy Logic: Theory and Applications;* Prentice-Hall of India Pvt. Ltd.: New Delhi, 2000.
[6] Berkan, R. C.; Trubatch, S. L. *Fuzzy Systems Design Principles;* Standard Publishers Distributors: New Delhi, 2000.
[7] Kosko, B. *Neural Network and Fuzzy Systems: A Dynamical Systems Approach to Machine Intelligence;* Prentice-Hall India, New Delhi, 2003.
[8] Ross, T. J. *Fuzzy Logic with Engineering Application;* ISBN-13: 9788126513376; Wiley-India: New Delhi, 2007, 2nd Edn.
[9] Yen, J.; Langari, R. *Fuzzy Logic: Intelligence, Control and Information;* Pearson Education: New Delhi, 2006.
[10] Dubois, D.; Prade, H. *Fuzzy Sets and Systems* 1979, *vol 2*, 327-348.
[11] Guruprasad, R.; Behera, B. K. *Ind. J. Fibre Text. Res.* 2010, *vol 35*, 75-84.
[12] Xu, B.; Dale, D. S.; Huang, Y.; Watson, M. D. *Text. Res. J.* 2002, *vol 72,* 504-509.
[13] Xu, B.; Fang, C.; Watson, M. D. *Text. Res. J.* 1999, *vol 69,* 656-662.
[14] Siddaiah, M.; Lieberman, M. A.; Hughs, S. E.; Prasad, N. R. *Proc. of the 8th International Fuzzy Systems Association World Congress;* Taipei, 1999, *vol 1*, 151–155.
[15] Sarna, E.; Wlochowicz, A.; Sarna, M. *Fibres Text. Eastern Eur.* 2006, *vol 14,* 88-91.
[16] Zaman, R.; Wunsch, D. C. *Prediction of Yarn Strength from Fibre Properties from Fuzzy ARTMAP;* 1997, http:// www.acil.ttu.edu/users/Raonak/papers/itc.htm
[17] Sette, S.; Boullart, L.; Langenhove, L. V. *Text. Res. J.* 2000, *vol 70,* 375-386.
[18] Sette, S.; Langenhove, L. V. *J. Text. Inst.* 2003, *vol 94,* 103-109.
[19] Kuo, C. F. J.; Hsiao, K. I.; Wu, Y. S. *Text. Res. J.* 2004, *vol 74,* 231-235.
[20] Ceven, E. K.; Özdemir, Ö. *Fibres Text. Eastern Eur.* 2007, *vol 15,* 55-59.

[21] Choi, H. T.; Jeong, S. H.; Kim, S. R.; Jaung, J. Y.; Kim, S. H. *Text. Res. J.* 2001, *vol 71,* 563-573.

[22] Lin, J. J.; Lin, C. H.; Tsai, I. S. *Text. Res. J.* 1995, *vol 65,* 697-709.

[23] Wang, X.; Yao, M. *Int. J. Clothing Sci. Techol.* 2000, *vol 12,* 114-123.

[24] Kuo, C. F. J.; Shis, C. Y.; Kao, C. Y.; Lee, J. Y. *Text. Res. J.* 2005, *vol 75,* 779-783.

[25] Kuo, C. F. J.; Shis, C. Y.; Lee, Y. H. *Text. Res. J.* 2004, *vol 74,* 107-111.

[26] Raheel, M.; Liu J. *Text. Res. J.* 1991, *vol 61,* 31-38.

[27] Park, S. W.; Huang, Y. G. *Text. Res. J.* 1999, *vol 69,* 19-24.

[28] Chen, Y.; Collier, B.; Hu, P.; Quebedeaux, D. *Text. Res. J.* 2000, *vol 70,* 443-448.

[29] Hwang, Y. G.; Park, S. W.; Kang, B. C.; Bae, C. K.; Choo, K. *J. Korean Fibre Soc.* 1998, *vol 35,* 119-124.

[30] Lau, T. W.; Hui, P. C. L.; Ng, F. S. F.; Chan, K. C. C. *Comp. in Industry* 2006, *vol 57,* 82-92.

[31] Huang, C. C.; Chang, K. T. *Text. Res. J.* 2001, *vol 71,* 987-992.

[32] Dorrity, J. L.; Vachtsevanos, G.; Daves, G.; Rim, S.; Kumar, A. National Textile Center Annual Report: 1994, September, 157-165.

[33] Kayacan, M. C.; Dayik, M.; Colak, O.; Kodaloglu, M. *Fibres Text. Eastern Eur.* 2004, *vol 12,* 29-33.

[34] Dayik, M.; Kayacan, M. C.; Acar, M.; Calis, H. *J. Text. Inst.* 2008, *vol 99,* 505-514.

[35] Semnani, D.; Sheikhzadeh, M. *World Aca. Sci. Eng. Techol.* 2007, *vol 29,* 77-82.

[36] Deng, X.; Vroman, P.; Zeng, X.; Koehl, L. *J. Info. Computing Sci.* 2007, *vol 2,* 93-102.

[37] Huang, C. C.; Yu, W. H. *Text. Res. J.* 1999, *vol 69,* 914-918.

[38] Biswas, A.; Ghosh, A. Silk Cocoon Grading by Fuzzy Logic, *In Soft Computing in Textile Engineering: Yarns, Fabrics, Garments and Technical Textile Applications;* Majumdar, A. Ed.; Woodhead Publishing Limited: Cambridge, 2010.

[39] Chen, S. J.; Hwang, C. L. *Fuzzy Multiple Attribute Decision Making Methods and Applications;* Springer: New York, 1992.

[40] Majumdar, A.; Ghosh, A. *J. Eng. Fibers Fabrics* 2008, *vol 3,* 61-68.

In: Modeling and Simulation in Fibrous Materials
Editor: Asis Patanaik
ISBN: 978-1-62100-116-4

Chapter 6

Application of Artificial Neural Network and Empirical Modeling in Yarn and Woven Engineering

Ashvani Goyal[1,*] and Harinder Pal[2]

[1]Department of Textile Technology, The Technological Institute of Textile and Sciences, Bhiwani, India

[2]School of Engineering and Sciences, B.P.S. Women University, Khanpur Kalan, Sonepat, India

Abstract

This chapter deals with the application of artificial neural network modeling as well as empirical modeling in the field of yarn and woven engineering. It covers various case studies about mentioned methodologies in the area of yarn manufacturing, yarn properties prediction, fabric manufacturing, and fabric properties prediction. Some of the case studies in yarn engineering includes bale selection, predicting spinnability, simulating spinning processes, card web defect analysis, sliver unevenness, yarn properties prediction like strength, elongation, hairiness, unevenness, yarn grading etc. While fabric engineering case studies include designing of worsted fabric expert system, airbag fabric material, canopy fabric, fabric properties prediction like strength, elongation, thermal resistance, handle end-use etc. apart from the wide scope the ANN can have in the field of textiles.

6.1. Introduction

In textile, there is high degree of variability involving the large number of variables in raw material as well as production processes. More ever the relation between these variables and final product properties are not well established. Lack of precise control on process parameters further makes the task difficult to produce desired product features [1]. A lot of

*E-mail: ashvanigoel@rediffmail.com

work has been done to determine the methods of predicting yarn and fabric properties by empirical and artificial neural network (ANN) modeling. The empirical model based on regression equation has the advantage of simplicity in describing the quantitative relationship between textile material properties. But their limitation is the applicability of these equations in predicting the properties outside the data used for deriving these equations. Further, deriving of these equations is expensive and time consuming. Also empirical model found difficulty in deriving equation for nonlinear relationship. So new approach based on ANN model is found popularity now-a-days. ANN models have ability to handle nonlinear relation and once designed ANN model can be updated for new input-output data after retraining the existing model with new data set. Also, ANN has the ability to solve control processes that are hard to deal with empirical modeling.

This chapter discusses the application of ANN modeling as well as empirical modeling in the yarn and woven engineering. Brief background about how different manufacturing process parameters in yarn and woven engineering affect some of the important properties is also discussed.

ANN has made its presence felt in all related aspects in textile from the fiber cultivation to garment manufacturing. Pattern recognition is one of the areas where ANN has played a significant role. Attempts have been made to identify fibers as well as trash using neural networks and NIR spectrophotometry. Even the animal fibers, merino and mohair have been recognized and classified by She et al. [2]. Cotton colors are also classified based on chromatic differences by the use of neural network. Classification of trash, such as leaf, bark and seed coat, the content of each of these trash particles is important in the sense to decide the cleaning process.

In yarn manufacturing and yarn properties prediction, ANN has been used in various areas like:

- Selecting cotton bales by spinning consistency index and micronaire
- Predicting the spinnability of a yarn
- Selection of best cotton blend in terms of price and quality
- Classification of card-web defects
- Control of sliver evenness
- Modeling the spinning process
- Predicting the strength and elongation of spun yarn
- Predicting the count-strength product (CSP) of ring-spun yarn
- Predicting the ring and rotor-spun-yarn hairiness from fiber properties
- Prediction of yarn shrinkage
- Predicting yarn unevenness (CV%) from fiber properties and yarn count
- Grading of yarn

In woven engineering, warp breakage rate has been predicted by the use of ANN. Fabric end-use even has been predicted by neural network. ANN embedded expert system has been tried for the design of canopy fabric. ANN application has also been realized in predicting following fabric properties:

- Evaluation of wrinkling of a fabric
- Handle evaluation of fabrics
- Predicting the pilling propensity of fabrics
- Predicting properties of worsted fabrics from yarn properties and weaving parameters
- Prediction of fabric handle and drape
- Airbag fabric material modeling
- Modeling the woven fabric strength using artificial neural network
- Prediction of initial load extension behavior of woven fabric

There are also various other areas like knitted fabric properties, deciding machine parameters for desired end use application, texturizing in man-made fibers, predicting performance of fabric during garment manufacturing, predicting drape, seam pucker, sewability, interlining properties, comfort rating of garments and fabric fault classification etc. where application of neural network has played in important role. So the application of ANN in above mentioned areas will be helpful in understanding some of the complexity of the process in simulating manufacturing process as well as predicting different properties.

6.2. Application of ANN in Yarn Engineering

6.2.1. Yarn Engineering

Yarn engineering, which implies the production of yarns with customer defined properties, has often attracted the attention of textile researchers. Further, from a yarn engineering point of view, a reverse model, that can predict the fiber properties and process parameters from yarn characteristics is even more desirable. Keeping this in mind, Majumdar et al. [3] have developed a yarn-to-fiber inverse model. 25 ring spun cotton yarn samples having linear densities ranging from 34 to 90 Ne were spun and tested for common yarn properties viz. tenacity, breaking elongation, unevenness and hairiness index. The yarn properties are taken as input parameters for the model to predict the spinning consistency index (SCI) and micronaire value of cotton as outputs. From the available 25 data sets, 20 were used for training and remaining 5 for testing. Only one hidden layer and log sigmoid transfer function was used for building the model. Training of model was done using back-propagation algorithm. From the model, it was found that mean absolute error of prediction was 4.38% and 1.92% for SCI and micronaire, respectively. Further, the correlation coefficient between actual and predicted value was 0.88 for SCI and 0.98 for fiber fineness. The model was also validated by three samples spun from cotton mix having predicted combination of SCI and micronaire and it is found that that engineered yarn have good agreement with target yarn in most of the properties.

6.2.1.1. Bale Selection to Meet Specified Ring Yarn Properties

Majumdar et al. [4] have tried a method of selecting cotton bales to meet the specified ring yarn properties using artificial neural networks. They have used five yarn properties and yarn count as input parameters and SCI and micronaire were the outputs to the neural network models.

Table 6.1. Prediction of SCI and micronaire by ANN models for different number of nodes [Source: Reference 4]

Number of	Mean error, %			Correlation coefficient R	
Nodes	SCI	Micronaire	Overall	SCI	Micronaire
2	4.56	6.79	5.68	0.786	0.422
4	4.99	4.87	4.93	0.760	0.845
6	4.08	5.35	4.72	0.800	0.853
8	4.31	5.12	4.72	0.791	0.844
10	4.40	5.16	4.78	0.782	0.784
12	4.32	5.16	4.74	0.783	0.869

SCI is a calculation for predicting the overall quality and spinnability of the cotton fiber. Authors have calculated SCI using the average fibers and yarn data of five consecutive years of USDA annual crop reports. The main use of SCI in selecting bales was to gain advantage of considering all major cotton properties. Within the SCI, there are various fiber properties to take the advantage of inherent correlation prevailing among the fiber properties.

They prepared 90 samples of three different counts of ring spun carded yarn from the cotton fibers of known SCI and micronaire values. All the bales of cotton fibers were tested with HVI before the spinning operation. 75 samples were used for training the neural networks, the remaining 15 sets of data were used for testing the trained networks. Six major yarn properties i.e. C.S.P (count strength product), tenacity, elongation, unevenness, hairiness and yarn count were used as the selected input parameters for the ANN model. The single hidden layer structure was selected for the model. The number of neurons in the hidden layer was varied from 2 to 16 by an increment of 2. The outputs from the neural networks were the SCI and micronaire.

It was observed that the ANN models with six nodes in the hidden layer exhibit the highest prediction accuracy (Table 6.1). Correlation coefficients between actual and predicted values are 0.800 and 0.853 for SCI and micronaire respectively. It was also observed that as the number of hidden nodes increases, the prediction performance becomes optimum when six nodes are used in the hidden layer. No improvement was noted after that. The reason for this was attributed to the memorization of training data by networks with too many nodes.

6.2.1.2. Modeling Spinning Process to Predict Spinnability

Pynckels et al. [5] have used neural nets for determining the spinnability of fibers. Determining spinnability is not an easy task because of the complex interaction involved in raw material characteristics and process parameters. Twenty types of cotton were evaluated for 29 fiber properties. Spinnability for ring and rotor yarn was investigated. In rotor spinning, machines parameters considered were the speed and the navel type. A criterion of spinnability chosen was that a yarn was considered to be non-spinnable if more than 5 breakages occurred during the first three minutes of the spinning process. Below is given some of the important parameters considered and measured:

- Mean length
- Length uniformity

- Strength
- Elongation at break
- Micronaire
- Mature fiber percentage
- Yellowness, brightness
- Wax content, sugar content
- Amount of trash
- Information about neps
- Percentage of short fibers
- Percentage of dust, number of seed particles

The authors claim that the spinnability of a given fiber quality on a rotor and ring spinning machine can be predicted with a reliability of 95% by means of a neural network. In another paper, Sette et al. [6] combined neural network and genetic algorithm to optimizing the fiber-to-yarn production process. The genetic algorithm is enforced with a sharing function and a Pareto optimization. They have selected fourteen fiber characteristics and five machine settings as input parameters. The output contains the tenacity and elongation of all constructed yarn. Machine parameters considered were yarn count, twist, spin tube, breaker speed and rotor speed. Fiber parameters were related to length, fineness, trash, color, strength, elongation and maturity.

The back-propagation network was set up with an input layer of nineteen nodes. Two hidden layers of twelve nodes each connect the input layer. The paper showed that optimization of yarn quality is easily achieved as a function of the necessarily input parameters. They have also compared neural network model with multiple regression analysis and the results showed that the network model can be considered more reliable than its statistical counterpart.

6.2.1.3. Card Web Defects

Shiau et al. [7] have done some work to classify web defects with a back-propagation neural network by color image processing. Manual inspection is time consuming and boring to perform. Also consistency of the results is questionable. Research is being done to introduce automatic inspection system to bring in the consistency. In card web analysis, neps and trash contents are criteria for assessment of yarn quality. The ANN is widely recognized as a useful classification technique for pattern recognition problems. Authors have tried to construct an appropriate topology for color image processing to automatically recognize neps and trash in a web. Neps and trash are similar to each other in shape, so color imaging helps in giving more information like hue, saturation and intensity. With the ANN, the RGB (red, green and blue) values corresponding with the image pixels are used to do the recognition. Three categories like normal web, neps and trash can be determined. They have used sigmoid logistical nonlinearity while developing the back-propagation neural network. The quality of color image impacts the recognition rate of a neural network. The parameters, contrast, brightness and background color influenced the recognition rate. So attempt was done to have ideal background color under moderate conditions of brightness and contrast to overcome the translucent problem of fibers in a web, specimens were reproduced in a color BMP image file format.

The experimental analysis showed that recognition rate could reach 99.63% under circumstances in which the neural network topology is 3-3-3. The method is suitable not only for cotton and man-made fibers of different lengths, but also for different web thickness as to a limit of 32.9 g/m^2. Authors suggest that since neps and trash in a web can be recognized, yarn quality not only can be assessed but also improved using a reference for adjusting manufacturing parameters.

6.2.1.4 Sliver Evenness

ANN can also be used to reduce variations in sliver linear density. Huang et al. [8] have used fuzzy self-organizing control and neural network control to develop autolevelling system. It was observed that neural network controller can reduce more CV% than the fuzzy self-organizing controller. The fuzzy self-organizing controller consists of a basic fuzzy logic control and a self-organizing scheme. The basic fuzzy logic control includes fuzzification, fuzzy knowledge base, decision making and a defuzzification. The self-organizing scheme corrects the control action and modifies the fuzzy control rules. In neural network controller, back-propagation learning algorithm was used. Unipolar sigmoid function was used as the activation function to compute the output of a node.

Linear density of the feed sliver and the desired linear density of the output sliver were used as the inputs and the ratio of front and back roller speed was the output. Authors claimed that the developed system combines the principles of both open and closed loop autoleveller. The experimental results showed the improved levelness of slivers by both controllers.

6.2.2. Yarn Properties Prediction

6.2.2.1. Prediction of Tensile Properties

Ramesh et al. [9] used ANN to build a model to predict the tensile properties of air-jet-spun yarns. In air-jet spinning, a drafted strand of fibers is passed through one or two fluid nozzles located between the front roller of a drafted system and take-up device. The properties of the yarns are greatly affected by the structure of the yarn which in turn is influenced by process variables as well as material properties. In air-jet spinning, the important process parameters are front and back nozzle pressure, the yarn delivery speed, the total draft, the ratio of the draft in the front and back zones of the drafting system, the thread-tension draft and the inter-jet distance [10]. Authors build a single hidden-layer neural network trained by using the back-propagation algorithm. The input parameters relevant with material and process variables, namely yarn count, blend and front and back-nozzle pressures on an air-jet spinning machines, are correlated with the experimentally determined yarn properties, breaking load and breaking elongation. Authors in his paper from observation claim that the errors of prediction were low despite the availability of only a small data set for training and in each case the prediction error was less than the standard deviation of experimentation.

The tenacity of air-jet spun yarn is also predicted using ANN by Zeng et al. [11]. They have taken the following process parameters as input viz. first and second nozzle pressure, spinning speed, distance between front roller nip and first nozzle inlet and the position of the jet orifice in the first nozzle to predict the yarn tenacity. The correlation coefficient (R) between predicted and measured value was 0.98 and it confirm that predicted values agree well with the measured one.

Majumdar et al. [12] tried to predict breaking elongation of ring spun cotton yarns using mathematical, statistical and ANN model. They have used the data from cotton crop study of 1997 and 1998 given by the International Textile centre, U.S.A. Seven cotton fiber properties taken from Uster HVI testing namely fiber bundle tenacity, elongation, upper half mean length (UHML), uniformity index, micronaire, reflectance degree and yellowness were used as input parameters along with yarn count. They have selected 87 samples for input – output data for carded ring spun yarn out of which 72 samples were used for training and remaining 15 samples for testing the predicted model. For ANN model prediction, they tried five different network structures with only one hidden layer and varied the number of neurons in hidden layer from 6 to 14 with increment of two in each step. Learning rate and momentum were optimized at 0.1 and 0.0 respectively and best prediction was found after 2500 iteration with 10 nodes in hidden layer. The correlation coefficient found between actual and predicted elongation was found very high i.e. R=0.938. They also ranked the fiber properties influencing breaking elongation of cotton yarns. Fiber elongation followed by fiber length uniformity and yellowness was the important fiber properties affecting the yarn breaking elongation the most.

6.2.2.2. Prediction of CSP

Cheng et al. [13] also tried to establish relationship between fiber properties and yarn strength. They used USDA cotton fiber and processing test results. The samples were taken from 169 bales of upland cotton. Fiber properties were measured using the HVI testing system, and included upper half mean length, length uniformity, short fiber content, strength, fineness, maturity ratio, greyness and yellowness. Fineness and maturity ratio were measured on the Shirley fineness maturity. Micronaire is not used in this study. 84 bales were chosen for training set and 85 bales were used in the testing set. All cotton samples spun into 22s ring spun yarns. Yarns were evaluated for skein strength, elongation, imperfections and appearance index. Count Strength Product (CSP) was used as the prediction target. The effect of the number of hidden neurons was also studied. Increasing the number of hidden-layer neurons always gave better results for the training set while four neurons in the hidden layer gave the best results for the testing set, yielding a correlation coefficient of 0.84 between predicted and actual values.

6.2.2.3. Modeling Spinning Process to Predict Ring and Rotor Yarn Properties

In another paper Pynckels et al. [14] have tried to predict yarn properties on rotor and ring spinning machine, given the relevant fiber properties and production conditions. These conditions were determined by the type of machine employed and the machine settings. In the investigation, coupled self-training system was chosen instead of single net. In this, a number of neural networks were trained and put in cascade i.e. the output of the first neural network was fed into several other neural networks. The fourteen fiber properties and the five machine parameters were studied, the list of which is given below:

Yarn properties:

- Linear regularity

- Elongation
- Tenacity
- Other irregularity per 1000 meter (yarn count)
- Twist
- Thin places per 1000 meters
- Thick places per 1000 meters
- Neps per 1000 meters
- Number of hairs per 250 meters

Machine parameters:

- Breaker speed
- Rotor speed
- Spin tube
- Twist
- Yarn count

Authors claim that neural network was able to simulate spinning process with an accuracy of over 95%.

6.2.2.4. Prediction of Yarn Hairiness

Yarn hairiness is defined as the fiber ends and loops protruding from the main yarn body [15]. It is undesirable property and directly responsible for pilling generation. It is mainly affected by fiber properties and spinning conditions. ANN algorithm was applied to estimate cause-effect relationship between fiber properties and yarn hairiness. Cotton fiber properties were measured using three different measurement systems – HVI and AFIS (Advanced Fiber Information System) as well as the traditional instruments including the fineness and maturity tester, the Stelometer, comb sorter and Shirley analyzer. Both Ring and rotor spun yarns were used. Total six neural network models were made based on each measurement system and for both, ring and rotor yarn. Each neural network is trained on known fiber properties and yarn hairiness. Results were compared using all these models. Also the impact of each fiber properties on ring and rotor spun yarn hairiness was analyzed. Back propagation algorithm was used. 100 samples were selected, 67 observations was used to train a neural network. 33 observations were used to test that network.

Comparing these models, the models based on the HVI data shows the highest prediction accuracy for both ring and rotor spun yarns hairiness. The coefficient of determination (R^2) values for training and testing data are 0.8779 and 0.842 for ring yarn hairiness and 0.8175 and 0.7716 for rotor yarn hairiness, respectively.

6.2.2.5. Prediction of Yarn Unevenness

Predicting the yarn irregularity from fiber properties is not new. Many studies have been done and different statistical models have been used. Zhu and Ethridge [16] have tried to develop neural network model for predicting the yarn irregularity based on inputs of fiber property measurements with the AFIS instrument. Back propagation algorithm and a nonlinear function called a sigmoid transfer function were used in the model. Five network configurations were selected as follows: 7-6-1, 7-8-1, 7-10-1, 7-12-1 and 7-14-1. These five neural network models were trained by using the 150 elements of experimental data. The six fiber properties used were: upper quartile length, mean fiber length, percentage short fibers, diameters, neps and total trash. The correlation coefficient between the experimental yarn irregularity (CV%) and the predicted yarn irregularity (CV%) is from 0.7984 to 0.8808 for five different neural networks. By comparing the correlation coefficient, the one layer and ten hidden neurons (7-10-1) prediction model give the best results.

6.2.2.6. Prediction of Yarn Grading

As we are aware of that the grading of textile yarn is an important parameter in deciding the market value of yarn, various methods like standard testing blackboard and modern electric inspection board by Lawson-Hemphill corporation are used to grade textile yarns based on yarn evenness. Though electronic inspection board improves the yarn grading system, but no evaluation of yarn strength by this system may allow grading a low strength yarn into a high grade yarn. Lien and Lee [17] used the back propagation neural network (BNN) and Karhunen-Loeve expansion method to build a neural network model who can grade the textile yarn effectively with considering different yarn properties. For the model, they have taken 30 data sets, out of which 20 were used as training data set and remaining 10 as testing data set. Six yarn properties are used as input nodes, whereas three output nodes represent the three yarn grades. Experimental results confirm that a back propagation neural network provides good grading accuracy and relearning ability. Also BNN system with proper comprehensive training samples or relearning procedure can give an error free textile yarn grading system.

6.3. Application of Empirical Modeling in Yarn Engineering

6.3.1. Relationship of Cotton Fiber Properties to Yarn Tenacity

Ramey et al. [18] have tried to establish the relationship of cotton fiber properties to yarn tenacity. 42 homogeneous samples of cotton were chosen. Various properties like reflectance, yellowness, micronaire reading, Arealometer measure A (specific area) and D (an indicator of immaturity), percentage elongation at break (E_1), Upper half mean (UHM) and mean length, Tenacity at 0 (T_0) and at 3.2 mm (T_1) gauge lengths were measured. Each sample was divided into eight subsamples. Two groups of subsamples were spun into 27 tex and two into 12 tex yarn at optimum twist. Two groups were spun into 27 tex yarn on open end spinning frame with a 52.6 machine – twist factor. The remaining two groups were spun into 27 tex yarn on the same open end frame with a 47.8 machine twist factor. The relative contributions of the fiber properties to yarn tenacity were analyzed by stepwise regression. At each step one independent variable was added that reduces the error sum of squares.

Correlations between the fiber properties were found. It was observed that Reflectance was negatively correlated with yellowness and T_1, micronaire reading and Uniformity ratio (UR) were positively correlated. Arealometer A was correlated negatively with T_0 and UR and positively with E_1. Arealometer D was negatively correlated with T_0, T_1 and UR and positively correlated with E_1. Tenacity measures were positively correlated with fiber-length measures. For detailed information, readers are suggested to refer article by Ramey et al. [18]

6.3.2. Relation of Cotton Fiber Properties to Yarn Breaking Elongation

Majumdar and Majumdar [12] tried to develop the mathematical and statistical model to predict the breaking elongation of cotton ring spun yarn. Along with the fiber properties from HVI testing, they also use yarn twist multiplier for mathematical model. In this study, they have concluded that prediction power of statistical model was better than the mathematical model. The correlation coefficient and mean absolute error was 0.870 and 6.696%, respectively for statistical model and 0.731 and 10.05%, respectively for mathematical model.

6.3.3. Relation of Fiber and Yarn Parameters to Yarn Characteristics

As we know, yarn properties are influenced by the fiber characteristics, roving properties, yarn count and twist. So this approach was used by Ureyen and Kadoglu [19] in their study to predict the important yarn properties of ring spun cotton yarns with the help of linear multiple regression analysis by using fiber, roving and yarn parameters. For their study, they have selected 15 different cotton roving samples from spinning mill. Fibers properties were measured on Uster HVI tester from the sliver collected from finisher draw frame. Each cotton samples were spun into ring yarns of count 20s, 25s, 30s and 35s and each count with three different twist coefficients (αc) of 3.8, 4.2 and 4.6. In total, 180 samples are prepared and tested for yarn strength, elongation, hairiness and unevenness value. A linear multiple regression analysis was adopted to establish a quantitative relationship of yarn properties with respect to fiber properties, roving properties, yarn count and yarn twist. Forward stepwise procedure was taken for all yarn properties except the elongation that was estimated by backward elimination. They have concluded with their curve fitting that the relationship between fiber properties were almost linear with high regression coefficient.

In an another study [20], the same authors predicted the important cotton ring spun yarn properties from fiber properties taken from AFIS fiber testing instrument using linear regression model. The same 180 samples were used to relate the important yarn properties with Uster AFIS fiber testing results, roving and yarn properties. The curve fitting using line multiple regression analysis shows that relationship between variable and yarn properties are nearly linear for each yarn properties. The results for multiple linear regression models for statistical analysis are also very good as indicated by very high R^2 value and low standard error values in their study.

Goodness of fit statistics and ANOVA analysis were also confirm the very good prediction power of their model at α=0.01 level. In their study, they also pointed out that fiber diameter is the most important parameters as indicated in regression analysis of all yarn properties and it is the only parameter that maintains significance in every equation. Their models confirm that roving properties have a great effect on all yarn properties along with yarn count and twist.

6.3.4. Prediction of Hairiness of Cotton/Polyester Blended OE Rotor Yarns

Baykal et al. [21] gave a statistical model to predict the hairiness of cotton/polyester blended rotor yarns for unobserved blend ratios and yarn count. In their study, they have prepared 50 yarn samples with five blend ratio of cotton and polyester with proportion of 0%, 25%, 50%, 75% and 100% for cotton and rest polyester in the corresponding blend and five different yarn counts with linear density 36.9, 29.5, 24.6, 21.1 and 18.5 tex with two replication at each design point. The hairiness test result was used to analyze mixture- process crossed design and cubic - quadric crossed model gave the best fitting model. The predicted model from study had strong capability of prediction as indicated by a high and positive correlation between predicted and observed H values.

6.4. Application of ANN in Woven Engineering

Fabric engineering involves a great deal of expertise and experience. It needs a thorough understanding of the functional properties and their key control construction parameters. When the relationship between a set of interrelated properties goes beyond the complete comprehension of the human brain, neural networks could be used to find the unknown function. The utility performance properties of woven fabric depend on the combined effect of the properties of the constituent fibers, yarn and fabric structure. According to Behera and Karthikeyan [22], the relationship between structure and property of the fabric is complex and inherently nonlinear and to create a predictive model, one must resolve the complexities.

6.4.1. Prediction of Woven Fabric Strength

Zeydan [23] has tried to model the multi-linear relationship between fiber, yarn and fabric parameters with strength of fabric using ANN- and Taguchi design of experiment (TDOE) methodologies. Jacquard woven fabrics were chosen for the study and the strength of jacquard woven fabric depends on several factors. The objective of the ANN model generated in this study was to help the engineers of planning department at the company to plan the manufacturing of fabric with a good estimation of fabric strength before the production order. The following parameters were identified as potentially important parameters affecting the strength of woven fabric:

- Number of warp yarns at fabric width
- Weft density
- Weft yarn count
- Fiber type of weft yarn
- Warp density
- Warp yarn count
- Fiber type of warp yarn

Because of the unavailability of yarn information like yarn strength and twist etc., they have not taken it into consideration. So parameters related to weaving process have been considered rather than yarn based parameters.

While modeling with ANN, "feed forward with a single hidden layer perception" form is adopted. The form is made up of one layer and the data propagate forward through the network from input to output. In this study, 120 training patterns and 10 testing patterns are used to train and test the network, respectively. The experimental study concluded that the design model enables to predict the strength of woven fabric with negligible deviation. On comparing TDOE and ANN, the fabric strength prediction using ANN can be obtained better than TDOE model. For more detail, readers are suggested to refer the article.

The tensile strength of woven fabric is also predicted by Majumdar et al. [24] with the use of ANN by feeding warp yarn strength, warp yarn elongation, ends per inch (EPI), picks per inch (PPI), weft count and tensile properties of warp yarns were used as input parameters and fabric strip strength as output for ANN model using back propagation algorithm. Out of available 33 samples results, 27 samples results were used for training data and remaining 6 for testing data. Only one hidden layer and log- sigmoid transfer function were used for the model. Five nodes in the input layer were selected because of five input parameters namely warp yarn strength, warp yarn elongation, EPI, PPI and weft count. The analysis showed that ANN model have a consistent and better prediction accuracy with much lower correlation coefficient between actual and predicted value of 0.92 with absolute error of prediction(7.83). Further, the model also informs that warp yarn strength and ends per inch are two dominant factors determine the strength of fabric in warp direction.

6.4.2. Prediction of Woven Fabric Handle

Basu et al. [25] developed an ANN model to predict the fabric handle characteristics from air-jet spun yarn properties. They also design an inverse model for predicting the air-jet yarn properties required for a particular fabric handle characteristics. In their study, they had spun 81 yarn samples of three different yarn linear densities of 19.68, 14.76 and 9.84 tex on air-jet spinning machine with different process variable using polyester fiber (40mm x 0.111.tex). Same 81 yarn samples were woven as weft into 81 woven fabric samples. An ANN model was designed on a single hidden layer feed-forward method based back propagation algorithm.

Among 81 data sets, 72 sets were used for training the model and remaining 9 for testing of model. The yarn properties namely flexural rigidity, compressional energy and hairiness were used as input parameters to predict the fabric handle properties namely bending length, kinetic frictional resistance, drape coefficient and compressional energy. The predicted model have good accuracy with high correlation coefficient of 0.92 for bending length, 0.93 for kinetic frictional resistance, 0.91 for drape coefficient and 0.90 for compressional energy. The inverse model designed for prediction of yarn properties for a particular fabric property was also designed successfully. For testing the accuracy of the derived inverses model, the authors have spun the yarns from the predicted process parameters for a particular fabric quality and again the fabric sample was constructed and tested. The difference in properties of fabric was within the statistical limit that ensures the success of inverse ANN model in fabric engineering.

6.4.3. Prediction of Woven Fabric Thermal Resistance

One of the common problems associated with statistical modeling is the nonlinear relationship between different fabric parameters and thermal comfort properties. Thus a model is required which can predict thermal comfort properties by considering all fabric parameters at the same time. In this regard, Bhattacharjee and Kothari [26] developed an ANN model to predict the steady state and transient thermal behavior of a fabric. In total, 86 samples of cotton woven fabric were taken, out of which 70 samples were considered for training and remaining 16 samples for testing of the model. Seven fabric parameters viz. type of weave, warp count, weft count, EPI, PPI, fabric thickness and aerial density were used as input parameters and thermal resistance along with maximum heat flow Q_{max} as outputs. Two cases of ANN models were considered. In the first model, two neural network works in tandem with one common set of input parameters and two networks had thermal resistance and Q_{max} as output whereas, in the second model, one neural network with seven input parameters were considered along with two output parameters namely thermal resistance and Q_{max}. In the first model, the first network has 10 and 8 neurons in first and second hidden layer whereas the second network had 5 and 7 neurons in first and second hidden layer. The second model had 7 and 15 neurons for first and second hidden layer. The mean error percentage from first and second model was 8.61% and 10.42% respectively. The coefficients of determination (R^2) for the first and second model were 0.94 and 0.69 for thermal resistance and 0.64 and 0.46 for Q_{max}. This shows that the first model have better prediction of steady state and transient thermal behavior of a fabric.

6.4.4. Prediction of Fabric End-Use

As we aware, establishing a relation between fabric properties and its end-use will certainly help the textile designer to integrate fabric properties into designing a garment. In a study, Chen et al. [27] have design a neural network model for predicting fabric end-use based on fabric properties tested on Kawabata instrument. Fabric properties of extension, shear, bending, compression and friction and roughness were tested on Kawabata KES instrument for all the 100 apparel fabric samples from different fabric and garment manufacturers. Including fabric weight, the total 16 variables were used for training data set. 100 samples are divided into three groups namely blouse, shirts and suiting (suit, jacket and coat). For the model, 90 samples were used for training and 10 for testing the model. Two number of hidden layer was selected along with mean square error of 0.01 for designing the model. The designed model has good accuracy in predicting the fabric end-use with 0.07 estimated error rate of prediction. The established predicted model can provide guidance to fabric manufacturer, fashion designer and makers-up in fabric design, fabric selection and fabric proper use.

6.4.5. Design of Canopy Fabric

Behera and Karthikeyan [22] described a method to design canopy fabrics with the use of an ANN-embedded expert system. The functional properties of a parachute is mainly influenced by the properties and performance of the canopy fabric, which in turn are influenced by major fabric constructional parameters, viz., fabric areal density, thread density, thread linear density, weave, type of yarn. The performance parameters chosen were breaking load, elongation, air porosity, tearing strength and bursting strength. Out of the mentioned parameters, the constructional parameters are used as input, the performance parameters as the desired output for forward engineering, and the parameters are interchanged for reverse engineering. Forward engineering is a methodology wherein the constructional parameters are known and accordingly performance parameters of the given fabric are predicted, whereas in reverse engineering, the required constructional parameters are predicted from the desired performance parameters. The radial basis function is used as the learning algorithm for the ANN. For the expert system, Microsoft Visual Basic was used as the front end for user interface while the ANN toolbox of MATLAB is used for the network fabrication; database is maintained using MS Access. The results concluded that the design prediction have excellent correlation with all the samples and the ANN based on RBF is found promising for a new domain of design prediction techniques.

6.4.6. Designing of Worsted Fabric Expert System

Designing of expert system requires sufficient knowledge, experience and data base to assist in fabric engineering. The process of fabric engineering is very complicated since it may differ from mill to mill and case to case. According to Fan and Hunter [28], the whole process of fabric engineering involves fabric requirement-fabric designing comprising fabric composition, structure, manufacturing parameters, sample trials considering evaluation in terms of cost, performance and characteristics modification of fabric design and processing parameters to improve the end product or reduce the cost.

There are four areas where an expert system can support in the fabric engineering process. These are:

- Fabric design
- Prediction of fabric properties and performance
- Interpretation of testing results
- Modification of fabric design and processing parameters

Based on above consideration, worsted fabric expert system was developed. The system has eight advisers to provide advice or guidance for determining fabric composition, weave, yarn count and sett, weaving details, yarn type, twist, fiber specification and finishing procedures [28].

In another paper [29], Fan and Hunter have tried to develop an ANN model for predicting the properties of worsted fabrics. Back propagation neural network with one single hidden layer was used. The input parameters were related to fiber composition, yarn type and fabric structural parameters like weave that are important to fabric properties and performance. The outputs were seam slippage, wrinkle performance, abrasion resistance, shear rigidity, bending rigidity and thickness. Authors claim that the evaluation of model shows very good agreement between the predicted and the experimental values.

6.4.7. Prediction of Properties of Airbag Fabric Material

Two important properties of airbag fabric material that contribute to the energy-absorbing capabilities are permeabilities and its biaxial stress-strain characteristics in the plane of the fabric. Keshavaraj et al. [30] have used neural network architecture for modeling of airbag fabric material of nylon and polyester fabrics. They investigated the modeling of fluid flow through woven airbag fabrics. Experimental permeability data for fabrics under biaxial strain conditions were obtained through a blister-inflation technique. The various properties of the fabrics can be incorporated while training the proposed network architecture.

The purpose of the experiment was to develop a model that can be used to predict the changes in permeability with inflation temperature and internal pressure drop. The effects of the type of the fabrics i.e. denier and weave type with nylons and calendaring in case of polyester, biaxial strain, biaxial stress and pressure drop were considered in the model. Authors compared the performance of nylon 66 and thermoplastic polyester fabrics. The effect of denier of the fiber on the performance, both in terms of permeability and biaxial stress-strain was considered in the case of the nylon 66 materials. In the case of polyester the effect of calendaring was investigated. The network used in the model was a: 3-input node, 4-hidden node, 1-output node architecture. Authors claimed that the predictions obtained from this neural network model agreed very well with the experimental data. So the neural nets can be considered as a serious design tool use in determining permeability and biaxial stress-strain relationships for textile fabrics used in airbags.

6.5. APPLICATION OF EMPIRICAL MODELING IN WOVEN ENGINEERING

Majumdar et al. [24] used the multiple regression model to predict tensile strength of plain woven fabric and found that correlation coefficient between actual and predicted value are 0.811 with 13.65% absolute error of prediction. They used five input parameters namely warp yarn strength, warp yarn elongation, EPI, PPI, weft count for the prediction. Apart from warp yarn strength and ends per inch, weft count is also dominant parameter affecting tensile strength of woven fabric.

The prediction of tensile strength of woven fabric is also discussed by Hussain et al. [31]. They have developed a statistical model for prediction of tensile strength of 52:48 polyester/cotton blended woven fabrics using empirical data. In total, 270 fabric samples of plain and twill weave were constructed by using polyester/cotton blended yarn of 15, 20 and 25 tex ring spun yarns. 234 samples were used for predicting the model and remaining 36 samples (18 plain weave and 18 twill weave) were used to check the validity of the model. Different statistical model was designed for predicting the warp way and weft way fabric strength and the model with warp strength, weft strength, ends/25mm, picks/25mm and float length gave the best prediction accuracy. Table 6.2. gave a comparison of actual and predicted fabric strip strength for 36 samples used for checking the validity of the developed model. The correlation between actual and predicted strength were found to be 0.997 and 0.993 for warp and weft way fabric strength that indicate the accuracy of the predicting model.

Table 6.2. Comparison of actual and predicted fabric strip strength [Source: Reference 31]

Plain weave (float length=1)						Twill weave (float length=3)					
Warp strip strength, N			Weft strip strength, N			Warp strip strength, N			Weft strip strength, N		
Predicted	Actual	Diff.%	Predicted	Actual	Diff.%	Predicted	Actual	Diff.%	Predicted	Actual	Diff.%
348.45	358.00	2.67	242.24	230.00	-5.32	318.66	320.00	0.42	214.22	219.00	2.18
415.52	423.00	1.77	306.96	305.00	-0.64	385.73	373.00	-3.41	278.95	266.00	-4.87
371.17	384.00	3.34	415.09	423.00	1.87	341.37	335.00	-1.90	387.07	379.00	-2.13
413.78	416.00	0.53	343.30	339.00	-1.27	383.98	383.00	-0.26	315.28	324.00	2.69
371.77	356.00	-4.43	448.18	429.00	-4.47	341.98	361.00	5.27	420.16	433.00	2.96
451.07	444.00	-1.59	675.94	680.00	0.60	421.28	425.00	0.88	647.93	644.00	-0.61
506.84	514.00	1.39	306.97	288.00	-6.59	477.05	484.00	1.44	278.95	272.00	-2.55
576.63	556.00	-3.71	259.18	247.00	-4.93	546.84	535.00	-2.21	231.17	219.00	-5.56
505.10	489.00	-3.29	343.30	343.00	-0.09	475.30	483.00	1.59	315.29	325.00	2.99
611.57	597.00	-2.44	512.29	552.00	7.19	581.78	553.00	-5.20	484.27	455.00	-6.43
530.16	544.00	2.54	566.30	580.00	2.36	500.37	493.00	-1.49	538.28	527.00	-2.14
612.18	628.00	2.52	574.77	582.00	1.24	582.39	589.00	1.12	546.75	510.00	-7.21
673.54	675.00	0.22	261.08	250.00	-4.43	643.75	637.00	-1.06	233.07	238.00	2.07
813.29	815.00	0.21	382.07	383.00	0.24	783.50	763.00	-2.69	354.05	364.00	2.73
696.26	708.00	1.66	433.93	453.00	4.21	666.46	671.00	0.68	405.92	418.00	2.89
811.55	806.00	-0.69	442.41	454.00	2.55	781.75	775.00	-0.87	414.39	422.00	1.80
696.86	704.00	1.01	467.03	466.00	-0.22	667.07	671.00	0.59	439.01	443.00	0.90
812.15	801.00	-1.39	475.50	479.00	0.73	782.36	782.00	-0.05	447.48	447.00	-0.11

Morino and Matsudaira [32] had predicted the mechanical properties and hand value of the fabric from the newly defined weave structure parameters namely the crossing over firmness factor) and the floating yarn factor. For their study, 5.9 tex cotton yarn were used for warp and 8.3 tex/24 polyester filament was used for the weft. Ten fabrics with three fundamental weave namely plain, 2/2 twill and 5- harness satin are constructed as mentioned below:

- Plain weave
- 2/2 twill weave
- 5 harness satin weave
- Crape weave
- Huck-a-back weave
- Special honeycomb weave
- Sponge Weave
- Granite weave
- Dice weave, and
- 10- harness weave

Multiple regression equations were calculated for different mechanical and primary hand value that are affected by weave structure parameter. They had found that weave structure mainly affecting the bending and shearing and also Hari and Shinayakasa (primary hand parameters) properties. They also concluded that mechanical parameter (tensile linearity LT and tensile energy WT), compression properties and surface properties are not affected by defined weave structure parameters.

6.6. A Comparative Analysis of ANN and Empirical Modeling

Statistical prediction methodologies usually involves linear model while the relation between the input variables which includes raw materials and process variables and the output properties is not linear and not well established. Ethridge et al. [33] noted that prediction could be improved with a nonlinear model. The optimum regression equation, which they obtained in their investigation, included a quadratic term in micronaire, the logarithm of strength and length uniformity, and the square of greyness time's yellowness. The main advantage of ANN approach over theoretical methods lies in the fact that it does not require the accurate mathematical model of the system while modeling complicated nonlinear processes [34].

Majumdar et al. [24] give a comparative analysis of prediction power of ANN model and linear regression model for predicting the tensile strength of woven fabric by using warp yarn strength, warp yarn elongation, EPI, PPI, weft count as input parameters and found that the prediction accuracy is better for ANN model than linear regression model (Table 6.3). Both models confirm the significance of warp yarn strength and ends per inch in predicting the tensile strength of woven fabric. Further, regression model confirm the weft count as third important parameters whereas ANN model confirm picks per inch as third important parameter in predicting the tensile strength of woven fabric.

Rajamanickam et al. [35] in his paper compare four modeling methodologies for predicting the strength of air-jet spun yarns. The methodologies analyzed were: mathematical model that are derived from first principles, empirical model that use statistical techniques, a computer simulation model that efficiently model the inherent randomness of the material and structural parameters determining yarn strength and ANN model. From their experience in developing these methodologies, authors conclude that an ANN model is best suited for predicting yarn tensile properties in day-to-day use, provided the model is not used for extrapolation. However, empirical models can also be used if sufficiently computing facility is not available. In other work, Guha et al. [36] have examined the relative performance of mechanistic model, simple statistical models and neural network models for predicting ring spun yarn tenacity from fiber properties and process parameters. The results showed that neural network models are capable of giving better results than mechanistic models and statistical model [36]. Majumdar and Majumdar [12] gave a comparative assessment of mathematical, statistical and ANN model for predicting the breaking elongation of ring spun cotton yarns from cotton fiber properties taken from Uster HVI test result and concluded that ANN model had the best prediction power followed by mathematical and statistical model.

Table 6.3. Comparison of ANN and regression model in predicting fabric strip strength [Source: Reference 24]

Statistical parameter	Training data		Testing data	
	ANN model	Linear regression model	ANN model	Linear regression model
Correlation coefficient (R)	0.92	0.81	0.95	0.94
Mean absolute error %	7.83	13.65	7.78	6.97

Zhu et al. [37] have compared the prediction of cotton yarn irregularity by neural network model and multiple-linear regression model. It was observed that the neural network model has a better prediction result than regression model.

6.7. Future Trend

As we are aware, there are large numbers of papers published in the area of application of empirical and ANN modeling for parameter prediction for both forward and backward yarn and fabric engineering, but it has been proven that ANN is more precise, accurate and has great potential for modeling complex and nonlinear processes of the textile industry. Apart from this, the ANN modeling techniques are still not used in textile industry as because of absence of user friendly software specially designed for textile industry. So there is a great need of designing of user friendly software so that there end-use can be easily and precisely explored for process control in textile industry.

Apart from ANN, new soft computing techniques like fuzzy logic, support vector machines and genetic algorithm etc. should also be explore for yarn and fabric engineering.

ACKNOWLEDGMENT

The authors acknowledge Indian Journal of Fiber and Textile Research, AUTEX Research Journal and Fibers and Polymers for granting permission to reproduce tables used in this chapter.

REFERENCES

[1] Chattopadhyay, R.; Guha, A. *Text. Prog.* 2004, *vol 35,* 1-46.
[2] She, F. H.; Kong, L. X.; Nahavandi, S.; Kouzani, A. Z. *Text. Asia* 2001, *vol 32(11)*, 44-47.
[3] Majumdar, A.; Majumdar, P. K.; Sarkar, B. *J. Text. Inst.* 2006, *vol 97(5)*, 429-434.
[4] Majumdar, A.; Majumdar, P. K.; Sarkar, B. *AUTEX Res. J.* 2004, *vol 4(1)*, 1-8.
[5] Pynckels, F.; Kiekens, P.; Sette, S.; Van Langenhove, L.; Impe, K. *J. Text. Inst.* 1995, *vol 86*, 425-437.
[6] Sette, S.; Boullart, L.; Langenhove, L. V.; Kiekens, P. *Text. Res. J.* 1997, *vol 67*, 84-92.
[7] Shiau,Y. R.; Tsai, I. S.; Lin, C. S. *Text. Res. J.* 2000, *vol 70,* 633-640.
[8] Huang, C.C.; Chang, K. T. *Text. Res. J.* 2001, *vol 71,* 987-992.
[9] Ramesh, M. C.; Rajamanickam, R.; Jayaraman, S. *J. Text. Inst.* 1995, *vol 86,*459-469.
[10] Lawrence, C. A.; Baqui, M. A. *Text. Res. J.* 1991, *vol 61,* 123-130.
[11] Zeng, Y. C.; Wang, K. F.; Yu, C. W. *Text. Res. J.* 2004, *vol 74,* 689-694.
[12] Majumdar, P. K.; Majumdar, A. *Text. Res. J.* 2004, *vol 74,* 652-655.
[13] Cheng, L.; Adams, D. L. *Text. Res. J.* 1995, *vol 65,* 495-500.
[14] Pynckels, F.; Kiekens, P.; Sette, S.; Langenhove, L. V.; Impe, K. *J. Text. Inst.* 1997, *vol 88*, 440-448.
[15] Zhu, R.; Ethridge, M. D. *Text. Res. J.* 1997, *vol 67*, 694-698.
[16] Zhu, R.; Ethridge, M. D. *J. Text. Inst.* 1996, *vol 87*, 509-512.
[17] Lien, H. C.; Lee, S. *Neural Comput. Appl.* 2004, *vol 13,* 185-192.
[18] Ramey, H. H.; Lawson, R. Jr.; Worley, S. Jr. *Text. Res. J.* 1977, *vol 47,* 685-691.
[19] Ureyen, M. E.; Kadoglu, H. *Text. Res. J.* 2006, *vol 76*, 360-366.
[20] Ureyen, M. E.; Kadoglu, H. *Fibers Text. East. Eur.* 2007, *vol 15*, 63-67.
[21] Baykal, P. D.; Babaarsian, O.; Ruzvan, E. *Fibers Text East. Eur.* 2007, *vol 15(4),* 46-49.
[22] Behera, B. K.; Karthikeyan, B. *J. Ind. Text.* 2006, *vol 36(2),* 111-123.
[23] Zeydan, M. *Int. J. Clothing Sci. Technol.* 2008, *vol 20(2),* 104-118.
[24] Majumdar, A.; Ghosh, A.; Saha, S. S.; Roy, A.; Barman, S.; Panigrahi, D.; Biswas, A. *Fibers Polym.* 2008, *vol 9,* 240-245.
[25] Basu, A.; Chellamani, K. P.; Kumar, R. *J. Text. Inst.* 2002, *vol 93*, 283-296.
[26] Bhattacharjee, D.; Kothari, V. K. *Text. Res. J.* 2007, *vol 77*, 4-12.
[27] Chen, Y.; Zhao, T.; Collier, B. J. *J. Text. Inst.* 2001, *vol 92*, 157-163.
[28] Fan, J.; Hunter, L. *Text. Res. J.* 1998, *vol 68,* 680-686.
[29] Fan, J.; Hunter, L. *Text. Res. J.* 1998, *vol 68,* 763-771.
[30] Keshavaraj, R.; Tock, R.W.; Haycook, D. *J. Appl. Polym. Sci.* 1996, *vol 60,* 2329-2338.
[31] Hussain, T.; Malik, Z. A.; Tanwari, A. *Ind. J. Fiber Text. Res.* 2010, *vol 35,* 243-249.

[32] Morino, H.; Matsudaira, M. *Text. Res. J.* 2005, *vol 75,* 252-257.
[33] Ethridge, M. D.; Towery, J. D.; Hembree, J. F. *Text. Res. J.* 1982, *vol 52,* 35-45.
[34] Hadizadeh, M.; Jeddi, A. A. A.; Tehran, M. A. *Text. Res. J.* 2009, *vol 79,* 1-11.
[35] Ramesh, M. C.; Rajamanickam, R.; Jayaraman, S. *J. Text. Inst.* 1995, *vol 86,* 459-469.
[36] Guha, A.; Chattopadhyay, R.; Jayadeva. *J. Text. Inst.* 2001, *vol 92,* 139-145.
[37] Zhu, R.; Ethridge, M. D. *J. Text. Inst.* 1996, *vol 87,* 509-512.

In: Modeling and Simulation in Fibrous Materials
Editor: Asis Patanaik
ISBN: 978-1-62100-116-4

Chapter 7

APPLICATION OF ANN, FEA AND EMPIRICAL MODELING IN PREDICTING FABRIC DRAPE

Ajit Kumar Pattanayak* and Ameersing Luximon
Institute of Textiles and Clothing, The Hong Kong Polytechnic University, Hung Hom Kowloon, Hong Kong

ABSTRACT

The important role of fabric drape in determining the aesthetics of garments is discussed in this chapter. Special attention has been made to discuss the classification, measurement, empirical modeling and simulation of the drape. For empirical modeling the relationship between the fabric drape and fabric mechanical properties is explored from various studies. The most used techniques such as regression analysis and artificial neural network model is discussed for the prediction of fabric drape. For the simulation of different fabric models, finite element model, geometrical model and physical models have been used and discussed. So a comprehensive review has been carried out from different studies for the deep understanding of fabric drape research as well as the application of fabric drape in the modern usage like computer graphics and animation industry.

7.1. INTRODUCTION

Textile materials have a unique characteristic to undergo 3-D draping deformations by buckling and formation of graceful folds of single and double curvature. Hence, fabric drape is a very important property due to its influence on the appearance of clothing. Textiles terms and definitions of the Textile Institute [1] defined drape as the ability of a fabric to hang limply in graceful folds. Cusick [2] defined drape as the deformation of the fabric produced by the gravity when part of the fabric is partially is directly supported. According to Stylios et

*E-mail: ajitpatnaik1@yahoo.com

al. [3], the fabric drapeability may be defined as a degree of deformation of fabric, which is produced by its own gravity when only supported partly by other objects.

Drape is an important factor that affects the aesthetics and dynamic functionality of fabrics determining the adjustment of clothing to the human silhouette and providing the description of the fabric deformation produced by gravity when the fabric is partially supported. This unique characteristic provides a sense of fullness and a graceful appearance and this property distinguishes fabrics from other sheet materials. Drape describes the ability of a textile material to orient itself into folds in more than one plane under its own weight. This unique characteristic offers sense of fullness and a graceful appearance, which distinguish textiles from other materials such as paper or polymer films.

Practically drape determines apparel design and influences fabric quality for specific end uses. When a fabric is draped; it can bend in one or more directions. Usually, curtains and drapes bends in one direction, whereas garments and upholstery exhibit a complex three-dimensional form with double curvature. Hence, fabric drape is a complex mathematical problem involving large deformation under low stresses. [4]

Fabric drape [5] as described above shows the crucial impact on apparel design. This is the reason for the through investigations by many researchers in this area. Many studies have also been carried out to quantify the drape. Hence a numerous instruments ranging from a simple cantilever bending tester to a dynamic drape tester developed for measuring fabric drape. "Drape coefficient (DC)" is the main parameter used to quantify fabric drape. Though useful, it is insufficient to characterize complex forms such as garments. Many researchers have studied the fabric drape and from their study, drape can be classified into two categories, static and dynamic drape. Apart from measurement of drape, a great deal of research has been directed towards the prediction of drape from fabric mechanical properties. Also many studies have been conducted to model, simulate and animation of fabric. A garment will have areas where fit is important whereas in other areas the shape will be determined by drape characteristics. A simulation of the overall appearance will be assessed and modified until the design is judged acceptable or otherwise. Major issues are the link between the density, material properties and the simulated shape, the scope of the user interface and the definition of the resulting patterns. Hinds et al. [6] tried to evaluate methods for specifying and modifying garment designs, modeled mathematically in a CAD system. In early computer animations, the clothes for mannequins or virtual humans were generated by attaching clothes to the skin layer. This restricted any independent motion for the cloth. In the past few years, researchers are able to animate human figures and cloth objects with independent movement making way for true-to-life animation of cloth on virtual figures. The computer animation and entertainment industry is focused significantly on the appearance of virtual actors. The increased demand for more realistic virtual actor's clothing has influenced development of faster, better and easier simulation tools. As a result, there are more and more applications of cloth modeling in animation movies.

7.2. Measurement of Drape

Fabric drape is one of the visual components in the aesthetic assessment of fabrics. The drape of a fabric is defined as a description of the deformation of the fabric produced by gravity when only part of it is directly supported. Since different end users require different amounts of drape, it would be desirable to develop a technique to predict the drape of a fabric from the knowledge of its other properties.

7.2.1. Two Dimensional Drape

As drape plays a very important role for the determination of aesthetic assessment of fabrics during use, a great deal of attention has been made to quantify the drape.

The quantification of drape behavior of fabrics has been started with Pierce [7]. Pierce developed the "cantilever method" for measurement of fabric bending properties and used bending as a measure of fabric drape. The Peirce Cantilever Test is performed on the commercially available Shirley Stiffness Tester (Figure 7.1a). The principle of the Cantilever Stiffness Test [8] is depicted in Figure 7.1b. The instrument allows a rectangular strip of fabric, 6 inches by 1 inch to bend under its own weight to a fixed angle when projected as a cantilever. The fabric sample in the instrument represents a cantilever beam that is uniformly loaded by its own weight and bends downward until it reaches 41.5^0 The longer the projected length, the stiffer the fabric.

The fabric sample is placed below the metal ruler (Figure 7.1a) and slowly moved forward. The movement of the fabric continues on the tapered surface until the tip of the specimen viewed in the mirror cuts both the index lines. Then bending length is read from the scale mark opposite to the zero line engraved on the platform. The mean value for the bending length is calculated by testing the specimen four times at each end and again with the strip turned over. Using the average value of bending length (c), Peirce calculated flexural rigidity (G), and bending modulus (q).

From the length (l) and angle (θ) bending length (c) can be calculated using equations 7.1 and 7.2. Flexural rigidity of the fabric can be calculated from the bending length (C) and weight using equations 7.3 or 7.4 where w_1 is the cloth weight in ounces per square yard, and w_2 the cloth weight in grams per square yard.

Bending modulus (q) is calculated from the equations 7.5 or 7.6 where g_1 is the cloth thickness in thousands of an inch, and g_2 the cloth thickness in centimeters.

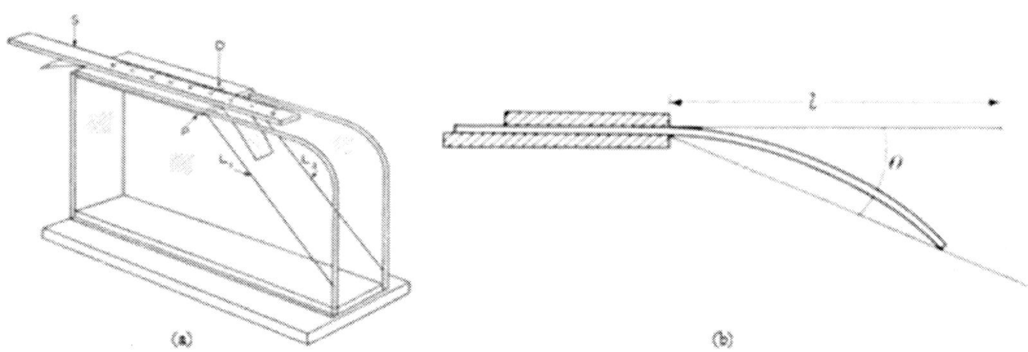

Figure 7.1. (a) Shirley Stiffness Tester (b) Principle of Cantilever Stiffness Test.

$$c = lf_1(\theta) \tag{7.1}$$

$$f_1(\theta) = \left[\frac{\cos\frac{\theta}{2}}{8\tan\theta}\right]^{\frac{1}{3}} \tag{7.2}$$

$$G = 3.39 \times w_1 c^3\ mg/cm \tag{7.3}$$

$$G = w_2 c^3 \times 10^3\ mg/cm \tag{7.4}$$

$$q = \frac{732G}{g_1^{\frac{3}{1}}}\ kg/cm^2 \tag{7.5}$$

$$q = \frac{12G \times 10^{-6}}{g_2^{\frac{3}{2}}} kg/cm^2 \tag{7.6}$$

This method is still considered as the standard method of the measurement of two-dimensional bending of fabrics.

Apart from Cantilever method, a number of stiffness testing instruments have been designed, developed and studied. The initial research shows stiffness as one of the factors determining drape characteristics of fabrics. The other instruments used for the measurement of the 2-D drape measurement are Heart Loop tester [9], Schiefer Flexometer [10], Planoflex [11, 12] and MIT Drapemeter [9, 13].

The instruments mentioned above are predominantly those measuring two dimensional distortions of fabrics. These instruments provided an acceptable correlation in terms of measuring stiffness. Due to two dimensional testing of samples, these instruments were incapable of differentiating fabrics from papers having the same bending length. The inability to discriminate between a piece of paper and a piece of fabric having the same stiffness value was the major disadvantage of characterizing drape using two dimensional distortions.

Abbott [14] compared results from subjective evaluation of stiffness with five laboratory methods of measuring stiffness. The Peirce Cantilever test, the Heart Loop test, the Schiefer Flexometer, the Planoflex and the MIT Drapemeter were the instrument methods selected for the study. He found that the Peirce Cantilever test provided the best correlation, and four of the five instrument methods provided significant correlation, with the subjective evaluation.

Sun [15] proposed a new tester which uses a cross-shaped specimen to measure fabric drape and stiffness where the warp and the weft strips of the specimen are supported at the central part and bend under its own weight. Then, he traced the X and Y coordinate of the hanging strips for the measurement of the drape angles. He also explained a single test for the measurement of the stiffness along warp and weft direction and calculated the flexural rigidity and bending length by using Peirce's formula. He found a strong correlation between the stiffness data measured by the Shirley stiffness tester and the Fabric Assurance by Simple Testing (FAST)-2 bending meter of FAST system [16].

Research in this area was continued by several researchers like Behre [17], Dahlberg [18], Grosberg [19], Lindberg [20] et al. and Oloffson [21]. Their work contributed to the understanding of drape, but two-dimensional drape assessment can't capture the complex three-dimensional double curvature deformations of 3D drape.

7.2.2. Three Dimensional Drape

A three dimensional fabric drape representation was necessary for making appropriate evaluation of drape characteristics quantitatively. To overcome the limitations of two dimensional measurement of stiffness as the estimating parameter of drape, Chu et al. [22] in Fabric Research Laboratories developed the F.R.L Drapemeter and introduced dimensionless drape coefficient as a measure of drape. Later Cusick [23] developed a drapemeter based on similar principles to quantify drape of the fabric. This Cusick Drapemeter became a standard instrument for the drape measurement.

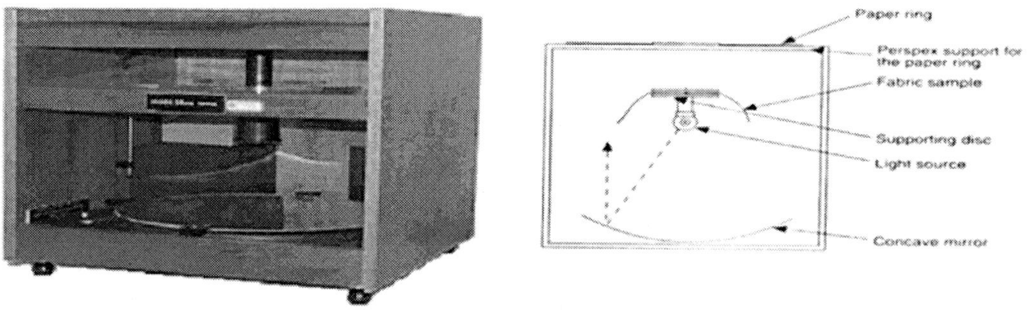

Figure 7.2. Cusick's drapemeter.

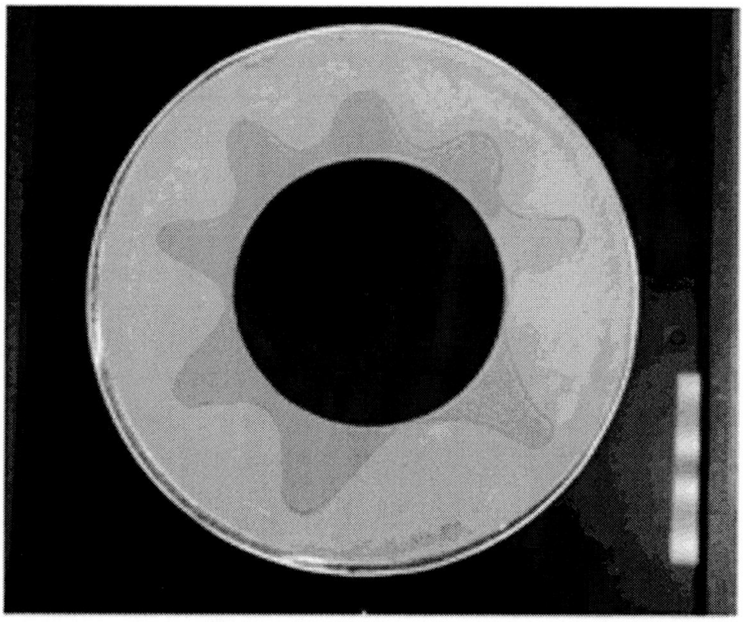

Figure 7.3. Drape configuration of fabric on drapemeter.

The schematic diagram of the Cusick Drapemeter is shown in Figure 7.2. Figure 7.3. shows the drape configuration of fabrics on the Drapemeter. This is the apparatus used extensively for the determination of fabric drape. In this instrument, a circular specimen of 36 centimeters in diameter is supported on a circular disk of 18 centimeters in diameter. The unsupported area drapes over the edges of the support disk to form the drape configuration of the fabric specimen. Equation 7.7 is used for the calculation of drape coefficient from instrument.

$$DC\% = \frac{Draped\ Area - Supporting\ Disc\ Area}{Specimen\ Area - Supporting\ Disc\ Area} \times 100 \qquad (7.7)$$

Again Cusick [24] introduced a much simpler method of calculating Drape Coefficient using a paper weighing method. In this method, a circular piece of paper, of radius R, is placed under the center of the tester. The perimeter of the shadow of the draped fabric is then drawn on the paper. The circle of paper is folded and weighed to give W_1. The paper is then cut along the perimeter of the shadow, and the paper in the shape of the shadow of the area is weighed to give W_2. Drape Coefficient (DC) is expressed as the ratio of W_1 and W_2 in percentage as shown in equation 7.8.

$$DC\% = W_2/W_1 \times 100 \qquad (7.8)$$

A measure of 100% on this instrument indicates a complete rigid fabric, while a value of 0% represents a completely limp fabric. The values in practice range from about 30% for a loose, open weave rayon fabric, to about 90% for a stretched cotton gingham, and about 95% for stiff nonwoven. [25]

Collier et al. [26] developed a modified version of a drapemeter. The instrument utilizes a bottom surface of photovoltaic cells to determine the amount of light blocked by a fabric specimen draped on a pedestal. The amount of light absorbed by photovoltaic cells is displayed digitally relating the amount of drape of the fabric.

7.3.1. Drape Measurement by Image Analysis

One recent significant development in the area of drape measurement is the use of an image analysis technique. Many researchers worked on improving the conventional technique of measuring drape using the Cusick Drapemeter. Typically the image analysis setup (Figure 7.4) consists of a Drapemeter to measure the drape of the fabric, a digital camera to capture the draped image of the mounted fabric sample, and a computer to analyze the captured image and translate it into appropriate output.

Vangheluwe and Kiekens [27] measured the drape coefficient using image analysis. A charge coupled device (CCD) camera was mounted centrally above the drapemeter. This camera sent the image to a monitor and a frame grabber in a personal computer, and the frame grabber digitally sent the image. Calibration of the drape coefficient was preceded by recording the image from the drape tester in terms of area. The image analysis system saves both time and paper. The drape coefficient can be evaluated accurately within a few seconds.

The measuring system is time-efficient and change of drape can be measured and comparisons made within a short time. Stylios et al. [28] developed a new generation of drapemeter which measured the drape of any fabric both statically and dynamically in three dimensions by using a CCD camera as a vision sensor. This system, called the Marlin Monroe Meter (M3), was used to measure the drape behavior of fabric (without being restricted to small circular fabric specimens) and to verify the theoretical prediction model. The draped profile of the specimen was taken and presented on a computer.

In addition to this, evaluation of three-dimensional drape on a real garment can also be carried out using the Moire Camera System. The system can convert images into digital data; for example the three-dimensional drape image of a flared skirt can be successfully predicted and presented on paper. Ruckman et al. [29] attempted to improve the optical method of Cusick's drapemeter by image and numerical analysis. They integrated Cusick's drape principle with image analysis technique to form a static and dynamic drape automatic measuring system (SDDAMS) which was used to obtain an image file allowing the calculation of static and dynamic drape coefficients of the fabrics and mathematical and statistical analysis. These values are not calculated using a ratio utilizing fabric mass but using fabric area; the calculation of fabric area is based on the number of pixels. Robson and Long [30] evaluated the fabric drape by automatic characterization of drape profiles using an image analysis technique. The study concludes by establishing a strong correlation between the traditional cut and weigh method of calculating the Drape Coefficient with the image analysis method. It is noted in the study that the three parameter combination of Drape Coefficient, number of nodes, mean node severity (a height/width gradient measure), and variation in node severity (standard deviation of gradient) is required for reasonable discrimination of drape profiles of fabrics. Kenkare et al. [31] proposed a modified method of measuring fabric drape using the Cusick Drapemeter. The modified method involves digitally capturing image and processing it in simple steps using image processing software. The study was conducted using a range of woven fabric samples. The fabrics were conventionally evaluated using the Cusick Drapemeter. Digital images of the draped fabrics were captured and processed as well using the modified method. Drape coefficient was selected to use as the comparative parameter to evaluate the results from conventional and digital method. The average drape coefficient of each fabric sample obtained through the conventional cut-and-weigh technique was compared statistically with that obtained through the modified digital technique. The study demonstrates that results of the modified digital method for evaluating drape were similar to that of conventional method of drape evaluation.

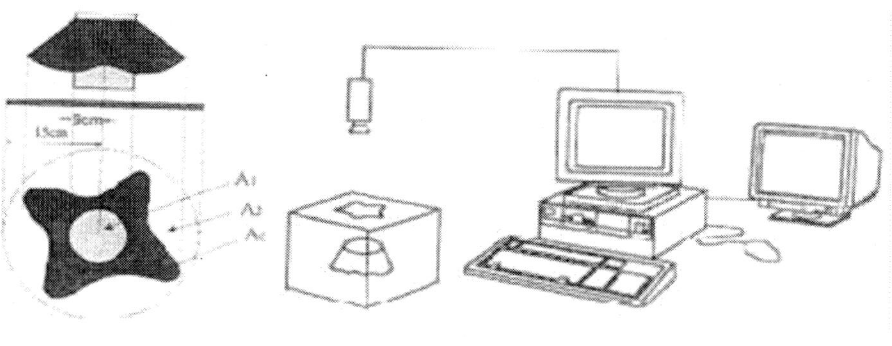

Figure 7.4. Diagram of image analysis system for the measurement of drapeability.

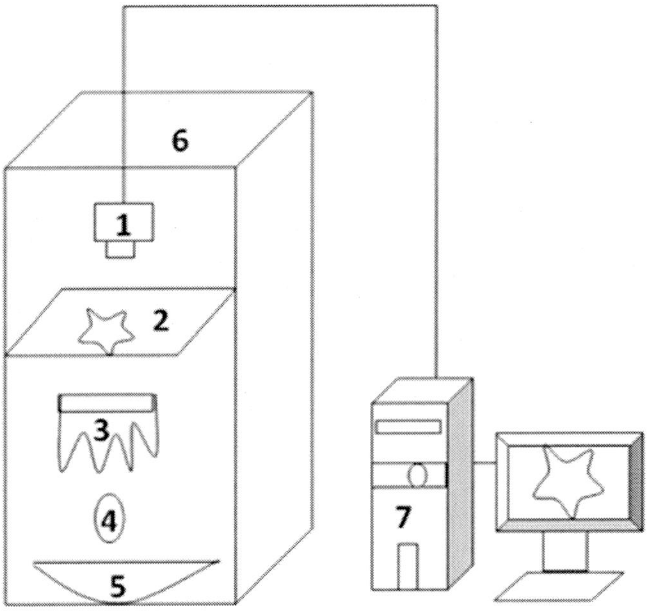

Figure 7.5. The schematic diagram of the digital drapemeter.

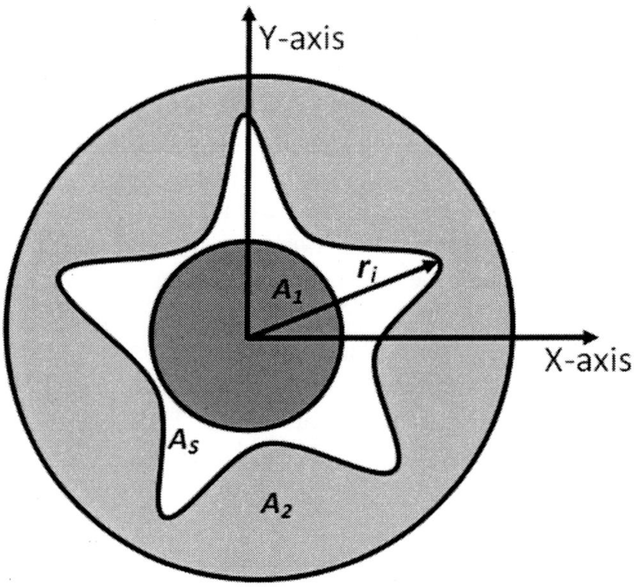

Figure 7.6. The outline of a drape profile.

According to Matsudaria and Yhang [32], drape constant is not always constant for each fabric. The coefficient changes mostly with the number of nodes. Therefore, more reliable and meaningful methods to measure the complete drape profile is necessary. The Drape Coefficient can provide an objective description of deformation, although it is not a complete description. Behera and Pandigadiya [33] explained four different drape parameters such as number of nodes, drape distance ratio, fold depth index and amplitude with drape coefficient for complete understanding of fabric drape profile. The five drape parameters of these fabrics were measured using a special developed drapemeter [34] based on image processing technique. The schematic diagram of the drapemeter is shown in Figure 7.5. In this design, the draped shadow image was captured at top glass plate (2). For this a bulb (3) was lightened up which throws the light on concave mirror (5) placed underneath. Then light reflected from the concave mirror, would be parallel rays. Thus gives an exact shadow of the draped fabric. For capturing the image without the background a thin uniform density paper was placed on the top glass plate. Thus the digital camera (1) mounted above, operated by the computer captures the clear image formed. This image was then saved in the computer (6 and 7) and processed by the program to give the various drape parameters such as drape coefficient, drape distance ratio, amplitude, fold depth index, and number of nodes. The schematic of the captured draped shadow is shown in Figure 7.6.

Definition of Drape Parameters

Drape Coefficient, $$DC = \frac{A_S - A_1}{A_2 - A_1} 100\% \quad (7.9)$$

Drape Distance Ratio, $$DDR = \frac{r_2 - r_s}{r_2 - r_1} 100\% \quad (7.10)$$

Fold Depth Index, $$FDI = \frac{r_{max} - r_{min}}{r_2 - r_1} 100\% \quad (7.11)$$

$$\text{Amplitude} = \frac{r_{max} - r_{min}}{2} \quad (7.12)$$

where A_S is the area of the draped fabric image
A_1 is area of the fabric supporting disc
A_2 is the area of the undraped fabric sample
r_1 is the radius of the fabric supporting disc
r_2 is the radius of the undraped fabric sample
r_{max} = maximum (r_i)
r_{min} = minimum (r_i)

$r_{avg} = \frac{1}{360}\sum_{i=1}^{360} r_i$, where, i = 1,......, 360 (7.13)

Thilagavathi and Natrajan [35] proposed a new method of measuring drape using a modified drape meter where the elegance of drape is more effectively observed in the elevation view of a draped fabric specimen. In this new method, apart from the top projection, the elevation projection along warp and weft directions of draped specimen mounted on the drape meter disc is taken using additional light source. The new three-dimensional drape coefficient (3-D DC) calculated from top projection and weft-way elevation projection is capable of differentiating the samples with different drape appearance but same conventional drape coefficient. This 3-D DC has better correlation with most of the drape describing parameters compared to the conventional drape coefficient.

Investigating drape on seamed fabrics can improve apparel design and fabric end-use applications. Moreover, it contributes to garment drape prediction for the clothing CAD system. Hu and Chung [36] analyzed the fundamentals of drape of seamed fabrics using Cusick's drapemeter. They put radial and circular seams to simple plain and twill woven fabrics with various fiber contents of cotton, linen, silk, wool, and polyester and investigated the effects of seam allowance and seam position experimentally. They determined and compared the drape behavior in terms of drape coefficient, node analysis, and drape profile. Ucar et al. [37] analyzes the effect of five-thread overlock seams (stitch type 516, seam type Ssa-1) on the drape behavior of heavy weight knit fabrics to provide prediction equations for drape by statistical analysis. They found from the regression analysis, the drape coefficients of seamed heavy weight knit fabrics and the rating values obtained from a subjective analyses are predicted with 0.80 and 0.86 correlation coefficients, respectively.

Jeong [38] proposed 'Drape Distance Ratio' (Figure 7.7) as an alternative to Drape Coefficient. The Drape Distance Ratio increases as fabrics become flexible, which is reverse of Drape Coefficient where the value decreases as fabrics become flexible.

$$R_d = \left[\frac{r_f - r_{ad}}{r_f - r_d}\right] \times 100 \tag{7.14}$$

where:

r_f = radius of fabric before being draped
r_d = radius of disk of drapemeter
r_{ad} = average distance to edge of draped fabric; r_i
r_i = radius and angle at i^{th} point
R_d = drape - distance ratio

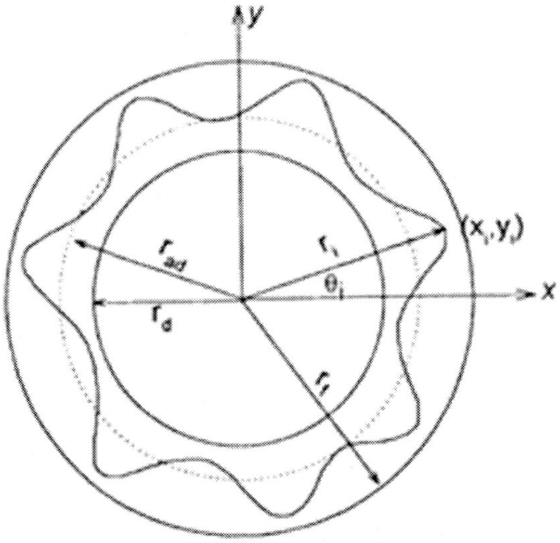

Figure 7.7. Diagram showing parameters of drape-distance ratio.

7.3. PREDICTION OF DRAPE

Drape of woven fabrics induced by gravitational force is influenced by a combination of certain deformation properties and structural characteristics. Stiffness, shear and extension are the relevant measure of deformation, while fabric weight and thickness are the appropriate structural characteristics. The type of woven construction (for example, plain, matt, twill, satin) is also a factor influencing the drape of woven fabric. Many researchers tried to find out the relationship between the fabric drape behaviors with the fabric mechanical properties. A brief of their findings are discussed below.

Chu et al. [39] found that drapeability is dependent on three basic fabric parameters: Young's modulus (Y), the cross-sectional moment of inertia (I), and fabric weight (W) and expressed drape coefficient in terms of these three parameters, DC= f (B/W) where, B= Y/I.

Hearle and Amirbayat [40] conducted a theoretical investigation of drape by using a different approach to study complex buckling. They found that the geometric form of deformation can be related to two dimensionless energy groups J_1 and J_2, which relate bending, membrane, and potential energies, and are defined in terms of sheet parameters and size. Their experiments showed that drape is not only a function of J_1 and J_2, but must also be influenced by other parameters such as the full set of anisotropic in-plane membrane and out-of-plane bending and cross term elastic constants, and perhaps the nonlinearity of response.

Sudnik [41] correlated fabric drapeability with bending length, and again confirmed the importance of bending length in predicting fabric drape. He determined that shear resistance was also a factor, although not as important. However, research by Collier [42] found shear properties to be more significant predictors of fabric drape than bending properties and shear hysteresis, and that shear hysteresis was closely related to DC.

Collier [43] presented a paper in which six parameters were measured: shear stiffness, bending hysteresis, bending stiffness, shear hysteresis at 0.5°, shear hysteresis at 5° from the KES testers, and bending rigidity from cantilever. He found that both bending stiffness from the KES and bending rigidity from the cantilever, and shear hysteresis and thickness were significant in the model for predicting the drape coefficient. Although, shear hysteresis and bending stiffness from the KES explained most of the variation, with the other two variables being less important. They concluded that shear hysteresis is very important.

According to Hearle [44], the major mode of deformation is the fabric bending, but due to the occurrence of the double curvature some shearing must also occur. There must be some tensile deformation and may be some compressive deformation, but these because of high stiffness along the threads are likely to be small and will be ignored. Buckling behavior will also be important in determining the form and magnitude of drape, but is related to bending stiffness. Thus expect bending and shearing properties to be the main factors influencing drape.

Frydrych et al. [45] examined the relationship between drape coefficient and mechanical properties of fabric. They found the correlation between the values of drape coefficient, and the values of the mechanical properties influencing formability viz. initial tensile modulus, bending rigidity, on the basis of experiment carried out. It was observed that influence of applied fabric weaves and their weight (W) on the fabric drape was significant, whereas the influence of fabric thickness (T) was not observed. The major mechanical properties influencing fabric drape are initial tensile modulus, shear rigidity and fabric formability.

Jeong and Philips [46] studied the effects of fabric structure and mechanical properties on the draping characteristics of fabrics. The fabric cover factor has a large impact on fabric drape because of its effect on the bending rigidity. Yarn interactions and crimp are determined by weave structure, also influences the fabric drape by modifying the shear rigidity.

Sang-Song [47] used the regression and neural network methods to develop new translation equations for calculating the drape coefficient of fabric using three fabric mechanical properties, namely bending rigidity, shear rigidity and formability, by the FAST system. He found that the fabric mechanical properties show significant correlations with drape coefficient. He observed that both regression models and neural network models have high correlation coefficients and low mean absolute values. He also observed that the neural network method has better ability of fitting than regression analysis.

7.3.1. Empirical Modeling of Fabric Drape

Cusick [24] proved by statistical inference that the fabric drape involves curvature in more than one direction, and the deformation is dependent on the shear angle (A) in addition to bending length (c). He used 130 fabrics for his multiple regressions. Regression equations formulated for the relationship between drape coefficient, bending length and shear angle is:

$$DC = 35.6c - 3.61c^2 - 2.59A + 0.0461A^2 + 17 \tag{7.15}$$

Mooraka and Niwa [48] generated an equation to predict fabric drape using data from the KES system and concluded that fabric weight and bending rigidity were the most important factors. In their study, DC% is found by $(B/W)^{1/3}$. The correlation coefficient r between DC% and $(B/W)^{1/3}$ is 0.767 which is greater than the value of 0.686 for DC% and B only. The use of bending rigidity from the warp, weft and bias direction on a regression equation allows for better prediction of DC% than by using a mean bending rigidity. Physical properties which contribute greatly to the DC% are bending properties followed by weight, thickness, and shear properties. When bending and shearing hysteresis is large, DC% would be large and unstable.

Niwa and Seto [49] published a paper concerned with the relationship between drapeability and mechanical properties of fabrics. They used mechanical parameters $(B/W)^{1/3}$, $(2HB/W)^{1/3}$, $(G/W)^{1/3}$ and $(2HG/W)^{1/3}$ as independent variables, where B, 2HB, W, G and 2HG are bending rigidity, bending hysteresis, weight per unit area, shear stiffness and shear respectively. These parameters were derived from the analysis of the bending of a fabric cantilever having hysteresis in bending and shear by applying the elastic theory. An equation to describe drape coefficient was then introduced.

$$DC = b_0 + b_1\sqrt[3]{\frac{B}{W}} + b_2\sqrt[3]{\frac{2HB}{W}} + b_3\sqrt[3]{\frac{G}{W}} + b_4\sqrt[3]{\frac{2HG}{W}} \tag{7.16}$$

Hu and Chan [50] considered the relationship between the fabric drape from the Cusick Drapemeter and sixteen values from the Kawabata Fabric Evaluation System. They used four mathematical models (equations 7.17 to 7.20) based on earlier studies to evaluate the influence of the Kawabata parameters on Drape Coefficient.

$$DC = b_0 + \sum_{i=1}^{n} b_i x_i \tag{7.17}$$

$$DC = b_0 + b_1\sqrt[3]{\frac{B}{W}} + b_2\sqrt[3]{\frac{2HB}{W}} + b_3\sqrt[3]{\frac{G}{W}} + b_4\sqrt[3]{\frac{2HG}{W}} \tag{7.18}$$

$$DC = b_0 + \sum_{i=1}^{n} b_i lnx_i \tag{7.19}$$

$$lnDC = b_0 + \sum_{i=1}^{n} b_i lnx_i \tag{7.20}$$

where:

b_0 and b_i are arbitrary constants
n is number of parameters closely related to the Drape Coefficient (DC)
x_i represents a mechanical property parameter

Matsudaira and Yang [51] found that there existed an inherent node number for any fabric, and the conventional static drape coefficient (D_s) could be measured accurately by an imaging system. Yang and Matsudaira [52] also derived regression equations from the static drape shape of isotropic and anisotropic fabrics using cosine functions and showed that the static drape coefficient (D_s) and the number of nodes (n) can be calculated from the following equations:

$$D_s = \frac{4a^2+2b^2+2a_m^2+b_m^2-4R_0^2}{12R_0^2} \tag{7.21}$$

$$n = 12.797 - 269.9\sqrt[3]{\frac{B}{W}} + 38060\frac{B}{W} - 2.67\frac{G}{W} + 13.03\sqrt{\frac{2HG}{W}} \tag{7.22}$$

where:

$$a = 35.981 + 1519\sqrt[3]{\frac{B}{W}} - 204300\frac{B}{W} - 23.27\sqrt[3]{\frac{G}{W}} + 0.0178G \tag{7.23}$$

$$b = 29.834 - 1.945n - 0.0188G - 91.84\frac{2HG}{W} \tag{7.24}$$

$$a_m = 9063\left(\frac{B_1-B_2}{W}\right)^{2/3} \tag{7.25}$$

$$b_m = 6224\left(\frac{B_1-B_2}{W}\right)^{2/3} \tag{7.26}$$

R_0 is the circular supporting stand of the drapemeter.
B = bending rigidity (mN.m^2/m)
G = shear rigidity (N/m/rad)
2HG = shear hysteresis at 0.0087 radian (N/m)
W= fabric weight (g/m^2)
$B_{1\,\&2}$ = bending rigidity in warp and weft directions.

Behera and Pattanayak [53] explored the relationship between the low stress mechanical properties measured by KES instrument and the five drape parameters determined using image processing technique. The regression equations derived are shown in equation 7.27 to 7.31.

$$DC = 57.47 + 0.47\,WT + 97\,B + 118\,(2HB) - 1.42\,G - 90\,W \quad (7.27)$$

$$DDR = 12.8 - 0.170\,RT + 5.1B - 6.4\,(2HB) + 13.7G + 1.23(2HG) - 0.52\,(2HG5) + 32.7W \quad (7.28)$$

$$FDI = 2.103 - 0.89\,EM - 0.28RT + 74B - 59\,(2HB) + 21.2\,G - 1.06\,(2HG5) \quad (7.29)$$

$$Amplitude = 1.579 + 82.7B - 0.4\,(2HG) + 0.39\,(2HG5) + 0.054\,RC \quad (7.30)$$

$$N = 7.569 + 0.044\,RT - 41.1\,B + 10.4\,(2HB) - 1.67\,G - 0.44(2HG5)\ (R^2 = 0.76) \quad (7.31)$$

where, DC, DDR, FDI and N are the drape coefficient, drape distance ratio, Fold depth index, and the number of nodes respectively. The other symbols used are the standard symbols for KES properties [62].

7.3.2. Artificial Neural Network

Artificial Neural Network (ANN) is an attractive predictive modeling tool used to solve problems in areas where a large number of factors contribute to the eventual outcome, but precise relationship between these various factors and their outcomes cannot be defined. A neural network is a massively parallel-distributed processor that has natural propensity for storing experimental knowledge and making it available for use. They are powerful non-linear regression algorithms and have proven to be an ideal tool to build models directly from non-linear data. Therefore, Artificial Neural Network can be trained, using fabric input-output data sets consisting of fabric structural details, yarn and fiber parameters as inputs and corresponding objectively measured fabric properties as outputs, to predict the properties of desired fabric [54]. Jedda et al. [55] investigated the relationship between the fabric drape coefficient from the drape meter and mechanical properties tested on the fabric assurance by simple testing system (FAST) and proposed a neural model and compared the neural model with the regression model. The relationship between the measured drape attributes and the subjective drape grades is a non-linear phenomenon. Consequently, it requires the use of modern artificial intelligence techniques to find the relationship. Therefore, a neural network was "trained" with the appropriate data.

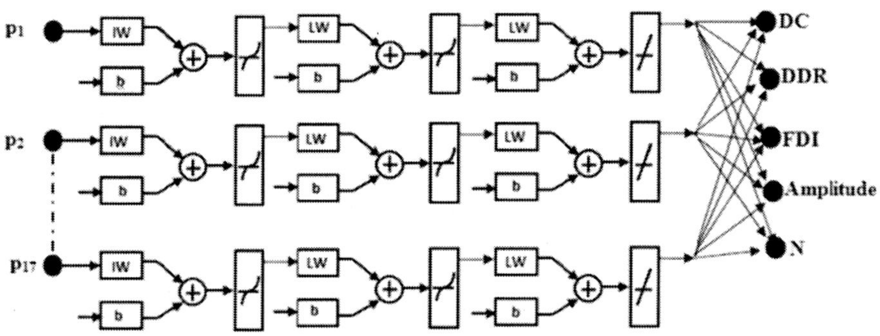

Figure 7.8. The architecture of neural network.

The predictive power of this model was compared with models derived from more tradition techniques of linear regression and discriminant analysis, and found to be more successful. The relationship between the measured drape attributes and the mechanical properties of the fabric was also established in the same way.

Pattanayak [56] explored the relationship between various drape parameters and fabric mechanical properties using an artificial neural network technique. These various drape parameters are drape coefficient (DC), drape distance ratio (DDR), fold depth index (FDI), amplitude and number of nodes (N). He used the fabric mechanical properties determined from KES as the input parameters where as the five drape parameters are used as the output. For this study sixty five cotton fabric samples are used with a spectrum of weave structures with a wide range of aerial densities.

In their study [57], a feed forward network trained with back propagation is used with a set of input vectors (p) and a set of associated desired output vectors called target vectors. This network architecture is shown in the Figure 7.8. The weights (W) and biases (b) can be updated. Back propagation is chosen because of its capability of multilayer network which are inter-connected by nonlinear differentiable transfer functions [57] and the input vectors and the corresponding target vectors were used to train a network until it could approximate a function, associate input vectors with specific output vectors in an appropriate way as it is defined. This feed forward network was created with 17-element input vectors and 5-element target vectors and this network was created with one hidden layers of tan-sigmoid neurons followed by an output layer of linear neurons. Multiple layers of neurons with nonlinear transfer functions allowed the network to learn nonlinear and linear relationships between input and output vectors. The linear output layer lets the network produce values outside the range -1 to +1. Levenberg-Marquardt optimization network training function was used to update the weight and bias values. Levenberg-Marquardt algorithm [58] is designed to approach second-order training speed without having to compute the Hessian matrix. When the performance function has the form of a sum of squares (*mse*), then the Hessian matrix (H) can be approximated as:

$$H = J^T \tag{7.32}$$

and the gradient(g) can be computed as:

$$g = J^T e \tag{7.33}$$

where J is the Jacobian matrix that contains first derivatives of the network errors with respect to the weights and biases, and e is a vector of network errors. The Jacobian matrix can be computed through a standard back propagation technique that is much less complex than computing the Hessian matrix. Before training, inputs and targets were normalized so that their values should be fall in the interval [-1, 1]. The normalized input values effectively become a part of the network, just like the network weights and biases. The output of the network was trained to produce outputs in the range [-1, 1] and the final output was converted back into the original units for further calculations. To avoid the over fitting problem during neural network training, we used early stopping method for improving generalization. In this technique the available data was divided into three subsets. The first subset (60%) was the training set, which was used for computing the gradient and updating the network weights and biases. The second subset (20%) was the validation set. The error on the validation set was monitored during the training process. The validation error decreased during the initial phase of training, as did the training set error. However, when the network began to overfit the data, the error on the validation set typically began to rise. When the validation error increases for a specified number of iterations, the training is stopped, and the weights and biases at the minimum of the validation error were returned. The last 20% i.e. the third subset was used as a completely independent test of network generalization. The performance function used for measuring this neural network was the mean sum of squares of the network errors (*mse*) which is defined in equation 7.34.

$$mse = \frac{1}{N}\sum_{i=1}^{N}(e_i)^2 = \frac{1}{N}\sum_{i=1}^{N}(t_i - a_i)^2 \tag{7.34}$$

where 'N' is number of samples, 'e' is the network error, 't_i' is the actual value and 'a_i' is the predicted value of the network. After the building of the network, the network efficiency was improved by changing the number of neurons per layer and number of layers. As an initial guess, the network created with five numbers of neurons and one hidden layer. The resulting network had five output neurons because the target vectors had five elements. Then the network has manually varied with the different number of neurons and the network is optimized with twenty numbers of neurons as per Figure 7.9. Then the network is again varied with different number of layers and the number of layers is optimized with two hidden layers as per Figure 7.9. So the optimized network was selected which contains two hidden layers and twenty neurons per hidden layer. Figure 7.9. and 7.10. show that the network performance increases with the increase in the number of neurons as well as with number of hidden layers but after a certain limit, it shows negative trend. The number of neurons up to twenty provides flexibility to the network but beyond this number the complexity of the connection in between the successive layers increases. Hence the network produces more errors. A two-layer feed forward network can potentially learn virtually any input-output relationship but feed forward networks with more layers might learn complex relationships more quickly. With increase in hidden layers, the complexity between the layers may cause more errors. Figure 7.11. describes the performance training algorithm. This is a useful diagnostic tool to plot the training, validation, and test errors to check the progress of training. This figure describes that the test set error and the validation set error has similar characteristics, and it doesn't appear that any significant over fitting has occurred and the training stopped after 11 iterations because the validation error increased. The next step is to perform some analysis of the network response where the entire data set is supplied to the network (training, validation, and test) and a linear regression between the network outputs (O) and the corresponding targets (T) are performed. The results are shown in Figure 7.12. This figure has four subfigures (a, b, c and d). The subfigures 'a', 'b', 'c' and 'd' explains the network response of the training, testing, validation and over all state respectively. It is observed from this figure that all the four outputs are tracking the targets reasonably well. The relation between the outputs and target for the sub data set are shown in the equations 7.35 to 7.38.

$$\text{Output (Training)} = 0.92\ \text{Target} + 0.0068\ (R^2 = 0.92) \tag{7.35}$$

$$\text{Output (Testing)} = 0.92\ \text{Target} + 0.0068\ (R^2 = 0.81) \tag{7.36}$$

$$\text{Output (Validation)} = 0.92\ \text{Target} + 0.0068\ (R^2 = 0.83) \tag{7.37}$$

$$\text{Output (Overall)} = 0.92\ \text{Target} + 0.0068\ (R^2 = 0.89) \tag{7.38}$$

Fan et al. [59] proposed to use a fuzzy-neural network system to predict and display the drape image of garments made from different fabrics and styles. The basic logic is to find and display a drape image from a database that is closest to the drape of the newly designed garment of the same style.

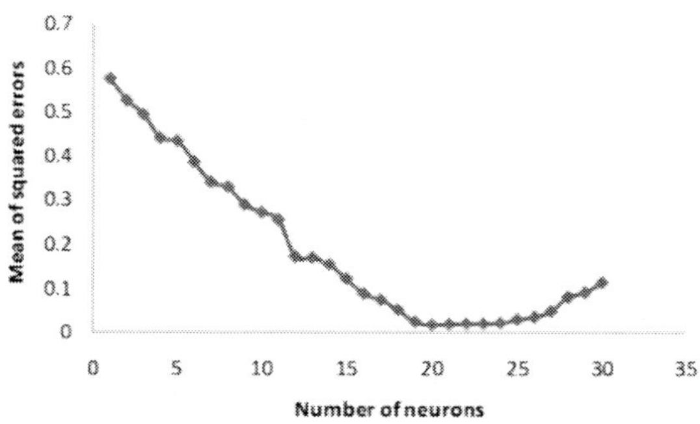

Figure 7.9. A plot between mean squared error and number of neurons.

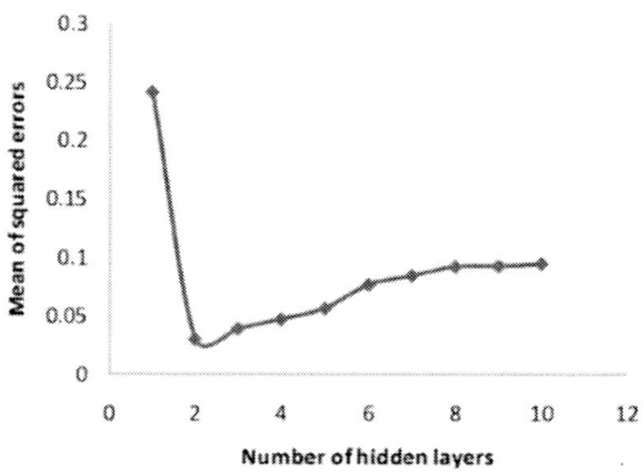

Figure 7.10. A plot between mean squared error and number of hidden layers.

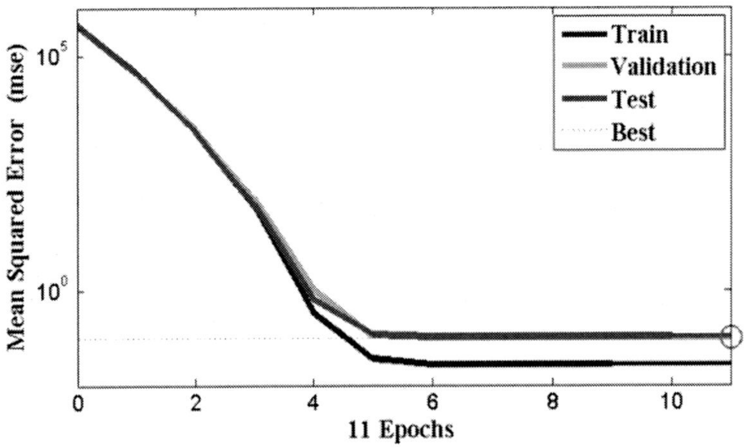

Figure 7.11. The performance plot of the drape parameters (Best validation performance is 0.091782 at epoch 11).

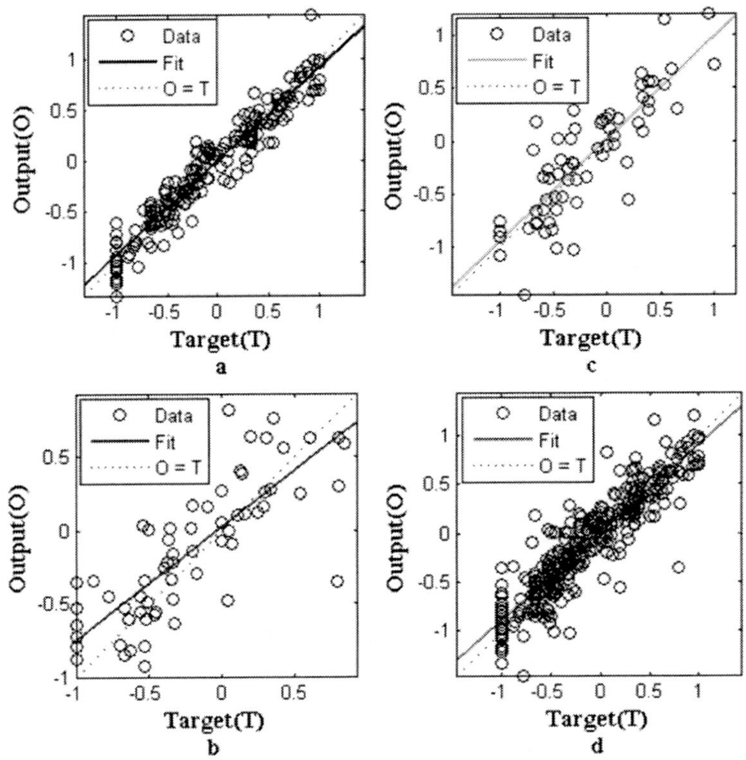

Figure 7.12. The network outputs of the drape parameters.

The advantages of the approach include very fast computation, avoid the difficulty of taking into account the effects of accessories, seams and styles on drape in conventional drape simulation and, if sufficient drape images are stored in the database, the predicted image can be very close to the actual once. A disadvantage is that only limited styles and changeable features dimensions can be allowed in the approach.

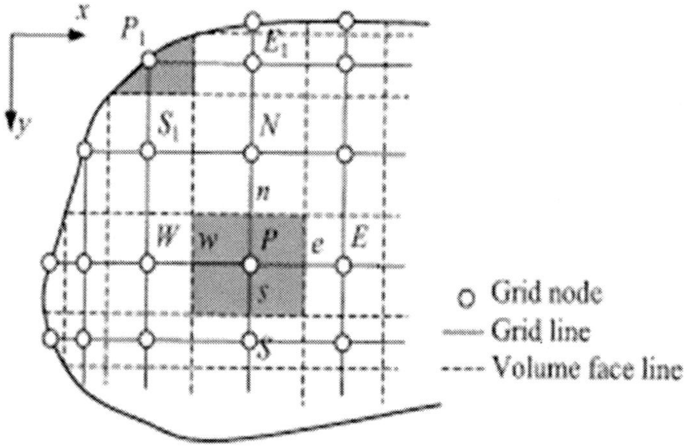

Figure 7.13. A non-uniform grid scheme for an arbitrarily shaped fabric sheet [Source: Reference 60].

7.4. DRAPE MODELING

The reasons for modeling fabric materials are numerous ranging from visualization to precise evaluation of fabric characteristics. Fabric simulation could assist in various steps of design, development, and manufacturing of fabric products. It could also help in speeding up business processes with increased speed in communicating design details electronically.

7.4.1. Finite Element Draping Model

The Finite element study of draping behavior can be undertaken by considering the deformation of a sheet material with certain properties. In order to model such behavior, it is important to know the material's load/deflection character. The simplest related case of finite element modeling is that involving only small strains and displacements and linear material properties. A fabric, when bend under its own weight, however, undergoes very slight stiffening and will bend through large displacements about two normal axes simultaneously. Large displacements must therefore be taken into consideration. This together with the fact that woven fabrics are not easily stretched, leads into classification of draping behavior as a small-strain/large-displacement problem in finite element analysis.

In brief, the basis of the finite element method is the representation of a body or a structure by an assemblage of subdivisions called finite element. Simple functions are chosen approximate the distribution or variation of the actual displacements over each finite element. The unknown magnitudes or amplitudes of the displacement functions are the displacements at the nodal points. Hence, the final solution will yield the approximate displacements at the discrete locations in the body, the nodal points. A displacement function can be expressed in various forms, such as polynomials and trigonometric functions. Since polynomials offer ease in mathematical manipulations, they have been employed in finite element applications.

A variation principle of mechanics, such as principle of minimum potential energy, is usually employed to obtain the set of equilibrium equations for each element. The potential energy of the loaded elastic body or structure is represented by the sum of the internal energy stored as a result of the deformations and the potential of the external loads. If the body is in a state of equilibrium, this energy is a minimum. The equilibrium equations for the entire body are then obtained by combining the equations for the individual elements in such a way that continuity of displacements is preserved at the interconnecting nodes. These equations are modified for the given displacement boundary conditions and then solved to obtain the unknown displacements.

Many researchers are tried to model the complex behavior of fabric drape using finite element analysis. Drape is a characteristic behavior of flexible fabric, so it is difficult to model fabric. They considered the fabric as different models and modeled accordingly. Some of their findings are discussed below.

Chen et al. [60] discussed a finite-volume method for deformation analysis of woven fabrics. They extend this finite-volume method to the prediction of drape behavior of woven fabrics and garments in contact with rigid objects or body forms. In their finite-volume method, the whole surface of an initially flat woven fabric sheet is taken to be the computational domain of the problem. The domain is then subdivided into a number of structured small patches called finite volumes (or control volumes) as shown in the Figure 7.13. Each control volume contains only one grid node. The curvatures and membrane strains and hence the out-of-plane bending and in-plane membrane strain energies of a typical control volume can be evaluated using the global coordinates of its grid node and several neighboring grid nodes surrounding it. The total strain energy of the whole fabric sheet is thus found by adding together the contributions from all control volumes. The equilibrium equations of the fabric sheet are derived employing the principle of stationary total potential energy. The fabric is assumed to be linearly elastic and orthotropic. Geometric non-linearity is considered in the formulation.

As the fabric is assumed to be an orthotropic elastic material, the corresponding stress resultants (force per unit length) for the section nPe is described by:

$$\begin{Bmatrix} N_1 \\ N_2 \\ S_{12} \end{Bmatrix} = \begin{bmatrix} E_1 & E_{12} & 0 \\ E_{12} & E_2 & 0 \\ 0 & 0 & G \end{bmatrix} \begin{Bmatrix} \varepsilon_1 \\ \varepsilon_2 \\ \gamma_{12} \end{Bmatrix} = \begin{bmatrix} E_1 & E_{12} & 0 \\ E_{12} & E_2 & 0 \\ 0 & 0 & G \end{bmatrix} \begin{Bmatrix} \varepsilon_{Pe} \\ \varepsilon_{Pn} \\ \gamma_{nPe} \end{Bmatrix} \tag{7.39}$$

where, G is the shear rigidity and E_1, E_2 and E_{12} are explained below:

$$E_1 = \frac{E_{warp}}{1 - \nu_{warp}\nu_{weft}} \tag{7.40}$$

$$E_2 = \frac{E_{weft}}{1 - \nu_{warp}\nu_{weft}} \tag{7.41}$$

$$E_{12} = \frac{\nu_{warp}E_{weft}}{1 - \nu_{warp}\nu_{weft}} = \frac{\nu_{weft}E_{warp}}{1 - \nu_{warp}\nu_{weft}} \tag{7.42}$$

here, E_{wrap} and E_{weft} are the membrane rigidities respectively determined from the tensile tests, and ν_{warp} and ν_{weft} are the corresponding Poisson's ratio. The authors have presented a number of numerical simulations of fabric sheets and simple garments by using the above discussed finite volume method. One of the simulated pictures is shown in Figure 7.14. The material properties used to create these simulations are tabulated in the Table 7.1. The Poisson's ratio was set to zero in all calculations. Dai et al. [61] introduced a novel method to model drape using a few shape parameters, predicted according to the pattern structure and mechanical properties of fabric. They used this technique to visualize the 3-D drapeability of fabric as well as to simulate of a skirt. The general shape of a flared skirt of large deformation predicted based on several shape parameters and used pre-draped initial shape for the popular physically-based model - particle system for the construction of skirt model. They used Kawabata Evaluation System (KES) [62] parameters for the input of the mechanical properties. Their simulated results show good agreement with actual fabric materials.

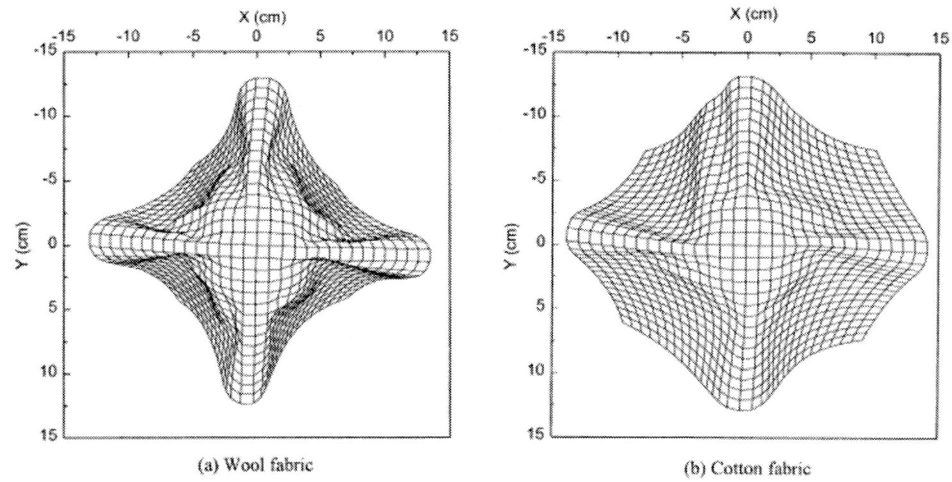

Figure 7.14. Drape shapes of square fabric sheets over a sphere: (a) wool fabric; (b) cotton fabric [Source: Reference 60].

Table 7.1. Material properties of fabrics

Property	Unit	Symbol	Polyester	Wool	Cotton
Tensile Rigidity(Warp)	gf/cm	E_{warp}	235724.7	1118.2	2531.6
Tensile Rigidity(Weft)	gf/cm	E_{weft}	145680.8	759.2	1413.5
Bending Rigidity(Warp)	Gf.cm^2/cm	D_{warp}	0.322	0.083	0.068
Bending Rigidity(Weft)	Gf.cm^2/cm	D_{weft}	0.199	0.063	0.030
Shearing Rigidity	gf/cm	G	2800	41.8	250.7
Weight	gf/cm^2	W	0.01715	0.019	0.0095
Thickness	cm	H	0.0254	0.0593	0.0469

7.4.2. Geometric Models

Geometric modeling was probably the first method devised to simulate fabric. Geometric models do not require a huge amount of computing time due to the fact that they do not consider physical properties of fabric. Geometrically based techniques are designed for capturing the static geometric appearance of fabric. These models represent folds and creases in the form of geometric equations providing specific solutions. Geometric models do not provide realistic simulation, but they are easy to compute.

Weil [63] used a simple geometric model based on the 'theory of cables'. His work explained a method for modeling fabric suspended from a finite number of points. The model was developed by fitting catenary curves between two diagonal points of attachment. The shape of the model is refined by dividing the fabric into increasingly smaller triangles and fitting catenary curves. As a final step, a spline approximation provides a smooth surface.

Hinds et al. [6] worked on interactive garment design. In developing a technique, they considered the human body in the form of a digitized mannequin and created a geometric model of a garment by specifying the outline of each garment panel. Then, they defined the panels in the form of edges and specified them by offsetting the panels from the body. The technique also provides designers with the option of placing textures and folds to enhance the model appearance. The folds in the model were based on sinusoidal offsets added in to the model. In the next stage of their research, authors devised a method for generating two-dimensional patterns from three-dimensional panels.

Bao et al. [64] investigated the effect of the mechanical properties on the MIT drape behaviors of fabrics by experiment and simulation. An improved MIT drape device enables them to detect the drape shapes of fabrics specimens with higher reproducibility than that of earlier authors. They simulated the large deformation of a fabric sample under drape by a non-linear FEM combined with an incremental method, where the sample is modeled by an elastic thin shell. They observed from the computed results that the simulation is valid for the estimation of the real MIT drape properties of fabrics, and that the MIT drape characteristics of fabrics depend appreciably on the bending rigidity and torsional one but little on the extensional rigidity and shearing one.

7.4.3. Physical Models

Physical techniques may be computationally more expensive than those of the geometric method. The time required for simulating a simple fabric sequence may range anywhere from a few minutes to days. One distinctive advantage of a physical model is its intuitive control of the parameter values. The input of an increased mass value in the model would result in depiction of heavier fabric in the model. Physical methods use mathematical tools such as differential equations to design models. Physical models can further be classified as; (a) physical models using energies and (b) physical models using forces.

Feynmann [65] proposed an 'elasticity-based' physical model. Feynmann simulated some of the mechanical properties of the fabric by defining a set of energy functions over a 2-D grid of 3-D points. The total energy of his fabric model contains tensile strain, bending and gravity terms. He minimized the energy of the grid with a stochastic technique and a multigrid method. Feynman assumed that the fabric is a continuous flexible material and derived his energy functions from the theory of elastic shells. His energy functions are only based on the distance between the points and a simple measure of curvature. The strain energy is defined as:

$$E_s = \frac{E}{1-v^2}\left(u_{xx}^2 - u_{yy}^2\right) + \frac{2vE}{1-v^2} u_{xx} u_{yy} \tag{7.43}$$

where, E is the Young's modulus, v is the Poisson's ratio and u_{ii} is the strain. The energy of bending is defined as:

$$E_b(S) = \int_0^{v_{max}} \int_0^{u_{max}} c_1 k^2 du dv \tag{7.44}$$

where k is the principal curvature of the surface and c_1 is the mechanical stiffness parameter.

Because his functions are based on the behavior of a deforming membrane, they can't maintain tight distance constraint between adjacent points, and yield a very stretchy fabric. Under normal loads, a real draping of fabric does not stretch significantly. Feynman's approach doesn't take into consideration the shearing behavior of fabric, as well as the self-intersection of the model and interaction with arbitrary solid geometric models. Since he assumed that fabric is a membrane, he also introduced a questionable energy of buckling that attempts to model cloth's differing behavior under compression and extension. In later work Ng et al. [67] developed a fabric visualization system based on Feynman's model which employs multigrid methods and searching techniques to rapidly calculate fabric drape simulation on a low-powered PC.

Taking continuum mechanics and differential geometry as their starting point, Terzopoulos et al. [67-69] developed a wide range of deformable physical based models for computer graphics. They created physical based models for computer animation useful for producing qualitatively familiar behaviors without requiring a prohibitive number of computations. For which, they presented a simplified set of equations based on elasticity theory that describe elastic and inelastic deformations, interactions with solid geometry, and fracture for flexible curves, surfaces and solids.

Lin and Stylios [70] investigated the non-linear behavior of fabric drape which is devoted to analysis of fabric pre-buckling large deflection deformation. They described an accurate analytical solution of the deflection height for fabric draping, which is applied to characterize the aesthetic of fabric drape feature and to calculate the fabric drape coefficient and also determined a non-linear equation to characterize the post-buckling mode of a draped fabric with n-folds, based on classical mechanics theory, incorporating the fabric anisotropic behavior. Fabric drape configuration, namely the wavelengths and amplitudes of ring-type buckling were predicted and

Dias et al. [71] presented a computational model for plain woven fabrics which can represent known elastic behavior in deformation such as planar extension, shearing and out-of-plane bending, drape, and buckling. They assumed the fabric to be an orthotropic linear elastic continuum and discretized by a mesh of triangles. Then each triangle links three particles which are capable to measure the stress and strain of the underlying medium. For the planar deformation, they assume the hypothesis of the plate under plane stress from the classical theory of elasticity. For the out-of-plane deformation, they allow linear elasticity and nonlinear displacement in bending, by modeling curvature interactions along the edges of neighboring triangles. To formulate the dynamic equilibrium of the mechanical system, they used Newton's second law and nonlinear elastic material behavior was made possible by piecewise linear approximation of measured data. They finally presented the results visually of the fabrics under various deformation conditions like tensile, draping, and buckling situations.

Rodel et al. [72] investigated the fundamentals of three-dimensional handling of fabrics and presented a model for the calculation by approximation of a drape test standardized in the textile industry. For that, a prerequisite is the implementation of algorithms for the simulation of draping of ready-made fabrics. They considered the woven fabric is a two-dimensional continuum as woven fabrics are of low thickness compared with the other dimensions and they considered shell theory for the for the simulation model. They also included significant material parameters for the more detailed treatment of physical and mechanical properties and their correct mathematical and physical formulation. Investigations concern several technical and technological parameters of woven fabrics, mainly the influence of the finishing process on the mechanical properties of the fabrics.

7.4.4. Hybrid Models

Hybrid based techniques integrate both the global geometric features and local physical behavior of fabric. The hybrid models usually use the geometric technique to determine the basic shape of the simulated fabric and then employ a physical technique to refine the simulation.

Rudomin [73] uses a geometric approximation as a starting condition to reduce the computation time for simulating fabric. The author suggests that this method could generate an initial configuration for a physically based fabric model. Rudomin creates a rough estimate of the general shape of a piece of fabric and produces a set of polygons as a '3D Fabric hull'. This '3D Fabric hull' provides the approximate shape of the draped fabric.

Yu et al. [74] developed an explicit–dynamic-analysis code to predict the shape of woven fabric as it deforms into a natural shape under gravity for the three-dimensional drape simulation of large samples of woven fabric. They incorporated a contact algorithm into the code, which is based on explicit algorithms for the non-linear dynamics of shells with a simple damping effect. They validated the model by performing various simulations and compared the results with those obtained from experiments.

Kang and Kim [75] developed a fast response three dimensional garment drape shape prediction system by establishing a human body model generator for the garment draping on it. For the mass production of different size of garments for the various sized body models, the cross-sectional value from the anthropometric data was used as the standard for the size accommodation to make a resizable human body model. To construct the fabric drape shape prediction system, the finite element analysis method has been utilized. The designed garment pieces were divided into fine quadrilateral elements using a specially coded mesh generating program, and then some appropriate sewing conditions were assigned to transform two dimensional patterns into three dimensional shapes. The final drape shape of the garment was determined from the solutions of the contact condition with human body, deformations, and the weights of the elements constituting the garment pieces, as well as the surface texture of the fabric.

Kunii and Gotoda [76] present a wrinkle modeling hybrid technique to simulate fabric. They initially follow a physical simulation to obtain the shape of the fabric using a gradient descent method of finding the energy minima. Then they identify a small set of characteristic deformations of wrinkled fabric. Based on the set of defined wrinkles, they calculate the response to changing geometric constraints. Since the method is based on approximation of wrinkles, it is useful for animation but not for actual representation of wrinkled fabric.

7.4.5. Discrete (Particle) System

Fabric is a composite element composed of intermeshed warp and/or weft yarns rather than a continuous sheet. Discrete models describing fabric behavior with spring mass or particle models have been proposed. In the particle modeling technique, the model is characterized by microscopic representation of fabric rather than macroscopic continuum approximation of the mechanical constraints between threads in the woven material. The warp and / or weft yarn crossings in fabric are treated as particles. The particle grid embodies all constraints arising due to mechanical interactions. Particle systems discretize the material itself as a set of point masses or particles that interact with a set of forces that approximately model the behavior of fabric. A particle system simulating small scale mechanics with Kawabata data, a particle system using a spring model for all metric, shearing and bending elasticity, and a particle system using Kawabata models on a square grid and Runge Kutta integration, illustrate the application of particle system models.

Breen et al. [77] developed a non-continuum particle model that attempted to directly capture the underlying fine grained mechanical structure of fabric. “Their model is based on the observation that fabric is best described as a mechanism of interacting mechanical parts rather than a continuous substance, and derives its macro scale dynamic properties from the micro-mechanical interaction between threads”. Breen et al. [78] followed two steps in simulating fabric with a particle model. First, the particles were allowed to fall freely to determine collision of objects and ground. This results in a rough shape of a draped fabric. In the second step, an energy minimization technique is applied to the inter-particle energy functions to generate fine details in the shape of fabric. The model simulates fabric considering fabric mechanical properties from the KESF.

7.5. Dynamic Drape

The drapebility of a fabric under known static conditions should be determined, fabric drapability under Dynamic Conditions, for example, under the influence of air twirling, is also important, especially now a days when garments are made with lighter-weight fabrics. In the case of dynamic drape, it is generally accepted that the apparel articles floats outwards (twirls) under the centrifugal acceleration, reversal, or other conditions imposed on it by the wearer or by the external forces such as air. Under these conditions, the ability to float gracefully is regarded as an important ability of the dynamic drape of the fabric.

There have been very few researchers involved in investigation of dynamic drape behavior of fabric. Yang and Matsudaira [79-82] of Kanazawa University, Japan, discussed various aspects of dynamic drape in a series of papers published in *Journal of Textile Machinery Society of Japan*. At first the researchers found that there existed an inherent node number for any fabric, and a conventional static drape coefficient *Ds* for a fabric could be measured by an image analysis system with high accuracy and reproducibility.

Yang and Matsudaira [83] defined the dynamic drape coefficient with swinging motion (*Dd*) that is considered to be similar to the human body's motion when walking. They derived a regression equation from the basic mechanical parameters of fabrics. The testing instrument built as a trial to measure dynamic drapeability of fabric. The instrument is composed of a circular supporting stand capable of rotating in the speed range of 0 to 240 rpm. The stand not only can rotate at various speeds, but also can turn in a reverse direction at an arbitrary angle. In order to get a stable drape for samples, the sample is rotated at 100 rpm for 1 minute. At this stage, two dimensional images are captured continuously at $1/30^{th}$ second time intervals. The images are captured by an image analysis system. The setup consists of a charge-coupled device camera (CCD) connected to a computer processor. To determine Drape Coefficient, an image taken by the CCD camera is put on the 512 x 512 dots picture plane and memorized in frame memory showing each dot as a monochrome tone by 256 steps of brightness. The image is divided into two grades, white and black, by an adequate threshold of the intensity of brightness. Later, an adequate image having the same direction to be compared is chosen from a frame memory and the projected area is calculated. From the measured area the dynamic drape coefficient is calculated. The dynamic drape coefficient with swinging motion is very similar to the human walking motion. The dynamic drape coefficient (*Dd*) is the change in degree of draping shape of fabrics at a swinging motion. If *Dd* is large, the fabric draping shape is changed easily by a small force such as light wind and or the swinging motion of human body.

$$D_d = \left[\frac{S_{max}-S_{min}}{\pi R_1^2-\pi R_0^2}\right] \times 100\% \tag{7.45}$$

$$D_d = 90.217 + 0.1183W - 720.3\sqrt[3]{\frac{B}{W}} - 41.3\sqrt[3]{\frac{G}{W}} \tag{7.46}$$

where, D_d is the dynamic drape coefficient
S_{max} is the maximum projected area at the turn around angle
S_{min} is the minimum projected area at the turn around angle
R_0 is the radius of the supporting disc
R_1 is the radius of the sample fabric disc
B is the bending rigidity of the sample fabric ($mN.m^2/m$)
G is the shear rigidity of the sample fabric (Nm/rad)
W is the areal density (g/m^2)

Shyr et al. [84] devised a new dynamic drape automatic measuring system integrating Cusick's drapemeter principle with the image analysis technique to automatically measure the static and dynamic drape coefficients of fabrics and investigated the relationship for four natural-fiber woven fabrics between the fabric drape coefficient and sixteen physical properties, based on the Kawabata Evaluation System for fabrics (KES-F). Their experimental data of the dynamic drape coefficient versus the rotating speed were well fitted to a Boltzmann function and the correlation coefficient analysis showed that the static drape coefficient and the dynamic drape coefficients of these four natural-fiber woven fabrics, at 100 and 125 r.p.m. did not have a very good correlation, apart from the wool fabric. The key parameters for the static drape coefficient and dynamic drape coefficient at 100 r.p.m. of each natural-fiber fabric were selected from sixteen physical properties using a stepwise regression method. Their results showed that the selected key parameters of different natural-fiber fabrics were not entirely the same, and that the static drape coefficient of a fabric could not show dynamic performance. However, the bending and shearing blocked properties were found to be most closely associated with the static and dynamic drape coefficient for the test fabrics. The tensile, compressional, surface and weight blocked properties were nuisance-blocked properties for the drape coefficient.

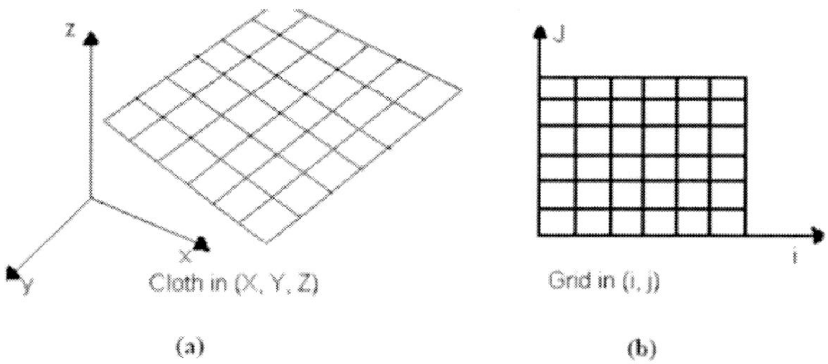

Figure 7.15. Feynmann's elasticity based model.

Conclusion

Fabric drape research started with the quantification of drape and a lot of attention has been made by researchers for the accurate measurement and empirical modeling. Several researchers beginning with Pierce have contributed to the field of drape measurement in the form of designing an instrument, studying the parameters influencing the drape, or by defining the relationship between the mechanical properties and drape. Drape characteristics are influenced by complex interactions between varying factors of bending, shear, fabric history, operating conditions, and finishing. Based on previous studies [48-50] it can be noted that there are some contradictory conclusions regarding the properties influencing fabric drape. This may be due to the fact that the kinds of fabrics used in the studies were quite different. The regression method and artificial neural network has been used recently for the better predictability.

Researchers also tried to simulate the drape behavior for computer animation industry. Their studies and the application area leads to the modeling of the fabric drape behavior using various techniques like finite element method, computer aided designs etc. Most recently much attention has been given for modeling of garment drape using various software tools. Although many studies haven made to simulate the fabric drape, it is still a focused area as all of the studies are based upon certain assumptions. Most of these assumptions are based upon the material properties like Poisson's ratio. For a better model the fabric mechanical properties should be properly characterized.

REFERENCES

[1] Denton, M. J.; Daniels, P. N. *Textile Terms and Definitions*; ISBN-13: 9781870372442; Woodhead Publishing Ltd: Cambridge, 2002, 11th Edn.

[2] Cusick, G. E. *J. Text. Inst. 1965*, *vol 59*, 596-606.

[3] Stylos, G. K.; Wan, T. R. *Int. J. Clothing Sci. Technol.* 1999, *vol 11,* 10-18.

[4] Hu, J. *Structure and Mechanics of Woven Fabrics*; ISBN: 0849328268; Wood head Publishing Ltd and CRC Press LLC: Cambridge, 2004, 1st Edn, pp 247-260.

[5] Bahtia, R.; Phadke, S. *J. Text. Assoc.* 2005, *vol 65,* 283.

[6] Hinds, B. K.; Mccartney, J.; Mclune, W.; SEOW, B. L. *Int. J. Clothing Sci. Technol.* 2000, *vol 12,* 4-5.

[7] Pierce, F. T. *J. Text. Inst.* 1930, *vol 21,* 377-416.

[8] Booth, J. E. *Principles of Textile Testing – An Introduction to Physical Methods of Testing Textile Fibers*, Yarns and Fabrics; ISBN: 0592063259; Chemical Publishing Company Inc.: NY, 1969, 3rd Edn, pp 282-288.

[9] Winn, L. J.; Schwarz, E. R. *Am. Dyestuff Reptr.* 1939, *vol 28,* 688-694.

[10] Schiefer, H. F. *Res. Natl. Bur. Std.* 1933, *vol 10,* 647-649.

[11] Derby, E. C. *Am. Dyestuff Reptr.* 1942, *vol 31,* 497-499.

[12] Winn, L. J.; Schwarz, E. R. *Am. Dyestuff Reptr.* 1940, *vol 29,* 400-403.

[13] Winn, L. J.; Schwarz, E. R. *Am. Dyestuff Reptr.* 1941, *vol 30,* 226-230.

[14] Abott, N. J. *Text. Res. J. 1951*, *vol 21,* 435-444.

[15] Sun, M. N. *Text. Res. J.* 2008, *vol 78,* 761-770.

[16] CSIRO division of wool technology; The FAST System for the Objective Measurement of Fabric Properties – Operation, Interpretation and Applications. CSIRO, Sydney.

[17] Behre, B. *Text. Res. J.* 1961, *vol 31,* 87-93.

[18] Dahlberg, B. *Text. Res. J.* 1961, *vol 31,* 94-99.

[19] Grosberg, P. *Text. Res. J.* 1966, *vol 36,* 205-11.

[20] Lindberg, J. *J. Text. Inst.* 1960, *vol 51,* T1475-1493.

[21] Oloffson, B. *J. Text. Inst.* 1964, *vol 55,* T541-57.

[22] Chu, C. C.; Cummings, C. L.; Teixeira, N. A. *Text. Res. J.* 1950, *vol 20,* 539-548.

[23] Cusick, G. E. *J. Text. Inst.* 1965, *vol 56,* 596-606.

[24] Cusick, G. E. *J. Text. Inst.* 1968, *vol 59,* 253-260.

[25] Sul, I.; Kang, T. J.; Kim, S. M.; Chi, Y. S. *Text. Res. J.* 2006, *vol 76,* 712-719.

[26] Collier, B. J.; Collier, J. R.; Scarberry, H.; Swearingen, A. *ACPTC Combined Proceedings*, 1998 35.

[27] Vangheluwe, L.; Kiekens, P. *Int. J. Clothing Sci. Technol.* 1993, *vol 5,* 5-8.

[28] Stylios, G. K.; Wan, T. R. *Int. J. Clothing Sci. Technol.* 1999, *vol 11,* 10-18.

[29] Ruckman, J. E.; Cheng, K. B.; Murray, R. 1998, *vol 10,* 56-61.

[30] Robson, D.; Long, C. C. *Clothing Text. Res. J.* 2000, *vol 18,* 1-8.

[31] Kenkare, N.; May-Plumlee, T. *J. Text. App. Tech. Man.*2005, *vol 4,* 123-128.

[32] Matsudaira, M.; Yang, M. *J. Text. Mach. Soc. Jap.* 2000, *vol 91,* 600-615.
[33] Behera, B. K.; Pandigadiya, A. *Text. Asia* 2003, *vol 34,* 45-50.
[34] Behera, B. K.; Mishra, R. *J. Tex. Inst.* 2006, *vol 97,* 142-153.
[35] Thilagavathi, G.; Natranjan, V. *Ind. J. Fibre Text. Res.* 2003, *vol 28,* 41-49.
[36] Hu, J.; Chung, S. *Text. Res. J.* 1998, *vol 68,* 913-919.
[37] Ucar, N.; Kalaoglu, F.; Bahtiyar, D.; Bilac, O. E. *Text. Res. J.* 2004, *vol 74,* 166-171.
[38] Jeong, Y. J. *J. Text. Inst.* 1998, *vol 89,* 59-69.
[39] Chu, C. C.; Platt, M. M.; Hamburger, W. J. *Text. Res. J.* 1960, *vol 30,* 66-67.
[40] Hearle, J .W. S.; Amirbayat, J. *Text. Res. J.* 1985, *vol 56,* 727-733.
[41] Sudnik, Z. M. *Text. Inst. Ind.* 1972, *vol 10,* 14-18.
[42] Collier, R.; Collier, B. J.; Toole, J. O.; Sargand, S. M. *J. Text. Inst.* 1991, *vol 10,* 96-107.
[43] Collier, B. J. *Clothing Text. Res. J. 1991, vol 10,* 46-52.
[44] Hearle, J. W. S. *Int. J. Clothing Sci. Technol.* 2004, *vol 16,* 141-152.
[45] Frydych, I.; Dziworska, G.; Matusiak, M. *Fibre Text. East. Eur.* 2003, *vol 11,* 31-37.
[46] Jeong, Y. J.; Philips, D. G. *J. Text. Inst.* 1998, *vol 89,* 70-79.
[47] Sang-Song, L. *Fibre Text. East. Eur.* 2003, *vol 28,* 270-274.
[48] Morooka, H.; Niwa, M. *J. Text. Mach. Soc. Jap.* 1976, *vol 22,* 67-73.
[49] Niwa, M.; Seto, F. *J. Text. Mach. Soc. Jap. 1986, vol 39,* 161-168.
[50] Hu, J.; Chan, Y. F. *Text. Res. J.* 1998, *vol 68,* 57-64.
[51] Matsudaira, M.; Yang, M. *J. Text. Mach. Soc. Jap. 1997*, *vol 50,* T242.
[52] Yang, M.; Matsudaira, M. *J. Text. Mach. Soc. Jap. 1998*, *vol 51,* T65.
[53] Behera, B. K.; Pattanayak, A. K. *Ind. J. Fibre Text. Res. J.* 2008, *vol 33,* 230-238.
[54] Chattopadhaya, R.; Guha, A. *Text. Prog.* 2004, *vol 35,* 1-21.
[55] Jedda, H.; Ghith, A.; Sakli, F. *J. Text. Inst.* 2007, *vol 98,* 219-225.
[56] Pattanayak, A. *Modeling and Analysis of Drape Behavior of Woven Fabrics under Static and Dynamic Conditions*; PhD Thesis, Indian Institute of Technology, New Delhi, 2010; pp 1-208.
[57] Rumelhart, D. E.; Mcclelland, J. *Parallel Data Processing;* ISBN-13: 9780262680530; The M.I.T. Press: Cambridge, 1986, 1st Edn, pp 318-362.
[58] Nocedal, J.; Wright, S. J. *Numerical Optimization;* ISBN:9780262680530; Springer: NY, 1999, 1st Edn, pp 262-269.
[59] Fan, J.; Newton, E.; Au, R.; Chan, S. C. F. *Text. Res. J.* 2001, *vol 71,* 605-611.
[60] Chen, S. F.; Hu, J. L.; Teng, J. G. *Finite Ele. Anl. Design* 2001, *vol 37,* 513-531.
[61] Dai, X.; Furukawa, T.; Mitsui, S.; Takatear, M.; Shimizu, Y. *Int. J. Clothing Sci. Technol.* 2001, *vol 13,* 23-37.
[62] Kawabata, S. *Text. Asia* 2000, *vol 31,* 30-32.
[63] Weil, J. *Comp. Graphics* 1986, *vol 20,* 49-55.
[64] Bao, L.; Takatera, M.; Sawada, K.; Sakurai, M.; Nakazawa, M.; Shinohara, A. *Sen'i Gakkaishi.* 2002, *vol 58,* 77-83.
[65] Feynman, C. R. *Modeling the Appearance of Cloth;* MSc Thesis, Massachusetts Institute of Technology, Cambridge, 1986; pp 74-98.
[66] Ng, H. N.; Grimsdale, R. L.; Allen, W. G. *Comput. Graphics* 1995, *vol 19,* 423-430.
[67] Terzopoulos, D.; Platt, J.; Fleischer, K. *Comput. Graphics* 1987, *vol 21,* 205-214.
[68] Terzopoulos, D.; Fleischer, K. *Comput. Graphics* 1988, *vol 22,* 269-278.
[69] Terzopoulos, D.; Fleischer, K. *Vis. Computer* 1988, *vol 4,* 269-278.

[70] Lin, H.; Stylios, G. K. *J. Text..Inst.* 2009, *vol 100,* 35-43.
[71] Dias, J. M. S.; Gamito, M. N.; Rebordao, J. M. *Text. Res. J.* 2000, *vol 70,* 285-297.
[72] Rodel, H.; Schenk, A.; Krzywinski, S. *Int. J. Clothing Sci. Technol.* 2000, *vol 12,* 10-11.
[73] Rudomin, I. J. *Simulating Cloth Using a Mixed Geometric-Physical Method*; PhD Thesis, University of Pennsylvania, Philadelphia, 1990; pp 23-41.
[74] Yu, W. R.; Kang, T. J.; Chung, K. *J. Text. Inst.* 2000, *vol 91,* 285-301.
[75] Kang, T. J.; Kim, S. M. *Int. J. Clothing Sci. Technol.* 2000, *vol 12,* 39-49.
[76] Kunii, T. L.; Gotoda, H. *Vis. Comput.* 1990, *vol 6,* 326-336.
[77] Breen, D. E.; House, D. H.; Getto, P. H. *Vis. Comput.* 1992, *vol 8,* 264-277.
[78] Breen, D. E., House, D. H.; Wongy, M. J. *Text. Res. J.* 1994, *vol 64,* 663-685
[79] Yang, M.; Matsudaira, M. *J. Text. Mach. Soc. Jap.* 1998a, *vol 51,* T65-T71.
[80] Yang, M.; Matsudaira, M. *J. Text. Mach. Soc. Jap.* 1998b, *vol 51,* T182-191.
[81] Yang, M.; Matsudaira, M. *J. Text. Mach. Soc. Jap.* 1999, *vol 52,* T167-175.
[82] Yang, M.; Matsudaira, M. *J. Text. Mach. Soc. Jap.* 2000, *vol 53,* T115-120.
[83] Yang, M.; Matsudaira, M. *J. Text. Mach. Soc. Jap.* 2001, *vol 54,* T57-64.
[84] Shyr, T. W.; Wang, P. N.; Cheng, K. B. *Fiber. Text. East. Eur.* 2007, *vol 15,* 81-86.

In: Modeling and Simulation in Fibrous Materials
Editor: Asis Patanaik
ISBN: 978-1-62100-116-4

Chapter 8

Applications of ANN and Statistical Modeling in Predicting Nonwoven Properties

Ting Chen* and Lili Wu

College of Textile and Clothing Engineering, National Engineering Laboratory for Modern Silk, Soochow University, Suzhou 215021, China

Abstract

Because of their superior and particular properties, uses of nonwoven products are continuing to expand. The properties of nonwoven fabrics are closely related with the processing technologies and their structural characteristics. As a nonlinear problem, predicting nonwoven properties from the processing parameters and structural characteristic can be realized by artificial neural network (ANN) and statistical models. There are two parts in this chapter. (1) The ANN and statistical models are established for predicting the fiber diameter of melt blown nonwoven fabrics from the processing parameters. The results show that the ANN model yields very accurate prediction and a reasonably good ANN model can be achieved with relatively few data points. Statistical models may be used for intuitive or qualitative analysis. (2) When applied in the field of textiles, ANN models often have to be established in the case of few samples but many input variables. Thus small-scaled ANN models are needed. Input variable selection is an effective solution to achieve small-scaled ANN models. An input variable selection method to help model the structure-property relations of nonwoven fabrics for filtration application was developed. The structural parameters are selected by utilizing this selection method which can deal with nonlinear relationships between input variables and outputs and no large number of data is required for running it. The ANN models of structure-property relations of nonwovens were established. The results show that the ANN model yields accurate prediction and a reasonably good ANN model can be achieved with relatively few data points by integrated with the input variable selecting method developed in this chapter. The results also show that there is a great potential for ANN in the field of computer-assisted design in nonwoven technology.

*E-mail: tingchen@suda.edu.cn

8.1. INTRODUCTION

Textile fabrics can be divided into woven, knitted, and nonwoven fabrics according to their processing methods. Nonwoven fabrics are such kinds of polymer materials that have sheet or web structures made by bonding or entangling fibers or filaments and perforating films with mechanical, thermal, or chemical methods. Because of the superior and particular properties of nonwoven products, their uses are continuing to expand. The major end-uses of nonwoven fabrics include disposable personal hygiene products, medical products, wipes and towels, filtration media, coated fabrics, geotextiles, roofing products, and interlinings.

The properties of nonwoven fabrics are directly related with the processing technologies and their structural characteristics. Therefore, investigating the relationship between the processing technologies, structural characteristics, and properties will be not only beneficial to better understanding of nonwoven products but also possible of achieving computer-assisted design of nonwoven fabrics. This relationship of nonwoven fabrics, however, is a complicated nonlinear problem that is sometimes difficult to be modeled physically.

As a nonlinear problem, predicting nonwoven properties from the processing parameters and structural characteristics can be realized by an empirical modeling method that includes the statistical model, artificial neural network (ANN) model and others. ANN models have been shown to provide good approximations in presence of noisy data and smaller number of experimental points and the assumptions under which ANN models work are less strict than those for statistical models [1].

There are two parts in this chapter. In part 1 (section 8.3), the above-mentioned models, i.e. the ANN model and statistical model are used to predict the fiber diameter of melt blown nonwoven fabrics from the processing parameters. The results are expected to give an indication of the relative roles of these models in predicting the fiber diameter of melt blown nonwoven fabrics. In part 2 (section 8.4), to meet the demand of establishing small-scaled ANN models, an input variable selection method was developed to help model the structure-property relations of nonwoven fabrics for filtration application. The structural parameters were selected by utilizing this method. The ANN models of structure-property relations of nonwovens were established. This section will establish a reasonably good ANN model that can generalize well and consider more structural parameters as the model inputs.

8.2. ANN AND STATISTICAL MODELS FOR PREDICTING FIBER DIAMETER OF MELT BLOWN NONWOVEN FABRICS

Melt blowing is used commercially as a one-step process for converting polymer resin directly into a nonwoven mat of superfine fibers. In the melt blowing process, high velocity hot air streams impact upon a stream of molten polymer as the polymer issues from a fine capillary. The result of this impact is that the polymer is rapidly (in about 50 μs) attenuated into fiber as fine as 1 μm in diameter.

The fiber diameter is a key property of the melt blown nonwoven fabrics and strongly affected by the processing parameters such as polymer flow rate, initial air velocity, and die-to-collector distance. In this section, ANN and statistical models were employed to predict the fiber diameter from the processing parameters.

8.2.1. Experiment

It is well known that empirical models including statistical model and ANN model are based on experimental data. Thus, in order to achieve these models, experiments were carried out on the melt blowing equipment at Nutrend Technology Developing Chemistry and Electronic Engineer Corporation Limited.

The dual slot die was used in the experiments. The spinneret hole diameter was 0.3 mm. The length of the die was 300 mm. The angle between the air slot and the spinneret axis was 30°, the slot width was 0.25 mm and the head width was 0.5 mm. The polymer used was S904 polypropylene with the melt flow index of 52. The processing parameters concerned were the polymer flow rate: 0.0017, 0.0025, 0.0033 g/s; initial air velocity: 87, 174, 261 m/s; and die-to-collector distance: 8, 11, 14 cm.

Because it is quite difficult to produce large numbers of samples for investigating the effects of the processing parameters on the fiber diameter in the factory test, the experiments are designed by using the orthogonal experimental method in order to reduce the number of experimental trials.

In this chapter, the orthogonal table of L_9 (3^4) was employed. To investigate the main effects of each processing parameter, the single factor experiments were also performed. The experimental program is shown in Table 8.1 [2-3].

In this table, experiments of No. 1 to 7 are for the single factor experiments and the orthogonal experiments cover No. 3, 5, 7 and No. 8 to 13. The image analysis method was employed to measure the fiber diameter. The images of nonwoven samples were acquired by the QUESTER three-dimensional video frequency microscope and then processed by the image analysis software named Image-Pro Plus to measure the fiber diameter. Further details about the experiment can be found in reference [3].

Table 8.1. Experimental program and results [Source: Reference 3]

No.	Polymer flow rate (g/s)	Initial air velocity (m/s)	Die-to-collector distance (cm)	Measured fiber diameter (μm)
1	0.0025	87	8	5.45
2	0.0025	174	8	4.31
3	0.0025	261	8	3.66
4	0.0017	174	8	3.69
5	0.0033	174	8	5.23
6	0.0025	174	11	4.13
7	0.0025	174	14	4.04
8	0.0017	87	8	5.44
9	0.0025	87	11	5.38
10	0.0033	87	14	5.82
11	0.0017	174	11	3.60
12	0.0017	261	14	4.04
13	0.0033	261	11	5.11

8.2.2. ANN Model

A feed forward artificial neural network is discussed in this chapter. In order to make the modeling work closely linked with industrial applications, the samples used for this work were collected from a nonwoven fabric manufacturer as mentioned above. For lack of plentiful samples, small-scaled ANN models were established in this chapter. Therefore, a three-layered ANN was built which consists of an input layer, an output layer and a hidden layer. The input layer has three neurons corresponding to the polymer flow rate, the initial air velocity and the die-to-collector distance. The output layer contains only one neuron corresponding to the fiber diameter. To determine the number of neurons in the hidden layer, all the experimental data were split into a training set (with 11 data points) and a testing set (with 2 data points). Then the number of hidden neurons was changed from 2 to 3 (more hidden neurons will cause too many unknown weights while the number of samples is quite limited). All combinations of 11 and 2 data points were used to train and test the ANN. Altogether there were 156 cases being trained and tested (78 for 2 hidden neurons and 78 for 3 hidden neurons). Comparing the average prediction error, it can be found that the average error of 2 hidden neurons was 0.031%, which was lower than that of 3 hidden neurons (0.050%). So the number of neurons in the hidden layer was determined to be two.

The transfer functions of the hidden layer and output layer neurons were the hyperbolic tangent function and pure linear function, respectively. The ANN was trained with the help of the error back propagation algorithm using the Matlab Neural Network Toolbox. The back propagation algorithm updates the network weights and biases in the direction in which the performance function decreases most rapidly. The training function used was "traingdx" which combines adaptive learning rate with momentum training and can increase the stability of the training procedure. Before training the ANN, the training data were normalized suitably.

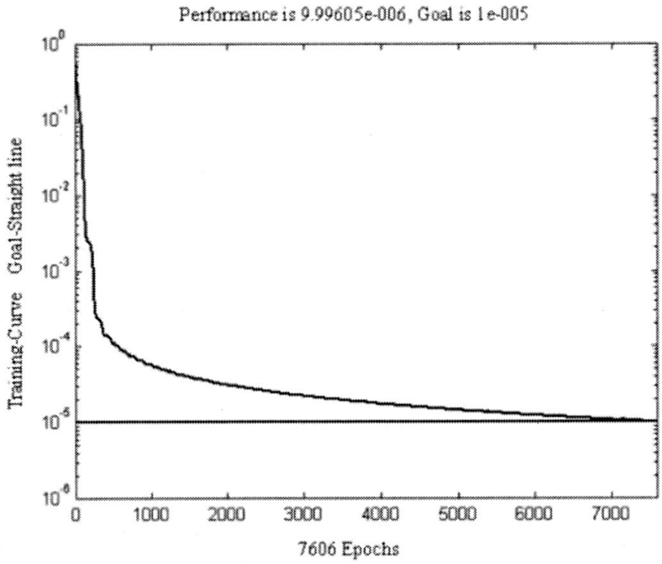

Figure 8.1. Evolution of training error [Source: Reference 3].

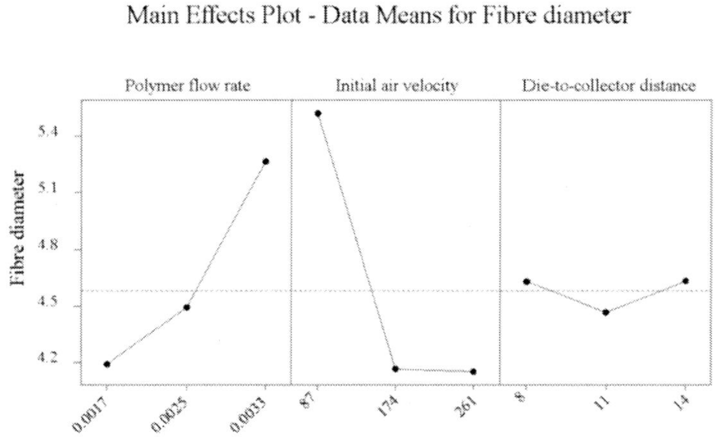

Figure 8.2. Main effects plots for fiber diameter [Source: Reference 3].

A key to successfully fit the ANN was to keep a testing set to test the prediction capabilities of the model. ANN models that are accurate to a high degree increase the confidence of an optimization procedure. In order to test the ANN model, all the experimental data were split into a training set (with 12 data points) and a testing set (with 1 data point). All combinations of 12 and 1 data points were used to train and test the ANN. Altogether there were 13 cases being trained and tested. Figure 8.1. illustrates the evolution of the training error in one of these cases [2-3]. Finally, the average of all the 13 ANN results was compared to the statistical model.

8.2.3. Statistical Model

As far as the statistical model is concerned, the input variables were the polymer flow rate (G), the initial air velocity (u_a) and the die-to-collector distance (L). The output was the fiber diameter (D). In order to study how the fiber diameter is affected by each of the variables, main effects plots and interaction plots are given in Figures 8.2. and 8.3 [2-3]. It can be found from Figure 8.2. that the effects of the polymer flow rate and initial air velocity upon the fiber diameter were large compared to the effect of the die-to-collector distance. Figure 8.3. shows the apparent interaction because the lines were not parallel, implying that the effect of each variable upon the fiber diameter depends upon other variables. All the experimental data were split into a fitting set (with 12 data points) and a testing set (with 1 data point). The 12 data points were used to establish the multivariate nonlinear regression equation, the remaining 1 data point for testing the equation, and this procedure was repeated for all combinations of 12 and 1 data points. Altogether there are 13 cases being fitted and tested. The average of all the 13 statistical results was compared with that from the ANN model. The average R^2 fit on the statistical model was 0.934. The error is taken as the difference between the predicted and measured fiber diameter, expressed as a percentage of the measured fiber diameter. The multivariate nonlinear regression equation is:

$$D = a + b_1 G + b_2 u_a + b_3 L + b_4 G^2 + b_5 u_a^2 + b_6 L^2 + b_7 G u_a + b_8 u_a L + b_9 L G \quad (8.1)$$

The parameters a and b_1, b_2, b_3 b_4, b_5, b_6, b_7, b_8, b_9 are constants.

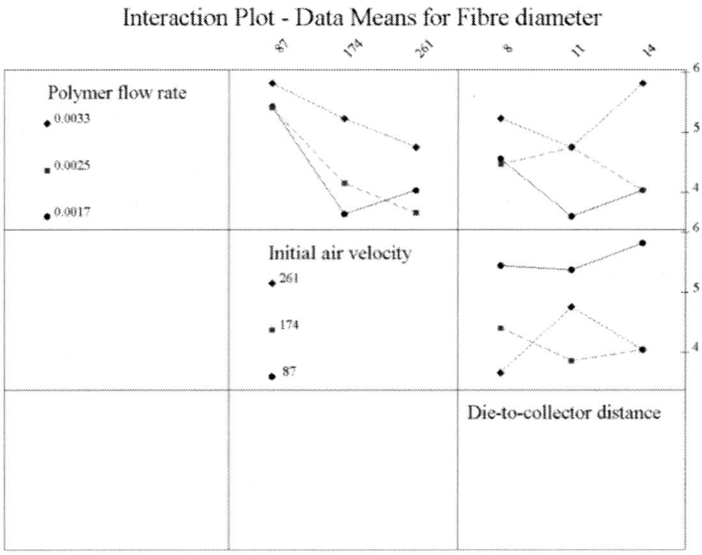

Figure 8.3. Interaction effect plots for fiber diameter [Source: Reference 3].

Table 8.2. Measured and predicted fiber diameters [Source: Reference 3]

No.	Measured fibre diameter (μm)	Physics model		Statistical model		ANN model	
		Predicted diameter	Error (%)	Predicted diameter	Error (%)	Predicted diameter	Error (%)
1	5.45	5.068	-7.009	5.634	3.384	5.446	-0.077
2	4.31	3.891	-9.722	4.167	-3.326	4.310	-0.005
3	3.66	3.322	-9.235	3.767	2.921	3.659	-0.018
4	3.69	3.358	-8.997	3.707	0.458	3.694	0.107
5	5.23	4.733	-9.503	5.225	-0.087	5.234	0.073
6	4.13	3.727	-9.758	4.101	-0.697	4.126	-0.095
7	4.04	3.592	-11.089	4.161	3.005	4.043	0.073
8	5.44	4.868	-10.515	5.345	-1.741	5.436	-0.070
9	5.38	4.922	-8.513	5.283	-1.806	5.389	0.169
10	5.82	5.348	-8.110	5.792	-0.488	5.819	-0.022
11	3.60	3.175	-11.806	3.718	3.283	3.600	0.010
12	4.04	3.604	-10.792	3.857	-4.531	4.039	-0.016
13	5.11	4.516	-11.624	5.140	0.593	5.112	0.037
Average			-9.744		0.074		0.013
Standard deviation			1.409		2.564		0.077

8.2.4. Results and Discussion

To make a comparison of the predicting performance, the prediction results of the physical model are also listed with those of ANN and statistical models. Further details about the physical model for predicting the fiber diameter from the processing parameters can be found in reference [4]. Table 8.2. shows the measured fiber diameters, predicted fiber diameters and prediction errors of the three models [2-3]. The predicted fiber diameters of the statistical and ANN model were the average of 13 results. It can be seen from this table that the performance of the ANN and statistical models were similar to each other (giving average errors of 0.013% and 0.074%, respectively) while the physical model (average error -9.744%) performs worse than both of them. It is also worthwhile to note that the standard deviation of error was lowest in the ANN model (0.077%). On the other hand, the standard deviation of error in the physical model was 1.409% while that in the statistical model was 2.564% which means that the prediction errors of the statistical model are more discrete than those of ANN and physical model. In all the thirteen samples, the absolute value of error in the ANN model was lower than those in other two models.

It is obvious that the ANN model has a better performance than the other two models. The reasons why the prediction error of the physical model is higher than the other two models may be as follows. (1) Before establishing the physical model, in order to simplify the establishing and solving procedure of the model, a number of assumptions have to be made some of which may influence the prediction result. For example, the constitutive equation of power-law fluid was introduced into the physical model as above mentioned. However, this rheological model only considers the viscosity of the fluid. In fact, the polymer melt in the melt blowing process is a kind of viscoelastic fluid. The elasticity of the polymer (for instance, the die swell near the spinneret) will undoubtedly affect its rheological behavior and thus the final fiber diameter. Another example is the ignoring of the surface tension in the momentum equation of the physical model. The air drawing of the polymer melts is a procedure of the gradual increase of specific surface area. In order to minimize the free energy of the system, the surface tension tends to minimize the fluid surface. So the surface tension is a kind of resistance force to the air drawing. Ignoring the surface tension may be the reason why the predicted fiber diameters are always smaller than the measured ones in all the thirteen samples. (2) Some physical model can have an analytical solution while many others such as the physical model in this paper have to be solved numerically. Numerical computation may cause some loss of information, which can also deteriorate the prediction result. (3) The physical model is independent of the experimental results while the ANN and statistical models strongly rely on the experimental results. The nature of the ANN and statistical models is to build a model with one part (the training sets) of the experimental results and then utilize this model to predict the other part (the testing sets) of the results. On the contrary, the physical model only considers the necessary inputs, initial and boundary conditions. It seems easier to find out the input-output relation by analyzing the existing input-output data sets but more difficult to generalize this relation to other data sets.

It can be found from Table 8.2. that the average value and standard deviation of prediction error of the statistical model were higher than the ANN model. Some probable reasons may be as follows. (1) The statistical model was probably unable to capture the nonlinearity that exists in the complex interaction between the processing parameters and the fiber diameter very accurately. (2) The statistical model was based on second order equations. It might be possible to capture the complex interaction between the processing parameters and the fiber diameter only with a high order equation- something that the ANN does. And because the physical model was built upon the inherent physical principles of phenomena of interest while the running mechanism of the statistical model was a 'black box', the prediction errors of the physical model (standard deviation of 1.409%) were less discrete than those of the statistical model (standard deviation of 2.564%).

As can be seen, the three models have their own characteristics and can be applied in different situations. The physical model is built upon the inherent physical principles of phenomena of interest and can produce reasonably good prediction results as long as the prediction error is tolerable when the experimental data is not available.

In fact, the prediction error may be lowered by improving the physical model in various ways. The ANN model is capable of making very precise predictions. When the required number of experimental data sets is available and only the final result (not the physical procedure) is of interest, the ANN model is undoubtedly a good choice. It is well known that increasing the number of training sets will decrease the prediction error. But it is quite difficult to obtain large quantities of experimental data sets especially in many engineering problems closely linked with industrial applications. However, this section gives an example of a situation where a reasonably good ANN model can be achieved with relatively few data points. The statistical model, which also needs certain number of experimental data sets, is easier to be established and understood than the physical and ANN models. So it may be used for intuitive or qualitative analysis.

8.3. ANN Model of Structure-Property Relations of Nonwoven Fabrics

It is known that many factors will influence the properties of nonwoven fabrics. On the other hand, in order to make research work closely linked with industrial applications, the nonwoven samples used are often collected from nonwoven manufacturers. It is impossible to obtain many samples made from different raw materials and different technologies from a nonwoven factory that is in stable production.

So ANN models have to be established in the case of few samples but many input variables. In practice, several factors may limit the successful application and development of an ANN model. One of the critical factors is the dimensionality of the input space. The problems associated with high input dimensionality in ANN modeling are recognized: (1) as input dimensionality increases, the computational complexity and memory requirements of the model increase; (2) learning becomes more difficult with irrelevant inputs; (3) poor model accuracy may result from additional unnecessary inputs; (4) understanding complex models is more difficult than understanding simple models that give comparable results. To establish small-scaled ANN models, several solutions can be adopted. (1) Because a three-layer (input layer-hidden layer-output layer) ANN can approximate to any nonlinear functions with limited discontinuity points, three-layer ANN models are preferred. (2) Do not include all the outputs in one ANN model. Establish one ANN model for each output solely. As a result, there is only one output in each ANN. (3) Decrease the dimensionality of the input space. Selecting the relevant input variables and removing irrelevant ones is one of the solutions to decrease the dimensionality of the input space and thus help establish small-scaled ANN models. In this section, an input variable selection method is developed to select the relevant input variables for ANN models. The ANN models of structure-property relations of nonwoven fabrics are established.

8.3.1. Input Variable Selection Method

In this section, an input variable selection method to rank the structural parameters was developed. This two-part method can deal with nonlinear relationships between input variables and output and no large number of data was required for running it. The first part takes the human knowledge on nonwoven products into account (VA_k). The second part was a data sensitivity criterion based on a distance method (S_k).

The ranking criterion was formulated as follows:

Let $X_s = (x_{s1}, x_{s2}, \ldots, x_{sk}, \ldots, x_{sn})^{\mathrm{T}}$ denotes the input vector of all the structural parameters and $Y_s = (y_{s1}, y_{s2}, \ldots, y_{sj}, \ldots, y_{sm},)^{\mathrm{T}}$ the output vector of properties. The subscript s indicates the sth sample ($s \in \{1, \ldots, i, \ldots, l, \ldots, z\}$). All the recorded data have been normalized to eliminate the scale effects and the series of data contains z samples. To rank the relevant inputs for a given output y_j, a criterion variable F_k was defined as follows:

$$F_k = g_1 \cdot VA_k(x_k, y_j) + g_2 \cdot S_k \; (k \in \{1, \ldots, n\}, j \in \{1, \ldots, m\}) \tag{8.2}$$

where g_1 and g_2 are two positive coefficients. The criterion was designed for searching the best compromise between the human knowledge and data sensitivity.

The first part (VA_k) of the ranking criterion was determined with the aid of the human knowledge [5]. As shown in Figure 8.4, the universe of discourse of y_j is divided into t equivalent intervals C_{jp} ($p \in \{1, \ldots, t\text{-}1\}$) [6-7]. The set A_{kp} was constructed with the set of input data x_k which corresponds to the output interval C_{jp} of y_j. The human knowledge shown in Table 8.3 is expressed with linguistic sentences [6-7], such as:

Rule 1: IF x_1 is increasing AND y_1 is increasing THEN $R(x_1, y_1) = +1$
Rule 2: IF x_1 is increasing AND y_1 is decreasing THEN $R(x_1, y_1) = -1$
Rule 3: IF x_1 is decreasing AND y_1 is increasing THEN $R(x_1, y_1) = -1$
Rule 4: IF x_1 is decreasing AND y_1 is decreasing THEN $R(x_1, y_1) = +1$

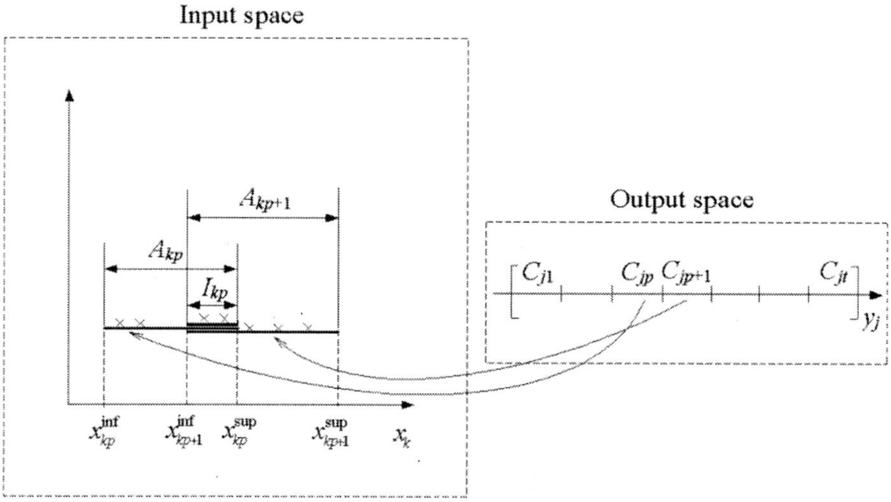

Figure 8.4. Relationship between input and output spaces [Source: Reference 7].

Then VA_k can be calculated using the following formula.

$$\begin{cases} VA_k(x_k, y_j) = \dfrac{1}{t-1}\sum_{p=1}^{t-1} va_p \\ x_{kp}^{\inf} = \min_{s\in\{1,\ldots,z\}} \{x_{sk} | y_{sj} \in C_{jp}\} \text{ and } x_{kp}^{\sup} = \max_{s\in\{1,\ldots,z\}} \{x_{sk} | y_{sj} \in C_{jp}\} \\ \text{if } I_{kp} = \phi, \begin{cases} va_p = \dfrac{1}{2}|R(x_k, y_j)| \times [1 + R(x_k, y_j)], \text{ if } x_{kp+1}^{\inf} \geq x_{kp}^{\sup} \\ va_p = \dfrac{1}{2}|R(x_k, y_j)| \times [1 - R(x_k, y_j)], \text{ if } x_{kp+1}^{\sup} \leq x_{kp}^{\inf} \end{cases} \\ \text{if } I_{kp} \neq \phi, \begin{cases} va_p = \dfrac{1}{2}|R(x_k, y_j)| \times [1 + R(x_k, y_j)] \times \left(1 - \dfrac{|I_{kp}|}{|U_{kp}|}\right), \text{ if } x_{kp+1}^{\sup} \geq x_{kp}^{\sup} \\ va_p = \dfrac{1}{2}|R(x_k, y_j)| \times [1 - R(x_k, y_j)] \times \left(1 - \dfrac{|I_{kp}|}{|U_{kp}|}\right), \text{ if } x_{kp+1}^{\inf} \leq x_{kp}^{\inf} \end{cases} \end{cases} \tag{8.3}$$

where R (x_1, y_1) is the relation index between the structural parameter x_1 and property y_1; va_p is the human knowledge criterion value in the interval C_{jp}; $x_{kp}^{\inf}$ and $x_{kp}^{\sup}$ are the lower bound (inferior limit) and upper bound (superior limit) of set A_{kp}, respectively; I_{kp} and U_{kp} are the intersection set and union set generated by A_{kp} and A_{kp+1}, respectively; ϕ is an empty set. The data sensitivity criterion S_k in equation (8.2) implies the following two hypotheses [7]:

(1) IF a small variation of an input variable corresponds to a big variation of the output, THEN this input is considered as a sensitive variable.
(2) IF a big variation of an input variable corresponds to a small variation of the output, THEN this input is considered as an insensitive variable.

Therefore, according to criterion S_k, an input variable is considered to be relevant if its small variation induces a great variation of an output.

$$T_k = \sum_{\substack{i\neq l \\ 1}}^{z} \frac{d(y_{ij}, y_{lj})}{d_k'(X_i, X_l)} \tag{8.4}$$

$$S_k = \frac{\max_{k\in\{1,\ldots,n\}}(T_k) - T_k}{\max_{k\in\{1,\ldots,n\}}(T_k) - \min_{k\in\{1,\ldots,n\}}(T_k)} \tag{8.5}$$

where $d_k'(X_i, X_l) = \sqrt{d^2(X_i, X_l) - d_k^2(X_i, X_l)}$, d (X_i, X_l) is the Euclidean distance between X_i and X_l in the input space. d_k (X_i, X_l) is the projection of d (X_i, X_l) on the axis x_k. And d (y_{ij}, y_{lj}) is the Euclidean distance between y_i and y_l of the jth output. The smaller the value T_k the more relevant to y_j will be the input x_k. So S_k is calculated by equation (8.5) to be standardized and have the same tendency as the human knowledge VA_k (larger VA_k means more relevant).

Table 8.3. Human knowledge on nonwoven products [Source: Reference 7]

Properties Structural Parameters	Air permeability	Strength at break (MD)	Elongation at break (MD)
Fiber			
Length	+1	+1	+1
Count	-1	+1	+1
Nonwoven			
Thickness	-1	+1	+1
Total pore volume	+1	-1	-1
Basis weight	-1	+1	+1
Basis weight uniformity	+1	+1	+1
Fiber volume density	-1	+1	+1

Two methods are employed to determine the weights g_1 and g_2 in equation (8.2). As shown in equations (8.6) and (8.7), method 1 uses the variation coefficient of VA_k and S_k as their weights g^*_{11} and g^*_{21}, respectively. The second subscript "1" of g^*_{11} and g^*_{21} indicates method 1. The principle of this method is as follows. A larger variation coefficient means that the corresponding criterion has stronger capability to differentiate between samples. So this criterion should be assigned a larger weight [8]. As shown in equations (8.8) and (8.9), method 2 utilizes the dispersion maximization decision principle to determine the weights g^*_{12} and g^*_{22} which means larger dispersion of a criterion corresponds to larger weight [9]. The second subscript "2" of g^*_{12} and g^*_{22} indicates method 2. The above weights are then normalized using equation (8.10). The final weights are the arithmetic average of weights determined by the two methods, as shown in equations (8.11) and (8.12).

$$g^*_{11} = \frac{\sqrt{\frac{1}{n-1}\sum_{k=1}^{n}\left(VA_k - \frac{1}{n}\sum_{k=1}^{n}VA_k\right)^2}}{\frac{1}{n}\sum_{k=1}^{n}VA_k} \tag{8.6}$$

$$g^*_{21} = \frac{\sqrt{\frac{1}{n-1}\sum_{k=1}^{n}\left(S_k - \frac{1}{n}\sum_{k=1}^{n}S_k\right)^2}}{\frac{1}{n}\sum_{k=1}^{n}S_k} \tag{8.7}$$

$$g^*_{12} = \frac{\sum_{\substack{i\neq k\\1}}^{n}\left|VA_i - VA_k\right|}{\sqrt{\left(\sum_{\substack{i\neq k\\1}}^{n}\left|VA_i - VA_k\right|\right)^2 + \left(\sum_{\substack{i\neq k\\1}}^{n}\left|S_i - S_k\right|\right)^2}} \tag{8.8}$$

$$g_{22}^{*} = \frac{\sum_{i \neq k,1}^{n} \left| S_i - S_k \right|}{\sqrt{\left(\sum_{i \neq k,1}^{n} \left| VA_i - VA_k \right| \right)^2 + \left(\sum_{i \neq k,1}^{n} \left| S_i - S_k \right| \right)^2}} \tag{8.9}$$

$$g_{11} = \frac{g_{11}^{*}}{g_{11}^{*} + g_{21}^{*}}, \quad g_{21} = \frac{g_{21}^{*}}{g_{11}^{*} + g_{21}^{*}}, \quad g_{12} = \frac{g_{12}^{*}}{g_{12}^{*} + g_{22}^{*}}, \quad g_{22} = \frac{g_{22}^{*}}{g_{12}^{*} + g_{22}^{*}} \tag{8.10}$$

$$g_1 = \frac{1}{2}\left(g_{11} + g_{12}\right) \tag{8.11}$$

$$g_2 = \frac{1}{2}\left(g_{21} + g_{22}\right) \tag{8.12}$$

After calculating VA_k, S_k and the weights g_1, g_2, the criterion variable F_k of each input x_k for a given output y_j can be determined. The larger the value F_k, the more relevant to y_j will be the input x_k. Then all the F_k are ranked in a descending order. Accordingly, the relevancies of all input variables are in the same order as the value of F_k ranked. That is, the input corresponding to the first F_k of this rank will be the most relevant input to output y_j, and the like.

8.3.2. ANN Model

Seven structural parameters and three properties of nonwoven fabrics will be used as the inputs and outputs of the ANN model. Each property was modeled respectively. For lack of plentiful samples, small-scaled ANN models were established in this section. As far as the model is concerned, five inputs and one output are preferred. A feed forward ANN is created using the Matlab Neural Network Toolbox. There are one hidden layer with two neurons (more hidden neurons will cause too many unknown weights while the number of samples is quite limited) and one output layer with one neuron in the ANN. The transfer functions of the hidden layer and output layer neurons were the hyperbolic tangent function and pure linear function, respectively. The ANN was trained with the help of the error back propagation algorithm. In order to avoid overfitting, the Bayesian framework was employed in the training procedure. In this framework the weights and biases of the network are assumed to be random variables with specified distributions. The regularization parameters are related to the unknown variances associated with these distributions. We can then estimate these parameters using statistical techniques [10]. The training function used was "trainbr" which updates the weight and bias values according to Levenberg-Marquardt optimization. It minimizes a linear combination of squared errors and weights, and then determines the correct combination so as to produce a network that generalizes well. The process is called Bayesian regularization. In order to test the ANN model, all the experimental data are split into a training set (with 17 data points) and a testing set (with 1 data point). All combinations of 17 and 1 data points are used to train and test the ANN. Altogether there are 18 cases being trained and tested. Finally, the average of all the 18 ANN results was compared to the experimental data.

8.3.3. Experiment

In order to make research work closely linked with industrial applications, the samples used were all collected from a nonwoven manufacturer in France. The method of nonwoven manufacturing was dry-laid for web forming and thermal bonding for web bonding. The fiber used was polyester with a round cross-section. There are seven structural parameters, i.e. the fiber length, fiber count, total pore volume (*tp*), basis weight uniformity, thickness (*tk*), basis weight (*bw*) and fiber volume density (*fvd*) of nonwoven fabrics. The total pore volume of nonwoven fabrics is defined as the percentage of all the pores in a nonwoven sample and is calculated with the following formula:

$$\mathrm{tp} = [1 - \mathrm{bw}/(\mathrm{tk} \times \text{fiber density} \times 1000)] \times 100\% \tag{8.13}$$

The basis weight uniformity is the patchiness or mass non-uniformity of nonwoven samples. A monochromatic LED light source (Philips Lumileds Lighting Company, San Jose, CA) and a digital camera (Uniq Vision Inc., Santa Clara, CA) are used to get the images of nonwovens. The basis weight uniformity was determined with the help of image analysis method [11]. The fiber volume density of nonwoven fabrics is defined as the total length (km) of fibers in a unit volume (m^3) of nonwoven fabrics:

$$\mathrm{fvd} = \mathrm{bw}/(\mathrm{tk} \times \text{fiber count} \times 0.0001) \tag{8.14}$$

The properties of nonwoven fabrics are the air permeability, strength at break along the machine direction (MD) and elongation at break along the machine direction (MD). The air permeability is measured on an air permeability tester (Textest AG, Schwerzenbach, Switzerland). The strength and elongation at break are measured on a universal tensile tester (Instron Deutschland GmbH, Darmstadt, Germany). All the structural parameters (inputs) and properties (outputs) are shown in Table 8.4. [7]. The technical computing software of Matlab was employed in the modeling process.

8.3.4. Results and Discussion

As mentioned above, there are altogether eighteen samples in this investigation. First, the structural parameters of nonwovens are selected with regard to each property using the method put forward in section 8.4.1. Table 8.5 to Table 8.7. gives the ranking results of structural parameters for the three properties of nonwovens [6-7].

As far as the air permeability is concerned, Table 8.5 shows that the weights $g_1 = 0.5585$ for VA_k and $g_2 = 0.4415$ for S_k, which shows that the two criterions have almost equal importance while the human knowledge criterion is a little more important. It can be seen from Table 8.5 that the most relevant structural parameter is the fiber volume density of nonwovens, followed by the thickness and basis weight uniformity of nonwovens. Note that the criterion S_k is insufficient to explain the ranking. For example, considering S_k (only the measured data), the thickness of nonwovens is the most irrelevant parameter. By adding a more general knowledge about the products (Human knowledge), the thickness increases to the second place, which coincides with our general knowledge on the close relationship between the air permeability and the thickness of nonwovens.

The ranking order for the strength at break (MD) is shown in Table 8.6. For this property, the weights $g_1 = 0.4542$ for VA_k and $g_2 = 0.5458$ for S_k. It can be found from Table 8.6 that the most relevant structural parameter is the fiber volume density, followed by the fiber count and basis weight uniformity of nonwovens.

The ranking order for the elongation at break (MD) is shown in Table 8.7. In this case, the weights $g_1 = 0.5644$ for VA_k and $g_2 = 0.4356$ for S_k. It can be seen from Table 8.7 that the most relevant structural parameter is the fiber count, followed by the basis weight uniformity and fiber volume density of nonwovens.

Table 8.4. Experimental results of structural parameters and properties of nonwoven fabrics [Source: Reference 7]

	Structural parameters							Properties		
Sample no.	Fiber length (mm)	Fiber count (dtex)	Total pore volume (%)	Basis weight unifor m-mity (%)	Thick -ness (mm)	Basis weight (g/m^2)	Fiber volume density (km/m^3)	Air permeab -ility (l/m^2/s)	Strengt -h at break (N)	Elongat -ion at break (%)
1	40	2.0	90.46	87.68	0.17	21	791855	1193.89	39.30	8.70
2	40	2.0	89.72	88.70	0.21	25	763126	765.28	51.30	8.60
3	40	2.0	89.71	85.59	0.25	29	743590	668.06	67.60	8.80
4	40	2.0	88.34	87.43	0.24	33	881410	488.61	69.90	10.10
5	40	2.0	87.94	87.98	0.35	48	879121	389.71	97.20	9.80
6	40	2.0	85.00	86.85	0.34	63	1187783	364.72	133.00	9.40
7	120	4.1	90.12	85.98	0.11	15	332594	1698.00	27.50	21.30
8	120	4.1	88.85	87.84	0.13	20	375235	1361.00	41.70	24.20
9	120	4.1	86.41	88.56	0.16	30	457317	930.00	68.70	15.10
10	120	4.1	84.30	90.64	0.18	39	528455	715.00	91.00	18.40
11	120	4.1	80.37	96.74	0.24	65	660569	267.00	186.00	15.60
12	120	4.1	85.19	99.14	0.45	92	498645	258.00	220.00	18.20
13	120	3.3	94.53	82.25	0.265	20	150943	756.00	27.60	21.80
14	120	3.3	88.50	83.12	0.189	30	317460	722.00	53.90	33.20
15	120	3.3	91.15	86.42	0.442	54	244344	360.00	99.10	34.90
16	120	3.3	89.52	82.04	0.498	72	289157	214.00	154.00	33.00
17	120	3.3	89.48	81.56	0.544	79	290441	187.50	172.00	33.80
18	120	3.3	87.83	80.05	0.643	108	335925	131.00	154.00	24.20

Table 8.5. Ranking of structural parameters for air permeability [Source: Reference 7]

	Input	Human Knowledge							Data Sensitivity						Ranking	
		R	VA	g^*_{11}	g^*_{12}	g_{11}	g_{12}	g_1	S_k	g^*_{21}	g^*_{22}	g_{21}	g_{22}	g_2	F_k	Rank
1	Fiber length	1	0						0.4559						0.2013	7
2	Fiber count	-1	0						0.5401						0.2385	6
3	Total pore volume	1	0.1721						0.4451						0.2926	5
4	Basis weight uniformity	1	0.3127	0.9684	0.7436	0.5905	0.5265	0.5585	0.7233	0.6716	0.6686	0.4095	0.4735	0.4415	0.4940	3
5	Thickness	-1	0.9820						0						0.5484	2
6	Basis weight	-1	0.6439						0.2202						0.4569	4
7	Fiber volume density	-1	0.4999						1						0.7207	1

Table 8.6. Ranking of structural parameters for strength at break (MD) [Source: Reference 7]

	Input	Human Knowledge							Data Sensitivity						Ranking	
		R	VA	g^*_{11}	g^*_{12}	g_{11}	g_{12}	g_1	S_k	g^*_{21}	g^*_{22}	g_{21}	g_{22}	g_2	F_k	Rank
1	Fiber length	1	0						0.5640						0.3078	5
2	Fiber count	1	0.1852						0.7710						0.5049	2
3	Total pore volume	-1	0.3490						0.5428						0.4548	4
4	Basis weight uniformity	1	0.4837	0.7200	0.5669	0.5007	0.4076	0.4542	0.4683	0.7180	0.8238	0.4993	0.5924	0.5458	0.4753	3
5	Thickness	1	0.5079						0.0921						0.2810	7
6	Basis weight	1	0.6667						0						0.3028	6
7	Fiber volume density	1	0.1249						1						0.6025	1

Table 8.7. Ranking of structural parameters for elongation at break (MD) [Source: Reference 7]

	Input	Human Knowledge							Data Sensitivity						Ranking	
		R	VA	g^*_{11}	g^*_{12}	g_{11}	g_{12}	g_1	S_k	g^*_{21}	g^*_{22}	g_{21}	g_{22}	g_2	F_k	Rank
1	Fiber length	1	0.6667						0						0.3763	6
2	Fiber count	1	1						0.1174						0.6156	1
3	Total pore volume	-1	0.1642						0.6871						0.3920	4
4	Basis weight uniformity	1	0.2664	1.1129	0.7010	0.6330	0.4957	0.5644	0.7961	0.6451	0.7132	0.3670	0.5043	0.4356	0.4971	2
5	Thickness	1	0.1332						0.7110						0.3849	5
6	Basis weight	1	0.0753						0.6774						0.3376	7
7	Fiber volume density	1	0						1						0.4356	3

Table 8.8. Results of ANN model: air permeability [Source: Reference 7]

Sample no.	Experimental value ($l/m^2/s$)	Predicted value ($l/m^2/s$)	Error (%)
1	1193.89	1079.78	-9.56
2	765.28	822.62	7.49
3	668.06	608.15	-8.97
4	488.61	442.18	-9.50
5	389.71	368.42	-5.46
6	364.72	336.18	-7.83
7	1698.00	1578.36	-7.05
8	1361.00	1493.41	9.73
9	930.00	969.49	4.25
10	715.00	679.42	-4.98
11	267.00	286.56	7.33
12	258.00	242.89	-5.86
13	756.00	825.37	9.18
14	722.00	767.87	6.35
15	360.00	324.87	-9.76
16	214.00	205.51	-3.97
17	187.50	204.74	9.19
18	131.00	137.96	5.31
Average error			-0.78

Table 8.9 Results of ANN model: strength at break (MD) [Source: Reference 7]

Sample no.	Experimental value (N)	Predicted value (N)	Error (%)
1	39.30	35.85	-8.78
2	51.30	55.37	7.93
3	67.60	72.04	6.57
4	69.90	66.45	-4.94
5	97.20	104.64	7.65
6	133.00	124.02	-6.75
7	27.50	24.79	-9.85
8	41.70	37.98	-8.92
9	68.70	72.19	5.08
10	91.00	98.82	8.59
11	186.00	174.71	-6.07
12	220.00	203.26	-7.61
13	27.60	30.06	8.91
14	53.90	49.77	-7.66
15	99.10	93.16	-5.99
16	154.00	168.05	9.12
17	172.00	183.26	6.55
18	154.00	139.24	-9.58
Average error			-0.88

Table 8.10. Results of ANN model: elongation at break (MD) [Source: Reference 7]

Sample no.	Experimental value (%)	Predicted value (%)	Error (%)
1	8.70	9.19	5.63
2	8.60	8.13	-5.47
3	8.80	9.22	4.77
4	10.10	9.26	-8.32
5	9.80	8.97	-8.47
6	9.40	8.53	-9.26
7	21.30	23.15	8.69
8	24.20	22.16	-8.43
9	15.10	15.99	5.89
10	18.40	16.79	-8.75
11	15.60	16.82	7.82
12	18.20	17.53	-3.68
13	21.80	20.36	-6.61
14	33.20	35.28	6.27
15	34.90	32.18	-7.79
16	33.00	35.92	8.85
17	33.80	31.85	-5.77
18	24.20	26.48	9.42
Average error			-0.84

For designing the ANN model of structure-air permeability relation, the five inputs selected are the fiber volume density, thickness, basis weight uniformity, basis weight, and total pore volume of nonwovens according to the result of Table 8.5. Table 8.8 gives the experimental values, predicted values and errors for the air permeability [6-7]. The predicted values are the average of 18 results. The average error -0.78% proves the effectiveness of the ANN model.

The five inputs selected for the strength at break (MD) are the fiber volume density, fiber count, basis weight uniformity, total pore volume, and fiber length according to the result of Table 8.6. Table 8.9 gives the experimental values, predicted values and errors for the air permeability [6-7]. The predicted values are the average of 18 results. The average error -0.88% confirms the validation of the ANN model. For the elongation at break (MD) prediction, with referring to Table 8.7 the five inputs selected are the fiber count, basis weight uniformity, fiber volume density, total pore volume and thickness of nonwovens. Table 8.10 gives the experimental values, predicted values and errors for the elongation at break (MD) [6-7]. The predicted values are the average of 18 results. The average error is -0.84% which indicates that the ANN model has a fine performance.

It can be seen from Table 8.8 to Table 8.10 [7] that some of the errors between the experimental value and predicted value are a little larger (the absolute value is nearly 10%) although the average error is small (the absolute value is less than 1%). The possible reasons may be as follows. (1) The number of samples for training is quite small. It is well known that more training samples will bring smaller prediction errors. However, just as mentioned above, in order to make research work closely linked with industrial applications, the samples used are all collected from a nonwoven manufacturer. It is impossible to obtain many samples made from different raw materials and different technologies from a textile mill that is in stable production.

In fact, it is exactly the aim of this section to establish small-scaled ANN models with few samples but tolerable predicting error for textile applications. (2) The structural parameters are selected and several parameters that are not very relevant to the properties investigated are excluded from the ANN model. This will cause information loss including loss of useful information, which will produce larger prediction errors.

A small number of samples require small-scaled ANN models that have few input neurons. In order to decrease the prediction errors, solutions are applied to the established ANN model.

For example, the Bayesian framework is employed in the training procedure to avoid over fitting. By comparison, it is found that the prediction errors can be much reduced than the conventional back propagation algorithm. There are several advantages of the input variable selection method put forward in this section. (1) This method takes not only the input variables but also the output into account. It takes not only the experimental data itself but also the human knowledge on the relationship between the input variables and output into account. (2) It uses the primitive variables to rank, not their transformations as those in the principal component analysis. So the variables have clear physical meanings. (3) No large number of data is required for running this method. It also does not require the data obeys any statistical distribution. (4) This method can deal with nonlinear relationships between input variables and output. (5) This method is easy to perform and does not lead to complex computation.

CONCLUSION

The ANN and statistical method are used to predict nonwoven properties from the processing parameters and structural characteristics of nonwoven fabrics. The ANN and statistical models are established for predicting the fiber diameter of melt blown nonwoven fabrics from the processing parameters. The results show that the ANN model yields very accurate prediction (average error of 0.013%) and a reasonably good ANN model can be achieved with relatively few data points. Statistical models may be used for intuitive or qualitative analysis.

When applied in the field of textiles, ANN models often have to be established in the case of few samples but many input variables. Thus small-scaled ANN models are needed. Input variable selection is an effective solution to achieve small-scaled ANN models. An input variable selection method to help model the structure-property relations of nonwoven fabrics for filtration use is developed. The structural parameters are selected by utilizing this selection method which can deal with nonlinear relationships between input variables and output and no large number of data is required for running it. The ANN models of structure-property relations of nonwovens are established. The results show that the ANN model yields accurate prediction and a reasonably good ANN model can be achieved with relatively few data points by integrated with the input variable selecting method developed in this chapter. The results also show that there is great potential for ANN in the field of computer-assisted design in nonwoven technology.

FUTURE TREND

Because of the advantages of ANN, many researches have concentrated on applications of ANN in predicting nonwoven properties. However, most of them used the error back propagation (BP) neural network in prediction. It is known that BP neural network has the following limitations: (1) The convergence obtained from back propagation learning is very slow. (2) The convergence in back propagation learning is not guaranteed. (3) The result may generally converge to any local minimum on the error surface since stochastic gradient descent exists on a nonlinear surface. Sometimes it is difficult to achieve good predictions using BP neural network. Therefore, ANNs other than BP neural network will be preferred in some occasions.

For example, the learning vector quantization (LVQ) neural network is one of the choices. LVQ neural network has the following advantages: (1) The neural network is trained significantly faster than other neural network techniques like Back Propagation. (2) It is able to summarize or reduce large datasets to a smaller number of codebook vectors suitable for classification or visualization. (3) It is able to generalize features in the dataset providing a level of robustness. (4) It can approximate just about any classification problem as long as the attributes can be compared using a meaningful distance measure. (5) It is not limited in the number of dimensions in the codebook vectors like nearest neighbor techniques. (6) Normalization of input data is not required (normalized may improve accuracy if attribute values vary greatly). (7) It can handle data with missing values. The LVQ neural network can be combined with the wavelet texture analysis to grade the uniformity of nonwoven fabrics.

Another example is the generalized dynamic fuzzy (GDF) neural network. The Fuzzy Neural Network (FNN) is the synergy of fuzzy logic and neural networks. The typical approach of FNN is to build standard neural networks which are designed to approximate a fuzzy inference through the structure of neural networks. Nevertheless, in most FNNs, structure identification is still difficult because the determination of hidden nodes in neural networks can be viewed as the problem of rule identification. On the other hand, almost all the existing FNNs are trained by the iterative back propagation algorithm so that real-time applications of these methods are not feasible. In order to cope with these problems, GDF neural networks have been developed. Based on the GDF neural network, both the structure and parameter identification can be achieved quickly and simultaneously by Hierarchical on-line self organizing learning paradigm without partitioning the input space and selecting proper parameters. Fuzzy rules can be automatically generated one by one to approximate the target mapping without considering the other rules and their positions. Moreover, the sensitivities of rules and input variables are analyzed based on the error reduction ratio. The GDF neural network can be combined with the wavelet transform to identify such nonwoven properties as defects, uniformity, fiber orientation and others.

ACKNOWLEDGMENTS

Projects 51076110 were supported by the National Natural Science Foundation of China. Project 200761 was supported by the Foundation for the Author of National Excellent Doctoral Dissertation of China. Project 111076 was supported by the Fok Ying Tung Education Foundation. Project BK2009123 was supported by the Natural Science Foundation of Jiangsu province and also sponsored by Qing Lan Project and by the Priority Academic Program Development of Jiangsu Higher Education Institutions.

REFERENCES

[1] Castro, J. C.; Ríos, M. C.; Mount-Campbell, C. A. *Modeling Simul. Mat. Sci. Eng.* 2004, *vol 12*, S121-S149.
[2] Chen, T.; Wang, J.; Huang, X. *J. Appl. Polym. Sci.* 2006, *vol 101*, 4275-4380.
[3] Chen, T.; Li, L.; Huang, X. *Modeling Simul. Mater. Sci. Eng.* 2005, *vol 13*, 575-584.
[4] Zheng, H. *The Study on the Process-Structure-Filtration Properties of PP Melt Blown Non-woven Fabric*; Master Thesis, Donghua University, Shanghai, 2002.
[5] Chen, T.; Huang, X. *Modeling Simul. Mater. Sci. Eng.* 2004, *vol 12*, 381-388.
[6] Zeng, X.; Koehl, L.; Sanoun, M.; Bueno, M. A.; Renner, M. *Int. J. Gen. Syst.* 2004, *vol 33*, 243-258.
[7] Chen, T.; Li, L.; Koehl, L.; Vroman, P.; Zeng, X. *J. Appl. Polym. Sci.* 2007, *vol 103*, 442-450.
[8] Zeng, X.; Koehl, L. *Int. J. Intell. Syst.* 2003, *vol 18*, 355-366.
[9] Hu, Y.; He, S. *Overall Evaluation Method*; Science Press: Beijing, 2000, pp 42-45.
[10] Wang, Y. *China Soft Sci.* 1998, *vol 13*, 36-38.

[11] MacKay, D. J. C. *Neural Comput.* 1992, *vol 4*, 415-447.
[12] Pourdeyhimi, B.; Koehl, L. *Text. Res. J.* 2002, *vol 72*, 1065-1072.

In: Modeling and Simulation in Fibrous Materials
Editor: Asis Patanaik
ISBN: 978-1-62100-116-4

Chapter 9

MODELING AND SIMULATION OF DIELECTRIC PERMITTIVITY AND ELECTROMAGNETIC SHIELDING EFFICIENCY OF FIBROUS MATERIAL

Kausik Bal[1,*] and V. K. Kothari[2]
[1]University of Calcutta, India
[2]Department of Textile Technology, Indian Institute of Technology Delhi
New Delhi, India

ABSTRACT

Dielectric and electromagnetic shielding properties of fibrous material are gaining importance in context to emerging technologies, material development and environmental, health and data safety and industrial processing and characterization requirements. Modeling of these properties and simulation of the behavior of the components or the system is integral parts of any research on these properties. Various methods, such as topology based analytical method, empirical and semi-empirical mixing laws, extrapolation techniques and computational methods in relation to dielectric permittivity of heterogeneous mixtures, particularly as fibrous materials, have been discussed. Basic concepts of electromagnetic interference shielding have been discussed and fundamental modeling approaches are outlined and scopes in these research areas particularly for fibrous materials have been discussed.

9.1. INTRODUCTION

Modeling is an integral part of research on electrical, dielectric and electromagnetic properties of any given system. The problem is often one of interaction of matter and energy, conversion of energy from one form to the other or transmission of energy from one point of interest to another and can be tackled at various dimensional scales from a macro to subatomic level. The measurement methods often limit the study at a particular level of

* E-mail: kb.iitd@gmail.com

hierarchy of the dimensions. The researcher measures the inputs and the outputs using the measuring devices. The material or system behaves like a black box, where only the inputs, outputs and transfer characteristics can be known without any knowledge of its internal construction or mechanism and the typical job of the researcher is often to speculate the contents and mechanisms present inside the black box from careful observation and analysis of the inputs and outputs. By varying the inputs in systematic ways and by recording the corresponding outputs carefully, maximum possible information is obtained by the researcher. As a result of these efforts, with experiences and analogies to other physical or abstract systems which behave closely, the speculations are zeroed-in to some specific mathematical expressions demonstrating what might be happening inside the black box. These expressions, relating the variables of the system, can be obtained analytically, empirically, statistically, or by using a number of available computational tools. This process is generally termed as modeling and these equations which relate the inputs and the outputs on the basis of speculations and assumptions are often termed as mathematical models of the system. The equations in most cases are simplified expressions of a complicated system and thus are analogous to physical models, often idealized and reduced or enlarged in size by several orders of magnitude, which are used to conveniently visualize a complicated life-size physical object for better understanding. Such models may vary in their attributes such as generality, robustness, accuracy, etc. and same problem may be analyzed with a number of models each offering a certain difference in attributes and advantages.

Heterogeneous materials and mixtures of two or more different components pose more challenges for modeling and simulation than a uniform single-component substance. However, most of the real materials encountered are many-component systems. In fact, heterogeneity may be encountered in any dimensional scale which is smaller than the physical material under study [1]. This heterogeneity may be due to domains of different materials or same material in different states. It can easily be understood that such domains can continue to be smaller even to atomic and subatomic levels. Therefore, it is a matter of choice, often limited by the measuring capabilities, to work in specific dimensional scales for modeling and simulation.

This chapter reviews various modeling approaches which have been used in relation to the dielectric and electromagnetic shielding properties of fibrous materials. Fibrous materials are heterogeneous materials even on a macroscopic scale and discussions in this chapter will be on modeling of fibrous structures and their dielectric and electromagnetic properties in the macroscopic scale. Each fiber has a microstructure involving crystalline and amorphous regions composed with long chain polymers and is often itself heterogeneous. This micro-structural heterogeneity of fibers will not be considered in this chapter. Simulation of dielectric properties of fibrous materials is quite unexplored research area. This chapter will make an attempt to discuss the research possibilities in this area on the basis of present state of the art. Modeling and simulation of electromagnetic shielding behavior of fibrous materials is also a very less explored research area. This chapter will briefly outline the various methods currently followed for electromagnetic shielding and the scope of research for modeling and simulation of the same. The underlying idea of this chapter is to introduce the diverse methods and approaches of modeling the permittivity of fibrous insulating materials and the scopes for various approaches for modeling the electromagnetic shielding properties of textile based materials. The readers are encouraged to go through the references cited for a more detailed understanding of the specific research area.

9.2. Physics of Dielectric and Electromagnetic Properties: Revisiting the Fundamentals

a) Permittivity and Dissipation Factor

A dielectric is ideally an insulating material which, when introduced between two parallel plates of a capacitor, increases the capacitance of the system [2-3]. The capacitance of a capacitor completely filled with a dielectric depends on its geometry and the property of the dielectric material. For example, the capacitance of a parallel plate capacitor of surface area *A* (m^2) with a plate separation of *L* (m) is given by:

$$C = \varepsilon_r \times \varepsilon_0 \times \frac{A}{L} \tag{9.1}$$

where ε_0 is permittivity of free space which is a universal constant ($\varepsilon_0 = 8.854 \times 10^{-12}$ F/m) and ε_r is relative permittivity of the dielectric medium occupying the entire volume between the parallel plates. In general for vacuum and for air $\varepsilon_r = 1$.

When the capacitor plates are charged with an alternating voltage source such as a sinusoidal voltage source $\mathrm{v} = \mathrm{v}_0 \times e^{j\omega t}$, where ω is the angular frequency of the signal, the total current through the capacitor becomes a vector sum of a charging current I_c which leads the applied voltage by 90° ($I_c = j \times \omega \times C \times \mathrm{v}$) and a loss current I_l ($I_l = G \times \mathrm{v}$, where *G* is the conductance of the dielectric) which is in-phase with the applied voltage. The total current, therefore, leads the voltage by an angle θ (0° < θ < 90°). The relative permittivity of the dielectric in such case is a complex quantity given by:

$$\varepsilon_r^* = \varepsilon_r^{'} - j \times \varepsilon_r^{''} \tag{9.2}$$

where $\varepsilon_r^{'}$ is the real part of permittivity, called 'real permittivity' or sometimes 'static permittivity' and $\varepsilon_r^{''}$ is the imaginary part of permittivity, also known as 'relative loss factor'. The loss tangent is given by:

$$\tan\delta = \frac{\varepsilon_r^{''}}{\varepsilon_r^{'}} \tag{9.3}$$

The loss angle δ is the angle made by the total current vector with the imaginary axis ($\delta = 90^\circ - \theta$). The loss tangent is also known as 'Dissipation Factor' and is conventionally denoted as 'D_f' or '*D*'.

b) Electromagnetic Field Vectors

Four vectors $\vec{E}$, $\vec{H}$, $\vec{D}$ and $\vec{B}$ are considered for a general description of electromagnetic field. All of these vectors are assumed to be finite throughout the entire field and are also assumed to be continuous functions of position and time with continuous derivatives at all ordinary points of the field. The discontinuities may occur on surfaces or the interfaces of two different mediums. $\vec{E}$ and $\vec{H}$ are the intensities of electric and magnetic field respectively, $\vec{D}$ is called electrostatic displacement and $\vec{B}$ is called magnetic induction. In its differential form, Maxwell's equations involving these vectors are considered as the fundamental governing rules for any electromagnetic phenomena in macroscopic scale and are given as:

$$\vec{\nabla}\times\vec{E}+\frac{\partial\vec{B}}{\partial t}=0 \qquad (9.4)$$

$$\vec{\nabla}\times\vec{H}-\frac{\partial\vec{D}}{\partial t}=\vec{J} \qquad (9.5)$$

where $\vec{J}$ is the electric current density vector. The fact that one can assume that the initial generation of the field took place at a finite time in past and that the electrostatic displacement is generated by the charge density ρ leads to two divergence equations which together with equation 9.4 and 9.5 are known as the four famous equations of Maxwell in the theory of electromagnetism:

$$\vec{\nabla}\cdot\vec{B}=0 \qquad (9.6)$$

and

$$\vec{\nabla}\cdot\vec{D}=\rho \qquad (9.7)$$

The field characteristics and the relative magnitude between the magnetic field ($\vec{H}$) and the electric field ($\vec{E}$) depend upon the nature of the source and the distance of the electromagnetic (EM) wave from its source. A region in space is said to have a 'near field' or induction field in which the field properties are determined mainly by the characteristics of the source. Beyond this, the space constitutes 'far field' or radiation field where the properties are mainly determined by the characteristics of the medium of propagation. The boundary between the near and far field is located at a distance of $\left(\frac{\lambda}{2\times\pi}\right)$ where λ is the wavelength of the radiating EM wave. The wave impedance Z is the ratio of the magnitude of electric and magnetic field intensities $(\frac{E}{H})$. For air and free space, the wave impedance is $Z_0=377\Omega$. If the source has high current and low voltage, i.e., low wave impedance, the near field is said to be predominantly magnetic. This may be caused due to a transformer or power lines. On the other hand, if the source has very high impedance, the near field is predominantly electric. As the distance from the source increases, both E-field and H-field gradually diminishes or 'attenuates' and eventually in far field region, the radiated waves lose their curvature and become plane.

9.3. MODELING PERMITTIVITY OF HETEROGENEOUS MEDIUM

In order to understand or interpret the results of capacitance measurement with a dielectric material occupying some space between the parallel plates of a parallel plate capacitor system, a suitable model of the system is required. Thus, for example, equation 9.1 represents a model of a uniform homogeneous isotropic dielectric medium occupying the entire space between the parallel electrode plates of a parallel plate capacitor system having effective area A and a gap L between the parallel plates. Materials such as vacuum, air, de-aerated water in liquid form would satisfy this condition and therefore in studies of dielectric properties of such materials, equation 9.1 can be used either to determine the material permittivity ε_r from measured value of capacitance C or to predict the capacitance of the parallel plate electrode system when such a uniform homogeneous isotropic medium having a known permittivity ε_r occupies the space between the parallel plates.

However, when the dielectric material is not homogeneous, the model of a homogeneous dielectric material of which equation 9.1 is a mathematical representation, does not fit. In such cases, the heterogeneity of the dielectric has to be taken into consideration and a new equation describing this has to be formulated. In discussions on the dielectric properties of heterogeneous mediums, a concept of homogenization is often used and the over-all dielectric property of the heterogeneous medium is described by a bulk property termed as 'effective relative permittivity' which can be used in the form of equation 9.1. The homogenization approach enables to assign this 'effective permittivity' to the medium which is the permittivity of a homogeneous medium which, when replaces the heterogeneous medium, does not affect the overall behavior of the system.

Two common models that are used in many of the complicated systems involving capacitors are 'series' and 'parallel' models. In the simplest of these cases, the heterogeneous dielectric medium can be modeled as two homogeneous dielectric mediums each occupying some distinct space between the electrode plates. In the simplest form of the 'series' model, both the dielectric materials cover the entire area of the electrode plates, but are stacked in series occupying a fraction of the gap between the electrode plates, as shown in Figure 9.1 (a) and in the simplest form of the 'parallel' model, both the dielectric materials cover the entire gap between the electrode plates, but are stacked side-by-side or in parallel formation occupying a fraction of the effective area of the electrode plates, as shown in Figure 9.1 (b). Figure 9.1 shows the equivalent circuit representations of the 'series' (Figure 9.1 (aa)) and 'parallel' (Figure 9.1 (bb)) models. The models are described mathematically as:

$$C_s = \left[(C_1)^{-1} + (C_2)^{-1} \right]^{-1} \tag{9.8}$$

and

$$C_p = C_1 + C_2 \tag{9.9}$$

where C_s is the capacitance of the parallel plate electrode containing the heterogeneous medium of two homogeneous dielectric arranged in 'series' and C_p is the capacitance of the parallel plate electrode containing the heterogeneous medium of two homogeneous dielectric arranged in 'parallel' and C_1 and C_2 in each case are corresponding capacitance of hypothetical parallel plate capacitors containing the homogeneous dielectric material 1 and 2 respectively between the parallel plates.

Textile materials are, however, much more complex than the simple models described above. Complexity in describing the model of textile or fibrous materials arise because of the facts that a) the material may itself be a composite of different fibers or filaments, impurities, moisture, other chemicals such as dye particles or pigments or finishing compounds, b) in addition to being heterogeneous, the textile materials are made of units called fibers, which are, in many cases, anisotropic and having variability, and c) the scaling up of organization from fibers to fabric normally involve many different processes and the configuration of the material takes different forms in such a way that practically modeling the system at fiber level and then up-scaling it to fabric is extremely challenging. The flexibility of textiles makes the exact geometrical description of position, orientation and conformation of the fibers constituting it under a dynamic condition very difficult. All these factors add up to indicate that predicting the dielectric properties of textile or fibrous materials with reasonable accuracy is extremely difficult.

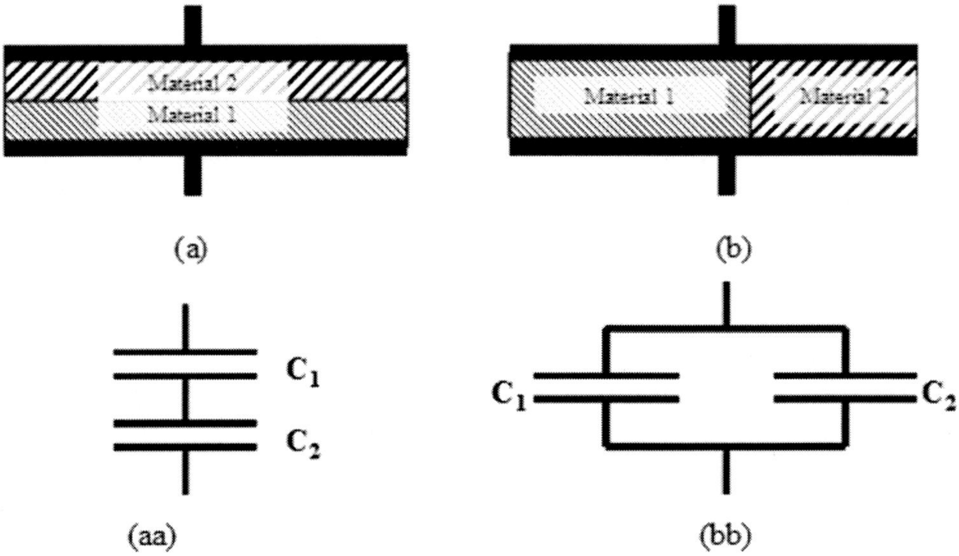

Figure 9.1. Models of heterogeneous dielectric materials a) Series b) Parallel and corresponding equivalent circuits of the series and parallel models of dielectric materials aa) series bb) parallel [Source: Reference 11].

Nevertheless, the dielectric properties of textile and fibrous materials have been studied for a long time [4]. Textile materials were some of the earliest ones to be recognized in relation to their abilities for electrical insulation [5].

In his experiments with electricity, Faraday used twines, which were probably textile materials, to insulate successive coils of conducting wires wound helically on dry wood or glass. Attempts to model the capacitor system with a textile or fibrous material in between the electrode plates have been made by few authors using analytical techniques involving the topology of the system. Some authors have tried to apply various empirical and semi-empirical formulas available for dielectric mixtures to the case of textile and fibrous materials. Some authors have used regression and extrapolation method to find out permittivity from measured capacitance. The following sub-sections briefly describe these attempts.

a) Mathematical Modeling

Mathematical consideration of capacitance was reported in literature by Boyd in relation to the description of an electronic instrument for measuring mass variation in slivers, rovings and yarns [6]. In his work, he did not consider any specific geometry of the linear fibrous structure, but started from a mathematical description of any arbitrary shape of solid fibers occupying portion of the space between the two parallel electrode plates. The situation is shown in Figure 9.2. The capacitance of an elementary area of the capacitor is given as:

$$dc = \varepsilon_0 \times \frac{dxdy}{\dfrac{L-u}{1.0} + \dfrac{u}{\varepsilon_r}} = \frac{\varepsilon_0}{L} \times \frac{dxdy}{\left\{1 - \left(1 - \dfrac{1}{\varepsilon_r}\right) \times \dfrac{u}{L}\right\}} \tag{9.10}$$

Hence the total capacitance of the system can be calculated as:

$$C = \frac{\varepsilon_0}{L} \times \int_{y_1}^{y_2} \int_{x_1}^{x_2} \frac{dxdy}{\left\{1 - \left(1 - \dfrac{1}{\varepsilon_r}\right) \times \dfrac{u}{L}\right\}} \tag{9.11}$$

Expanding the above equation and using $K = 1 - \dfrac{1}{\varepsilon_r}$, the integration as shown above leads to an expression as:

$$C = \frac{\varepsilon_0 \times A}{L} + \frac{\varepsilon_0 \times K}{L^2} \times v + R \tag{9.12}$$

where v is volume of solid fibrous material between the electrode plates and

$$R = \frac{\varepsilon_0}{L} \times \int_{y_1}^{y_2}\int_{x_1}^{x_2}\sum_{i=2}^{\infty}\left\{K^i \times \left(\frac{u}{L}\right)^i\right\} dxdy \quad (9.13)$$

For slivers and wool tops or any other similar highly porous material, $\frac{u}{L}$ may have very low value. In such cases, the higher order term R may be neglected and then the relationship between measured capacitance C and the volume v is effectively linear. For denser material such as yarns and when the gap between the plates is smaller, it was shown that the capacitance can be expressed as:

$$C = \frac{\varepsilon_0 \times a \times (b-L)}{L} + \frac{\varepsilon_0 \times a}{K \times \lambda} \times \left[\frac{2}{\sqrt{1-K^2 \times \lambda^2}} \tan^{-1}\sqrt{\frac{1+K\times\lambda}{1-K\times\lambda}} - \frac{\pi}{2}\right] \quad (9.14)$$

where λ is the porosity of the yarn itself. The relationship between C and λ was demonstrated graphically by putting numerical values of the other parameters of the above equation. It was shown that the relationship was very linear for all values of λ below 0.5 and it was argued that most practical yarns except the monofilaments fall into this category.

Capacitance of a cylindrical solid such as a yarn between two parallel plates of the capacitor was studied by Mack using mathematical modeling on the basis of purely electrostatic considerations of the system [7]. In this work, the change in capacitance δC which takes place due to the insertion of the yarn or solid cylinder was expressed in terms of the permittivity of the solid, the yarn diameter and the distance of the yarn axis from the nearest electrode plate for a fixed gap between the parallel plates. It was theoretically demonstrated that the position of the cylinder or yarn with respect to the nearest electrode plate was important, especially when the solid was very close to the plate. It was also shown that the yarn cross section, yarn density and the anisotropy of permittivity of fibers had certain influence on the measured value of capacitance.

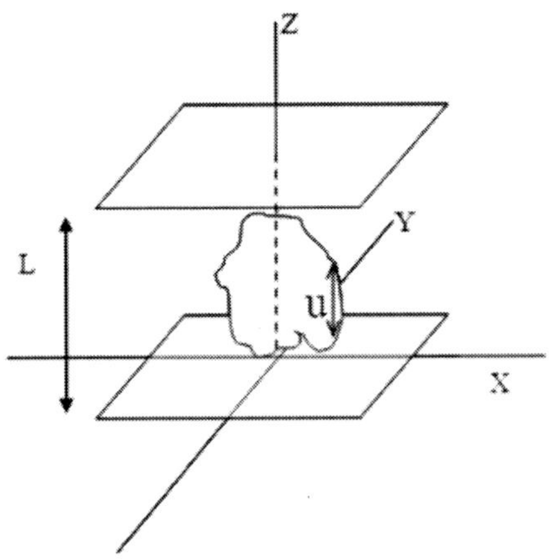

Figure 9.2. Any arbitrary shape and volume of dielectric between two parallel capacitor plates [Source: Reference 11].

The capacitance of a parallel plate capacitor with a woven fabric in between the electrode plates has been investigated recently and few models were compared for this purpose [8]. The models varied in simplicity of the assumptions, starting from a basic one in which the fabric is substituted by an equivalent polymer sheet and simple series capacitance model such as the one given by equation 9.8 could be used.

Another model considered the fabric to be substituted by an equivalent perforated sheet which effectively was a solid equivalent polymer sheet as the previous case, but this time covering a fraction of area of the electrode plates. The authors argued that this fraction could be approximated by the fractional cover of the fabric. The results indicated slight improvement for the second model over the first one for tests carried out with plain woven fabrics made from high density polypropylene monofilaments [8]. Hearle also reported similar models and applied them to calculate the permittivity and loss factor of fiber materials from capacitance measurements of yarns using a pair of specially designed conical electrode plates [9-10].

A more detailed model of capacitance of the parallel plate system with a non-crimped fabric has been developed recently [8, 11]. The unit cell of the fabric was divided into four zones as shown in Figure 9.3.

The total capacitance is the algebraic sum of the capacitance of the four zones. The capacitances of the individual zones were calculated by following the mathematical consideration given by Boyd [6] as described above. Thus, the capacitance of zone 00 is:

$$C_{00} = \frac{\varepsilon_0}{L} \times (P_1 - d_1) \times (P_2 - d_2) \tag{9.15}$$

where P_1, P_2 are yarn spacing in warp and weft directions respectively and d_1, d_2 are diameters of warp and weft respectively and L is the gap between electrode plates. The capacitance of the parallel plates with a cylindrical yarn in between the plates can be calculated from the geometry shown in Figure 9.4. From Figure 9.4, we have:

$$u_1 = 2 \times \sqrt{\left(\frac{d_1}{2}\right)^2 - x^2} \tag{9.16}$$

Similarly it can be obtained that:

$$u_2 = 2 \times \sqrt{\left(\frac{d_2}{2}\right)^2 - y^2} \tag{9.17}$$

Hence it can be shown that the capacitance of zone 01 is given as:

$$C_{01} = \frac{\varepsilon_0}{L} \times \int_{\frac{d_1}{2}}^{\left(P_1 - \frac{d_1}{2}\right)} \int_{-\frac{d_2}{2}}^{+\frac{d_2}{2}} \frac{dxdy}{\left\{1 - \frac{\sqrt{d_2^2 - 4 \times y^2}}{L} \times \left(1 - \frac{1}{\varepsilon_r}\right)\right\}} \tag{9.18}$$

Similarly the capacitance of zone 10 and zone 11 can also be calculated so that the total capacitance is expressed as:

$$C = \left(A \times n_{wp} \times n_{wt}\right) \times \frac{\varepsilon_0}{L} \times \Bigg[(P_1 - d_1) \times (P_2 - d_2) + \int_{\frac{d_1}{2}}^{\left(P_1 - \frac{d_1}{2}\right)} \int_{-\frac{d_2}{2}}^{+\frac{d_2}{2}} \frac{dxdy}{\left\{1 - \frac{\sqrt{d_2^2 - 4 \times y^2}}{L} \times \left(1 - \frac{1}{\varepsilon_r}\right)\right\}} +$$

$$\int_{-\frac{d_1}{2}}^{\frac{d_1}{2}} \int_{+\frac{d_2}{2}}^{\left(P_2 - \frac{d_2}{2}\right)} \frac{dxdy}{\left\{1 - \frac{\sqrt{d_1^2 - 4 \times x^2}}{L} \times \left(1 - \frac{1}{\varepsilon_r}\right)\right\}} +$$

$$\int_{-\frac{d_1}{2}}^{\frac{d_1}{2}} \int_{-\frac{d_2}{2}}^{+\frac{d_2}{2}} \frac{dxdy}{\left\{1 - \frac{\left(\sqrt{d_1^2 - 4 \times x^2} + \sqrt{d_2^2 - 4 \times y^2}\right)}{L} \times \left(1 - \frac{1}{\varepsilon_r}\right)\right\}} \Bigg] \tag{9.19}$$

Non-crimped fabrics are being used to make polymer composites. The above method can be used to calculate the capacitance of such materials. However, the authors did not study any such material experimentally and therefore the accuracy of this model is yet to be established in practice for real composite structures containing non-crimped fabric reinforcement. The authors used this model as an approximation in order to simplify the structure of a crimped woven fabric and it was found that the results gave better accuracy than the previously described models based on equivalent polymer films [11].

A further detailed model of capacitance of a capacitor system with woven fabrics was developed considering fabric geometry involving yarn interlacement. In a similar way as the previous case, the unit cell was divided into four capacitance zones and the capacitance value for each zone was calculated. The detailed mathematical treatment is reported elsewhere [8, 11]. It was shown that this model gave better results compared to the previous three models for plain woven fabrics made from high density polyethylene monofilament yarns [11]. However, it is expected that such detail of the structure would be even more important for much higher frequencies where the wavelength of the electromagnetic wave becomes close to the dimension of a unit cell or even less. For higher wavelengths, homogenization concept should be applicable [12]. This aspect is further discussed in details in the subsequent section.

Theoretical investigation to develop model of dielectric properties of fibrous materials has been reported for fiber reinforced or fabric reinforced composite materials by few authors which follows a different route.

For example, Chin and Lee have developed an electrical R-C circuit model to predict the dielectric properties of fiber reinforced polymer composites [13]. The R-C circuit has resistance components (R) to account for the dielectric loss and capacitance components (C) to account for the permittivity. They considered the dielectric anisotropy of fibers and derived binary mixture rules for different orientation directions of the parallel fibers with respect to the applied direction of the electrical field. Their modeling approach is depicted briefly in Figure 9.5. Figures 9.5 (a) and (b) shows the representative volume elements (RVE) with the fiber axis along and perpendicular to the electric field respectively. The equivalent circuits for permittivity modeling are shown in Figures 9.5 (aa) and (bbb) respectively. The first case is a simple one of parallel connection of the capacitance corresponding to the matrix (C_m) and that corresponding to the fiber (C_{f1}). The second case is schematically broken down into a network of capacitors as shown in Figure 9.5 (bb) which leads eventually to the circuit shown in Figure 9.5 (bbb). The final model of capacitance for any given orientation of fibers with respect to the applied electric field is obtained by arranging the respective RVEs accordingly in either series or parallel or a combination of both.

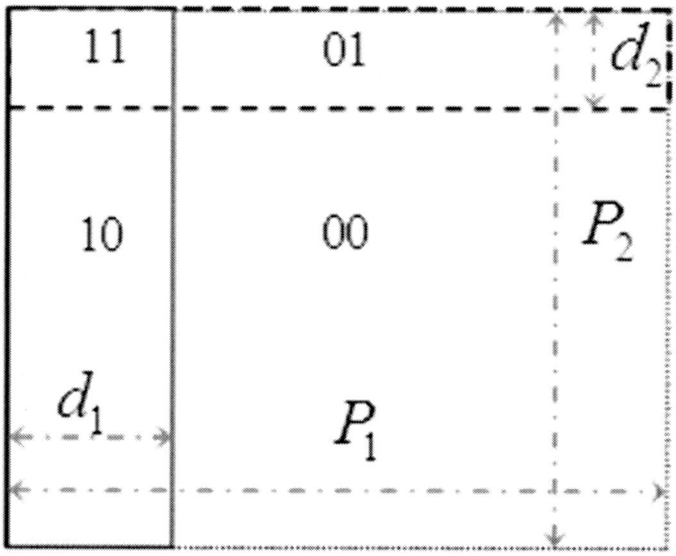

Figure 9.3. Schematic diagram of four capacitance zones (11, 10, 01 and 00) of unit fabric cell [Source: Reference 11].

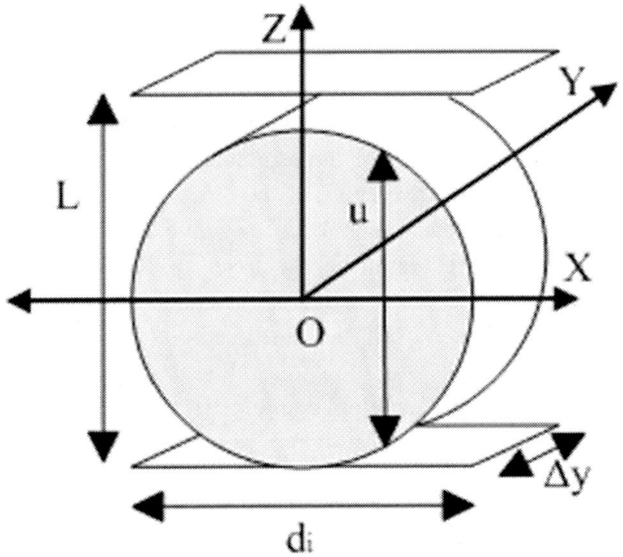

Figure 9.4. Geometry of cylindrical yarn between capacitor plates [Source: Reference 11].

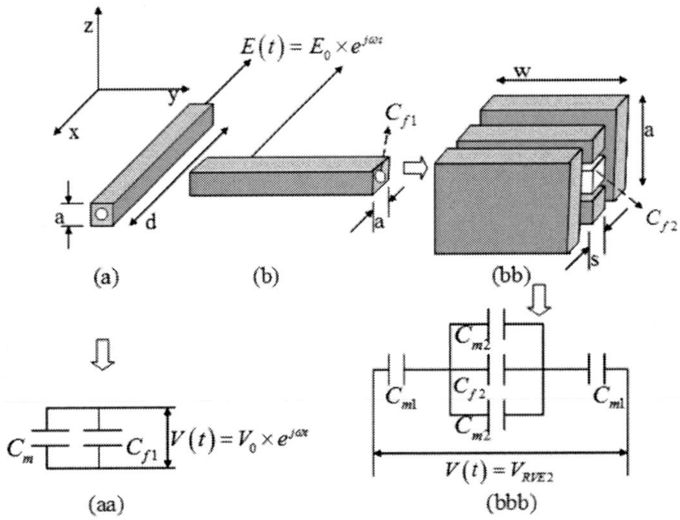

Figure 9.5. RVE of E-glass/epoxy composite and equivalent circuits: (a) schematic diagram of electric field along fiber axis, (b) schematic diagram of electric field perpendicular to fiber axis, (bb) schematic diagram of the second case with the RVE broken down in 5 parts, (aa) equivalent circuit model for the first case composed of parallel capacitances and a time-varying voltage source, (bbb) equivalent circuit model of second case [Redrawn from Reference 13]

The above approach is valid for unidirectional reinforced composites but in other cases such as woven fiber reinforced composites different approaches may be required. One approach has already been discussed above which incorporates the weave structural details either in non-crimped or crimped configurations. Yao et al. have reported another approach which is in line with the one just described [14]. In case of 3D woven composites, the general description of the model is pictorially depicted by the RVE and the corresponding circuits shown in Figures 9.6 (a) and (b), respectively. The RVE is assumed to have smaller cell elements arranged in n columns, m rows and l layers as shown in Figure 9.6 (a). The cross section of the RVE perpendicular to the electric field consists of $(n \times m)$ sub-areas denoted by A_{ij}, $i = 1,2,3,...$ and $j = 1,2,3,...$ and the thickness of the layers is denoted by d_k, $k = 1,2,3,...$ thus the entire RVE consists of a total of $(n \times m \times l)$ cells which may be represented by a capacitance C_{ijk}. This leads to the design of the equivalent circuit shown in Figure 9.6 (b). Using this approach, the authors studied the cases of single fiber type and hybrid composites and showed that the theory gave good agreement with experimental results for single fiber type composites.

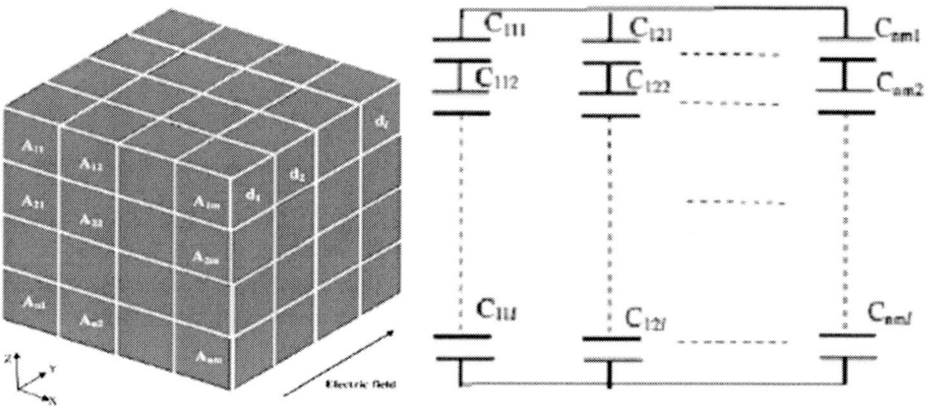

(a) (b)

Figure 9.6. (a) Schematic diagram of the RVE of 3D orthogonal woven composite and (b) its equivalent circuit [Source: Reference 14].

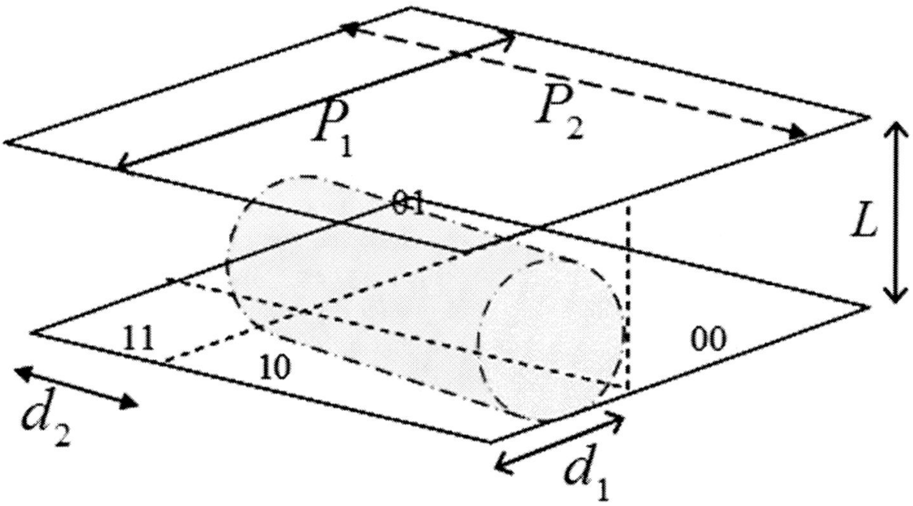

Figure 9.7. Schematic of solid volume in zone 10 of the unit cell [Source: Reference 11].

b) Empirical and Semi-Empirical Modeling

Textile materials, in general, are heterogeneous mixtures of fibers and air. Therefore the capacitance measurement method in most cases involves measurement of the capacitance of this air-fiber mixture. How the individual dielectric permittivity of air and fiber (in case of air-fiber mixture), vacuum and fiber (in case of studies in a vacuum chamber) or resin and fiber (in case of fiber and fabric reinforced composites) contribute to the overall capacitance of the mixture is a particular problem which belongs to a more general problem of dielectric properties of multi-component mixtures. The prediction of dielectric properties of multi-component mixtures has remained a topic of interest throughout the development of dielectric theories [1]. The subject is important because in reality most of the actual materials which are used for various purposes consist of more than a single component. In studies of dielectric properties of colloids and solutions, powders and crystals, food grains and food materials, soils and rocks and other geological materials, fiber reinforced composites, biological substances such as cells and tissues, polymer blends, polymers with carbonaceous fillers including carbon nanoparticles, hygroscopic materials such as wood and cellulosics, porous materials, laminated materials, papers and textiles, one of the main challenges in engineering is to predict the effective permittivity of the mixture from the knowledge of permittivity of the constituents. Alternatively, the problem may be one of interpretation and estimation of the permittivity of at least one of the constituents from the measured effective permittivity of the mixture.

In general, the problem is related to the dimension scale in which the material is being studied. When the particle size is sufficiently lower in dimension than the electromagnetic wave length, a homogenization approach can be made [12]. The homogenization approach is more convenient and practical at times especially because it requires very few inputs regarding the material organization, the volume fraction being the principal of them, and it yields to a relatively macroscopic description of the dielectric property, namely, the effective permittivity of the mixture.

The problem of predicting the overall dielectric permittivity of a heterogeneous mixture from the permittivity of its constituents, which is more commonly known as 'mixing law', has been approached by various methods ranging from purely analytical treatments for idealized topologies to empirical approximations [2]. Recently, numerical simulations using finite elements, finite differences and boundary integrals have also been explored [1, 2, 12, 15-20]. A lot of literature is there concerning each of these methods and with the advancement of our observational capabilities from micro to nano-scale and the advancement of our computational power exceeding tera-hertz speed, the old problem of predicting dielectric permittivity of heterogeneous mixture is getting new dimensions and new solutions.

The various methods used by researchers to predict the permittivity of the mixture from the permittivity of the components are often broadly divided into the following: bounding methods, effective medium approximation, percolation theory and numerical simulation. The various formulas obtained using many of these methods have been reviewed in a number of articles [2, 12, 15].

The most general bounds were given by Wiener on the basis of an analysis of laminated structures in which case the two phases or components are arranged in alternating order [21-22]. The system thus corresponds to capacitors connected in series and in parallel in a circuit. The lower bound is obtained in the case when the laminates are stacked with the normal to their contact surfaces parallel to the applied electric field and is given by equation 9.20 and the upper bound is obtained in the other extreme case when the same normal is aligned in perpendicular to the applied electric field and is given by the equation 9.21:

$$\varepsilon_{WL} = \frac{\varepsilon_r}{f + (1-f) \times \varepsilon_r} \tag{9.20}$$

$$\varepsilon_{WU} = (1-f) + f \times \varepsilon_r \tag{9.21}$$

Here f is the volume fraction of the solid dielectric material (for an air-solid mixture) and ε_r is the relative dielectric permittivity of the solid inclusion. For any arrangement of the components within the space between the parallel electrode plates, the permittivity value of the mixture lies within these two limits. A more strict set of limits was proposed by Hashin and Shtrikman [23]. Their bounds are given by equations 9.22 and 9.23 in case of an air-solid mixture:

$$\varepsilon_{HSL} = 1 + \frac{f}{\frac{1}{(\varepsilon_r - 1)} + \frac{1}{3} \times (1-f)} \tag{9.22}$$

$$\varepsilon_{HSU} = \varepsilon_r + \frac{(1-f)}{\frac{f}{3 \times \varepsilon_r} - \frac{1}{\varepsilon_r - 1}} \tag{9.23}$$

Bounding methods are sometimes useful especially in situations where no details regarding the topology of the dielectric components of the mixture are known. The bounding limits can be further narrowed down if further information regarding the topology of the inclusions is available. The effective medium approximation assumes that each constituent is surrounded by an 'effective medium' and the dielectric property of the effective medium is such that it is macroscopically equivalent to the real heterogeneous medium [1]. Various formulas have been proposed using this approach such as Maxwell Garnett formula [24], Rayleigh mixing formula [25], Böttcher mixing formula [26], Polder-van Santen formula [27], Taylor's formula [28], Bruggeman's EMT rule [29] etc. Almost all of these formulas can be generalized to the same expression shown in equation 9.24.

$$\frac{(\varepsilon_m - 1)}{(\varepsilon_m + 2) + \nu \times (\varepsilon_m - 1)} = f \times \frac{(\varepsilon_r - 1)}{(\varepsilon_r + 2) + \nu \times (\varepsilon_m - 1)} \tag{9.24}$$

where ε_m is the dielectric permittivity of the mixture. Sihvola [30] has shown that for integer values of (ν = 0, 1, 2, 3), equation 9.24 takes the form of Rayleigh mixing formula $(\nu = 0)$,

Bruggeman's formula $(\nu = 1)$, Böttcher's formula and Polder-van Santen formula $(\nu = 2)$ and coherent potential formula $(\nu = 3)$.

Several authors have derived formula for dielectric permittivity of a two component mixture from differential analysis. Most of them considered the excess polarization due to a small sphere introduced to the effective medium where the volume of the sphere can become infinitesimal enabling establishment of a differential equation for the effective permittivity of the mixture [30]. The particular solutions obtained for a 3D case by three of the most notable authors contain a fractional power of a third. These are Looyenga formula [31], Bruggeman [29] and Hanai formula [32] and formula by Sen et al. [33] given in equations 9.25, 9.26 and 9.27, respectively.

$$\varepsilon_m^{\frac{1}{3}} = (1-f) + f \times \varepsilon_r^{\frac{1}{3}} \tag{9.25}$$

$$\frac{(\varepsilon_r - \varepsilon_m)}{(\varepsilon_r - 1)} = (1-f) \times \varepsilon_m^{\frac{1}{3}} \tag{9.26}$$

$$\frac{(\varepsilon_m - 1)}{(\varepsilon_r - 1)} = f \times \left(\frac{\varepsilon_m}{\varepsilon_r}\right)^{\frac{1}{3}} \tag{9.27}$$

Formulas similar to the above are also there with different fractional powers. For example, formula given by Birchak et al. [34] use a square root form shown in equation 9.28.

$$\sqrt{\varepsilon_m} = (1-f) + f \times \sqrt{\varepsilon_r} \tag{9.28}$$

Calame [16] has proposed a numerical simulation method to solve an equation of the similar form as Hanai equation with power of n-th fraction, as shown in equation 9.29, and has described a method of empirically determining the value of n using least-squares fitting. Equation of the same form as equation 9.29 has also been discussed by Tuncer et al. [17] in context of effect of inclusion shape factor on the permittivity of the mixture.

$$\frac{(\varepsilon_m - 1)}{(\varepsilon_r - 1)} \times \left(\frac{\varepsilon_r}{\varepsilon_m}\right)^{\frac{1}{n}} = 1 - f \tag{9.29}$$

Another mixing formula was given by Lichtenecker [35-37] which can have two forms given by equations 9.30 and 9.31. Although earlier it was considered to be an empirical equation, but now it has been established that this equation can be theoretically derived from effective medium theory using a beta-function distribution [38]. It has also been shown that Lichtenecker's equation describes a specific topology of the composite mixture which may be represented as a spheroid system and it is different from the topology described by

Bruggeman [39]. This equation is reported to predict successfully the behavior of a disordered porous material [40-42].

$$\varepsilon_m^k = (1-f) + f \times \varepsilon_r^k \qquad (9.30)$$

where $-1 \leq k \leq 1$ and for $k \rightarrow 0$ it can be expressed as:

$$\ln \varepsilon_m = f \times \ln \varepsilon_r \qquad (9.31)$$

It can be seen from equations 9.25, 9.28 and 9.30 that for a single dielectric mixture in air, the form of Lichtenecker's equation becomes same as that of Looyenga's formula and Birchak's formula for *k* equal to 1/3 and 1/2, respectively. It has been pointed out by various authors that the value of k may be related to the inclusion topology [39-41, 43].

Some of the researchers who have worked on dielectric characterization of textile fibers have used few of the mixture formulas available at their time. Hearle described the problem of interpretation of results from measurement of air-fiber mixture and reviewed few of the relationships between fiber volume fraction and effective permittivity of the mixture [9, 10]. He mentioned that a combination of two formulas, one being that for the case where the solid dielectric covers the entire area of the electrodes but only a fraction of the gap between the plates (for an air-solid two phase system this would be similar to the lower bound proposed by Wiener) and the other being Lichtenecker's logarithmic equation. However, in his paper [9], he did not give information regarding exactly how these two formulas were combined mathematically. He claimed that results obtained by him were similar to, although not in exact quantitative agreement with, the combination of the two equations. A preference towards Lichtenecker's equation was obtained by Shaw and Windle [44] while working on microwave cavity resonator techniques with glass fibers in comparison with classical Lorentz equation [45].

Studies on the woven fabrics using plain weaves made from high density polypropylene monofilaments were carried out using ASTM D150 standard method [46]. Four fabrics were used, each of which had different filament diameter, yarn spacing, thickness and areal mass density. Each of these fabrics was kept between the capacitor plates and the corresponding volume fraction was calculated from the fabric areal density, density of the fiber and the electrode gap. The effective permittivity of the capacitor system with the fabric was calculated from the recorded capacitance value and the over-all capacitor dimensions using equation 9.1. From this effective permittivity (ε_m) and the effective volume fraction (f), the relative permittivity (ε_r) of the fiber was calculated following the various mixture equations and power-law equations described in this section above. The average values for results obtained from all the mixture equations ranged between 2.028 and 2.4. The dispersion of standard deviations of the calculated permittivity for almost all the dielectric mixture equations was of the same order as the standard deviation obtained in case of measurement with the Teflon sheets. Apparently, the difference of fabric structures for the four different woven fabrics did not affect the results significantly under the measurement set-up. It appeared that instead of using the exact fabric geometry, the fabric volume fraction, calculated on the basis of fabric areal mass density and the fiber density alone could be used to calculate the permittivity of the fiber from the effective permittivity of the air-fabric mixture [11, 46]. However, which of the dielectric mixture formulas should be chosen could be decided.

A hybrid model to predict the effective permittivity of plain woven fabrics was developed recently on the basis of mixing law equations and the fabric geometry [11]. In this model a fabric with plain weave was considered and the unit cell or repeat unit of such fabric from the point of view of its effective permittivity was the same as the unit shown in Figure 9.3. The homogenization was introduced in this model in each of the four individual zones of the unit cell. Thus each of the four zones denoted as 00, 01, 10 and 11 had its own effective permittivity ε_{ij} , where $i, j \in \{0,1\}$, 1 indicated presence of yarn, 0 indicated absence of corresponding yarn and i, j denoted warp and weft yarns respectively. The effective permittivity ε_{ij} was calculated from the solid volume fraction of fibers f_{ij} in the respective zone using a mixing equation. The over-all effective permittivity of the fabric was then calculated by adding the individual zones following a parallel capacitance model.

The volume fraction of solid fibers in each zones were calculated as follows:

Volume fraction of fiber in zone 10:

The case is schematically shown in Figure 9.7 where the shaded volume represents the solid cylindrical fiber (monofilament yarn).

$$f_{10} = \frac{\text{volume of fibre in zone 10}}{\text{volume of zone 10}} = \frac{\pi \times \left(\frac{d_1^2}{4}\right) \times l_{10}}{d_1 \times (P_2 - d_2) \times L} \tag{9.32}$$

where l_{10} is the length of the cylindrical yarn in the zone, d and P denotes yarn diameter and yarn spacing of the corresponding yarn, suffix 1 denotes warp and suffix 2 denotes weft. The value of l_{10} is estimated in the following way.

With reference to Figure 9.8, it is assumed that the portion of the yarn axis length l within interlacement zone 11 is approximately equal to the length of BC where:

$$BC = \frac{1}{2} \times (d_1 + d_2) \times 2\theta = (d_1 + d_2) \times \theta \tag{9.33}$$

Using the standard relation [8]:

$$\theta_i = 1.849 \times \sqrt{c_i} \tag{9.34}$$

We have:

$$BC=(d_1+d_2)\times 1.849\times\sqrt{c_1} \tag{9.35}$$

The total length of warp yarn in the unit cell is given by:

$$l=(1+c_1)\times P_2 \tag{9.36}$$

Hence, the length of warp yarn in zone 10 is:

$$l_{10}=BD-BC=l-BC=P_2\times(1+c_1)-(d_1+d_2)\times 1.849\times\sqrt{c_1} \tag{9.37}$$

Therefore:

$$f_{10}=\frac{\pi\times d_1\times\left\{P_2\times(1+c_1)-(d_1+d_2)\times 1.849\times\sqrt{c_1}\right\}}{4\times(P_2-d_2)\times L} \tag{9.38}$$

Similarly, it can be shown that the solid fiber volume fraction in zone 01 is given by:

$$f_{01}=\frac{\pi\times d_2\times\left\{P_1\times(1+c_2)-(d_1+d_2)\times 1.849\times\sqrt{c_2}\right\}}{4\times(P_1-d_1)\times L} \tag{9.39}$$

The volume fraction of solid fibers in zone 11 can be calculated as:

$$f_{11}=\frac{\text{volume of warp in zone 11}+\text{volume of weft in zone 11}}{\text{volume of zone 11}}$$

Volume of warp in zone 11 is the volume of the curved cylinder of diameter d_1 having length equal to BC as shown in Figure 9.8. The length BC is given by the equation 9.35. The volume of weft in zone 11 can also be calculated analogically. Therefore, it can be obtained that:

$$f_{11}=\frac{\pi\times 1.849\times(d_1+d_2)}{4\times L}\times\left\{\frac{d_1}{d_2}\times\sqrt{c_1}+\frac{d_2}{d_1}\times\sqrt{c_2}\right\} \tag{9.40}$$

And obviously, the volume fraction of solid fibers in zone 00 is given as:

$$f_{00}=0 \tag{9.41}$$

Now, according to the various mixing laws for effective permittivity of a mixture of solid and air, the effective permittivity of the mixture is a function of the solid volume fraction. In other words:

$$\varepsilon_{ij} = F\left(\varepsilon_r, f_{ij}\right) \tag{9.42}$$

where ε_{ij} is the effective permittivity of zone ij, $i, j \in \{0,1\}$, and i denotes warp, j denotes weft as already mentioned before. The function F depends on the choice of mixing equation.

Hence, the overall effective permittivity of the fabric is given as:

$$\varepsilon_m = \frac{(P_1 - d_1)\times(P_2 - d_2) + (P_2 - d_2)\times d_1 \times \varepsilon_{10} + (P_1 - d_1)\times d_2 \times \varepsilon_{01} + d_1 \times d_2 \times \varepsilon_{11}}{(P_1 \times P_2)} \tag{9.43}$$

The results of applying the hybrid model, when compared to the results obtained by the mixing laws straight-forwarded, did not show any remarkable difference although the values obtained using the hybrid model were marginally higher in each case [11].

However, due to the nature of the assumptions made in this model, particularly due to the approximation made in calculating the length of yarn in the individual zones, the calculated volume fraction was under-valued to some extent. But on the other hand, the dispersion of values for each mixing rule had increased remarkably. It appeared that the weave structure of the fabrics was not appreciably influencing the measured values under the measuring conditions.

A concept of homogenization is therefore valid. This is also logical as the equivalent wavelength of 1 KHz signal is approximately 3×10^5 meters which is much higher than the size of the repeat unit of the fabrics. At very high frequencies (> 10 GHz), the effects of weave may start to show more.

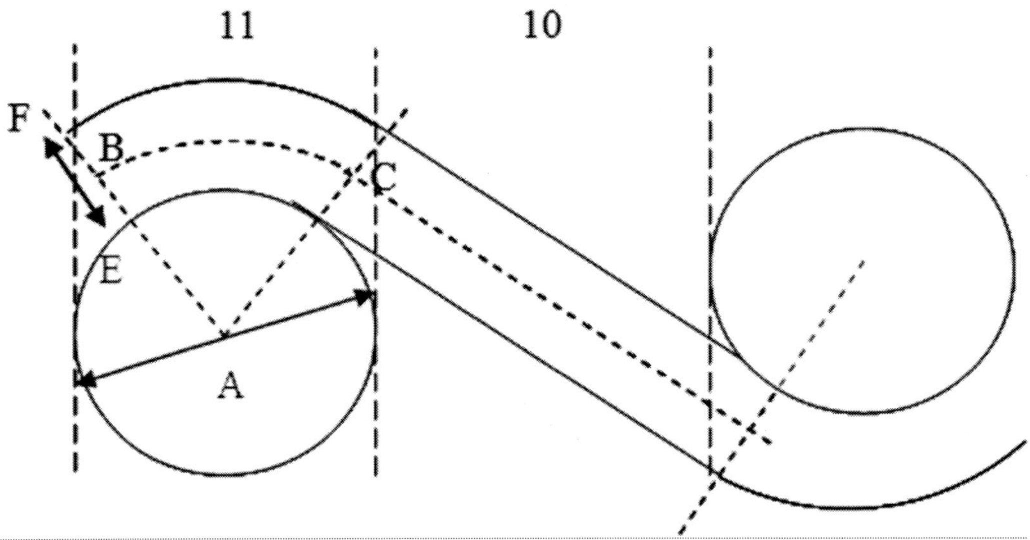

Figure 9.8. Geometry of the structure (longitudinal cross section at the plane containing warp yarn axis) at zone 10 [Source: Reference 11].

Few other models are also there on literature to predict permittivity of composites having fiber or fibrous inclusions. For example, Davies studied this problem theoretically and derived approximate formulas for both isotropic and axially symmetric cases [47-48]. Various formulas were derived corresponding to different geometries of the fiber and matrix phases. However, experimental validation was not given in the study by the author.

In general, for fiber or fabric reinforced composite materials, mixing law equations can be applied provided the homogenization concept is valid for the system. In such cases, the volume fraction of fibers in a two phase composite of which the polymer matrix forms the other phase can be related to the effective permittivity in a similar manner as the case of mixture of fibers and air and similar results may be expected for same geometrical configuration of the solid fibers provided the permittivity of the matrix material is also very low. Interface in such fiber reinforced polymer composites may become important in certain cases, especially for very high frequency measurements such as microwave dielectric measurements near X-band frequency. For example, an interlayer model has been developed for such composites by considering that a third component representing the interface has enveloped the inclusions in a separate layer [49]. Various shapes of the inclusion were considered including long needle or fibers. The formula which was derived describes the complex permittivity of the composite as a function of the volume fraction of the components, their permittivity values and the shape factor of the solid inclusion. The effect of the interface layer on the static permittivity of the composite was discussed. The special case of a conducting interface layer such as a layer of adsorbed water was also discussed in the light of this model.

The role of absorbed water is very important for the dielectric properties of fibrous materials. Most of the textile commodity fibers, natural or manmade, contain moisture in varying quantities. Moisture or water molecules may also be trapped within the porous structure of the fibrous material due to condensation or incomplete drying. The dielectric properties of water are markedly different from those of fiber forming polymers. A remarkable increase in measured value of permittivity has been reported when the fibrous material is measured at humid condition [10]. It is natural, therefore, that the role of adsorbed water is investigated in a number of papers on dielectric properties of fibrous materials. In fact, the role of moisture in porous materials has been studied by researchers in context of many other fields too. Among them, geological materials such as soil and rock have been studied by many authors [50-57]. Most of them were at microwave or radio-frequencies. Food materials containing moisture have also been studied by many authors at microwave frequencies [58-59]. Wood [60-62], inorganic and organic powders or crystals [63-64] and paper [60, 65-66] are some other porous materials containing moisture which have been studied at various frequencies by many authors. The role of moisture in fibrous porous materials such as yarns and fiber bundles and also in single fibers has also been studied by few authors at various frequencies [9, 10, 44, 67-77]. A number of authors have given models for dielectric properties of wet porous materials but it has been recognized that arriving at a general model in such case which would apply to all type of porous materials under various testing conditions is not likely due to the complicated nature of the system. De Loor [50] has discussed a bounding method and concluded that such method may be improved with more detailed information of the structure and interaction of the components of mixture.

It has been acknowledged that the absorption and adsorption of water by hygroscopic materials lead to attachment of water molecules under different conditions. For example, in an interesting work, Teschke et al. [78] have measured the dielectric permittivity profile of interfacial water using charged mica surfaces and found that the permittivity increases with distance from the solid surface. Some models have been proposed based on a bound and unbound layer whereas some models distinguished between the bound, mobile and liquid (or capillary) water in the porous structure. Accordingly, three phase and four phase models have been proposed and applied to explain results. The effect of bound layer becomes more significant in high frequency measurements such as microwave measurement. In case of wool fibers, Windle and Shaw [70-71] have shown that a concept of localized water and mobile water (similar to the bound and unbound layers) based on Cassie's multilayer absorption theory [79] can be successfully applied to explain the behavior of wool-water system at microwave frequencies.

At low frequencies the dielectric behavior of mixtures containing water may be different and perhaps more complicated as has been indicated and discussed by de Loor [50]. In context with fibrous structures such as yarns, Hearle has demonstrated that the dielectric behavior at low frequencies differ from those reported by Shaw and Windle at microwave frequencies [9].

c) Curve Fitting and Extrapolation Technique

Some researchers have reported techniques involving least square fitting with experimental data and extrapolation method in order to calculate permittivity from the experimental plots of effective permittivity vs. solid volume fraction. Nelson [80] has mentioned some of these approaches involving complex permittivity in case of granular materials. These include a quadratic least square fit used by Kent [81] and linear extrapolation of $\sqrt{\varepsilon_m}$ with bulk density as observed by Klein [82] and also a linear extrapolation of $\sqrt[3]{\varepsilon_m}$ with bulk density for air-solid mixture as obtained by Nelson [80]. In case of air-fiber mixture, extrapolation technique has also been used by Hearle [9] and by Kirkwood et al. [83]. Kirkwood et al. worked with various cotton fibers and observed that existing dielectric mixture formulas including Lichtenecker's equation gave unsatisfactory results. However, while Hearle had studied the dielectric properties with a wider material volume fraction range, Kirkwood et al.'s work was limited to low volume fraction of fiber strand only.

The porosity of a fibrous material can vary between the limits 0 and 1. However, when this fibrous material is put between the parallel plates of a capacitor, additional air may also be present depending on the gap between the plates and the sample size and this can decrease the effective volume fraction of solid in the space between the capacitor plates. In fact it is almost impossible to achieve a very high value of volume fraction in a capacitance measuring system with a fibrous material without deforming the original material. In case of woven fabrics, the porosity becomes even higher due to the pores between the yarns and in the interlaced regions. The problem with such high porosity material in case of extrapolation is that the permittivity of the fiber material, which can be obtained by extrapolating the curve to a value of 1 in the X axis showing the solid volume fraction, may have errors due to absence of real data in most part of the curve over a wide range of volume fractions between the maximum volume fraction actually obtained and the theoretical maximum value 1. Nevertheless, as was demonstrated in another study involving various plain woven fabrics made from high density polypropylene monofilaments, nylon monofilaments and staple fiber yarns, the effective permittivity of the fabric material had strong correlation with the solid volume fraction [11]. However, unlike the linear fit of data reported by Kirkwood and others, in this case a very good fit was obtained with a second order polynomial curve. Comparison with results obtained using other methods such as the mixing law equations showed that the least square fit method using second order polynomial equation gave reasonable accuracy [46].

9.4. Simulation of Dielectric Permittivity of Heterogeneous Mixtures

Simulation is imitation or reconstruction of existing or proposed system or process and may cover diverse array of subjects such as science, economics, entertainment, etc. This can be done in a virtual environment using computers or in a physical environment using specially engineered components to artificially create a state which provides more convenient means to study a real system. The convenience is sometimes in terms of time, cost, measurement and control of different variables or parameters influencing the input or outcome or a combination of these. The field of computer simulation has undergone an immense growth during the past few years and today a number of specialist software tools are available in the market for various applications in scientific and engineering research. Specialized research areas such as computational fluid dynamics, computational electrodynamics etc. have attracted a huge number of projects and research funds.

Computational simulation of the dielectric properties of heterogeneous mixtures or composites is useful for the material scientists involved in various fields. Designers and material engineers can use such simulations to try new structures with a combination of different dielectric materials to achieve a specific required property or performance in a much shorter time without actually producing the composites physically. A number of computational modeling and simulation techniques have been reported for dielectric hetero-structures [1, 2, 16-20, 84-92]. Finite Element Modeling (FEM) and Finite Difference methods are most commonly used for such purpose although other methods such as Fourier Expansion technique have also been used. Both 2D and 3D simulation have been carried out by the researchers.

A full discussion of various work done on computational modeling and simulation of dielectric properties of heterogeneous composites is beyond the scope or size of this chapter, however, a few introductory illustrations will be given here to draw the reader's interest in this subject. For example, in one of the studies, Cheng et al. [91] have considered a two phase composite material where the inclusion particles are assumed to be randomly distributed inside the matrix. The composite is visualized by a $10\times10\times10$ 3D cellular matrix structure. Figure 9.9 (a) shows such a structure for a volume fraction of 0.3 for the inclusions. The reader can recall a similar cellular division of structure reported by Yao et al. [14] which has been described earlier. The inclusion positions, shown with dark cubes, are randomized using a pseudo-random number generator. Provision of inclusion overlapping even at a very low inclusion concentration was there. FEM was used to solve the problem with the help of standard computational software. Figure 9.9 (b) shows the boundary conditions imposed on the three dimensional structure when the material was exposed to an electrostatic field. The calculation of the potential distribution for each element was carried out by dividing the domain into finite elements. The region of interest was meshed using an adaptive meshing technique by means of the software and once the potential distribution was known, the effective permittivity of the composite was calculated by the energy balance method. The results of the simulation showed that the disordered model could be applied to high permittivity contrast and large volume fractions of inclusions, while the Maxwell-Garnett rule and the conventional periodic models fail to predict the effective permittivity because they are not able to consider the overlapping of inclusions. The numerical results obtained by the disordered model were close to those obtained by the Bruggeman rule due to the clustering effects.

A procedure for numerical calculations of distributions of the time-harmonic electric field and energy losses and the effective complex permittivity of a composite dielectric with a periodic structure has been reported by Serdyuk et al. [92]. The frequency-dependent parameters of the constituents of the composite were obtained experimentally and were taken as input for the 3D numerical modeling performed using finite-volume and finite-element methods and the calculated frequency-dependent effective complex permittivity of the composite was compared with experimentally obtained data. The structure of the composite and the unit cell used for the simulation are shown together in Figure 9.10 (a) and the boundary conditions of the field problem in the unit cell are depicted in Figure 9.10 (b). An equivalent circuit was developed for solving the problem using Finite Volume Method (FVM). The result of simulations using FEM and FVM were compared by the authors and was shown to have good match with experimental data. Computational methods have been used in the field of fibrous materials mainly in relation to mechanical properties and processes [93-96]. FEM methods have been used in relation to ballistic properties of flexible and rigid textile composites, fabric drape, tensile properties of fabrics, thermal and liquid transport properties of fibrous materials, etc.

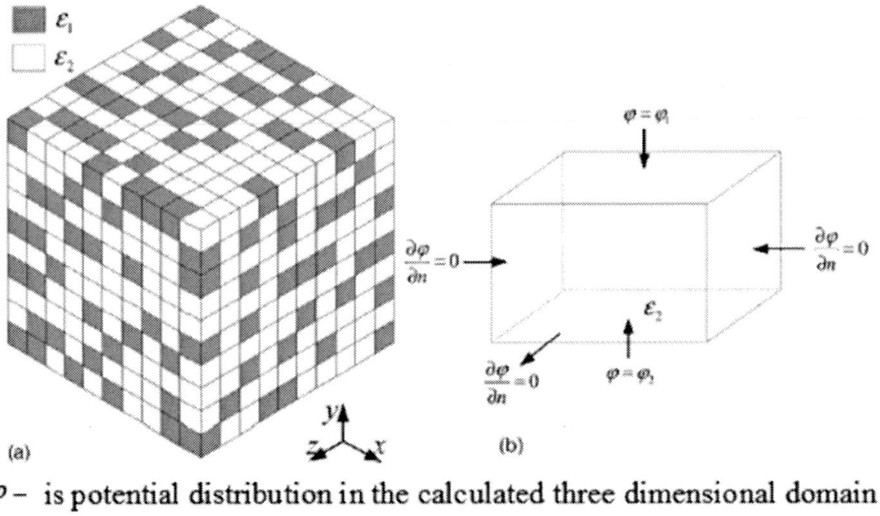

Figure 9.9. (a) A three-dimensional cubic structure with 30% volume fraction of inclusions. It is composed of $10 \times 10 \times 10$ cube cells in which the inclusions occupy the positions randomly in the matrix material, (b) Boundary conditions related to the three-dimensional structure [Source: Reference 91].

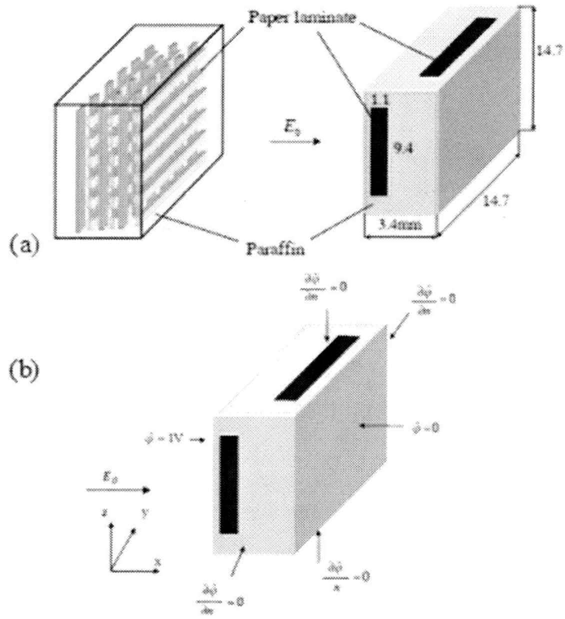

Figure 9.10. (a) The structure of composite material and the unit cell used for simulation (E_0 is the applied external electric field), (b) Boundary conditions related to the unit cell structure [Source: Reference 92].

However, literature on computational modeling and simulation of dielectric or electrostatic properties of fibrous textile materials is not yet available. The obvious challenges involved in this are tackling the mechanical flexibility of the fibers, the complex hierarchical architecture of textiles, identifying or developing a suitable meshing technique to generate the 3D model of the structure having several degrees of freedom, etc. Relating established analytical and empirical models to the computational models for dielectric properties of textile structures would also be an interesting research option.

9.5. Modeling and Prediction of EMI Shielding Behavior of Fibrous Materials

a) Fundamental Concepts of Electromagnetic Shielding

Electromagnetic interference (EMI) shielding has been defined in a broad way as any means used for the reduction of the radiated electromagnetic field in a prescribed region [97]. Electromagnetic compatibility (EMC) is the capability of electronic equipment or systems to be operated in the intended electromagnetic environment at design levels of efficiency and in general, electromagnetic shielding represents a way to improve the EMC performance of electronic equipments and devices. A similar need of electromagnetic shielding is to protect biological systems from unwanted health hazards caused by electromagnetic interference. Electromagnetic shielding is also used to ensure information security over various communication channels and networks through wired and wireless media.

The few basic ways in which electromagnetic interference may be shielded have generally been categorized into the following strategies [97]: using a barrier between the source of unwanted electromagnetic (EM) waves and the target object or space to be protected diverging the EM field from the victim object or space of interest using a counter source to reduce the effective level of intensity of EM field vectors in the prescribed space.

EMI shield in the form of a barrier material may be used to cover either the source of EMI or the object susceptible to EMI (which requires protection) or both. A number of different varieties of materials have been used as shielding materials with a wide range of electric conductivity and magnetic permeability and various type of geometry of such shield have been tried for this purpose. In most cases, the shield constitutes a shaped conducting material. It either reflects the incident EM wave or transmits it to the ground. Michael Faraday first developed this kind of shielding principle with respect to electric field, which is now known as 'Faraday cage' [98]. A magnetic material with very high magnetic permeability is required to shield low frequency magnetic field.

The shielding effectiveness depends on the frequency of the radiation, the distance from the source, the thickness of the shield and the shielding material. However, most of the shields in practice have to allow attachments to cables, wires or other inserts and joints and seams which lead to having apertures and gaps in the shield, thus breaching the shield integrity. Therefore, the overall design of the shield also affects the actual efficiency of the shield.

The shielding effectiveness (SE) is expressed as a function of the logarithm of the ratio of two electric or magnetic field or plane wave field intensities or power, usually expressed in decibels (dB). The most common definition of SE at any given point of the region of interest is that it is the ratio between the absolute value of the electric (or magnetic) field $\vec{E}$ (or $\vec{H}$) that is present at that point in space when the shield is present and the absolute value of the electric (or magnetic) field $\vec{E}_i$ (or $\vec{H}_i$) that would have been present at the same point in the absence of the shield, expressed in dB. In order to obtain positive values, normally the reciprocal of this definition is considered for calculation. Thus the following equations are used for calculating SE in electric and magnetic shielding, respectively:

$$SE_E = 20\log\frac{|\vec{E}_i|}{|\vec{E}|} \tag{9.44}$$

$$SE_H = 20\log\frac{|\vec{H}_i|}{|\vec{H}|} \tag{9.45}$$

The total shielding effectiveness of a shield is the sum of absorption factor (*A*), reflection factor (*R*) and a correction factor (*B*) to account for multiple reflections in case of thin shields. Thus:

$$SE = A + R + B \tag{9.46}$$

The absorption factor is due to the absorption losses which depend on the physical properties of the shield material. The distance required for an electrical wave to be attenuated to 37% (or e^{-1}) inside the material is defined as the skin depth δ of the material which can be expressed as:

$$\delta = \frac{1}{\sqrt{\pi \times f_{Hz} \times \mu \times \sigma}} \tag{9.47}$$

where f_{Hz} is frequency of the incident wave (in Hz), μ is relative permeability (compared to copper) and σ is conductivity relative to copper in IACS. Thus, for a material with thickness t, the absorption factor is given as:

$$A = 131 \times t \times \sqrt{f_{MHz} \times \mu \times \sigma} \tag{9.48}$$

The reflection loss factor is computed by considering the electric, magnetic and plane wave reflection losses separately according to the following set of equations:

$$\left.\begin{aligned} R_E &= 321.8 + 10\log\frac{\sigma}{f_{Hz}^{\ 3} \times r^2 \times \mu} \\ R_H &= 14.6 + 10\log\frac{f_{Hz} \times r^2 \times \sigma}{\mu} \\ R_P &= 168 - 10\log\frac{f_{Hz} \times \mu}{\sigma} \end{aligned}\right\} \tag{9.49}$$

where R_E, R_H and R_P are reflection losses for electric, magnetic and plane wave fields respectively, expressed in dB, f_{Hz} is frequency in Hz and r is the distance of the shield from the source in m.

The computation of the correction factor B can be carried out as [98]:

$$B = 20\log\left|1 - \frac{\left(1.3 \times \sqrt{\dfrac{\mu}{f_{Hz} \times r^2 \times \sigma}} - 1\right)^2}{\left(1.3 \times \sqrt{\dfrac{\mu}{f_{Hz} \times r^2 \times \sigma}} + 1\right)^2} \times \left(10^{-\frac{A}{10}}\right) \times e^{-j \times 227 \times A}\right| \tag{9.50}$$

b) Modeling of Practical Electromagnetic Shields

The topology or geometric design of the shield affects the shielding effectiveness and modeling this effect of the shield geometry is important in designing better EMI shielding devices and enclosures. In practice, most of the electromagnetic shields are far from a continuous block of solid material. Therefore, mathematical description of such materials must take into account the major imperfections and leakages. Conventionally, the basic problem can be analyzed by calculating the shielding effectiveness of a metallic enclosure which obviously indicates the best-case shielding effectiveness since no leakages or aperture effects are considered in such cases. In case of shielding enclosures with cable inserts, EMI gaskets are used, and a second modeling problem may be to model such gaskets. For more practical considerations, the shield with apertures needs to be modeled. Brief accounts of such models are given in references 98-99. The cases of rectangular holes, circular holes, metallic mesh or wire screens and waveguide or honeycomb structures have been discussed by the authors. It was generally observed that in case of perforated conducting surfaces, more number of holes, each having smaller dimensions, are more effective in shielding than less number of holes of bigger diameter even for same degree of optical cover.

Substantial work has also been done on modeling the EMI and EMC phenomena itself. Unlike the models of the shields, these models try to describe and predict the various EM noise and emission spectra. These are also useful for designing shielding as these models provide easy understanding of the frequency and intensity of the EMI which is to be shielded. For example, statistical physical approaches based on probability theory have been discussed and reported in literature in this context.

Mathematical models based on probability have been developed to analyze the system as a whole on the basis of the statistical behavior of the EMI [100]. These models are useful in developing computer aided design and analysis procedures to solve EMI problems. Computational models of EMI and EMC problems from circuit level to a more complex system levels such as aircraft EMC have also been extensively studied. Techniques such as FEM, method of moments (MoM), Transmission Line method (TLM), Finite Difference Time Domain method (FDTD), Finite Difference Frequency Domain method (FDFD), Partial Element Equivalent Circuit model (PEEC) and a number of such methods have been used for this purpose for various applications. Interested readers can see reference 100 for a nice review of such approaches.

c) Modeling of Textile Based EMI Shields

Textile based shielding materials are interesting options for the equipment designers. Such materials may be woven or knitted fabrics or braids in single or multiple layers. The shielding is achieved in such materials mainly by incorporating metallic yarns, metal-coated yarns, yarns made from conducting fibers, by conductive coating of fabrics, etc. Although a number of literatures have been reported on the development of such materials, most of these works only relied on the experimental or direct measurement of shielding effectiveness and some empirical study of the effects of few design parameters.

From a structural point of view, fabrics may come close to the cases of sheets with many perforations and also to the wire mesh [99]. One can, therefore, apply the models for these simple structures to the case of knitted and woven fabrics as an approximation. However, it was shown in a study that the behavior of woven fabric made from yarns having metal fibers was markedly different from that of a metallic foil having similar thickness and matching level of perforations [101]. This type of difference may be due to imperfect contact between the conducting fibers within a yarn as well as between the interlaced yarns. In that case, a suitable model incorporating such contact resistance within the wire elements might be necessary.

The author, however, attributed the observed mismatch to a combination of aperture shielding and waveguide shielding, arguing that a textile fabric was unlike a metal thin foil, the fabric having a thickness which renders the pores between the yarns have somewhat waveguide-like elements [101]. Using this line of thought, an analytical model was developed and applied to predict successfully the better weave for shielding efficiency with the given type of yarns.

CONCLUSION

Modeling and simulation of dielectric permittivity and electromagnetic shielding properties of fibrous material is important for various applications. Such applications include developing functional materials for protection and insulation e.g. EMI shielding materials and insulated printed circuit boards, developing processing equipments for industry e.g. radio frequency drying and microwave heating and also in developing instruments for quality control and testing e.g. capacitance based mass unevenness tester, etc.

Fibrous materials pose a number of challenges in modeling due to its very unique shape and properties and the typical complex hierarchical architecture in any textile structure. Models based on the topology of textile to predict the effective permittivity of the structure have been discussed in this chapter. Applications of various mixing law equations, empirically and semi-empirically, between macroscopic variables and the permittivity have been discussed. The extrapolation technique is also useful for relating the permittivity of constituent materials and the fibrous assembly. Although not yet directly applied to textile materials, computational modeling offers a fertile ground to explore new techniques for studying fibrous structures as dielectrics and a few relevant examples of such modeling techniques have been discussed.

With ever increasing intensity of research in material science coupled with computational capabilities, with tremendous increase of commodity products involving electronic circuits and signals and with more awareness of environmental, health, security and safety hazards due to electromagnetic emissions, research on EMI, EMC and EMI shielding is intensifying like never before. Fibrous materials as a substrate or component of effective shield have interesting potential due to the unique features of textile structure.

However, many of the existing efforts are apparently concentrated on areas like developing conducting polymers and fibers, doping polymers and fibers with nano-materials for increasing conductivity, conductive coatings, incorporating conductive nano-fibers into composites, etc. Developing a modeling framework on physical and mathematical basis for such complex systems is still not very widely or deeply investigated in case of fibrous materials.

However in an age of wearable electronics, the interaction of textiles and electromagnetic fields are being encountered more frequently and it is logical to expect that serious need of such models as a design tool for the future textile engineers and material scientists will be felt by various people, industry and governments throughout the world and therefore this area is definitely a very fertile ground for doing research in coming years.

REFERENCES

[1] Brosseau, C. *J. Phys. D: Appl. Phys.* 2006, *vol 39*, 1277-1294.
[2] Tuncer, E.; Serdyuk, Y. V.; Gubanski, S. M. *IEEE Trans. Dielect. Electric. Insul.* 2002, *vol 9(5),* 809-828.
[3] Von Hippel, A. R. *Dielectric Materials and Applications;* The Technology Press of MIT and John Wiley Sons Inc: NY, 1961, 3rd Edn; Von Hippel, A. R.; Ed.
[4] Bal, K.; Kothari, V. K. *Ind. J. Fibre Text. Res.* 2009, *vol 34*, 191-199.
[5] Hearle, J. W. S. *Moisture in Textiles;* The Textile Institute, Butterworths Scientific Publications: London, 1960, Hearle, J. W. S.; Peters, R. H.; Eds.
[6] Boyd, G. N. *J. Text. Inst.* 1949, *vol 40*, T407-T423.
[7] Mack, C. *J. Text. Inst.* 1955, *vol 46*, T500-T511.
[8] Bal, K.; Kothari, V. K. *J. Electrostatics*, 2009, *vol 67(5),* 751-758.
[9] Hearle, J. W. S. *Text. Res. J.* 1954, *vol 24 (4),* 307-321.
[10] Morton, W. E.; Hearle, J. W. S. *Physical Properties of Textile Fibers;* ISBN: 1845692209; Woodhead Publishing Limited and CRC Press LLC: Cambridge, UK, 2008, 4th Edn, pp 1-792.
[11] Bal, K. *Studies on Dielectric Properties of Woven Fabrics;* PhD Thesis, Indian Institute of Technology, New Delhi, 2009; pp 1-187.
[12] Sihvola, A. H. *Electromagnetic Mixing Formulas and Applications;* ISBN-10: 0852967721; Institution of Electrical Engineers: London, UK, 1999.
[13] Chin, W. S.; Lee, D. G. *Compos. Struct.* 2006, *vol 74*, 153-162.
[14] Yao, L.; Wang, X.; Liang, F.; Wu, R.; Hu, B.; Feng, Y.; Qiu, Y. *Compos. Sci. Techol.* 2008, *vol 68*, 1794-1799.
[15] Greffe, J. L.; Grosse, C. *Dielectric Properties of Heterogeneous Materials;* Progress in Electromagnetics Research, Elsevier: Amsterdam, 1992, vol 6, Priou, A.; Ed.; pp. 41-100.
[16] Calame, J. P. *J. Appl. Phys.* 2008, *vol 104*, 114108.
[17] Tuncer, E.; Gubański, S. M.; Nettelblad, B. *J. Appl. Phys.* 2001, *vol 89 (12),* 8092-8100.
[18] Kärkkäinen, K., Sihvola, A.; Nikoskinen, K. *IEEE Trans. Geosci. Remote Sensing* 2001, *vol 39(5)*, 1013-1018.
[19] Sareni, B.; Krähenbühl, l.; Beroual, A.; Nicolas, A.; Brosseau, C. *IEEE Trans. Magnetics*, 1997, *vol 33 (2),* 1580-1583.
[20] Brosseau, C.; Beroual, A. *Prog. Mater. Sci.* 2003, *vol 48*, 373-456.
[21] Wiener, O. *Bericht. Ver-handlungen Königlich-Sächs. Ges. Wiss. Leipzig*, 1910, 256-277.
[22] Wiener, O. *Abh. Sächs. Ges. Akad. Wiss. Math.-Phys. Kl.* 1912, *vol 32 (6)*, 509-604.
[23] Hashin, Z.; Shtrikman, S. *J. Appl. Phys.* 1962, *vol 33 (10),* 3125-3131.
[24] Maxwell-Garnett, J. C. *Trans. Royal Soc. London*, 1904, *vol 203*, 385-420.
[25] Rayleigh, L. *Philos. Mag.* 1892, *vol 34*, 481-502.
[26] Böttcher, C. J. F. *Theory of Electric Polarization;* Elsevier: Amsterdam, 1952.
[27] Polder, D.; van Santen, J. H. *Physica*, 1946, *vol 12(5)*, 257-271.
[28] Taylor, L. *IEEE Trans. Antennas Propagat.* 1965, *vol AP-13 (6),* 943-947.
[29] Bruggeman, D. A. G. *Ann. Phys.* 1935, *vol 24 (5),* 636-664.

[30] Sihvola, A. H. *IEEE Trans. Geosci. Remote Sensing*, 1989, *vol 27 (4),* 403-415.
[31] Looyenga, H. *Physica*, 1965, *vol 31*, 401-406.
[32] Hanai, T. *Kolloid-Z*, 1960, *vol 171*, 23-31.
[33] Sen, P. N.; Scala, C.; Cohen, M. H. *Geophysics*, 1981, *vol 46 (5),* 781-795.
[34] Birchak, J. R.; Gardner, L. G.; Hipp, J. W.; Victor, J. M. *Proc. IEEE*, 1974, *vol 62 (1),* 93-98.
[35] Lichtenecker, K. *Phys. Zeitschr.* 1926, *vol 27*, 115.
[36] Lichtenecker, K. *Phys. Zeitschr.* 1929, *vol 30*, 805-809.
[37] Lichtenecker, K.; Rother, K. *Phys. Zeitschr.* 1931, *vol 32*, 255-260.
[38] Zakri, T.; Laurent, J. P.; Vauclin, M. *J. Phys. D: Appl. Phys.* 1998, *vol 31*, 1589-1594.
[39] Goncharenko, A. V.; Lozovsky, V. Z.; Venger, E. F. *Optics Commun.* 2000, *vol 174*, 19-32.
[40] Neelakantaswamy, P. S.; Turkman, R. I.; Sarkar, T. K. *Electronics Lett.* 1985, *vol 21(7),* 270-271.
[41] Neelakantaswamy, P. S. *Electronics Lett.* 1989, *vol 25(12),* 800-802.
[42] Brosseau, C. *J. Appl. Phys.* 1994, *vol 75(1),* 672-674.
[43] Zhou, P. H.; Deng, L. J.; Wu, B.-I.; Kong, J. A. *Prog. Electromag. Res.* 2008, *vol 85*, 69-82.
[44] Shaw, T. M.; Windle, J. J. *J. Appl. Phys.* 1950, *vol 21*, 956.
[45] Lorentz, H. A. *Theory of Electrons;* Teubner: Berlin, 1916, 2nd Edn.
[46] Bal, K.; Kothari, V. K. *IEEE Trans. Dielectr. Electr. Insul.* 2010, *vol 17(3)*, 887-895.
[47] Davies, W. E. A. *J. Phys. D: Appl. Phys.* 1974, *vol 7*, 120-130.
[48] Davies, W. E. A. *J. Phys. D: Appl. Phys.* 1974, *vol 7*, 1016-1029.
[49] Steeman, P. A. M.; Maurer, F. H. J. *Colloid Polym. Sci.* 1992, *vol 270 (11)*, 1069-1079.
[50] de Loor, G. P. *IEEE Trans. Geosci. Remote Sensing*, 1983, *vol GE-21(3)*, 364-369.
[51] Wang, J. R.; Schmugge, T. J. *IEEE Trans. Geosci. Remote Sensing,* 1980, *vol GE-18(4)*, 288-295.
[52] Dobson, M. C.; Ulaby, F. T.; Hallikainen, M. T.; El-Rayes M. A. *IEEE Trans. Geosci. Remote Sensing*, 1985, *vol GE-23(1),* 35-46.
[53] Boyarskii, D. A.; Tikhonov, V. V.; Komarova, N. Y. *Prog. Electromag. Res.* 2002, *vol 35*, 251-269.
[54] Robinson, D. A.; Cooper, J. D.; Gardner, C. M. K. *J. Hydrol.* 2002, *vol 255*, 39-49.
[55] Rusiniak, L. *Geophys. J. Int.* 2002, *vol 148*, 313-319.
[56] Jones, S. B.; Or, D. *Physica B*, 2003, *vol 338*, 284-290.
[57] Behari, J. *Microwave Dielectric Behavior of Wet Soils;* ISBN: 9781402032714; Anamaya Publishers and Springer: New Delhi, 2005.
[58] Datta, A. K.; Anantheswaran, R. C. *Handbook of Microwave Technology for Food Applications;* ISBN: 0824704908; Marcel Dekker Inc: NY, 2001.
[59] Henry, F.; Gaudillat, M.; Costa, L. C.; Lakkis, F. *Food Chem.* 2003, *vol 82*, 29-34.
[60] Torgovnikov, G. I. *Dielectric Properties of Wood and Wood-based Materials;* ISBN: 3540553940; Springer-Verlag: Berlin, 1993.
[61] Takeda, M. *Bull. Chem. Soc. Jpn.* 1951, *vol 24(4)*, 169-173.
[62] Komarov, V. V. *J. Comm. Tech. Electron.* 2007, *vol 52(11)*, 1245-1251.
[63] Bayley, S. T. *Trans. Faraday Soc.* 1951, *vol 47*, 518-522.
[64] Takashima, S. *J. Polym. Sci.* 1962, *vol 62*, 233-240.

[65] Kukharchik, P. D.; Serdyuk, V. M.; Titovitskii, I. A.; Belyachits, A. Ch.; Kurilo, N. I. *J. Comm. Tech. Electron.* 2001, *vol 46(11),* 1264-1269.

[66] Habeger, C. C.; Baum, G. A. *J. Appl. Pol. Sci.* 2003, *vol 28(3),* 969-981.

[67] Balls, W. L. *Nature*, 1946, *vol 158*, 9-11.

[68] Algie, J. E. *Text. Res. J.* 1964, *vol 34(6),* 477.

[69] Hearle, J. W. S. *Text. Res. J.* 1956, *vol 26(2),* 108-111.

[70] Windle, J. J.; Shaw, T. M. *J. Chem. Phys.* 1954, *vol 22(10),* 1752-1757.

[71] Windle, J. J.; Shaw, T. M. *Text. Res. J.* 1955, *vol 25(10),* 865-870.

[72] Kumar, A. *J. Text. Inst.* 1978, *vol 69(4)*, 116-117.

[73] Kumar, A. *J. Text. Inst.* 1980, *vol 71(4),* 210-212.

[74] Bailey, B.; Phelps, E. L. *Text. Res. J.* 1939, *vol 9(3),* 101-113.

[75] Algie, J. E. *Koll. Z. u. Z. Pol.* 1969, *vol 234(2)*, 1069-1078.

[76] Algie, J. E.; Gamble, R. A. *Koll. Z. u. Z. Pol.* 1973, *vol 251(8),* 554-562.

[77] Leveque, J. L.; Garson, J. C.; Pissis, P.; Boudouris, G. *Biopolymers* 1981, *vol 20*, 2649-2656.

[78] Teschke, O., Ceotto, G.; de Souza, E.F. *Phys. Rev. E*, 2001, *vol 64*, 1-10.

[79] Cassie, A. B. D. *Trans. Faraday Soc.* 1945, *vol 41*, 450-458.

[80] Nelson, S. O. *IEEE Trans. Instrum. Meas.* 2005, *vol 54(5),* 2033-2040.

[81] Kent, M. J. *Microwave Power*, 1977, *vol 12(4),* 341-345.

[82] Klein, A. J. *Microwave Power* 1981, *vol 16(3&4),* 289-304.

[83] Kirkwood, C. E.; Kendrick, Jr. N. S.; Brown, H. M. *Text. Res. J.* 1954, *vol 24(9),* 841-847.

[84] Hallouet, B.; Wetzel, B.; Pelster, R. *J. Nanomaterials*, 2007, *vol 2007*, Article ID 80814, doi:10.1155/2007/80814.

[85] Simoes, R.; Silva, J.; Vaia, R. *J. Nanosci. Nanotechnol.* 2010, *vol 10(4),* 2451-2457.

[86] Sareni, B.; Krahenbuhl, L.; Beroual, B.; Brosseau, C. *J. Electrostatics* 1997, *vol 40 & 41*, 489-494.

[87] Shen, L. C.; Liu, C.; Korringa, J.; Dunn, K. J. *J. Appl. Phys.* 1990, *vol 67, no. 11*, 7071-7081.

[88] Brosseau, C.; Beroual, A. *J. Phys. D: Appl. Phys.* 2001, *vol 34*, 704–710.

[89] Myroshnychenko, V.; Brosseau, C. *Phys. Rev. E*, 2005, *vol 71*, Article ID: 016701.

[90] Krakovsky, I.; Myroshnychenko, V. *J. Appl. Phys.* 2002, *vol 92, no. 11*, 6743-6748.

[91] Cheng, Y.; Chen, X.; Wu, K.; Wu, S.; Chen, Y.; Meng, Y. *J. Appl. Phys.* 2008, *vol 103*, Article no. 034111, doi: 10.1063/1.2838465.

[92] Serdyuk, Y. V.; Podoltsev, A. D.; Gubansky, S. M. *J. Electrostatics* 2005, *vol 63*, 1073-1091.

[93] Tan, P.; Tong, L.; Steven, G. P. *Compos. Part A.* 1997, *vol 28(11),* 903-922.

[94] Lomov, S. V.; Huysmansa, G.; Luoa, Y.; Parnasa, R. S.; Prodromoua, A.; Verpoesta, I.; Phelan, F. R. *Compos. Part A.* 2001, *vol 32(10),* 1379-1394.

[95] Rajagopalan, D.; Aneja, A. P.; Marchal, J- M. *Text. Res. J.* 2001, *vol 71(9),* 813-821.

[96] Wang, Y.; Sun, X. *Compos. Sci. Techol.* 2001, *vol 61(2),* 311-319.

[97] Celozzi, S.; Araneo, R.; Lovat, G. *Electromagnetic Shielding;* ISBN: 9780470055366; John Wiley & Sons: NJ, 2008.

[98] Tong, X. C. *Advanced Materials and Design for Electromagnetic Interference Shielding;* ISBN: 9781420073584; CRC Press, Taylor & Francis Group: Fl, 2009.

[99] White, D. R. J.; Mardiguian, M. *Electromagnetic Shielding;* Interference Control Technologies Inc: Gainesville, 1988.

[100] Kodali, V. P. *Engineering Electromagnetic Compatibility;* ISBN-10: 0780347439; IEEE Press: NY, 2001.

[101] Ghosh, S. *Radio Frequency Shielding with Conductive Textiles;* Master's thesis, Indian Institute of Technology, New Delhi, 2005; pp 1-87.

In: Modeling and Simulation in Fibrous Materials
Editor: Asis Patanaik
ISBN: 978-1-62100-116-4

Chapter 10

MODELING AND SIMULATION OF HEAT AND MASS TRANSFER PROPERTIES OF TEXTILE MATERIALS

D. Bhattacharjee[1,*] and B. Das[2]

[1]Terminal Ballistics Research Laboratory, Defence Research and Development Organization, Sector 30, Chandigarh, India

[2]Regional Silk Technological Research Station, Central Silk Board, Muga Farm, Khanapara, Guwahati, India

ABSTRACT

There are many factors that contribute to the well being of a human being. Most important of all is the feeling of comfort or satisfaction. At physiological level, comfort is obtained when there is a balance between the energy produced by a human being and the exchange of the energy with the environment i.e. the total amount of energy produced by the body should be equal to the energy spent by the body and the energy stored by the body.

Clothing creates a microclimate between the skin and the environment. This supports the body's thermoregulatory system to keep its temperature within a safe range, even when the external environment temperature changes to a large extent. This creates a sensation of "thermal or thermo-physiological comfort" to the wearer. Thermal comfort as defined by ASHRAE 55 is "that condition of mind which expresses satisfaction with the thermal environment". Clothing is therefore an integral part of the human life. Its primary function is to form a barrier against unsuitable environmental conditions and chemical and biological injuries. The factors which influence the choice of clothing depend upon its end use. For extreme climatic conditions or special environments e.g. space, underwater etc., the functional aspects and the performance of the clothing are most important. For everyday civilian use, apart from comfort and durability, fashion and aesthetics become prominent factors in the choice of the clothing. In both the cases, one of the important functions of clothing is to provide thermal comfort to the wearer.

The thermo-physiological comfort characteristics of the clothing assembly depends upon the heat and moisture/moisture vapor transfer properties of the textile materials and environmental factors like temperature, relative humidity, wind speed etc. Prediction of

*E-mail: debarati.bhattacharjee@gmail.com

thermal and moisture transmission properties have always been an issue of concern, as it facilitates to get the knowledge of the thermo-physiological behavior of the material/clothing assembly, before it has been manufactured. It is also useful to characterize the clothing comfort as well as it helps to design the fabric assembly as per specific requirement especially in cases of extreme climatic clothing.

The concept of clothing comfort and the factors influencing the same have been investigated by various researchers since 1930s till present date. Many investigations have been carried out to study the thermal and moisture properties of textile materials. Researchers have worked on the effect of various properties of textile materials on their transport values.

To understand the heat and moisture flow characteristics of textile fabrics, many mathematical models have been propounded. Many computational tools like Computational Fluid Dynamics (CFD), artificial neural networks, fuzzy logic and many more are also being used to understand the complex relationships between the clothing parameters and the perception of comfort. This chapter deals with the studies on heat and mass transfer properties of textile assemblies. The phenomena covered here are dry steady state heat transfer, transient heat transfer, moisture vapor and liquid moisture transfer and coupled heat and moisture transfer properties of fibers, fiber bundles, fibrous materials and other textile structures. The processes involved in each and the work done on modeling and simulation of the transfer processes till date, from the point of view of clothing comfort have been discussed.

10.1. Introduction

10.1.1. Thermoregulatory Properties of Clothing Assemblies

The factors that contribute to the well being of a human being are many and complex. Most important of all is the feeling of comfort or satisfaction. At physiological level, comfort is obtained when there is a balance between the energy produced by a human and the exchange of this energy with the environment i.e. the total amount of energy produced by the body should be equal to the energy spent by the body and the energy stored by the body. Clothing creates a microclimate between the skin and the environment. This supports the body's thermoregulatory system to keep its temperature within a safe range, even when the external environment temperature changes to a large extent. This creates a sensation of "thermal or thermo-physiological comfort" to the wearer.

The thermo-physiological comfort is dependent on the thermal as well as moisture related properties of the fabric. Thermal comfort is defined as 'that condition of mind which expresses satisfaction with the surrounding environment' [1]. Clothing is the integral part of the human life. It is used to protect the body from different environmental conditions by maintaining the body temperature. The primary function of clothing is to form a barrier against unsuitable environmental conditions and chemical and biological injuries. The factors which influence the choice of clothing depend upon its end use.

For extreme climatic conditions or special environments e.g. space, underwater etc., the functional aspects and the performance of the clothing are most important. For everyday civilian use, apart from comfort and durability, fashion and aesthetics become prominent factors in the choice of the clothing. In both the cases, one of the important functions of clothing is to provide thermal comfort to the wearer.

10.1.2. Heat and Moisture Vapor Transfer

In a regular atmospheric condition and during normal activity level, the heat produced by the metabolism is liberated from body to atmosphere by conduction, convection and radiation and perspiration in vapor form to maintain the body temperature. At higher activity levels and at higher atmospheric temperatures, the production of heat is very high and transmission of heat from the body to atmosphere is less, the sweat glands are activated to produce liquid perspiration as well [2]. The vapor form of perspiration is known as insensible perspiration and the liquid form as sensible perspiration. When the perspiration is transferred to the atmosphere, it carries heat (latent as well as sensible) thus reducing the body temperature. The perception of discomfort in the active case depends on the degree of skin wetness. It is also important to reduce the reduction of thermal insulation caused by moisture build-up. If the ratio of evaporated sweat and produced sweat is very low, moisture accumulates in the inner layer of the fabric system, thus reducing the thermal insulation of clothing and causing unwanted loss in body heat [3].

The thermo-physiological comfort characteristics of the clothing assembly depends upon the heat and moisture/moisture vapor transfer properties of the textile materials and environmental factors like temperature, relative humidity, wind speed etc. The concept of clothing comfort and the factors influencing the same have been investigated by various researchers since 1930s till present date. Many investigations have been carried out to study the thermal and moisture properties of textile materials. Prediction of thermal and moisture transmission properties have always been an issue of concern, as it facilitates the knowledge of the thermo-physiological behavior of the material/clothing assembly, even before it has been actually manufactured. It is also useful to characterize the clothing comfort as well as it helps to design the fabric assembly as per specific requirement especially in cases of extreme climate clothing.

10.1.3. Modeling and Simulation of Heat and Moisture Transfer Properties

Influence of various factors and prediction of heat and moisture transmission behavior of the clothing is recognized as an important research arena among the research fraternity in recent years. Theoretical prediction of heat and moisture transmission properties of textile materials is useful to characterize the clothing comfort and also it helps to design the fabric as per specifications. To understand the heat and moisture transmission characteristics of textile fabrics, many mathematical models have been advocated. Prior to mathematical modeling of heat and moisture transmission through fabrics (porous media), the dependency of the relevant material characteristics and the interface phenomenon among solid, liquid, gaseous vapor and air and liquid-vapor equilibrium must be considered. The coupled heat and moisture transfer in textile fabrics is important for understanding the dynamic thermal comfort of clothing. Moisture transmission through textiles is interrelated with the heat transfer phenomenon. Clothing heat and vapor resistances are important inputs for standards and models dealing with thermal comfort, heat- and cold-stress. A general mathematical model on heat and mass transfer through porous media is constructed by using the conservation equations of heat and mass transfer. So, the models developed on heat transfer also contemplate the vapor and liquid transmission. Heat and mass conservation equations in porous media are coupled and, in general, solved by using the values of temperature and moisture content from previous iterations to calculate the source terms. Various mathematical treatments of heat, moisture and combined heat and moisture transfer given by different scientists to predict the transmission behavior of the textile materials have been discussed in the latter part of this chapter.

This chapter discusses the basic principles of heat and mass transfer, separately and in combination, transfer phenomenon through textile and fibrous structures and various models and simulation tools available to predict the same.

10.2. Heat Transfer through Textile Structures

10.2.1. Principles of Heat Transfer

Heat transfer through textile structures comprises of (a) conduction, (b) convection, (c) radiation and, (d) evaporation [4-8]. The focus of this section is on 'dry' heat transfer. The term 'dry heat' is used to distinguish it from heat loss by evaporation of moisture, which takes place either within the skin, or at the surface of the skin. The moisture transfer and moisture based heat transfer will be discussed in latter sections.

Dry heat transfer due to conduction, convection and radiation depends upon a number of clothing and fabric parameters like fabric structure, thickness, porosity, fabric layers as well as environmental parameters. However, one mode may or may not be present under different circumstances e.g. in a still atmosphere where the fabric is kept between two static bodies without any kind of air flow between them, heat transfer due to natural and forced convection can be neglected and in this case heat transfer is only due to conduction and radiation. In an evacuated chamber/vacuum conductive heat transfer through still air can be neglected and heat transfer takes place mostly due to radiation and conduction through the solid material.

Thermal insulation of the clothing assembly is one of the most important parameters in characterization of thermo-physiological comfort of the wearer. It is a complex and synergistic relationship between the man, material and the environment. Fourt and Hollies [4] have given a detailed description of the basic thermo-physiological interactions between man, material and environment. The main modes of heat transfer in clothing are:

a) Conduction through the fibers
b) Conduction through the entrapped air between the fibers, yarns and various fabric layers
c) Convection through the free spaces between the yarns and fabric layers (only possible in case of air flowing at a velocity over and through the clothing)
d) Radiation across the free spaces between the yarns and fabric layers (only possible if the temperature gradient between the environment and the body is high)

To arrive at a suitable approximation of the thermal resistance it is important to understand all the modes of heat transfer and their physical phenomena.

10.2.1.1. Conduction

The governing equation of heat conduction in steady state is the Fourier's equation given by:

$$\dot{q} = \frac{dQ}{d\tau} = -kA\left[\frac{\partial T}{\partial x} + \frac{\partial T}{\partial y} + \frac{\partial T}{\partial z}\right] \tag{10.1}$$

where $\dot{q}$ is the rate of heat transfer; dQ is the quantity of heat conducted in time $d\tau$; T is the temperature gradient; A is the area of the specimen and k is the coefficient of thermal conduction (W/mK). Confining the system to a medium limited by two isothermal, parallel and infinite surfaces (Figure 10.1), the boundary conditions become:

$$\frac{\partial T}{\partial x} = \frac{\partial T}{\partial y} = 0$$

therefore, Equation 10.1 can be rewritten as:

$$\frac{dQ}{d\tau} = -kA\frac{\partial T}{\partial z} \tag{10.2}$$

Integrating equation 10.2 with the limits provided in Figure 10.1.

$$\dot{q} = -kA\frac{(T_1 - T_2)}{z} = -kA\frac{(\Delta T)}{z} \tag{10.3}$$

The negative sign shows the direction of the flow of heat which is opposite to the direction of increasing temperature, i.e. the heat flow is from the hot surface to cold surface. Equation 10.3 is analogous to the Ohm's law of electricity and the heat flow can be expressed as:

$$\dot{q} = -\frac{\Delta T}{\left(\frac{z}{kA}\right)} \tag{10.4}$$

The thermal resistance $R_{conduction}$ is given as:

$$R_{conduction} = \frac{d}{kA}\text{, where } d \text{ is the thickness of the material} \tag{10.5}$$

The above calculations are based on the steady state heat transfer i.e. the state where the temperature of a point in the body remains constant with time. However, before reaching the steady state, temperature is also a function of time (τ) i.e.

$$T = f(x, y, z, \tau) \tag{10.6}$$

The event till the heat flow is dependent upon time is known as the transient heat conduction. A fabric surface can, be assumed to be a semi-infinite body where at a given time, there's always a portion of the body where the temperature remains unchanged when there is a sudden temperature change in one of the boundaries. Consider a unidirectional flow through a semi infinite solid plate bounded by the y-z plane and extending to infinity in x-direction. Initially the entire body is at temperature T_0. At time τ_0 and distance x_0, face temperature at x is suddenly raised to T_s. For a unidirectional heat flow the governing equation (without any internal heat generation) is given by the Laplace law of conduction:

$$\frac{1}{\alpha}\frac{\partial T}{\partial \tau} = \frac{\partial^2 T}{\partial x^2} \tag{10.7}$$

where α is the thermal diffusivity which is $\frac{k}{c\rho}$ where c is the specific heat and ρ is the density of the material. The boundary conditions for the above equation for transient state are given by (i) $T(x, 0) = T_0$; and (ii) $T(0, \tau) = T_s$ for all $\tau > 0$.

Surface heat flow Q_0 at any instant of time is given by:

$$Q_0(\tau) = \frac{kA(T_s - T_0)}{\sqrt{\pi \alpha \tau}} \tag{10.8}$$

In case of steady state heat conduction, the material property is the conductivity 'k' which can be calculated once the heat loss from the body is known and the boundary temperature is measured. In case of transient heat flow, the main factor is the diffusivity 'α' which is equal to the ratio of the conductivity and the heat content of the body. Hence equation 10.8 is the basic governing equation of transient heat transfer with boundary conditions relevant for textile materials. Transient state heat conduction is related to instantaneous conduction of heat from the surface of the body to the clothing. Instantaneous heat transfer can be related to the warmth or coolness to touch and the warm-cool feeling of any clothing can be quantified.

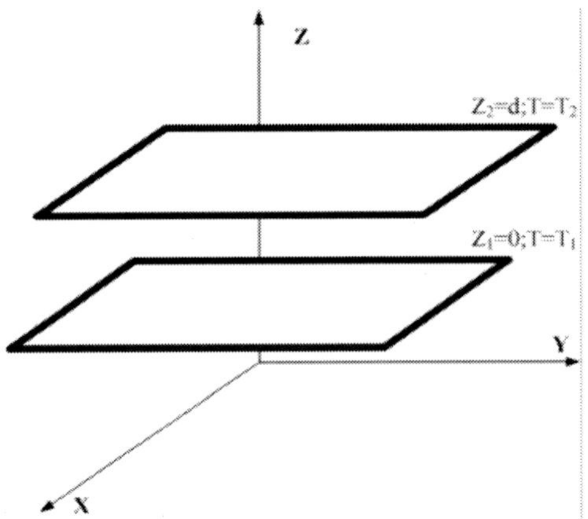

Figure 10.1. Conduction heat transfer through isothermal, infinite plane configuration of anisotropic medium [Source: Reference 8].

10.2.1.2. Convection

The basic equation for convection heat transfer is given by Newton's Law of Cooling:

$$\dot{q} = hA\Delta T \tag{10.9}$$

where *h* is convective coefficient of heat transfer. Here the material property is '*h*' which can be derived from the Nusselt number *Nu*. For forced convection *Nu* is:

$$Nu = \varphi(Re)\psi(Pr) = \frac{hL}{k} \tag{10.10}$$

For natural convection *Nu* is given by:

$$Nu = \varphi(Gr)\psi(Pr) = \frac{hL}{k} \tag{10.11}$$

where *Re* is Reynold's number; *Gr* is Grashof number; *Pr* is Prandtl number and *L* is characteristic length. The analytical and boundary layer methods are most convenient to use for prediction of the value of '*h*'. However, analytical methods involve a number of experiments to derive the empirical constants which can be time consuming as well as economically unviable. Most of the solutions of heat transfer through convection are approximate and based on empirical dimensionless numbers. These solutions are valid only for a specific set of conditions, so for every new situation, a new set of empirical coefficients have to be formulated. In such cases, the boundary layer technique gives the best approximations of the convection heat transfer. The boundary layer analysis for solving convection heat transfer equations involve the simultaneous solutions of the three basic Navier Stokes equations of fluid flow viz. the equation of continuity, the conservation of momentum equation and the conservation of energy equation all of which characterize the hydrodynamic and the thermal boundary layer build up over or within the material (Figure 10.2). A reference control volume of the fluid flow is considered while deriving these equations (Figure 10.3). The τ_{ij} in the figure represent the shear forces on the ith surface in the jth direction. ρu_i is the mass flow term and $\rho u_i u_j$ is the momentum carried into the control volume by the flow. The flow of fluid in terms of continuity, momentum and energy are given by the Navier Stokes equations. Equation for continuity is given by:

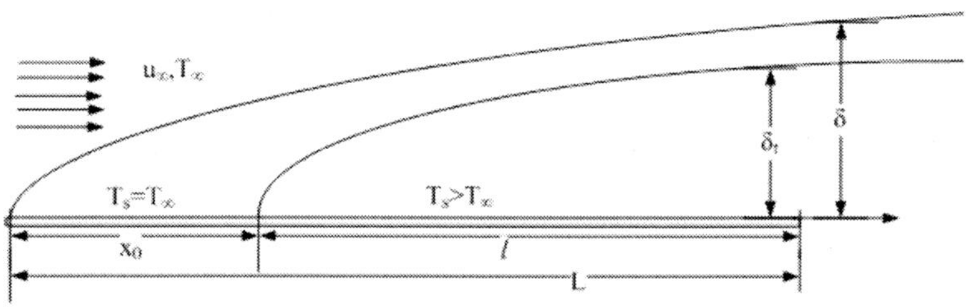

Figure 10.2. Hydrodynamic and thermal boundary layers with unheated starting length. T_∞ is the ambient temperature: T_s is the temperature of the plate; *l* is the characteristic length; x_0 is the unheated length; and, *L* is the specified length; δ_t is the thermal boundary layer and δ is the hydrodynamic boundary layer [Source: Reference 8].

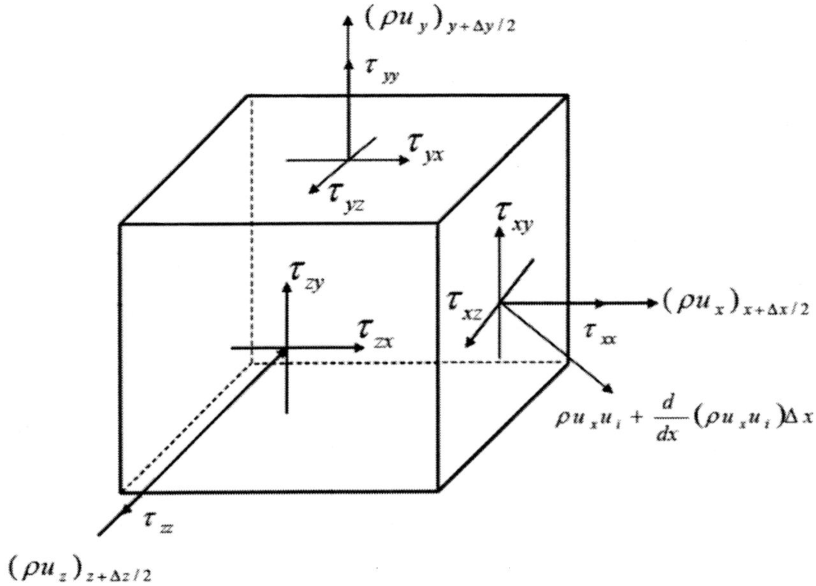

Figure 10.3. Reference control volume for fluid flow analysis [Source: Reference 8].

$$\frac{\partial \rho}{\partial t} = \frac{\partial}{\partial x_i}(\rho u_i) = S \tag{10.12}$$

where ρ is the fluid density, t is time, x_i is the direction tensor, u_i is the velocity in the direction i, and the S represents a source term. The source term is included for use with user-defined boundary conditions such as inlet and outlet mass flows.

The momentum equation is given as:

$$\begin{aligned}&(\textit{Rate of momentum entering control volume}) + (\textit{Summation of pressure and viscous forces}) \\ &= (\textit{Rate of momentum leaving control volume}) + (\textit{Rate of accumulation of momentum})\end{aligned}$$

$$\frac{\partial}{\partial t}(\rho u_i) + \frac{\partial}{\partial x_i}(\rho u_i u_j) = \frac{-\partial P}{\partial x_i} + \frac{\partial}{\partial x_j}\tau_{ji} + B_i + F_i \tag{10.13}$$

B_i is the buoyancy force term; F_i is the body force term; P is pressure; and μ is the dynamic viscosity. On the left side of Equation 10.13, the first term is the rate of mass flow increase in the control volume while the second term is the net momentum carried out of the control volume by fluid flow. The right hand side of equation contains the pressure force, net viscous forces, buoyancy force, and body force.

The conservation of energy equation states that the summation of heat flowing inside a body and the heat generated by the body is equal to the heat lost due to conduction through the body, convection from the surface and radiation from the body to the atmosphere and the heat retained by the body in terms of enthalpy and can be written as:

$$\frac{\partial}{\partial t}(\rho E)+\frac{\partial}{\partial x_i}[u_i(\rho E+P)]=\frac{\partial}{\partial x_i}\left(k_{eff}\frac{\partial T}{\partial x_i}+u_j(\tau_{ji})_{eff}\right) \tag{10.14}$$

where, $E=h-\frac{P}{\rho}+\frac{u_i^2}{2}$ and $h=cT$

k_{eff} is the effective conductivity of the material; c_p is the specific heat and τ_{ij} is the shear force. The two terms on the left side of the equation physically represent the time rate change in energy found in the control volume and net removal of energy out of the control volume. The right side of the equation gives the energy transfer due to conduction and viscous dissipation. Equations 10.12, 10.13 and 10.14 are solved simultaneously for different boundary conditions and an approximate solution is reached. Once the coefficient of connective heat transfer is obtained, the Newton's law of cooling can be equated to the Fourier's law of conduction and the resistance due to convection heat transfer can be obtained i.e.

$$R_{convection}=\frac{1}{hA} \tag{10.15}$$

10.2.1.3. Radiation

The easiest way to consider radiation heat exchange is by assuming transparent media where radiative exchange takes place only between surfaces. Net radiation heat exchange is calculated by. If two gray plates are considered in a transparent medium, the net radiation heat transfer $Q_{radiation}$ between them can be given by:

$$Q_{radiation}=\frac{\sigma(T_i-T_j)^4}{R_{ij}} \tag{10.16}$$

where subscripts i,j denote hot and cold surfaces respectively, σ is the Stefan-Boltzmann's constant and R_{ij} is the total resistance encountered during net radiation transfer through two surfaces. This value has two components (i) Surface resistance which depends upon the emissivity and the area of the surface. This is given by $R_{surface}=\frac{1-\varepsilon_i}{A_i}$ and (ii), Space resistance which depends upon the view factor of the two surfaces. This is given by $R_{space}=\frac{1}{A_iF_{ij}}$ where ε is the emissivity of the surface and F is the view factor. Considering all these parameters, Equation 10.17 becomes:

$$Q_{radiation}=\frac{\sigma(T_i-T_j)^4}{\left[\frac{1-\varepsilon_i}{A_i\varepsilon_i}+\frac{1}{A_iF_{ij}}+\frac{1-\varepsilon_j}{A_j\varepsilon_j}\right]} \tag{10.17}$$

The net radiation exchange can be considered as electrical resistance analogy (Figure 10.4) where resistance to radiation heat transfer is given by:

$$R_{radiation} = \left[\frac{1-\varepsilon_i}{A_i\varepsilon_i} + \frac{1}{A_iF_{ij}} + \frac{1-\varepsilon_j}{A_j\varepsilon_j}\right] \tag{10.18}$$

Radiation through a fibrous material, however, is not the same as through transparent media. Depending upon the shape and structure, the fibrous material can be considered as an absorbing scattering medium. As an electromagnetic wave hits a medium, depending upon the shape and size of the medium, it transmits part of its energy into it, part is reflected back and part scattered or redirected in various directions. This complicates the governing equation of radiation heat transfer or the Radiative Transfer Equation (RTE). A simple linear differential equation by combining one set of variables (spatial location) is converted into an integro-differential one by integration over another set of variables (solid angle). The final RTE is given as:

$$\frac{dI(\hat{s})}{dl} = -(\kappa + \sigma_s)I(\hat{s}) + \kappa I_b + \frac{\sigma_s}{4\pi}\int_{4\pi} \phi(\hat{s}', \hat{s})I(\hat{s}')d\Omega' \tag{10.19}$$

where I is the energy of radiation propagating in direction ŝ, I_b is the energy emitted by black body, κ is the absorption coefficient, σ_s is the scattering coefficient, and $d\Omega$ is the solid angle. Many mathematical models have been advocated which consider fibers as absorbing, scattering medium and depending upon the size of the fibers, Mie scattering has been considered [5].

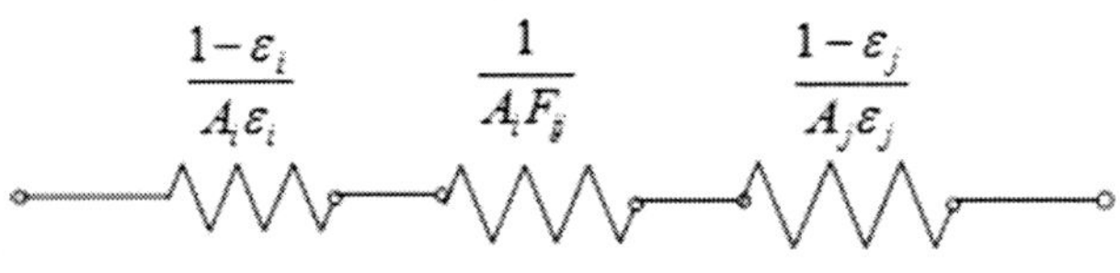

Figure 10.4. Electrical analogy of net heat transfer by radiation [Source: Reference 8].

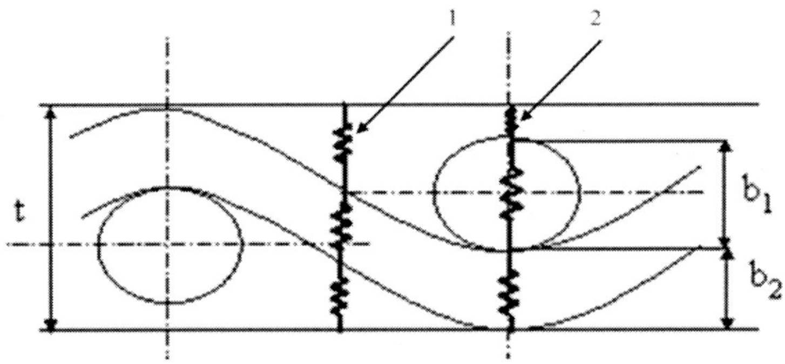

Figure 10.5. Transverse view of the fabric seen as system of resistances (1) Series resistance between unsupported length of weft and air (2) Series resistance between air and intersection zone. The resistance through air pore cannot be shown in this view [Source: Reference 8].

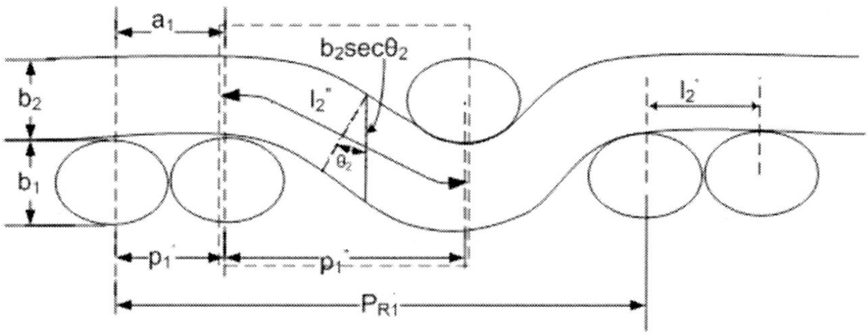

Figure 10.6. Cross section of a 1/2 twill fabric. The area inside the dashed region is assumed to follow Peirce's geometry for plain woven structure with circular/ flattened threads [Source: Reference 8].

10.2.2. Modeling of Heat Transfer through Textile Structures

10.2.2.1. Prediction of Thermal Resistance Using Effective Conductivity

Studies on heat transfer through various materials has been going on since a long time, these include heat transfer through various engineering materials and structures like stack of rods and fins, porous materials which include granular (pumice, concrete etc.), cellular (foam, cork etc.) as well as fibrous insulators. In the early thirties, Black and Mathew [6] were one of the first to provide a comprehensive review of the work done to measure the thermal transmission through fabrics. It was observed that the most important fabric factor which determines the thermal resistance of the fabric is fabric thickness. The simplest way to characterize the heat resistance is by considering the effective conduction heat transfer through the structure. Heat transfer due to conduction through porous materials and regular structures has been well documented. Kaviany [7] has given a comprehensive list of formulae to calculate the effective thermal conductivity through a porous material. For practical purposes a rough and ready estimate of k_m is provided by the weighted geometric mean i.e.

$$k_{eff} = k_S^{1-\phi} + k_f^{\phi} \tag{10.20}$$

where k_{eff} is the effective thermal conductivity, k_f and k_s are the thermal conductivities of fluid and solid parts in the porous material, and ϕ is the porosity.

Textile weaves have regular geometries which can be utilized to describe the flow of heat across the thickness. Fabric geometry for different weaves can be commonly defined in terms of its thickness [8]:

$$t = \frac{4}{3I}\sqrt{R^2 p_2^2 c_1 - Rp_2 c_1 (R-I)\frac{d_2}{e}} + d_1 \cdot e \tag{10.21}$$

where subscripts 1 and 2 denote warp and weft respectively, t is the thickness of the fabric, R is the number of ends or picks in a repeat, I is the number of intersections in a repeat, p is the average thread spacing, c is the crimp in the yarn, d is the diameter of the yarn and e is the coefficient of yarn flattening. When yarns are woven into fabric, the yarn during interlacement tends to flatten and they do not have the same flattened diameter throughout the fabric. Diameter of yarn is more at the crown of the interlacement and less in the rest of the body. The effective diameter in this case is the average of the flattened diameter a_{yarn} and normal diameter i.e.

$$d_{eff} = \frac{a_{yarn} + d_{yarn}}{2} = d_{yarn}\left[\frac{1+e}{2e}\right] \tag{10.22}$$

The fabric structure can be considered to be a woven mesh comprising of solid and fluid regions and the effective conductivity can be measured as a network of resistances across the fabric thickness (Figure 10.5). The effective thermal conductivity can be predicted using such a structure. Ismail et al. [9] gave a theoretical model to predict the thermal conductivity of a plain woven textile fabric using the simple Fourier's equation of conduction to obtain the effective conductivity of different sections in the fabric repeat. Kothari and Bhattacharjee [10] further simplified the problem and incorporated the effect of different weave structures into the calculation of the total heat loss due to conduction and the total resistance due to thermal conduction only. From the fabric geometry given in Figures 10.6 and 10.7, it can be seen that any basic weave can be represented in terms of a system of intersections, lengths of warp and weft yarns between the respective cross yarns and air pores where the total thread spacing on the warp side P_{R1} is given by

$$P_{R_1} = (R - I)p_1' + Ip_1'' \tag{10.23}$$

where 1, 2 subscripts denote warp and weft respectively, P_R is the total length of one repeat, p' is the thread spacing in float region, p'' is the thread spacing in intersection region.

In real situations, two cases arise (Figure 10.7) viz. case 1 where the yarns are apart from each other and air pore is present and case 2 where the warp yarns are close to each other and there is no air pore

Based on Fourier's equation, conduction heat loss through unsupported lengths of warp/weft $Q_{m_{1/2}}$ is given by $Q_{m_{1/2}} = \dfrac{A_{m_{1/2}}\Delta T}{\left(\dfrac{b_{1/2}\sec\theta_{1/2}}{k_{1/2}} + \dfrac{t - b_{1/2}\sec\theta_{1/2}}{k_a}\right)}$ where

$$A_{m_{1/2}} = R^2 a_{1/2}(p_{2/1} - a_{2/1}) \tag{10.24}$$

A_m is the total area of the unsupported length of warp or weft, a is the flattened major diameter, b is the flattened minor diameter, θ is the weave angle, $k_{1/2}$ is the conductivity of the warp/weft thread; k_a is the conductivity of the air. Conduction heat loss through the intersection Q_{st} is given by:

$$Q_{st} = \frac{A_{st}\Delta T}{\left(\dfrac{b_1}{k_1} + \dfrac{b_2}{k_2} + \dfrac{t - b_1 - b_2}{k_a}\right)} \quad \text{where } A_{st} = R^2 a_1 a_2 \tag{10.25}$$

Conduction heat loss through air pore Q_a is given by:

$$Q_a = \frac{A_a \Delta T}{\left(\frac{t}{k_a}\right)} \quad \text{where } A_a = R^2(p_1 - a_1)(p_2 - a_2) \tag{10.26}$$

The total heat loss due to conduction, therefore is:

$$Q_{conduction} = Q_{m_1} + Q_{m_2} + Q_{st} + Q_a \tag{10.27}$$

In case 2 (Figure 10.7.b) the conduction heat loss through air pore is zero i.e $Q_a = 0$. From the above equation, total thermal resistance due to conduction per unit area is given by:

$$R_{conduction} = \frac{\Delta T.(R^2 p_1 p_2)}{Q_{conduction}} \quad \text{K.m}^2\text{/W} \tag{10.28}$$

Rees [11] in early seventies claimed that even though textile fibers have higher conductivities as compared to air, the fibers occupy only a small fraction of the fabric and almost 90% of the bulk of the fabric is occupied by air, hence differences in the conductivities of various fibers are not of much relevance. Furthermore, the convective and radiative insulation provided by the fibers are much higher than the heat loss by conduction. The heat transfer through fabrics occurs by two mechanisms viz. (1) Convection: This depends upon temperature difference, density of the fibrous material and rate of movement of air and (2) Radiation: This depends upon the surface of clothing and temperature difference with the environment. Total heat energy from clothing to environment is given by:

$$Q_{clothing\ to\ env.} = A + B\sqrt{\frac{PV}{760}} \tag{10.29}$$

and $$R_a = \frac{1}{\left(A + B\sqrt{\frac{PV}{760}}\right)} \tag{10.30}$$

where A is the contribution from radiation; B is the contribution from conduction; P is pressure; V is wind velocity and R_a is the resistance to heat transfer. Convection was altogether neglected by Bankvall [12] who considered the heat loss due to conduction and radiation through fibrous insulations by assuming the total volume to contain layers of sheets having the same density and thickness as that of fibers and calculating the total heat loss due to radiation through a number of screens. The effective thermal conductivity due to radiation was given as:

$$\lambda_R = \frac{4\sigma L_0 T_m^3}{\left[\frac{1}{\beta} + \frac{L_0}{d} \cdot \left(\frac{2}{\varepsilon_0} - 1\right)\right]} \tag{10.31}$$

where $\beta = \frac{\varepsilon}{2-\varepsilon} = f(T, D, \varepsilon)$; λ_R is the radiative conductivity; L_0 is the mean free path; d is the diameter of the fiber; ε_0 is the emissivity of the surface; ε is the emissivity of the fiber. He concluded that conduction due to gas contributes the largest part of the thermal conductivity.

In a regular fabric structure understanding the radiation heat exchange between the various factors in the material can be somewhat complicated, as with the change in the shape and the size of the fibers, the radiation exchange properties change to a great deal. In the fabric geometry as given in Figure 10.8, the radiation heat transfer can be assumed to comprise of heat exchange between the body and the environment through the air-pore and heat exchange through the fibers and the yarns [13].

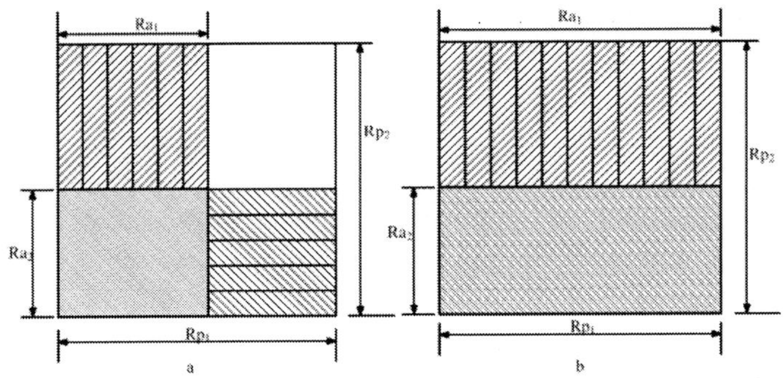

Figure 10.7. Simplification of the fabric geometry into a system of unsupported warp and weft lengths, intersecting zones and air pores. Two cases of such system can be possible (a) Presence of air pore (b) Absence of air pore [Source: Reference 8].

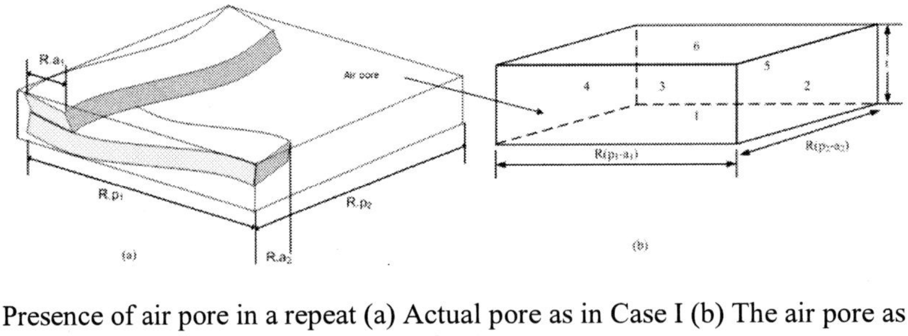

Figure 10.8. Presence of air pore in a repeat (a) Actual pore as in Case I (b) The air pore as an enclosed surface used for net radiation method [Source: Reference 8].

10.2.2.1.1. Radiation Heat Exchange through Air Pore

In one repeat weave of the fabric the air gap is enclosed on four sides by yarns, hence it can be assumed to be a problem of enclosed surfaces. Also, in absence of convection, the upper surface can be assumed to be a still black body. The basic equation of radiation heat transfer between two plates is given in Equation 10.16 Calculations for enclosure radiation through air pore can be carried out using net radiation formulation. This formulation was originally proposed by Siegel and Howell [14] and considers N discrete surfaces between which the radiation is exchanged. The basic equation is given as:

$$\sum_{j=1}^{N}\left(\frac{\delta_{kj}}{\varepsilon_i}-F_{k-j}\frac{1-\varepsilon_j}{\varepsilon_j}\right)\frac{Q_j}{A_j}=\sum_{j=1}^{N}\left(\delta_{kj}-F_{k-j}\right)\sigma T_j^4 \tag{10.32}$$

In the above equation subscripts represent surfaces, ε is the emissivity of a surface, F is the view factor, Q is the heat transfer from a surface, A is the area of a surface, σ is Stefan Boltzmann's constant, and T is the surface temperature. Also δ_{kj} is defined such that:

$$\delta_{kj}=\begin{cases}1 & when\ k=j\\ 0 & when\ k\neq j\end{cases}$$

Figure 10.8.(b) shows an enclosure consisting of 6 surfaces. The energy flux along the sides of the yarn is considered uniform. The lower surface (j = 1) is the hot plate and the upper surface (j = 6) is the atmosphere whose temperatures T_1 and T_6 respectively are known. The equation 10.34 can be converted into a matrix equation in the following way which can be solved using matrix inversion method with the unknown values being Q_1, T_2, T_3, T_4, T_5, Q_6.

$$\begin{bmatrix}
\frac{1}{\varepsilon_1 A_1} & F_{12}\sigma & F_{13}\sigma & F_{14}\sigma & F_{15}\sigma & \frac{-F_{16}(1-\varepsilon_6)}{A_6\varepsilon_6}\\
\frac{-F_{21}(1-\varepsilon_1)}{\varepsilon_1 A_1} & -\sigma & F_{23}\sigma & F_{24}\sigma & F_{25}\sigma & \frac{-F_{26}(1-\varepsilon_6)}{A_6\varepsilon_6}\\
\frac{-F_{31}(1-\varepsilon_1)}{\varepsilon_1 A_1} & F_{32}\sigma & -\sigma & F_{34}\sigma & F_{35}\sigma & \frac{-F_{36}(1-\varepsilon_6)}{A_6\varepsilon_6}\\
\frac{-F_{41}(1-\varepsilon_1)}{\varepsilon_1 A_1} & F_{42}\sigma & F_{43}\sigma & -\sigma & F_{45}\sigma & \frac{-F_{46}(1-\varepsilon_6)}{A_6\varepsilon_6}\\
\frac{-F_{51}(1-\varepsilon_1)}{\varepsilon_1 A_1} & F_{52}\sigma & F_{53}\sigma & F_{54}\sigma & -\sigma & \frac{-F_{56}(1-\varepsilon_6)}{A_6\varepsilon_6}\\
\frac{-F_{61}(1-\varepsilon_1)}{\varepsilon_1 A_1} & F_{62}\sigma & F_{63}\sigma & F_{64}\sigma & F_{65}\sigma & \frac{1}{\varepsilon_6 A_6}
\end{bmatrix}
\begin{bmatrix}Q_1\\ T_2^4\\ T_3^4\\ T_4^4\\ T_5^4\\ Q_6\end{bmatrix}
=
\begin{bmatrix}
\sigma T_1^4 & -F_{16}\sigma T_6^4\\
-F_{21}\sigma T_1^4 & -F_{26}\sigma T_6^4\\
-F_{31}\sigma T_1^4 & -F_{36}\sigma T_6^4\\
-F_{41}\sigma T_1^4 & -F_{46}\sigma T_6^4\\
-F_{51}\sigma T_1^4 & -F_{56}\sigma T_6^4\\
-F_{61}\sigma T_1^4 & \sigma T_6^4
\end{bmatrix} \tag{10.33}$$

Once these values are obtained, the total net radiation heat exchange between the hot plate and the atmosphere is equal to:

$$Q_{radiation}=Q_1-Q_6 \tag{10.34}$$

The resistance due to radiation through the air gap per unit area is given by:

$$R_{radiationair} = \frac{\Delta T.R^2 p_1 p_2}{Q_{radiation}} \quad (10.35)$$

10.2.2.1.2. Radiation Heat Exchange through Fibrous Assembly

Radiation through a fibrous assembly can be solved in many different ways. One way is to consider the total contribution of radiation as a system of resistances.

Yuen et al. [15] used Mie scattering theory to find a first-principle based solution to the heat transfer by combined conduction and radiation for high porosity silica insulation materials where the radiation flux is given as:

$$q_r = \sigma\left(T_L^4 - T_R^4\right)s_L s_R + \sum_{i=1}^{N} \sigma\left(T_i^4 - T_R^4\right)g_i s_R \quad (10.36)$$

where T_L is temperature of the hot side; T_R is the temperature of the cold side; s is the surface exchange; g is the anisotropic scattering factor; and i is surface element.

Another method of modeling the radiation is by using the Linear Anisotropic Scattering (LAS) method. The LAS method assumes that when an infinite cylinder of length L and radius a, with its axis pointed into the direction of $\hat{s}_f$, is irradiated obliquely by electromagnetic waves, the scattering will be confined to a conical surface instead of all 4π angles (Figure 10.9). Hence anisotropic scattering takes place.

Tong and Tien [16-17] argued that even under moderate temperatures of 300K-400 K, conduction and radiation are two dominant modes of heat transfer in fibrous insulations. Radiation energy flux in lightweight fibrous insulations can be predicted using the equation:

$$q_T = \frac{\sigma\left(T_1^4 - T_2^4\right)}{\left(1 + \gamma\tau_0\right)} \quad (10.37)$$

where $\gamma = \frac{(3 - \omega\,\alpha)}{4}$; $\omega = \frac{\sigma_s}{\sigma_e}$; $\alpha = 2(1 - 2b)$ and $\tau_0 = \sigma_e.L$

q_T is the radiation heat flux; T_1 and T_2 are the temperatures of hot and cold surfaces respectively; σ_e, σ_s are extinction and scattering coefficients respectively; b is the back scatter fraction; τ_0 is the optical thickness of the material; L is the thickness of the material.

Many researchers have carried out investigations in interaction of radiation and conduction heat loss in determining the total thermal insulation of the porous medium. Lee [18] considered the fibers as infinitely long cylinders and obtained the effect of fiber orientation in the radiative properties of the fibrous media and found that the highest insulation is obtained when the fibers are oriented parallel to the direction of heat flow and the highest radiative heat transfer takes place when the fibers are perpendicular to the heat flow direction.

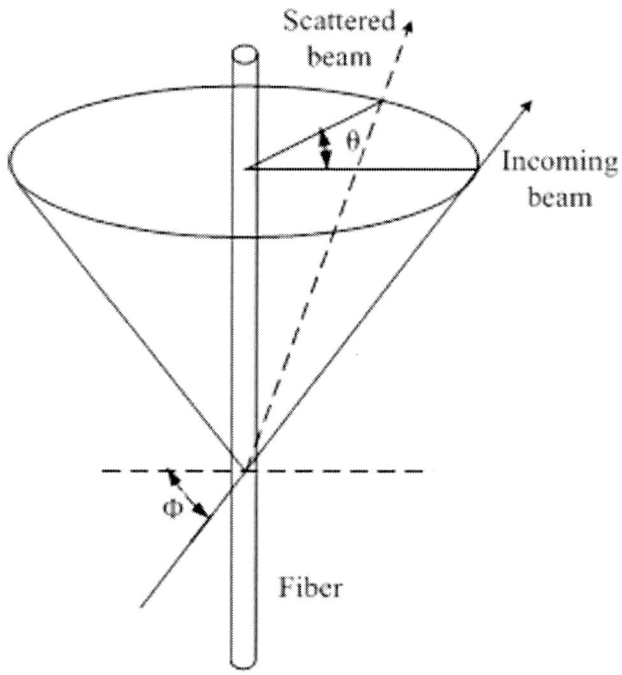

Figure 10.9. Linear anisotropic scattering of infinitely long cylinders [Source: Reference 5].

10.2.2.2. Convection Models

Convection heat transfer through fabric structures in particular has not been extensively investigated. In most cases, researchers like Yuen et al. [15], and Ismail et al. [9] have chosen to ignore the effect of heat transfer due to convection. Some of the researchers like Rees [11], and later Nield and Bejan [19] have considered the combined effect of conduction and convection. Studies on convection through porous media have increased significantly during recent years. This is because the applications of porous media can range from filtration to heat exchangers, energy storage units, ceramic processing, packed bed chemical reactors etc. Textile structures, being inherently porous in nature, have been put to use for many of these applications.

Extensive reviews of heat transfer and flow of Newtonian fluids in porous media were provided by Nield and Bejan [19] and Kaviany [7]. Modeling of forced convection in porous media involves several important physical phenomena. The momentum equation for low velocity, fully developed, one dimensional flow in a porous medium is given by the Darcy law for incompressible fluids:

$$\frac{\Delta p}{\Delta x} = \frac{\mu}{K} u \tag{10.38}$$

where p is the pressure gradient, u is the average velocity in the flow direction x, K is the constant known as hydrodynamic permeability and μ is the dynamic viscosity of the fluid. This equation is valid mainly for low velocity flows and do not consider the convective inertial drag. The inertial component is added in the Forchheimer equation:

$$\frac{\Delta p}{\Delta x} = \frac{\mu}{K} u + \frac{c}{\sqrt{K}} \rho u^2 \tag{10.39}$$

where c is a dimensionless constant and the second term in Equation 10.39 is called Forchheimer or the inertial resistance term [20]. The values of c and K are given as:

$$c = \frac{3.5}{d_p} \frac{(1-\varepsilon)}{\varepsilon^3} \quad \text{and} \quad K = \frac{d_p^2 \varepsilon^3}{150(1-\varepsilon)^2}$$

where ε is the porosity of the porous material, d_p is the characteristic dimension (in case of spherical particles, particle diameter). The hydrodynamic permeability 'K' is a geometric characteristic related to the structural features of the material and various length scales. When modeling laminar flow through a packed bed, the second term in the Equation 10.39 may be dropped, resulting in the Blake-Kozeny equation [21]. The equation over a packed bed then becomes:

$$\frac{\Delta p}{L} = \frac{150\mu(1-\varepsilon)^2}{d_p^2 \varepsilon^3} u \tag{10.40}$$

where μ is the viscosity, d_p is the mean particle diameter, L is the bed depth, and ε is the void fraction, defined as the volume of voids divided by the volume of the packed bed region. Jackson et al. [22] gave an extensive literature review and compilation on the values of hydrodynamic permeabilities of fibrous media calculated by various researchers. These were restricted to low Reynolds number flow through the fibrous media. The permeability of fibrous medium can be described by the porosity, size and arrangement of the fibers. The porosity can be given by the solid volume fraction φ. The relationship between permeability, size of the fiber and the solid volume fraction for a particular fiber arrangement is given by:

$$\frac{K}{a^2} = f(\phi) \tag{10.41}$$

where a is the fiber diameter and ϕ is the porosity.

10.2.3. Computational Fluid Dynamics

Computational Fluid Dynamics (CFD) is the analysis of systems involving fluid flow combined heat transfer and associated phenomena such as chemical reactions by means of computer-based simulation. Because of the mathematical complexities involved, theoretical fluid mechanics can only solve simple problems. Experimental fluid mechanics, on the other hand, is relatively expensive, and it is often difficult to observe the details in some flow fields. CFD, which has been developing very fast since 1970's, provides a tool that can simulate flow phenomena in some flow fields that cannot be solved by theoretical fluid mechanics and are difficult to study by experimentation. Most of the CFD tools are based on either finite volume or finite element methods which can solve modeling equations of continuity, momentum, and energy that have been temporally and spatially discretized. The basic process of formulation of a CFD model consists of preparing the model in a pre processing software, providing the required boundary conditions, discretization parameters and convergence parameters and finally using these conditions to solve the equations of continuity, conservation of momentum and energy. The flow chart of the simulation process is given in Figure 10.10.

There are many commercial CFD tools available viz. FLUENT, ANSYS/FLOTRAN, CFD-ACE+ etc. The choice of the tool depends upon a few aspects viz. (i) The type of flow and the model that is to be designed; (ii) ability to combine different mode of heat transfer along with the flow behavior; (iii) post processing features. CFD has been extensively used to understand the flow field and distributions of temperature around a human body in different environmental conditions. The fabric can be modeled as a porous material consisting of fibers and air. The equation of permeability for flow through perpendicular to an array of rods by Kuwabara [23] can be applied to find the permeability of the fabric, given by:

$$\frac{K}{a^2} = \frac{1}{8\phi}\left(-\ln\phi - \frac{3}{2} + 2\phi\right) \tag{10.42}$$

where K is the hydrodynamic permeability; a is the fiber diameter and ϕ is the porosity of the fibrous material. Table 10.1. gives the values of permeability for different fabrics.

The temperature profile when the fabric is placed over the guarded hot plate is given in Figure 10.11 The majority of the heat loss takes place from the top of the fabric surface. With increase in the air velocity over the fabric surface, the coefficient of heat transfer increases (Figure 10.12).

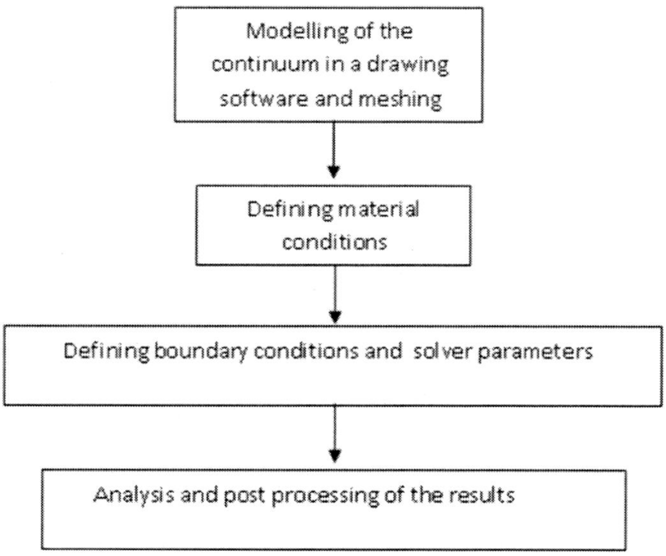

Figure 10.10. Flow chart of solution process by computational fluid dynamics [Source: Reference 8].

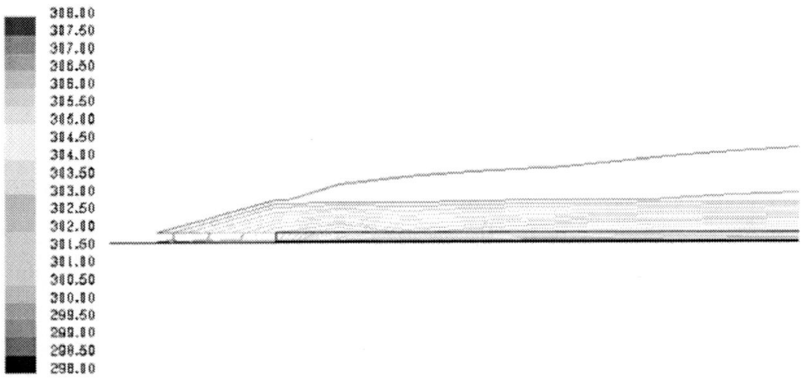

Figure 10.11. Temperature profile of forced convection when a fabric is placed over the guarded hot plate [Source: Reference 8].

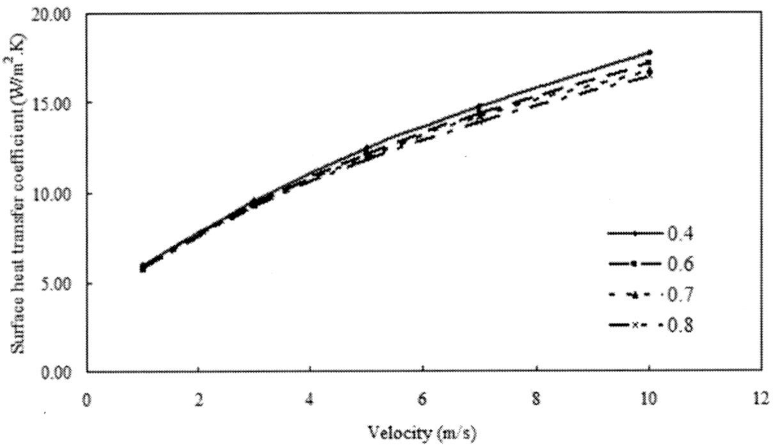

Figure 10.12. Effect of velocity of air on surface heat transfer coefficient for different porosities of fabrics [Source: Reference 8].

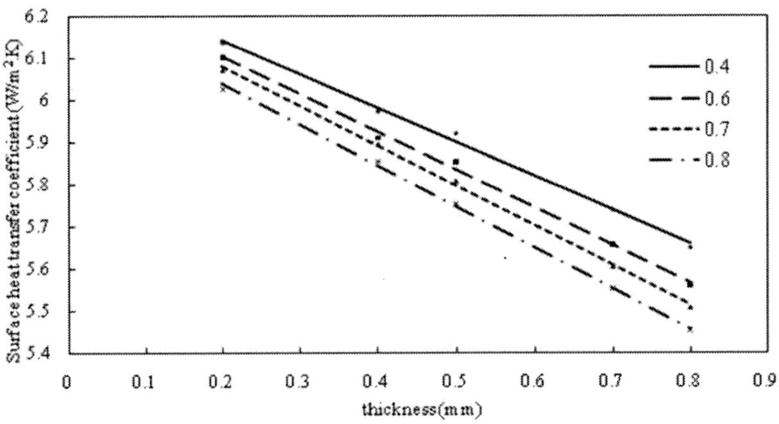

Figure 10.13. Effect of thickness of the porous fabric material of different porosities on the coefficient of surface heat transfer in case of forced convection [Source: Reference 8].

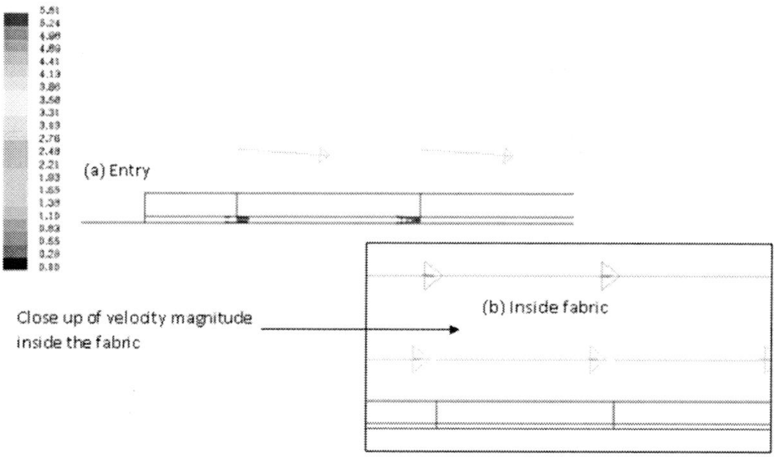

Figure 10.14. Velocity vectors inside the fabric region and the underlying air layer when the air velocity through the wind tunnel is 5 m/s (a) At the entry point, the air velocity is approximately equal to an average of 0.29 m/s (b) Inside the fabric, the magnitude of the velocity is equal to 0 [Source: Reference 8].

Table 10.1. Values of permeabilities of set of fabrics with different porosities [Source: Reference 8]

S.No	Type of flow	$f(\phi)$
1.	Flow parallel to an array of rods	$\frac{1}{4\phi}\left(-\ln\phi - K + 2\phi - \frac{\phi^2}{2}\right)$
2.	Flow perpendicular to an array of rods	$\frac{1}{8\phi}\left(-\ln\phi - \frac{3}{2} + 2\phi\right)$
3.	Flow through 3-dimensional arrays	$\frac{3}{20\phi}\left(-\ln\phi - 0.931 + O(\ln\phi)^{-1}\right)$

Thickness of the material is a major contributor in deciding the convective heat transfer coefficient of the fabric as compared to the porosity of the material. An interesting observation can be seen in Figures 10.12. and 10.13 where the coefficient of surface heat transfer decreases with increasing porosity which is contrary to the general expectations. This means, that for a particular thickness, the heat transfer is more when the porosity is less and vice versa. This happens because the increase in porosity means increase in still air spaces within the fabric which effectively increase the thermal resistance of the fabric (The thermal conductivity of air is 0.024W/m.K as compared to the thermal conductivity of the solid fiber which is approx.0.06W/mK). Hence, within increase in porosity, the heat transfer coefficient of the fabric is decreased [24]. The velocity vectors within the fabric structure, though existent are negligible in magnitude (Figure 10.14)

A similar kind of trend was also reported by Sung et al. [25]. In their case, they deduced that when the Darcy number decreases (with decrease in permeability), the porous material acts more like a solid and generates a higher recirculation zone, thereby increasing the heat transfer. One more reason for the insignificant effect of porosity is the position of the fabric with respect to the air flow. The pressure gradient between two surfaces of the fabrics is not as high in horizontal configuration as it is when the fabric is placed perpendicular to the direction of the air flow. The air pores within the fabric, in such configurations, do not play a significant role in the convection heat transfer and mostly act as solid insulating materials

10.3. Mass Transfer through Textile Structures

10.3.1. Principles of Mass Transfer

Mass transmission through textile material takes place both in the form of liquid water and water vapor. The process of moisture transport through clothing under transient humidity conditions is an important factor to influence the dynamic comfort of the wearer in practical use. Moisture may transfer through textile materials in vapor and in liquid form, as outlined below.

10.3.1.1. Water Vapor Transmission

Water vapor can pass through textile layers by the following mechanisms:

- Diffusion of the water vapor through the layers
- Absorption, transmission and desorption of the water vapor by the fibers
- Transmission of water vapor by forced convection

10.3.1.1.1. Diffusion Process

In the diffusion process, the vapor pressure gradient acts as a driving force in the transmission of moisture from one side of a textile layer to the other. The governing equation for diffusion is given by Fick's Law [26]:

$$J_{AX} = D_{AB}\frac{dC_A}{dx} \tag{10.43}$$

where, J_{Ax} is the rate of moisture flux; $\frac{dC_A}{dx}$ is the concentration gradient; and D_{AB} is the diffusion coefficient or mass diffusivity of one component, diffusing through another media.

For three dimensional mass transfers through the element, the governing equation is:

$$D_{AB}\frac{d^2C_A}{dx^2} + D_{AB}\frac{d^2C_A}{dy^2} + D_{AB}\frac{d^2C_A}{dz^2} + R_A = \frac{dC_A}{dt} \tag{10.44}$$

where R_A is the mass produced due to some chemical reaction and C_A is the moisture concentration. The diffusion coefficient of water vapor in air can be given as a function of temperature and pressure by the following equation [27]:

$$D_a = 2.20\times 10^{-5}\left[\frac{\theta}{\theta_0}\right]^2\left[\frac{P_0}{P}\right] \quad (10.45)$$

where D_a is the diffusion coefficient in bulk air (m^2/sec), θ is the absolute temperature (K), θ_0 is the standard temperature, P is the atmospheric pressure and P_0 is the standard pressure (bar).

Water vapor can diffuse through a textile structure in two ways, simple diffusion through the air spaces between the fibers and yarns and along the fiber itself [28]. In the case of diffusion along the fiber, water vapor diffuses from the inner surface of the fabric to the fibers' surface and then travels along the interior of the fibers and its surface, reaching the outer fabric surface. The diffusion flux through the textile material also depends on the concentration of water vapor in the inter-fiber void space [29].The equation of conservation of mass is given as:

$$K\frac{\partial^2 C_A}{\partial x^2} = \frac{\partial C_F}{\partial t} + \varepsilon\frac{\partial C_A}{\delta t} \quad (10.46)$$

where C_A and C_F are moisture concentrations both in void space and fibers respectively and ε is the porosity of the material.

In the case of hydrophilic fiber assemblies, vapor diffusion is governed by a non-Fickian, anomalous diffusion [30]. This is a two stage diffusion process. The two stage sorption has been first expressed by Nordon and David [31]. The first stage corresponds to a Fickian diffusion with a constant diffusion coefficient; but the second stage is non-Fickian diffusion associated with viscoelastic relaxation of the fiber structure [32].

$$\frac{\partial C_F}{\partial t} = (1-p)R_1 + pR_2 \quad (10.47)$$

where R_1 is the first stage moisture sorption rate of fibers, R_2 is the second stage moisture sorption rate of fibers, p is proportion of uptake that occurs during second stage.

The first stage moisture sorption rate of fibers R_1 can be obtained by considering that the sorption process is a radial diffusion of moisture in a cylindrical medium, governed by the following relationship [33]:

$$\frac{\partial C_F}{\partial t} = \frac{1}{r}\frac{\partial}{\partial r}\frac{\partial (rD_F C_F)}{\partial r} \quad (10.48)$$

where, D_F diffusion coefficient of fiber and r is the radius of the cylinder.

Second stage is much slower than the first, following an exponential relationship between the concentration gradient and the vapor flux [34]. The second stage sorption rate (R_2) is related to the local temperature, humidity, and sorption history of the fiber at each point in the fabric, which is assumed to have the following form:

$$R_2 = s_1 \frac{H_A - H_F}{|H_A - H_F|} \exp\left(s_2 / |H_A - H_F|\right) \tag{10.49}$$

where, s1 and s2 are constants for each fiber, HA and HF is the relative humidy of air and fiber respectively.In general, the diffusion coefficient of fibers increases with the increase in moisture concentration. In case of blended fabrics absorption of moisture by hygroscopic fibers will affect the response of fabrics to the moisture gradient [35]. Hence for simplification it is better to discuss the diffusion process in non-hygroscopic and hygroscopic cases separately. Bejan et al. [36] explained non-hygroscopic material as an inert solid phase. The effective diffusivity in this case has been determined by the following equation:

$$D_{eff} = \frac{\varepsilon D_a}{\tau} \tag{10.50}$$

where, ε and τ are porosity and tortuosity respectively. In case of non-hygroscopic material the predicted effective moisture diffusivity depends only on the geometrical arrangement of fibers and the moisture diffusivity in non-woven fabrics is given as:

$$D_{eff} = D_a P + D_a\left(1 - V_f - P\right)\left(1 - P\right) / \left(1 + sV_f - P\right) \tag{10.51}$$

where, P is the optical porosity corresponding to the air fraction s is the fiber shape factor.

Diffusion coefficient in case of hygroscopic material, due the influence of the absorption process has been discussed in the subsequent section.

10.3.1.1.2. Sorption-Desorption Process

Sorption-desorption process is an important process to maintain the microclimate during transient conditions. A hygroscopic fabric absorbs water vapor from the humid air close to the sweating skin and releases it in dry air. This enhances the flow of water vapor from the skin to the environment comparatively to a fabric which does not absorb and reduces the moisture built up in the microclimate [37-38]. In the absorption-desorption process an absorbing fabric works as a moisture source to the atmosphere. Considering hygroscopic fiber as a dynamic sink, the transient behavior of fabric is analyzed:

$$\varepsilon \frac{\delta C_A}{\delta t} = \frac{D_a \varepsilon}{\tau} \frac{\delta^2 C_A}{\delta x^2} - (1 - \varepsilon) \frac{\delta C_F}{\delta t} \tag{10.52}$$

Sorption kinetics in a diffusion process is described as:

$$\begin{cases} \dfrac{\delta C_F}{\delta t} = \dfrac{1}{r}\dfrac{\delta}{\delta r}\left(rD_f\dfrac{\delta C_F}{\delta r}\right) \text{for cylindrical fibres} \\ C_{fs} = f(C_a, T) \qquad \text{at fibre surface} \end{cases} \tag{10.53}$$

where, C_{fs} is the moisture concentration on the fiber surface.

Barnes and Holcombe [37] examined the magnitude of difference in moisture transport caused by fabric sorption, and its effect to wearer perception. Adsorption of water molecules takes place below a critical temperature, due to the Van der Waal's forces between the vapor molecules and the solid surface of the structure. In a thermodynamic equilibrium the chemical potential of vapor is equal to that of the absorbed film. An increase in vapor pressure causes an imbalance in chemical potential, and more vapor transfers to the absorbed layer to restore the equilibrium. The amount of water vapor which can be absorbed by the materials is dependent on the fiber regain and the humidity of the atmosphere (%). In the case of absorbent fibers, e.g. cotton, rayon, the moisture sorption is not only dependent on regain and humidity, but also on the phenomena associated with sorption hysteresis, the effect of heat, dimensional changes and elastic recovery effects, due to the reduced swelling of the fibers. During swelling, the fiber macro-molecules or micro-fibrils are pushed apart by the absorbed water molecules, reducing the pore size between the fibers as well as the yarns, thus reducing the water vapor transmission through the fabric. As swelling increases the capillaries between the fibers get blocked, resulting in lower wicking. Also, the distortion caused by swelling sets up internal stresses which influence the moisture sorption process. By the use of thermodynamics Morton and Hearle [30] have given a quantitative relation between swelling, moisture absorption and mechanical properties of the fibrous material:

$$\frac{V}{x}\left(\frac{\partial x}{\partial p}\right)_{XYZ} dX + \frac{V}{y}\left(\frac{\partial y}{\partial p}\right)_{XYZ} dY + \frac{V}{z}\left(\frac{\partial z}{\partial p}\right)_{XYZ} dZ = vdp - v\left(\frac{dp}{d\rho}\right)_{XYZ} d\rho \tag{10.54}$$

where V is the volume of the fibrous material, ∂p is change in force, $\partial \rho$ is change in moisture regain and $\partial x, \partial y, \partial z$ are displacements in x, y and z directions respectively.

10.3.1.1.3. Convection Process

Convection is a mode of moisture transfer that takes place while air is flowing over a moisture layer. The mass transfer in this process is controlled by the difference in moisture concentration between the surrounding atmosphere and the moisture source. The process is governed by following equation [39]:

$$Q_m = -Ah_m(C_a - C_\alpha) \tag{10.55}$$

where Q_m is the mass flow by convection through area A of the fabric along the direction of the flow, C_a is the vapor concentration on the fabric surface and C_α is the vapor concentration in the air. The flow is controlled by the concentration difference (C_a - C_α) and the convective mass transfer coefficient h_m, which depends on the fluid properties as well as on its velocity.

In the steady state, the latent heat lost by water due to evaporation is equal to the heat that comes to the water from the surrounding air, making it cooler. In that case the energy balance equation at the air-water interface is as follows:

$$q_{conv} = q_{evap} \tag{10.56}$$

where q_{conv} is the convective heat transfer from the surrounding air to the water and q_{evap} is the heat taken from the water due to evaporation. When internal ventilation coefficients are known at the skin, then the steady periodic time averaged sensible and latent heat losses per unit area from the skin is calculated, respectively, as [40]:

$$Q_s = h_{c(skin-air)}\left\{\frac{1}{\tau}\int_t^{t+\tau}(T_{sk} - T_a)dt\right\} + h_r\frac{1}{\tau}\int_t^{t+\tau}(T_{sk} - T_o)dt \tag{10.57}$$

$$Q_L = h_{fg}h_{m(skin-air)}\left\{\frac{1}{\tau}\int_t^{t+\tau}(P_{sk} - P_a)dt\right\} \tag{10.58}$$

where, Q_s and Q_L is the sensible and latent heat of ventilation respectively, T_{sk} is the skin temperature, T_a is the air temperature, P_{sk} is the skin vapor pressure, P_a is the air vapor pressure, $h_{c(skin-air)}$, $h_{m(skin-air)}$ is the mass transfer coefficient between the skin and the air layer, h_{fg} is the heat of evaporation of water.

10.3.1.2. Liquid Water Transmission

Liquid transfer through a porous structure involves two sequential processes– wetting and wicking. Wetting is the initial process involved in fluid spreading. In this process the fiber-air interface is replaced with a fiber-liquid interface. This process is followed by wicking. Capillary pressure and permeability are the two fundamental properties used to predict the overall wicking performance of a fabric [41]. The liquid flow through textiles is caused by fiber-liquid molecular attraction at the surface of the fiber materials, which is mainly determined by the surface tension and the effective capillary pore distribution and pathways [42]. Capillary action can be defined as the macroscopic motion or flow of a liquid under the influence of its own surface and interfacial forces. It is based on the intermolecular forces of cohesion and adhesion. If the forces of adhesion between the liquid and the tube wall are greater than the forces of cohesion between the molecules of the liquid, then capillary motion occurs. During wetting the forces at equilibrium at a solid-liquid boundary are commonly described by the Young-Dupre equation, given below [43]:

$$\gamma_{SV} - \gamma_{SL} = \gamma_{LV}\cos\theta \tag{10.59}$$

where, γ represents the tension at the interface between the various combinations of solid (*S*), liquid (*L*) and vapor (*V*), and θ is the contact angle between the liquid drop and the surface of the solid to be wetted (Figure 10.15). In the case of a textile material, the fiber represents the solid portion. When the liquid wets the fibers, it reaches the spaces between the fibers and capillary pressure develops. The liquid is forced by this pressure and dragged along the capillary due to the curvature of the meniscus in the narrow confines of the pores. The spontaneous flow of liquid or wicking occurs due to a pressure differential or capillary action. The magnitude of the capillary pressure is given by the Laplace equation [44]:

$$P = \frac{2\gamma_{LV} \cos\theta}{r_c} \tag{10.60}$$

where P is the capillary pressure developed in a capillary tube of radius r_c. Difference in capillary pressure in the pores causes the fluid to spread in the media. Hence, a liquid that does not wet the fibers cannot wick into the fabric [45]. In fibrous structure the pores are open channel capillaries, formed by the fibers' walls unlike the regular capillary channel [46]. The size and shape of the capillaries are dependent on the size and shape of the fibers as well their packing in yarn. Fiber type, under certain circumstances, can drastically change the structure of yarn, in turn changing the wicking properties of fabric. For example, changes in fiber properties when wet can significantly affect liquid movement and retention behaviors through fiber swelling. Yarns spun with natural fibers have irregular capillaries due to various factors such as fiber roughness, cross-sectional shape and limited length, which interrupt the flow along the length of the yarn [47]. The tortuosity of the pores also has a significant influence on the wicking process. It depends on the alignment of fibers as well as on irregularities in fiber diameter or shape along the pores [48]. With an increase in tortuosity of the pores, its wicking potential is reduced. The tortuosity in the capillary path increases with the increase in yarn twist.

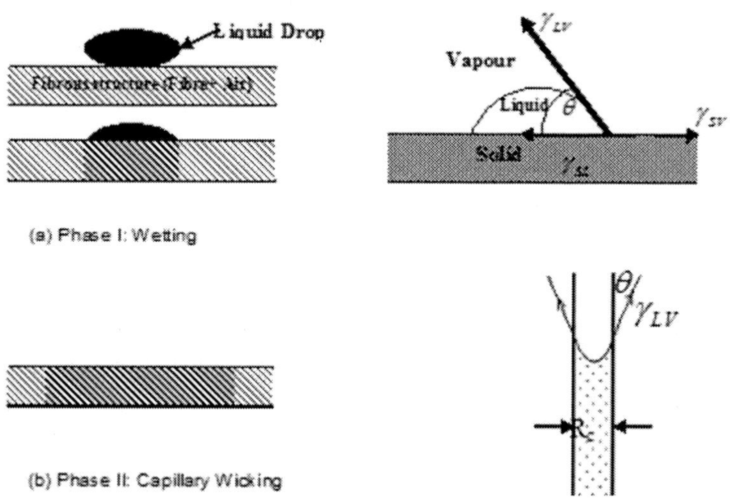

Figure 10.15. The liquid transfer processes through a porous media. γ_{SV} is the surface tension between solid and vapor interface, γ_{SL} is the surface tension between solid and liquid interface (dyne/cm), γ_{LV} is the surface tension between liquid and vapor interface, θ is the contact angle, Rc is the radius of the capillary [Source: Reference 63].

The density and geometry of fabric pores, which can be varied according to woven fabric structure, has a significant influence on the liquid flow pattern both in the interstices and downstream characterize porous media by analyzing the kinetics of capillary penetration. Ichikawa and Satoda [49] have described the interface dynamics of capillary flow in horizontal tube as:

$$\gamma \cos\theta.\pi D + \rho g z_0 \pi.\frac{D^2}{4} - 8\pi\mu.s\frac{ds}{dt} = \rho\pi.\frac{D^2}{4}.\left\{(s+x)\frac{d^2 s}{d^2 t} + \left(\frac{ds}{dt}\right)^2\right\} \tag{10.61}$$

where γ is the surface tension, θ is the contact angle, D is the inner diameter of a tube, ρ is the density of the liquid, g is the gravitational acceleration, z_0 is the hydrostatic height in the liquid container, μ is the viscosity, t is the time, s is the liquid column length from the inlet of the tube, and x and m are correction factors.

Capillary flow in a vertical circular capillary with the radius varying axially has been described by the following equation by Staples and Shaffer [50]:

$$t = \int_0^L \frac{8\eta[R(L)]^3 \int_0^L \frac{dz}{[R(z)]^4}}{2\gamma\cos\theta - R(L)\rho g L} dL \tag{10.62}$$

L is the length of the wetted liquid column.

It has been reported that the most important mechanism of fabric wicking is the motion of liquid in the void spaces between the fibers in a yarn [51]. Due to the laws of capillarity, the much larger pores between yarns do not contribute much to the long-range motion of liquid. It has also been seen that yarn intersections act as new reservoirs and feeds all branches equally and the rate of travel of liquid water is governed by the fiber arrangement in yarns which control capillary size and continuity [47]. Wicking mechanism in yarns is the key to fabric wicking performance.

10.3.1.3. Combined Vapor and Liquid Transmission

In a humid transient condition, moisture is transported through textiles both in liquid and vapor form. Li and Zhu [42] identified that the transportation of moisture in humid transient conditions happens in three different stages. The first stage is dominated by two fast processes - water vapor diffusion and liquid water diffusion in the air filling the inter fiber voids, which can reach a new steady state condition within fraction of a second. During this period, water vapor diffuses into the fabric due to the concentration gradient across the surfaces. Meanwhile, the liquid water starts to flow out of the regions of higher liquid content to the drier regions, driven by surface tension. During the second stage, the moisture sorption of the fibers is much slower than during the first stage, and takes a few minutes to a few hours to complete, depending on the heat transfer processes. Finally in the third stage steady state is reached in which all modes of moisture transfer and heat transfer become steady and the coupling effect between them becomes less significant. In this condition, distributions of temperature, water vapor concentration, fiber water content, liquid fraction volume and evaporation rate become independent of time. Combined liquid water and water vapor transmission along the fabric is very important in sweating condition. The liquid transport (i.e. liquid diffusion or capillary wicking) is very small compared with the vapor diffusion at low moisture content, whereas at saturation, capillary wicking is the major mechanism of moisture transport.

10.3.2. Modeling of Mass Transfer through Textile Structures

Prediction of moisture transmission properties of textile materials has always been the major concern before the clothing comfort can be characterized. This kind of work find immense use in the clothing industry as it offers to estimate of the comfort level of the fabric, even before actually it has been manufactured.

10.3.2.1. Modeling of Moisture Vapor Transmission

Several models have been developed by many researchers for better understanding of the diffusion phenomenon through textile materials. Wehner et al. [35] presented two mathematical models of the diffusion of vapor through a textile material, considering that: (1) the moisture diffusion within fibers is very rapid; (2) the moisture diffusion within fibers is very slow. Li et al. [52] have considered the two stage sorption process to develop mathematical model of moisture vapor transmission. They have explained two sets of variable of diffusion coefficient: a moisture content dependent coefficient for the first stage and a time dependent coefficient for the second stage. Li and Zhu [42] have developed a mathematical model on the simultaneous heat and moisture transfer, considering, moisture sorption, condensation, and capillary liquid diffusion in porous textiles. They have taken a series of complex coupling effects between heat transfer and moisture transfer processes, including heat transfer by conduction and moisture transfer by water vapor diffusion, liquid capillary action, and moisture sorption of fibers. They describe the diffusion coefficient of the moisture vapor in the fibers of the fabric; the liquid diffusivity has been determined by the following equation:

$$D_l(\varepsilon_l) = \frac{3\sigma \cos\phi \sin^{2\beta} d_c \varepsilon_l}{20\eta\varepsilon} \tag{10.63}$$

where d_c is the largest effective radius of the capillary, determined by the pore size distribution in the fabrics, σ is the density of the fiber and η is the viscosity of the liquid.

The heat released or absorbed within a hygroscopic textile layer with the change in humidity or temperature, is often physiologically very significant.

The equilibrium amount of moisture absorbed by a fiber assembly is a function of relative humidity and temperature. Le et al. [53] developed a model for the interactive heat and mass transfer in the time of forced convection of steam through absorbing fibrous media. They have calculated the changes in temperature in the condensing front due to flow of steam. They have used Darcy's law of fluid flow to describe the steam penetration through the media, but only the forced convectional flow of moisture, which is in the gas phase, has been considered.

Luo and Xu [54] have presented a 2D model based on Fick's second law and mass conservation law to understand the internal process of heat and moisture transfer in fibrous media.

$$\frac{\partial C^*}{\partial t^*} = D^*_a \left(\frac{\partial^2 C^*}{\partial^2 x^*} + \frac{\partial^2 C^*}{\partial^2 z^*} \right) \tag{10.64}$$

$$\frac{\partial C^*}{\partial t^*} = D^*_f \left(\frac{\partial^2 C^*}{\partial^2 x^*} + \frac{\partial^2 C^*}{\partial^2 z^*} \right) \tag{10.65}$$

Gibson and Charmchi [55] have developed a mathematical model of the convection and diffusion processes in textiles with inclusion of humidity dependent air permeability. They developed a set of partial differential equations to describe time dependant heat and mass transfer through porous hygroscopic materials; incorporating different important factors such as swelling of the fiber due to water imbibitions and the heat of sorption evolved when the water is absorbed by the polymer matrix, in the appropriate conservation and transport equations. They used the controlled volume finite difference method to solve the system of equations representing the convection/diffusion process and for the experimental purpose the Dynamic Moisture Permeable Cell (DMPC) has been used to validate the theoretical results. Their approximation for the humidity dependant permeability works well for the situation where the high humidity flow is being forced through the fabric but does not match the experimental data as closely in case of low humidity flow.

10.3.2.2. Modeling of Liquid Moisture Transmission

Darcy's law is used to describe a linear and slow steady state flow through a porous media, and is given by Equation 10.38.

Yarlagadda and Yoganathan [56] developed a mathematical model for fluid spreading in a composite web structure. They have used the mass equation for this purpose. They have considered an analogous three-dimensional transient heat transfer model with varying thermal conductivities and heat capacities, to represent the actual phenomenon of fluid spreading in the composite web. They have used three different models in their study to predict the dynamic fluid advancement in a web section:

$$K_x \frac{\partial^2 T}{\partial x^2} + K_y \frac{\partial^2 T}{\partial y^2} + K_z \frac{\partial^2 T}{\partial z^2} = \rho C_p \frac{dT}{dt} \tag{10.66}$$

where, K_x, K_y and K_z are the thermal conductivity of the material in the corresponding directions, c_p is the effective specific heat of the material.

With an idealized tube structure, the Hagen-Poiseuille law [57] for laminar flow through a circular tube is employed, as per the following equation:

$$Q\left(\frac{dh}{dt}\right)=\frac{r^2}{8\eta}\times\frac{\Delta P}{h} \tag{10.67}$$

where, Q is flow rate through the channel, η is viscosity of the liquid, r is capillary radius, ΔP is pressure gradient along the flow and h is the wetted length. Forces, acting during the vertical capillary flow through a capillary channel are the upward acting capillary force and the downward acting gravimetric force. The pressure gradient ΔP is given as:

$$\Delta P=\frac{2\pi r\gamma_{LV}\cos\theta-mg}{\pi r^2}$$
$$=(P_c-h\rho g) \tag{10.68}$$

where, ρ is the density of the liquid, m is the mass of the liquid, wicked along the channel and g is the gravitational acceleration. At the equilibrium height ($h_{\max}$), ΔP will be zero when

$$h_{\max}=\frac{P_c}{\rho g}$$

Hagen-Poiseuille's equation is also used for the steady state flow through a porous media. In such cases, to define the pore diameter, hydraulic diameter of the pore is considered and calculated on account of non-circularity and irregularity in the pore structure and spacing [58-59]. The capillary rise between the time of initial contact and the final equilibrium was obtained by Lukas-Washburn [60-61]:

$$\ln\left(1-\frac{\mathrm{h}}{\mathrm{h}_{\max}}\right)^{-1}-\frac{\mathrm{h}}{\mathrm{h}_{\max}}=\mathrm{C}_1\mathrm{t} \tag{10.69}$$

where h_{max} is the maximum wicking height or equilibrium wicking height and C_1 is a constant equal to $\frac{r_c^2\rho g}{8\eta\eta_{max}}$

At low values of t, where h is very small comparative to the h_{max}, the above equation has been approximated as follows to establish direct relation between wicking height (h) and the wicking time (t), commonly known as Washburn equation [61].

$$h=\sqrt{\frac{r\gamma\cos\theta}{2\eta}\times t} \tag{10.70}$$

Washburn equation is the only available equation till the date which directly provides the wicking height through a capillary channel for specified time t. This equation suits for the horizontal flow, where gravity effect is not present but the limitation in case of vertical wicking of this equation is that it is only valid for very low values of t and not for the complete wicking profile. The relation between the liquid front position and the time of the droplet wicking has been given by Lukas-Washburn equation (Equation 10.70). From this equation, it is expected that capillary rise at a specific time will be faster in a medium with larger pore size.

However, Miller, using a comparative wicking study, showed that this is not always the case [41]. He found that higher initial wicking through the capillaries with bigger diameter has been overtaken with time by the capillaries with smaller diameter.

A larger amount of liquid mass can be retained in larger pores but the distance of liquid advancement is limited. This may be explained by the Laplace equation, as the radius of the capillary decreases, the pressure generated in the capillary will be higher, causing faster flow through the capillary. When a large capillary *R* is connected to a small one [62], the net pressure acting to move liquid out of the large capillary into the small one is:

$$\Delta p = p_A - p_R = \left(\frac{2\gamma\cos\theta_A}{r_A}\right) - \left(\frac{2\gamma\cos\theta_R}{r_R}\right) \tag{10.71}$$

Combining the Laplace equation for capillarity and Hagen-Poiseuille law for laminar flow through pipe, Das [63] developed a model to explain the vertical wicking through yarn taking different fiber and yarn parameters into account. The factors have been taken into consideration in their model are fiber contact angle, number of filaments in a yarn, fiber denier, fiber cross sectional shape, yarn denier and twist level. This model not only determines the equilibrium wicking height, rather it provides the complete wicking profile. A mathematical function "*Lambertw*" [64-65] has been incorporated in the model, which helps to predict vertical wicking height at any given time, considering the gravitational effects.

$$\mathrm{h} = \mathrm{k}\left\{\mathrm{lambertw}\left(-\mathrm{e}^{-\frac{\mathrm{k+Bt}}{\mathrm{k}}}\right) + 1\right\} \tag{10.72}$$

$$\text{Where, } h = \text{Wicking height}, B = \frac{\left(\frac{D_H}{2}\right)^2}{8\eta}\rho g,\ k = \frac{\text{Capillary pressure}}{\rho g} = h_{max}$$

where, ΔP is pressure gradient, η is viscosity of the liquid, Q is flow rate (wicking height/time), D_H is hydraulic diameter of capillary (cm), h is wicking height at a time t, h_{max} is equilibrium wicking height, t is time of wicking, ρ is density of liquid, g is gravitational acceleration.

$$\text{Capillary pressure} = \frac{\text{Equivalent capillary force}}{\text{Area of the capillary}} \tag{10.73}$$

For twisted yarn, equivalent capillary force:

$$F_c = \frac{4\pi\mu\gamma_{LV}}{R_f}\int_0^{R_y}\frac{r_y\cos(\theta+\psi)}{\sqrt{1+4\pi^2 r_y^2 T^2}}dr_y - [\frac{2(\pi-\phi)\times 3.14}{180}\times R_f\times n\times\gamma_{LV}\times\cos(\theta+\cos^{-1}(\frac{1}{\sqrt{1+4\pi^2 r_y^2 T^2}}))$$

$$+R_f\left[\frac{(F+1)-\sin\phi}{\sin(\phi+\theta)}\right]\times\frac{2\pi}{180}\times(\pi-\theta-\phi)\times n\times\gamma_{LV}] \qquad (10.74)$$

where, $\varphi = \cos^{-1}\left[\frac{-\sin\theta\cos\theta\pm\sin\theta\sqrt{(F+1)^2-\sin^2\theta}}{(F+1)}\right]$ and $F = \frac{x}{2R_f}$

and n is the number of fibers at the outer layer of yarn, R_f is fiber radius, μ is yarn packing coefficient, x is distance between two fibers, a is the distance between nearest fibers' axis, γ_{LV} is surface tension between liquid and vapor interface (dyne/cm), θ is the contact angle, T is twist/cm, ψ is twist angle, R_y is yarn radius in cm.

In case of non-circular cross-sectional shaped fibers, R_f is replaced by:

$$\frac{1}{\chi}\sqrt{\frac{d}{\pi\times\rho_f\times 9\times 10^5}}$$

where d is fiber denier, χ is fiber shape factor and ρ_f is density of fiber. The new model has been verified conducting vertical wicking test of the following yarns using video metric method. The details of the used yarn samples have been given in Table 10.2

The model has been experimentally verified using multifilament polyester and polypropylene yarns. The details of the used yarns have been given in Table 10.3. Figure 10.16. gives the comparison between the equilibrium wicking height obtained from theoretical model and experimental results.

Table 10.2. Parameters of polyester and polypropylene yarns used for wicking studies [Source: Reference 63]

Sample code	Fiber type	Cross-sectional shape	Fiber shape factor	Contact angle	Fiber denier	Number of filaments /yarn	Yarn denier	Yarn radius (cm)
Trilobal	Polyester	Trilobal	1.27	75.75	4.72	32	151	0.0073
Triangular	Polyester	Triangular	1.16	75.75	4.72	32	151	0.0073
Circular-normal denier	Polyester	Circular	1	75.75	4.72	32	151	0.0072
Micro denier	Polyester	Circular	1	70	0.84	200	167	0.0085
PET1	Polyester	Circular	1	70.96	5.26	220	1158	0.019
PET2	Polyester	Circular	1	63.28	5.26	190	1000	0.018
PP1	Polypropelene	Circular	1	60.45	9.5	115	1100	0.022
PP2	Polypropelene	Circular	1	45.20	10	86	862	0.020

For all these samples the vertical wicking profile (from initial to equilibrium wicking) has been measured experimentally and also has been predicted from the theoretical model. It can be seen from Figures 10.17. and 10.18. that the developed model can predict the vertical wicking through yarn at any point of time with reasonable accuracy.

Table 10.3. Sample no. vs. yarn type [Source: Reference 63]

Sample No.	Yarn type
1	PET1
2	PET2
3	PP1
4	PP2
5	Microdenier-twist- 0 Turns/cm
6	Microdenier- 1 Turns/cm
7	Microdenier- 2 Turns/cm
8	Microdenier- 5 Turns/cm
9	Microdenier-10 Turns/cm
10	Trilobal
11	Triangular
12	Circular-normal denier-0 Turns/cm
13	Circular-normal denier-1 Turns/cm
14	Circular-normal denier-2 Turns/cm
15	Circular-normal denier-3 Turns/cm
16	Circular-normal denier-4 Turns/cm
17	Circular-normal denier-5 Turns/cm
18	Circular-normal denier-10 Turns/cm

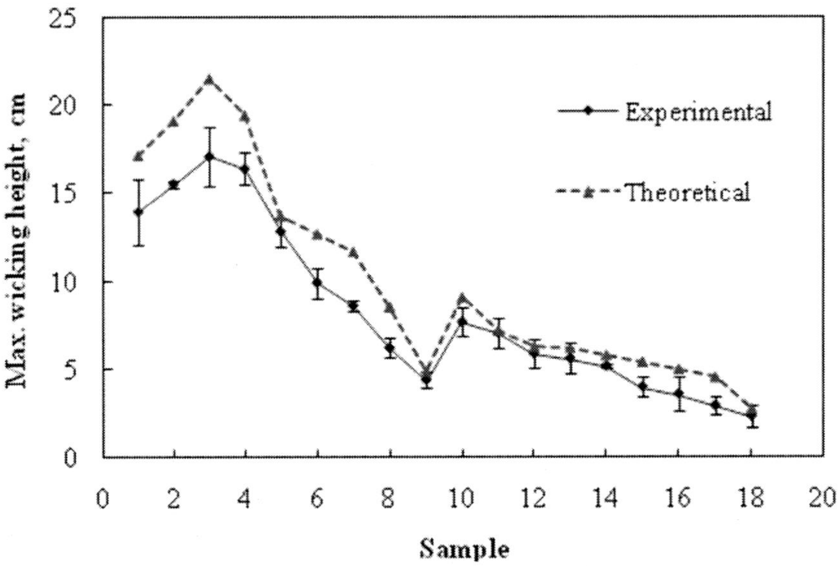

Figure 10.16. Comparison of maximum wicking height obtained from the vertical wicking test and the theoretical model [Source: Reference 63].

Based on the developed yarn model Das et al. [66], have modeled the vertical wicking through fabric. In order to model the vertical flow through fabric, along with the vertical flow through liquid carrying thread, the horizontal flow through the transverse threads has also been taken into account. A simplified fabric geometrical concept (Inclined tube geometry) and Peirce geometry for plain woven fabric was used to define the fabric structure. Warp and weft linear density, fabric sett and yarn crimp was considered in the fabric modeling. In virtue of the simplification the work only plain woven fabric structure has been considered for the mathematical calculation. For the determination of flow, initially the fabric has been defined considering a simplified geometrical concept, named as inclined tube geometry. In this concept the capillary channel geometry through the fabric has been defined considering the zigzag path of yarn, due the presence of crimp in the yarn, as shown in Figure 10.19. The angle of inclination of the tube is equal to the crimp angle of the thread. In case of vertical wicking through fabric along the warp direction, the warp threads are considered as the liquid carrying thread and the weft threads are the transverse threads which are in contact with warp threads and liquid will be transferred to a certain distance along this cross threads from the warp thread in the contact points. In case of vertical wicking through fabric along the weft direction, the role of the threads will be interchanged. Following equation describes the vertical movement of liquid along the fabric:

$$\mathrm{h}_{\mathrm{fab}} = \mathrm{k}' \times \{\mathrm{lambertw}(\text{-}\mathrm{e}^{(-(\mathrm{k}' + \frac{\mathrm{B}}{\mathrm{V_f}} \times \mathrm{t})/\mathrm{k}')})\} \tag{10.75}$$

where, $\mathrm{h}_{\mathrm{fab}}$ is fabric wicking height, k' is equilibrium wicking height along the fabric, V_f has been defined as volume factor, as it has replaced the term related with the structural parameters of the fabric and t is time of wicking. V_f has been used to replace the following term:

$$\frac{\cos\alpha_1^2}{Ac_1^{\,2}} \times \frac{1+c_1}{\cos\alpha_{Avg}} \times \left((1+c_1) + \frac{p_1}{p_2} \times (1+c_2) \times \frac{Ap_{weft}}{Ap_{warp}} \times \frac{Ac_2^{\,2}}{Ac_1^{\,2}} \times \frac{\cos\alpha_1^2}{\cos\alpha_2^2} \right) \tag{10.76}$$

where, Ap_{warp} and Ap_{weft} is the cross-sectional area of pore in warp and weft way respectively, p_1 and p_2 are warp and weft spacing, c_1 and c_2 are warp and weft crimp and α_1 and α_2 are warp and weft crimp angle respectively. Ac_1 and Ac_2 have replaced the terms $\left(1+\frac{D\alpha_1}{2l_1}(\cos\alpha_1 - 1)\right)$ and $\left(1+\frac{D\alpha_2}{2l_2}(\cos\alpha_2 - 1)\right)$ respectively; D is the summation of warp and weft diameter.

In Figure 10.20 the values of wicking height and the wicking values obtained from the experimental result for polyester fabric have been plotted.

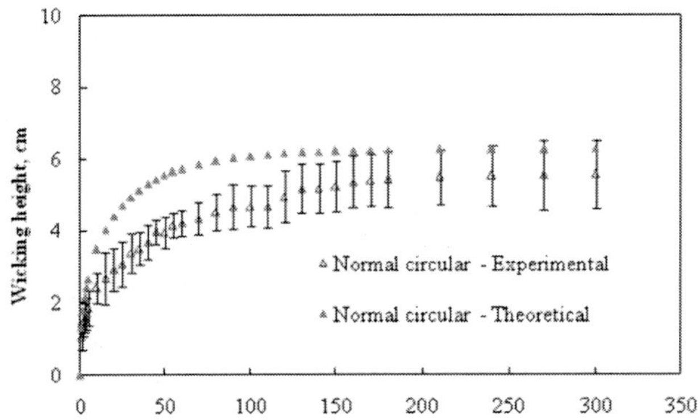

Figure 10.17. Comparison of theoretical and experimental wicking height through circular normal denier multifilament polyester yarn [Source: Reference 63].

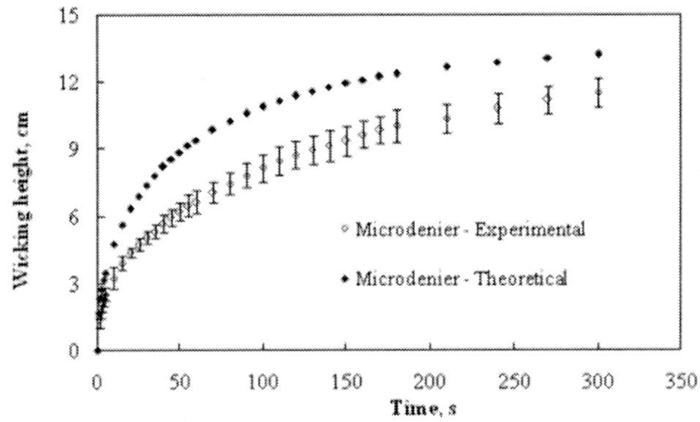

Figure 10.18. Comparison of theoretical and experimental wicking height through circular micro denier multifilament polyester yarn [Source: Reference 63].

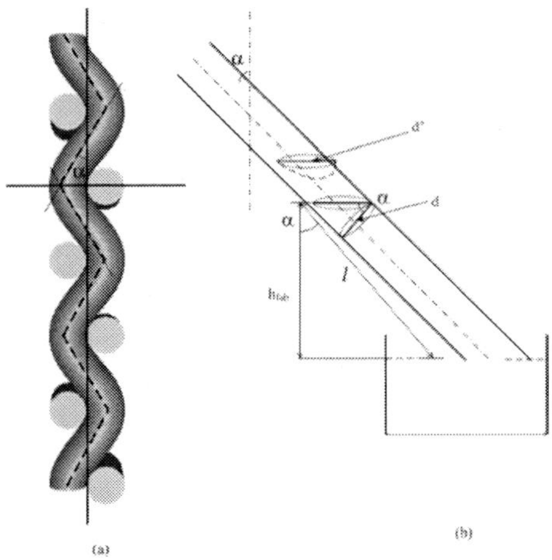

Figure 10.19. Yarn position in fabric, considering inclined tube geometry. α is crimp angle, d is diameter of the tube and d' is the diameter of the liquid meniscus, hfab is fabric wicking height and l is wicking length along the fabric [Source: Reference 63].

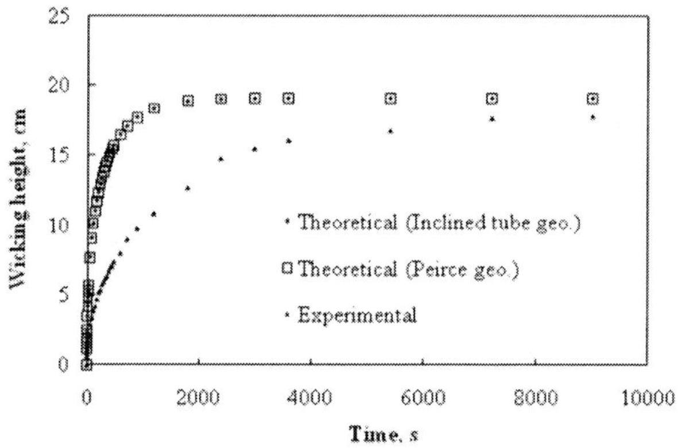

Figure 10.20. Comparison of wicking through polyester fabric as the result obtained from experiment and the proposed model [Source: Reference 63].

10.3.3. Coupled Heat and Moisture Transfer

Moisture transmission through a textile material is not only associated with the mass transfer processes, but heat transfer as well. Heat and moisture absorption in hygroscopic materials are interrelated. During the transmission of water molecules through textile materials, they are absorbed by fiber molecules due to their chemical nature and structure. The mechanism of the transient diffusion of heat and moisture into an assembly of textile fibers was first proposed and analyzed by Henry [29]. He developed a system of differential equations to describe the processes involved, derived from the conservation of mass and heat transfer:

$$\varepsilon \frac{\partial C_a}{\partial t} + (1-\varepsilon)\frac{\partial C_f}{\partial t} = K\frac{\varepsilon}{\tau} \cdot \frac{\partial C_a^{\ 2}}{\partial x^2} \tag{10.77}$$

$$C_v \frac{\partial T}{\partial t} - \lambda \frac{\partial C_f}{\partial t} = Q\frac{\partial T}{\partial x^2} \tag{10.78}$$

where, C_a and C_f are the water vapor concentration in the air filling the interfibrous pores and in the fibers of the fabric respectively. C_v is the volumetric heat capacity of the fabric, ε is the porosity of the fabric and Q the thermal conductivity of the fabric. Using these two basic equations Nordon and David [31] developed a model on coupled diffusion of moisture and heat in hygroscopic textile materials. They proposed an exponential relationship between the rate of change of water content of the fibers and the absolute difference between the relative humidity of the air and that with which the fiber would be in equilibrium. They considered a parameter *S* which controls the pseudo equilibrium towards which the first stage of sorption tends:

$$\frac{1}{\rho(1-\varepsilon)} \cdot \frac{\partial C_F}{\partial t} = k\left(y_A - y_F\right) \tag{10.79}$$

$$\frac{1}{\rho\left(1-\varepsilon\right)} \cdot \frac{\partial C_F}{\partial t} = \left[k_2 + k_1\left(S - \frac{y_F}{y_A}\right)^2 \frac{C_F}{\rho\left(1-\varepsilon\right)}\right]\left(y_A - y_F\right) \tag{10.80}$$

where y_A and y_F are the humidity of the air and that of the fibers in the fabric respectively. k_1, k_2 and k_3 are rate constants for mass transfer, ρ is the density of the fiber and S is a parameter controlling pseudo equilibrium. Absorption of water is followed by the liberation of heat, known as heat of absorption. It is the heat evolved when one gram of water is absorbed by the infinite mass of the material at given moisture regain. It is expressed in joules per gram (of water absorbed). The water may be absorbed from water vapor to give value Qv, or from liquid water to give value Ql. The heat evolved because of absorption of moisture by the fiber assemblies has a huge effect on the thermodynamics of the system.

It influences the diffusion process of the moisture transmission. The heat of sorption produced increases the temperature of the fibrous assemblies. As the temperature of the system increases the diffusion into the fibers coupled with the heat transfer process are imposed by the ability of fibers to absorb moisture. If a textile material is engrossed in humid atmosphere, the time required for the fibers to come to equilibrium with this atmosphere is negligible compared to the time required for the dissipation of heat generated or absorbed when the regain changes [67]. With an increase in humidity, the heat transfer efficiency of the material increases.

The heat transfer process also comes into play during the moisture transportation, under dynamic conditions, due to phase change of the water molecules. Thus, during the transient stages of moisture sorption and diffusion, the heat transfer process is coupled with four different forms of moisture transfer due to the heat released or absorbed during sorption/ desorption and evaporation/ condensation which in turn are affected by the efficiency of heat transfer and the length of the transient stage is dependent on the heat transfer process [68]. The coupling effect, between moisture diffusion and heat transfer depends on number of fiber properties, such as moisture absorption capacities (isotherm), water vapor diffusion coefficient, fiber diameter, fiber density and heat of sorption [69]. The heat of wetting of cellulosic fibers depends to some extent on the moisture regain and the crystalline structure, and it decreases proportionally with an increase in the degree of crystallinity of the fibers [70]. Two transient phenomena, buffering and chilling, are associated with the simultaneous heat and moisture vapor transport through fiber assemblies. The cooling effect or buffering effect is experienced due to perspiration in hot climates and the chilling effect is associated with the after exercise sweating in cool climates. At a sudden increase in relative humidity in the climate, fabrics absorb moisture maintaining a microclimatic condition and generating heat. This gives rise to a thermostatic or buffering action for the person wearing the fabric in clothing [71]. The cooling effect was first given by Spencer-Smith, who postulated that there would be a cooling effect at the onset of perspiration in hot climates, whereas in the case of cold climates it would result in a 'post exercise chilling effect' [72]. It reduces the working performance, even causing hypothermia. When water vapor (vapor perspiration) comes in contact with a cold wall (clothing), it condensates and reduce the thermal insulation of clothing. Both these phenomena are extremely dependent on atmospheric temperature and humidity conditions.

Conclusion and Future Scope

The phenomenon of heat and mass transfer through textile structures has been widely investigated. In spite of the inherent variability of textile materials, much of the physics behind the thermo-physiological comfort characteristics is already known. A number of researchers and scientists have tried to predict the same in terms of heat and mass transmission properties through fibers, yarns fabrics and other textile assemblies. However, owing to the complicated relationships between the various factors involved, most of the prediction still remains empirical in nature and are specific in nature. New tools like computational fluid dynamics, artificial neural networks, fuzzy logic, genetic algorithms are being currently used to simulate the actual conditions. CFD has been used to study the combined heat and mass transfer from manikin [73, 74]. Moghazy et al. [75] have worked on empirical modeling of the fabric comfort phenomenon using a combination of physical, neural-network, and fuzzy logic analysis. These tools can incorporate any amount of intricacies and inter-relationships between man-clothing-environment factors. However, these tools have to be used with care and experience as they are also prone to give misleading results if used with improper initial and boundary conditions and inappropriate functions.

References

[1] ANSI/ASHRAE 55 – 2004, Thermal Environmental Conditions for Human Occupancy.
[2] Havenith, G.; Holmer, I.; Den Hartog, E. A.; Parsons, K. C. *Ann. Occup. Hyg.* 1999, *vol 43*, 339-346.
[3] Chen, Y. S.; Fan, J.; Zhang, W. *Text. Res. J.* 2003, *vol 73*, 152-157.
[4] Fourt, L.; Hollies, N. R. S. *Clothing Comfort and Function*, *In Fiber Science Series;* Rebenfeld, L. Ed.; ISBN-10:0824712145; Marcel Dekker Inc.: New York, 1970.
[5] Modest, F. M. *Radiative Heat Transfer*; ISBN: 0125031637; Academic Press; 2003.
[6] Black, C. P.; Mathew, J. A. *J. Text. Inst.* 1934, *vol 25,* T197-T224.
[7] Kaviany, M. *Principles of Heat Transfer in Porous Media*; ISBN: 0387945504; Springer – Verlag: New York, 1995, 2nd Edn.
[8] Bhattacharjee, D. *Studies on Thermal Transmission Properties of Fabrics;* PhD Thesis, Indian Institute of Technology Delhi, New Delhi, 2007.
[9] Ismail, M. I.; Ammar, A. S. A.; El-Okeily, M. *Appl. Math. Modell.* 1988, *vol 12,* 434-440.
[10] Kothari, V. K.; Bhattacharjee, D. *J. Text. Inst.* 2008, *vol 99*, 421-432.
[11] Rees, W. H. *In The Third Shirley Conference Proceedings on Comfort*, Rees, W. H. Eds.; Shirley Institute, Manchester, UK, 1971.
[12] Bankvall, C. *J. Test. Eval.* 1973, *vol 1,* 235-243.
[13] Bhattacharjee, D.; Kothari, V. K. *Int. J. Heat Mass Transfer* 2009, *vol 52,* 2155-2160.

[14] Siegel, R.; Howell, J. R. *Thermal Radiation Heat Transfer*; ISBN-10: 0070573182; McGraw-Hill Kogakusha Ltd: Tokyo, 1972.

[15] Yuen, W. W.; Takara, E.; Cunnington, G. *The 6th ASME-JSME Thermal Engineering Joint Conference*, 2003.

[16] Tong, T. W; Tien, C. L. *ASME J. Heat Transfer* 1983, *vol 105*, 70-75.

[17] Tong, T. W.; Yang, Q. S.; Tien, C. L. *ASME J. Heat Transfer* 1983, *vol 105*, 76-81.

[18] Lee, S. C. *Intl. J. Heat Mass Transfer* 1989, *vol 32*, 311-319.

[19] Nield, D. A.; Bejan A. *Convection in Porous Media;* ISBN-10:0387984437; Springer-Verlag: New York, 1992.

[20] Angirasa, D.; Peterson, G.P. *Intersoc. Conf. Ther. Pheno.* 1996, 203-210.

[21] Ergun, S. *Chem. Eng. Prog.* 1952, *vol 48*, 89-94.

[22] Jackson, G. W.; James, D. F. Can. *J. Chem. Eng.* 1986, *vol 64*, 364-374.

[23] Kuwabara, S. *J. Phys. Soc. Japan* 1959, *vol 14*, 527-532.

[24] Bhattacharjee. D.; Kothari, V. K. *J. Text. Inst.* 2008, *vol 99*, 1754-2340.

[25] Sung, H. J.; Seo Y. K.; Hyun, J. M. *Int. J. Heat Fluid Flow* 1995, *vol 16,* 527-535.

[26] Sachdeva, R. C. *Fundamentals of Engineering Heat and Mass Transfer;* ISBN-13: 9788122421385; New Age International (P) Ltd: India, 2008, 3rd Edn, pp 595-597.

[27] Jost, W. *Diffusion in Solid, Liquids and Gases;* ISBN: 0123907500; Academic Press: New York, 1960, 3rd Edn., pp 423-425.

[28] Fohr, J. P. *Text. Res. J.* 2002, *vol 72*, 1-12.

[29] Henry, P. S. H. *Proc. R. Soc.* 1939, *vol 171A*, 115-141.

[30] Morton, W. E.; Hearle, J. W. S. *Physical Properties of Textile Fibers;* ISBN-10: 1870812417; The Textile Institute: Manchester, 1997, 3rd Edn.

[31] Nordon P. and David H. G. *Int. J. Heat Mass Transfer* 1967, *vol 10*, 853-865.

[32] Li, Y.; Holcombe, B. V. *Text. Res. J.* 1998, *vol 86,* 389-397.

[33] Crank, J. *The Mathematics of Diffusion;* ISBN-13:9780198534112; Oxford University Press: USA, 1980, 2nd Edn, pp 36-75.

[34] Nordon, P.; Mackay, B. H.; Downes, J. G.; Mcmahon, G. B. *Text. Res. J.* 1960, *vol 30*, 761-770.

[35] Wehner, J. A.; Miller, B.; Rebenfeld, L. *Text. Res. J.* 1988, *vol 58*, 581-592.

[36] Bejan, A.; Dincer, I.; Lorente, S.; Miguel, A. F.; Reis, A. H. *Porous and Complex Flow Structures in Modern Technologies;* ISBN-10:0387202250; Springer-Verlag: NY, 2004, 1st Edn.

[37] Barnes, J. C.; Holcombe, B. V. *Text. Res. J.* 1996, *vol* 66, 777-786.

[38] Suprun, N. *Int. J. Clothing Sci. Technol.* 2003, *vol 15*, 218-23.

[39] Incropera, F. P.; DeWitt, D. P. *Fundamentals of Heat of Mass Transfer;* ISBN: 9780471517290; John Wiley and Sons: NY, 1996, 4th Edn.

[40] Ghaddar, N.; Ghali, K.; Jones, B. *Convection and Ventilation in Fabric Layers, In Thermal and Moisture Transport in Fibrous Materials;* Pan, N.; Gibson, P. Eds.; ISBN: 1845690575; Woodhead Publishing Ltd; Cambridge, England, 2006, pp 271-307.

[41] Miller, B. *Int. Nonwovens J.* 2000, *vol 9,* 5-40.

[42] Li, Y.; Zhu, Q. *Text. Res. J.* 2003, *vol 73,* 515-524.

[43] Kissa, E. *Text. Res. J.* 1996, *vol 66*, 660-668.

[44] Chatterjee, P. K. *Absorbency*; ISBN-13:978-0444423771; Elsevier Science Publication Company: New Jersey, 1985.

[45] Wong, K. K.; Tao, X. M.; Yuen, C. W. M.; Yeung K. W. *Text. Res. J.* 2001, *vol 71*, 49-56.

[46] Mao, N.; Russell, S. J. *J. Text. Inst.* 2000, *vol 91*, 235-258.

[47] Hollies, N. R. S.; Kaessinger, M. M.; Bogaty, H. *Text. Res. J.* 1956, *vol 26*, 829-835.

[48] Perwuelz, A.; Mondon, P.; Caze, C. *Text. Res. J.* 2000, *vol 70,* 333-339.

[49] Ichikawa, N.; Satoda, Y. J. *Coll. Inter. Sci.* 1994, *vol 162*, 350–355.

[50] Staples, T. L.; Shaffer, D. G. *Coll. Surf. A: Physic. Eng. Asp.* 2002, *vol 204*, 239–250.

[51] Marchal, J. M. *Text. Res. J.* 2001, *vol 71*, 813-821.

[52] Li, Y.; Holcombe, B. V.; Scheider, A. M.; Apcar, F. *J. Tex. Inst.* 1993, *vol 84*, 267-273.

[53] Le, C. V.; Ly, N. G. *Text. Res. J.* 1995, *vol 65*, 203-212.

[54] Luo, X.; Xu, Q. *Appl. Math. Comp.* 2006, *vol 174,* 1135-1150.

[55] Gibson, P. W.; Charmchi, M. *Int. Comm. Heat Mass Transfer* 1997, *vol 24*, 709-724.

[56] Yarlagadda, A. P.; Yoganathan, A. P. *Text. Res. J.* 1990, *vol 60*, 23-32.

[57] Hagen, G. *Ann. Phys, Chem.* 1839, *vol 46,* 423-442.

[58] Millionshchikov, M. D. *Atom. Ener.* 1970, *vol 29,* 16-18.

[59] Lekner, J. *Eur. J. Phys.* 2007, *vol 28,* 521-527.

[60] Lukas, R. *Kolloid Z.* 1918, *vol 23,* 15-22.

[61] Washburn, E. W. *Phys. Rev.* 1921, *vol 17*, 273.

[62] Minor, F. W.; Schwartz, A. M.; Wulkow, E. A.; Buckles, L. C. *Text. Res. J.* 1959, *vol 29*, 931-939.

[63] Das, B. *Studies on Moisture Transmission through Fabrics*; PhD Thesis, Indian Institute of Technology Delhi, New Delhi, India, 2009.

[64] Weisstein, E.W. Lambert W-Function From *MathWorld* - A Wolfram Web Resource. http://mathworld.wolfram.com/LambertW-Function.html

[65] Fries, N.; Dreyer, M. *J. Coll. Inter. Sci.* 2008, *vol 320*, 259-263.

[66] Das, B.; Das, A.; Kothari, V. K.; Fangueiro, R. *J. Text. Inst.* 2010, *Accepted.*

[67] Cassie, A. B. D.; Atkins, B. E.; King, G. *Nature* 1939, *vol 143*, 162.

[68] Chen, Y. S.; Fan, J.; Zhang, W. *Text. Res. J.* 2003, *vol 73,* 152-157.

[69] Li, Y. *Text. Asia* 1999, *vol 6,* 39-41.

[70] Mizutani, C.; Tsujii, Y.; Bertoniere, N. *Text. Res. J.* 1999, *vol 69*, 559-564.

[71] Kim, J. O. *Text. Res. J.* 1999, *vol 69*, 193-202.

[72] Woodcock, A. M. *Text. Res. J.* 1966, *vol 36*, 855-856.

[73] Murakami, S.; Zeng, J.; Shinsuke, K. *Build. Environ.* 2000, *vol 35*, 489-500.

[74] Sørensen, D. N.; Voigt, L. K. *Build. Environ.* 2003, *vol 28*, 753-762.

[75] Mogahzy, A.Y.; Gupta, B. S.; Parachuru, R.; Broughton, R.; Abdel-Hady, F.; Pascoe, D.; Slaten, L.; Buschle-Diller, G. Design Oriented Fabric Comfort Model, Project No. S01-AE32, National Textile Centre Annual Report, 2003.

In: Modeling and Simulation in Fibrous Materials
Editor: Asis Patanaik
ISBN: 978-1-62100-116-4

Chapter 11

APPLICATION OF MODELING AND SIMULATION IN SMART AND TECHNICAL TEXTILES

Rajkishore Nayak[1, 2,*] and Rajiv Padhye[2]
[1]School of Fashion and Textiles, RMIT University, Melbourne and CSIRO Materials Science and Engineering, Melbourne, Australia
[2]School of Fashion and Textiles, RMIT University, Melbourne, Australia

ABSTRACT

This chapter includes the applications of modeling and simulation in the field of smart and technical textiles. Brief descriptions of different models for predicting the properties and performance of smart, intelligent, electronic and technical textiles are also included. The application of various types of models such as artificial neural network, computational fluid dynamics, finite element analysis, model of colloidal computing, fuzzy logic and magnetic resonance imaging techniques have been discussed. Different smart and technical textiles covered include phase change materials, shape memory materials, monofilament technical textiles, protective textiles and medical textiles. The models simulate various properties of smart and technical textiles such as comfort, tensile properties, water permeability, air permeability and performance. Some of the lacking areas which needed further developments and future directions are also highlighted.

11.1. INTRODUCTION

The application of modeling and simulation in textiles can be traced back to the work of Pierce [1] in 1930 for fabric drape simulation. Now as the textile industries around the globe are becoming extremely competitive, it is becoming more and more important for textile engineers to rely on modeling and simulation for automatisation of production processes and management procedures. Smart and technical textiles are gaining impetus day-by-day because of their enhanced properties and performance. Although numerous modeling tools have been

---

[*]E-mail: rkn.nayak@gmail.com

developed in case of textiles intended for clothing and apparel, the technical textile sector is still in infancy.

In this chapter the application of modeling and simulation in various areas of smart and intelligent textiles such as phase change materials (PCM), shape memory alloys (SMA), shape memory polymers (SMP), technical textiles such as automotive, medical, and protective clothing are highlighted. The other main area of smart textiles includes the electronic textiles. The application of modeling in the field of electronic textiles such as acoustic beam-former and shape-sensing pants has been discussed. The application of artificial neural network to model the complex physical behavior of airbag fabric has also been discussed. Various mathematical modeling developed for the technical textiles containing PCM such as diver's dry suit, protective clothing for cold environments and smart wall has been illustrated. The mathematical model used to calculate the values of heat and moisture transfer of porous textiles applied with PCM is also addressed.

Numerical models developed for other areas of technical textiles such as monofilament technical textiles, textile solar collectors coated with fluorescent dyestuff and ballistic fabrics are covered. The properties of technical textiles which are covered by the models include the water permeability, air permeability, liquid spreading, tensile properties and viscoelastic behavior. Model of colloidal computing supporting local computation and inexpensive communication among computational elements has also been addressed. The application of computational fluid dynamics (CFD) for modeling the airflow phenomenon in porous technical media containing fibers and fluid flow through the microstructure of hydroentangled fabric has been explored. The application of magnetic resonance imaging (MRI) technique to obtain 3D images which are used for CFD simulations has also been covered. The investigation on the initiation and development of damage during impact on the dynamic response of a composite motorcycle safety helmet by computational modeling is highlighted.

11.2. Application of Modeling and Simulation

11.2.1. Smart and Intelligent Textiles

There is a substantive difference between smart and intelligent textiles. Smart textiles can be defined as the textiles which can sense the environmental conditions or stimuli, whereas intelligent textiles can sense as well as react and respond to environmental conditions or stimuli. These stimuli could be thermal, chemical, mechanical, electric, magnetic or from other sources. Intelligent textiles may keep us warm in cold environments or cool in hot environments or provide us with considerable convenience and even fun in our normal day-to-day lives. Many intelligent textiles already feature in advanced types of clothing, mainly for protection, safety and added convenience.

11.2.1.1. Shape Memory Materials

Shape memory materials are those that can revert from the current shape to a previously held shape, usually due to the action of heat. The incorporation of shape memory materials such as alloys, polymers, bi-material film laminates and encapsulated bigels into garment confers greater versatility in the protection against extremes of heat or cold. When the shape memory materials are activated in garments, the air gap between adjacent layers of clothing are increased providing better insulation. Shape memory alloys (SMA) such as nickel-titanium and cuprous-zinc possess different properties below and above the activation temperature i.e. the temperature at which they are activated. Below the activation temperature the alloy is easily deformed, whilst once the activation temperature is reached, internal forces within the alloy cause it to return to a previously adopted shape and to become much stiffer. The temperature of activation can be chosen by altering the ratio of components in the alloy. The alloys also can react to temperature changes brought about by variations in physical activity levels.

Shape memory polymers (SMP) are known to be able to store up to three different shapes in memory and in general they cover wide property-ranges from stable to biodegradable, soft to hard and elastic to rigid, depending on the structural units that constitute the SMP. The shape memory effect is observed at a particular temperature when a polymer conforming to one shape returns to a previously adopted shape. For wearable electronic textile applications, the desirable temperatures for the shape memory effect to be triggered should be close to the skin temperature. The advantage of SMP over SMA is that being polymers, they will potentially be more compatible with textiles. As SMPs are thermoplastic polymers which melt at a much lower temperature than the alloys, they are not used in the case of hot liquids.

Polyurethanes (PU) with a segmented structure have been observed to have shape memory functions i.e. when they reach certain temperature, they deform into the memorized shape from any temporary one. The PUs are block-copolymers that consist of both soft and hard segments. A few reports have been published in the constitutive modeling of SMPs [2]. Hong et al. [3] evaluated the shape memory performance of segmented polyurethane fibers and utilized the viscoelastic theory for mathematical modeling of the thermo-mechanical behavior of the polymers. In this study, two materials were used to investigate the thermo-mechanical behavior of SMP and to access the feasibility of utilizing SMP smart fibers with thermo-responsive function. The two samples were synthesized by one- and two-step processes and were spun into fibers using conventional extrusion processes. The thermo-mechanical behavior of the SMP fibers was investigated using the tensile mode in dynamic mechanical thermal analysis (DMTA).

In the model, in order to describe the frozen stress and its activation in SMP, a linear viscoelastic theory was used as the first trial. Since the linear theory is limited to small deformations, subsequently a nonlinear viscoelastic theory was adopted for large deformations. The linear viscoelastic model was found to predict the characteristics of SMPs, especially the strain fixity and recovery properties for small deformations with some discrepancy between the experimental and calculated values. The main source for the error was found to be the reduced rigidity of SMPs due to the thermal treatment. This should be avoided for better shape memory performance of SMPs.

In another research, the thermo-mechanical behavior of SMPs was described by both linear and nonlinear viscoelastic theories [4]. In this work four element mechanical units consisting of spring, dashpot and frictional device were used to derive a constitutive differential equation. In order to determine the material properties by a constitutive differential equation; the modulus, viscosity and other parameters were assumed to decay exponentially with temperature. Liu et al. [5] developed a constitutive equation for SMPs based on thermodynamic concepts of entropy and internal energy. They adopted the concept of frozen strain and demonstrated the utility of the model by simulating the stress and strain recovery under various flexible constraint levels. Bi-material film laminates act similarly to bi-metallic strips by relying on differing coefficient of thermal expansion to produce a reversible bending effect. The temperature dependant bending movement can be harnessed to provide the reversible fabric separation required for variable insulation. Encapsulated bigels act by absorbing liquid at differing rates according to the temperature causing them to bend. This property offers possibilities for their use as actuators in a variable insulation system. There is limited application of modeling and simulation in these areas.

11.2.1.2. Phase Change Materials

One class of intelligent textiles include phase change materials (PCMs) which can absorb, store and release a large amount of latent heat over a defined temperature range when the material changes phase. PCMs are considered as substances with a high heat of fusion which melt and solidify at a certain temperature. They are also capable of storing and releasing a large amount of energy. Heat is absorbed or released when the material changes from solid to liquid phase and vice versa. PCMs are solid at room temperature, but as the temperature increases, they absorb the heat and melt, thus cooling the surroundings. Conversely, when the temperature drops, the material re-solidifies giving off heat, so warming the surroundings. Recently, the application of PCMs to improve the heat and moisture transfer properties of textiles and clothing has gained considerable attention. The thermal insulation of cold protective clothing may be significantly improved by incorporating micro-PCMs, that is a capsule containing small amounts of PCM. Numerical simulations based on mathematical modeling have been developed to describe the process of coupled heat and moisture transfer through porous technical textiles. These are effective ways to investigate the thermal properties of textiles and clothing. After incorporating PCMs into porous textiles, their dynamic thermal properties change significantly and the process of coupled heat and moisture transfer through these textiles become more complicated. Hence, the thermoregulatory effect of such textile assemblies with PCMs needs to be studied by mathematical modeling. Gear et al. [6] investigated the thermal storage and insulation properties of protective garments enhanced with PCMs using a finite difference procedure. They investigated the heat flow and thermal protection of diver's dry suit constructed from foam enhanced with microencapsulated PCM by using a one-dimensional finite difference procedure. The diver's dry suit with 0.0127m thickness was constructed from foam and the PCM applied was a mixture of octadecane (C_{18} paraffin) and a micro-fibrous insulating material called Thinsulate (a trademark of 3M Corporation). The outer layer of the suit was Thinsulate (40%) and the inner layer was PCM foam. The initial temperature of the suit was 32°C and the skin temperature of the diver was considered to be constant at 32°C. The experiment was carried out by immersing the diver's dry suit in water at 4°C.

In the study a mathematical model was developed to calculate the values of heat and moisture transfer through the porous textiles applied with PCMs. These values were compared with the porous textiles without PCMs [7]. The distribution of temperature, liquid water content and moisture concentration were numerically computed on the basis of a finite volume difference model. To validate the mathematical model, the surface temperature of the fabric with PCM was measured by a series of experiments. The diameters of the microcapsules were assumed to be identical and they were embedded in the porous textile structure. The mechanics of coupled heat and moisture transfer and thermal regulation by the PCM were discussed by considering the phase change process as a moving boundary problem and the phase temperature as a point. Polyester fabric was applied with a PCM which was mixed octadecane (99%) and impurities (1%) with a melting point between 28 to 29°C. The base polyester fabric was cast with a foam emulsion of micro-PCM which was applied to the back of the fabric by a coating method followed by drying and curing. A box plot chart of the experimental results was plotted (Figure 11.1) showing the comparison of computed temperature changes at upper surface of the fabric with experimental measurements. The box plot chart showed an obvious difference between the temperature of the samples with PCM and without PCM from 26 to 31.5°C. There was a reasonable agreement between the predicted value and the measured value for both the fabrics indicating the model is satisfactory. The computational and experimental results illustrated the complex multiple effects among different moisture transport and heat transfer processes. This model can be used to calculate the amount of microcapsules dispersed throughout the fabric or deposited in clumps, to study the orientation of the PCM layer to the body and the amount of body surface area covered by PCM garments. Some valuable numerical results in the design of new fabrics and smart garment products can also be obtained from this model. Although the effects of the amount of applied PCM on the heat and moisture transfer in textiles were discussed in the model, the influence of fiber hygroscopicity was not included. An analytical model of a diver's dry suit enhanced with microencapsulated PCMs was established by Nuckols [8]. The analytical model was used to assess the relative thermal performance of candidate Comfortemp (trade names used by Frisby Technologies of Clemmons, NC) foams with different weight percentages of micro-encapsulated PCMs in order to maximize their thermal storage capabilities. As steady-state heat transfer analyses are not helpful in assessing the dynamic behavior of the PCMs, an analytical model using a finite-difference technique was developed to assess the transient behavior. The numerical methods adopted for the study fall into the generic category of the Euler (Explicit) method [9] for solving a system of ordinary differential equations. A similar approach was presented by Hittle and Andre [10] to study the influence of PCM on a dry fabric. The effect of phase change temperature range which is related to the heating/cooling rate, on the thermoregulation process of clothing containing PCM was not considered in the study.

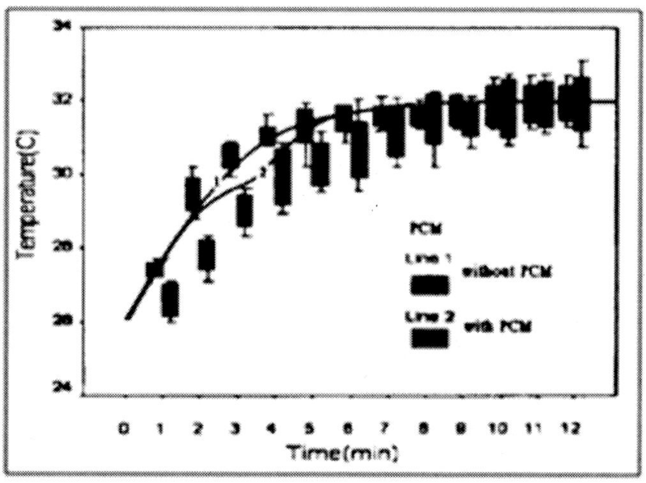

Figure 11.1. Comparison of computed temperature changes at fabric surfaces with experimental measurements (thickness = 2mm, $\varepsilon_m = 0.085$) [Source: Reference 7].

Wang et al. [11] described a simulation technique for the physical processes of coupled heat and moisture transfer in a clothing assembly containing PCM in cold environment (-15°C). The clothing system used for the study consisted of four layers, namely cotton, nonwoven polyester treated with PCM, nonwoven polyester fabric and waterproof breathable fabric. The PCM used was bromo-octadecane (BrC_{18}) with a melting point of 26.4°C which was enclosed in small polymer spheres with diameters of a few micrometers. A bionic skin model simulator operated in a climatic chamber was employed for the study. The surface temperature of the apparatus was maintained at 33°C and the sweating rate was kept equivalent to the insensible perspiration state (350 g/m^2/24hr).

The temperature and humidity were dynamically recorded by a computer using five groups of temperature and humidity sensors. The simulation can be used for predicting the temperature of the layer coated with PCM. Although there was some agreement between the experimental and simulated results, they don't fit very well. Both sets of results show that the PCM can delay the temperature change of the assembly when the environmental temperature becomes cold. This mathematical model can be used to predict the temperature of the assembly and the effective time of the PCM when the temperature of the environment changes.

Li et al. [12] analyzed the thermal characteristics of a three layer wall with microencapsulated PCM polyester fabric by numerical simulations. The simultaneous heat and moisture transfer through a porous building material containing PCM was taken into account by the mathematical model. The study provided an effective way to investigate the thermal properties of porous materials containing PCM, but the simulations were very complicated. The wall consisted of a wooden middle layer to which porous textile material was fixed at the inner and outer side as shown in Figure 11.2.

To identify the best wall structure, walls with and without PCM were implemented in this simulation. The thermal performance of PCM on the wall temperature variations was compared and examined. The temperature and relative humidity of the inner and outer environment were kept constant for the simulation. A series of numerical simulations with sets of properties and conditions were implemented in order to examine the thermal effects of the multi-layered wall. The model successfully predicted the behavior of the wall during the melting and solidifying process of PCM.

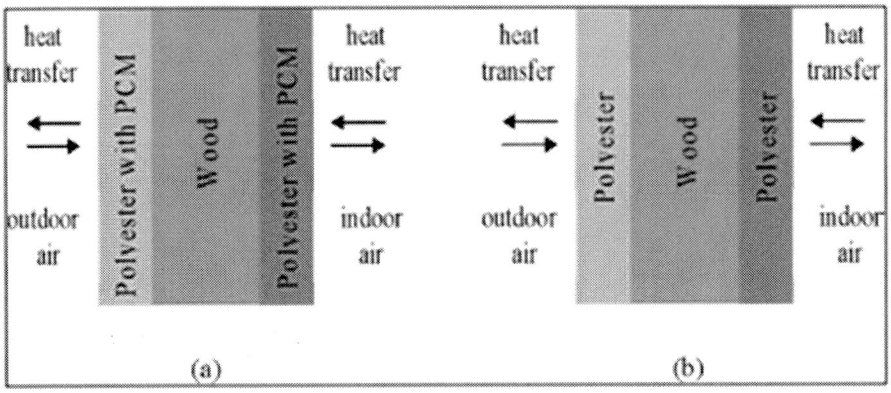

Figure 11.2. Three layer wall (a) with PCM and (b) without PCM [Source: Reference 12].

For the analysis of the interaction between PCM and self-heating material (i.e. the material, when in contact with air and without an energy supply, is liable to self-heat), the simulation of the physical processes of coupled heat and moisture transfer was studied by Li et al. [13]. Using the mathematical model on fabric incorporated with PCM and self-heating elements, the progressive coupled thermal and moisture transfer through the fabric was simulated according to the different initial conditions and boundary conditions. The simulation results indicated that the effective time of the PCM with self-heating was longer than the one without self-heating. This model can be used for designing and developing smart clothing by considering the interaction between PCM and self-heating material.

In the numerical model, the heat-flux from a temperature control system was treated as constant which was positive while the heating system works and negative while the cooling system works. The fabric used for the simulation was a single-layer fabric with 10% PCM (melting point 28°C) and a self-heating system based on polyester. When the temperature regulating function of the PCM disappears, the self-heating system is switched on or off to recharge the PCM by re-melting or re-condensing it. When the PCM is functional, the self-heating is switched off to save battery power. The simulation results had shown that the PCM can be recycled to maintain constant temperature in the fabric for longer time and the method can be used to optimize the design of the smart clothing. The experimental results and the calculated results showed good agreement.

In order to consider the properties of a PCM and the moisture sorption/desorption of the fibers more closely, a new model of heat and moisture transfer in hygroscopic textiles was developed by Fengzhi and Li [14]. The moisture sorption of fibers, the effects of the phase transition temperature range and the heating/cooling rate on the phase change processes of the PCM were taken into consideration in the model. In addition, the model was used to investigate how the fiber hygroscopicity influenced the heat and moisture properties of the textiles containing PCM microcapsules. The theoretical predictions were validated by experimental data. A series of three-dimensional diagrams illustrated the moisture transfer process in the inter fiber void spaces and within the fibers, and also their interaction with heat transfer and regulation of PCM microcapsules during ambient transient conditions.

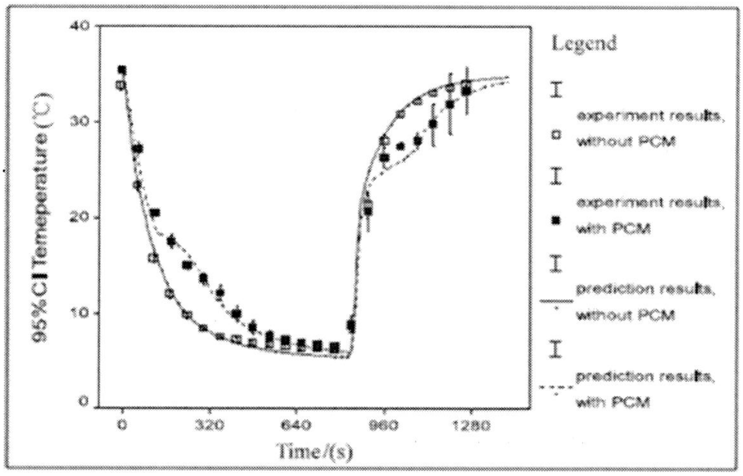

Figure 11.3. Comparison of the results from the predictive model and experimental data [Source: Reference 14].

The base material fabrics made of polyester (4.4 mm thickness) were used in the model. One fabric contained the PCM microcapsules (5 μm diameter) of bromo-octadecane (BrO_{18}) with a volume fraction of 0.35 and the other (control) fabric was without PCM. Two different climate chambers (hot and cold) were used and the temperature sensors were set in the middle of the fabrics. The relationship between the heating/cooling rates, the characterizing temperatures and latent heats were derived by a curve fitting method from the DSC data of the PCM. Figure 11.3 shows the comparative plots of the results from the predictive model and the experimental data. It can be observed that the theoretical predictions and the experimental data show good agreement during the PCM melting and freezing.

In another work, the thermoregulatory effects of PCM textile assembly have been carried out using a mathematical model of heat and moisture transfer through multilayered PCM textile assemblies [15]. This model comprised a three-layered polyester based PCM textile assembly having a 2.5 mm thick inner layer, a 5 mm thick middle layer, and an outer layer same as the inner layer, each separated by 1 mm layer of air. The temperature distribution throughout the PCM textile assembly was numerically computed with different PCM add-on levels, and compared for both warming and cooling cycles. This computational investigation could be used to design and develop smart functional clothing by applying PCMs with different range of transition temperature, add-on levels and temperature distribution.

11.2.1.3. Electronic Textiles

Electronic textiles or e-textiles are textile fabrics with interconnections and electronics (consisting of both sensing and processing elements) woven into them. These are the class of smart textiles which brings together specialists in information technology, microsystems, materials and textiles. E-textiles rely on simple computing elements embedded into fabric or directly into active yarn structure. E-textiles are an extreme form of distributed computing, physically spread over a relatively smaller space, but having a greater dependence on physical locality of computation, lower bandwidth requirement for communication and less available energy. E-textiles have potential applications in wearable computing, medical monitoring, military uniforms and sensor networks.

As the designing of e-textiles is still in its infancy, it is clear that prototyping alone cannot adequately explore the design space for most applications [16]. A simulation environment is necessary for e-textiles that allows to adequately model a range of areas such as the physical environment in which they will be operating, the behavior of sensing and processing elements and the software execution.

Also the simulation results are not useful unless the simulation environment has been validated against experimentally measured data. Any simulation tool for e-textiles must be able to model the physics of the environment and the behavior of the sensors used to monitor that environment. An e-textile simulation tool for wearable computing applications must also be able to capture aspects of the human body such as body size and limb motion in order to model the dynamic location of sensors.

The draping and motion of the cloth in response to the underlying movement of the wearer may also have to be modeled to fully account for the movement of the sensors because most garments do not fit tightly on the human frame. For the successful simulation of e-textiles, the areas to be encompassed include physical environment, sensor behavior, human body and motion, motion/draping of clothing, manufacturability (weaving and piece work), networking, power consumption and software execution. Sensing and processing elements must be interconnected through an on-fabric network.

Martin et al. [16] designed a large-scale acoustic beam-former to locate passing vehicles based on their acoustic emissions by using the Ptolemy II software as a means of integrating a wide range of simulation capabilities. The simulation environment has a representation of the number and location of sensors, processors, communicating wires and power lines. The model can simulate the location of power and ground lines and also the energy consumption. The results from the large-scale acoustic beam-forming fabric were used to illustrate the utility of the simulation environment in the exploration of e-textile design space.

A pair of shape-sensing pants were also developed by Martin et al. [16] for context-awareness in particular the current activity of the user. Before building the hardware prototype, the garments were simulated using recorded motion data and modeling the physical behavior of two types of sensors.

The simulation environment was used to train the neural network for classifying user motion into walking, running, jumping and standing using similar approach as described in the paper [17]. The simulation of the system before building a prototype has several advantages over prototyping alone, as mentioned below:

1. There is a higher confidence level that the system will function for a broad segment of the population by using the motion data for a wider set of individuals. As the motion database covers a wide range of body sizes and shapes, the testing of each iteration of the hardware prototype on individuals covering the same range of body sizes is eliminated.
2. The possibility of studying the type of sensors used and the location of the sensors on the body without building a hardware prototype exist. Thus the feasibility studies can be performed more rapidly and the number of iterations that a prototype must go through before it has the desired functionality is decreased.
3. The use of recorded human motions makes the experiments more repeatable. During the testing of a hardware prototype, the subject will not make exactly the same motion each time the experiment is run, introducing variability in the results.

In the work, physical models for the garment with two types of sensors, accelerometers and piezoelectric film strips were developed. Physical models for each of these sensor types were developed with data from a publicly available human motion database [18]. The human motion database consisted of three-dimensional position data for various locations on a subject's body for each video frame. The high frequency noise in the data was smoothened using the Matlab spline toolbox. The acceleration of each monitored location on the body was calculated from a sliding window of three position samples and the sampling frequency of the video system by using the smoothed position data. The resulting accelerations were used to drive models of an accelerometer and a piezoelectric film strip. The simulation model for the piezoelectric film was developed by sensing the angle of a joint due to the lack of force data in the motion database.

The simulation model consisted of inputs from three points forming a joint angle such as the hip, knee and heel. The best results were obtained by using individuals who were near the mid-range of the population in terms of maximum heel acceleration while the worst results were obtained for individuals who were outliers at either end of the population. The results of the simulation environment for the context-aware pants were verified against ground truth data, showing a close agreement between the simulation and actual data.

Conductive textiles are one class of e-textiles which are produced with the addition of nickel, copper and silver coatings to textiles to provide versatile combination of physical and electrical properties. These materials provide designers with new opportunities in developing contoured and portable products for both existing and emerging product markets. Several models have been developed on conductive textile characteristics. Cottet et al. [19] studied the conductive textile characteristics and modeled for high frequency signal transmission. A mathematical model was developed with different modules connected to the bus through stingers [20]. The network model based on conductive textile transmission lining (CTTL) eliminated the need for traditional cables and helped for data acquisition and simulation.

In order to study the deterioration of frequency dependent per-unit-length parameters of CTTL for wearable textiles, Chedid et al. [21] applied the combination of equivalent circuit modeling and three-dimensional Transmission Line Matrix (TLM) modeling. Besides the geometrical data, a TLM model requires the electrical properties such as conductivity, permittivity and permeability of the materials used. The deterioration of the electrical properties of the CTTL was investigated by measuring the per-unit-length parameters when CTTL was subjected to environmental stress during washing cycles. The transmission line of Nickel/Copper plated polyester fabric was used for the study. With the measured data of per-unit-length parameters (resistance and inductance), an equivalent circuit was created to model the degradation of the line. Both the equivalent circuit and TLM model described the degradation of the conductive textile when washed.

A commercially available electromagnetic simulation tool (Flo/EMC) from Flomerics was used in the TLM simulations and the computational mesh was composed of cuboids of different sizes. Convergence tests were performed to determine the effect of the open boundary location on the simulation results. In order to properly define thin plates in TLM simulation, their equivalent thickness (t_{eq}) was calculated from the surface resistivity. The effective conductivity of the conductive textile was calculated using the method proposed by Henn [22]. The conductive textile in the CTTL was modeled as thin plates of isotropic dielectric material with fixed frequency, independent relative permittivity, fixed permeability and a fixed conductivity. The efficiency of the model was confirmed by good agreement between the measured and simulated data and this model can assist during virtual prototyping in design phase of wearable electronics.

Carysse et al. [23] integrated textile sensors in a wireless monitoring suit for monitoring the electrocardiogram (ECG) and respiration rate of children in a hospital environment. The suit consisted of: a) the sensors integrated in a prototype belt of the suit, b) interface, c) data handling and storage devices, and d) transmission electronics. Two types of stainless steel electrodes (namely woven and knitted) were used as an alternative to the conventional gel electrodes. The model consisted of various factors such as the skin impendence, electrode impendence and a DC offset potential. A read-out amplifier was used to overcome the problems associated with the high skin-electrode impedance. A fabric sensor called as 'Respibelt' was simulated using different software packages for respiration measurement. This research concluded that textile materials have a strong potential to be used as sensor elements, antennas and interconnections in a wearable bio-instrumentation system.

Model of Colloidal Computing (MC^2) [24] concept borrowed from physical chemistry [25] with some of its characteristics resembling the features of e-textile based computational model, supports local computation and inexpensive communication among computational elements. As shown in Figure 11.4, simple computation particles are dispersed in an inexpensive communication medium which is sufficiently fast but unreliable. Colloidal particles in the case of unstable colloidal suspensions tend to coalesce or aggregate together due to Vander wall and electrostatic forces. Operational longevity and effective usage of e-textiles has to be supported by an entirely new computational and architectural model.

In MC^2 model, the application is viewed as a set of nano particles loosely communicating so as to exchange information and the underlying architecture has local computation as well as remote communication capabilities when needed. The MC^2 model supports a wide range of computational granularities from computation embedded in the yarn up to simple or clusters of sensors.

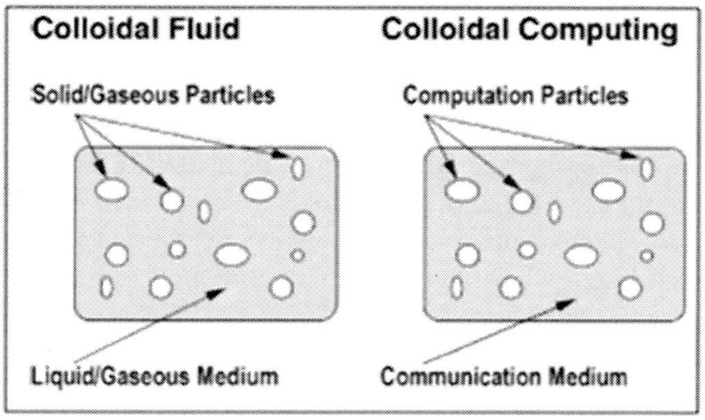

Figure 11.4. Model of Colloidal computing [Source: Reference 24].

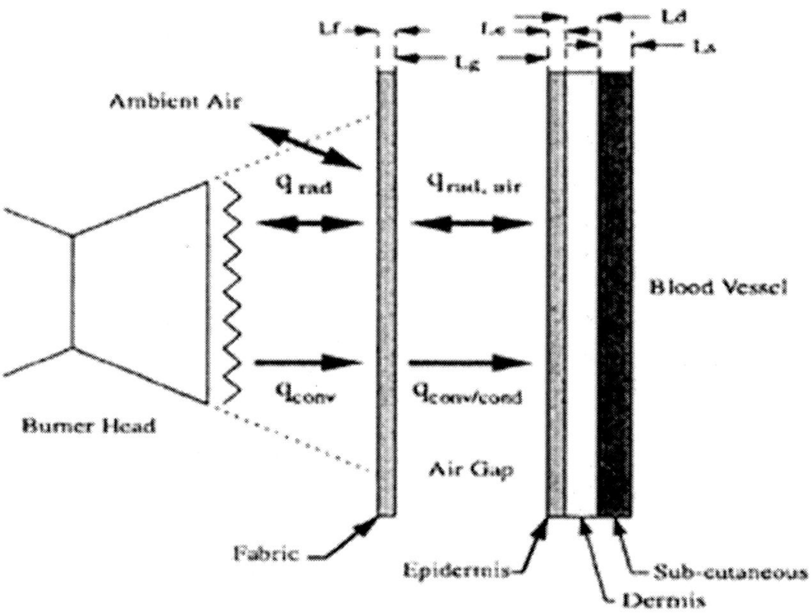

Figure 11.5. Schematic diagram of the one-dimensional heat transfer model [Source: Reference 30].

11.2.2. Technical Textiles

Technical textiles are the textiles manufactured for non-aesthetic purposes, where function is the primary criterion. Technical textiles can be manufactured from different types of fibers such as nylon, polyester and carbon with various structures such as woven (mainly plain, twill and sateen), knitted or braided. The required mechanical properties of technical textiles can be achieved by selecting suitable fiber and changing the geometric parameters of the fabric (yarn spacing and thickness).

As technical textiles are highly heterogeneous structures, the experimental and analytical tasks for engineering approach are managed by the mechanical characterization within the theoretical framework of homogenization for periodic media. In order to simplify the management of both experimental and analytical tasks of these heterogeneous structures, constitutive models were developed upon evaluation of stress and strain at macroscopic level. The reference scale for description of technical fabric properties are obtained from the tightness of the weave or the intensity of linkage among the yarns. The stress-strain behavior of technical textiles is non-linear over a wide range of stresses/strains. One of the most frequent concerns associated with technical textiles is the inability of researchers to accurately model the onset of failure under complex biaxial loading conditions [26].

The mechanical characterization of technical textiles should also account for the complex multi-axial loading conditions which they are subjected in most of the applications. In order to obtain a full mechanical characterization of orthotropic technical textiles, it is not enough to study only the axial properties. Different properties exhibited by technical fabric must be included in a numerical model of any approach such as the Finite Element Method (FEM). In the modeling of technical textiles, the most promising shell elements of several commercial finite element systems are not able to properly address the change of thread direction during deformation. Hence, a special type of membrane element must be built for successful modeling. The models for shear behavior of technical textiles are more complicated as they must include particular aspects such as friction between fibers and yarns, bending and variable yarn cross sections.

The two main assumptions normally made for modeling of technical textiles are the plane stress condition and orthotropic behavior. In case of technical textiles with low fabric thickness (of the order of a few yarn diameters), the stress components along the direction of thickness can be neglected and hence most of the technical textiles are modeled as 2-dimensional membranes undergoing plane stress state [27]. The common two-dimensional woven fabrics such as plain, twill and sateen are characterized by a very regular pattern with yarns arranged along two orthogonal directions. Therefore, a common assumption that exists is the symmetry of mechanical properties consistent with the symmetry of the geometrical weave pattern.

11.2.2.1. Medical Textiles

In recent years, technical textiles used in medical and related healthcare and hygiene sectors are an important and rapidly growing segment of the textile field. Computational models are increasingly playing major roles in the design of medical textiles. One class of medical textiles includes the absorbents such as incontinence pads and wound dressing materials. The use of validated computational models to predict liquid spreading in medical absorbents is an important tool for understanding and optimizing the function of absorbent products. In case of medical absorbents fluid tends to be driven along the sheets, rather than through the thickness. As the thickness of the absorbent is less than 1 cm, use of a flow cell to measure the permeability introduces significant interfacial effects. Landeryou et al. [28] reviewed different modeling strategies applied to describe fluid infiltration and transport in medical absorbents. It was demonstrated that as the porous structure of fibrous medical absorbents is inhomogeneous, new closure rules need to be developed for computational models of fluid spreading in them.

An empirical characterization model has been established to assess the performance of a cosmetic textile agent and cosmetic textiles [29]. The new terminology 'cosmetic textiles', has now opened up new target groups and sustainable markets in the textile industry. Upon contact with skin, skin caring fibrous materials are designed to transfer an active substance for cosmetic purposes. The idea is achieved by simply imparting the bioactive ingredients into wearable textiles so that with the natural movement of the body, the skin is slowly supplemented and revitalized. To achieve this functional effect, microencapsulation technology appeared to be an alternative method for providing satisfactory performance with increased durability

The characterization model consisted of three major categories, namely: 1) material characterization and ingredient identification, 2) fabric performance testing, and 3) biological safety and biological response to human skin. The first category of the characterization system successfully identified the cosmetic textile agent/cosmetic textiles in terms of surface morphologies, identification of ingredients and their particle size distribution. The second category mainly focused on the study of fabric performance properties after the treatment of the cosmetic textile agent. For the third category, the biological response and biological safety of the cosmetic textile agent with respect to human skin were investigated.

It was concluded that, the systematic characterization model established in the study can provide a guideline for commercial sectors and researchers to identify and evaluate the performance of both cosmetic textile agents and cosmetic textiles in safe and objective manner. The controlled-release properties of the cosmetic textile agent and cosmetic textiles, as well as subjective wearer trials, are the main parameters that should be taken into consideration.

11.2.2.2. Protective Clothing

Song et al. [30] developed a numerical model to simulate the heat transfer through a single- layer protective garment exposed to flash fire conditions. The approach used the thermo-physical properties of fabric as a function of fabric temperature accounting for the air layer distribution between the garment and manikin, and incorporating characteristics of flash fire generated in the chamber. Exposure to intense heat generally produce nonlinear changes to the thermo-physical and optical properties of the fabric [31-32]. Therefore, a heat transfer model cannot predict protective performance of clothing accurately without accounting the complex changes in the material properties from which they are made.

The PyroMan manikin fire test system used in the study was a size 40 regular manikin made from flame resistant polyester resin reinforced with fiberglass. The manikin was instrumented with 122 individual heat-flux sensors distributed all over the surface. The manikin was suspended from the ceiling of a fire resistant burn chamber surrounded by eight industrial burners capable of producing a large volume simulated flash fire that can fully engulf the manikin in flames. Two typical industrial thermal protective garments namely

Kevlar/PBI coverall and Nomex IIIA coverall were selected to demonstrate the model. Both the garments were laundered five times in an industrial laundering machine before testing. The changing temperatures of human tissue at two skin depths were calculated from the incident of heat-flux measured by these sensors during and after the exposure. A computer system controlled the data acquisition and calculated the surface heat-flux and the histories of skin temperature distribution.

The model (Figure 11.5) illustrated the mechanism of thermal energy transfer to the protective fabric both in radiative and convective modes. A one-dimensional finite difference model [33- 34] was used to simulate the heat transfer through the protective garment, intermediate air gap and human skin. Due to the nonlinear radiation terms the Gauss-Seidal point-by-point interactive scheme was employed to solve the equations of the model. The parameter estimation method proposed by Beck [35] was used to determine the variations of fabric thermal conductivity and volumetric heat capacity as they change considerably during exposure to intense heat.

The three-dimensional body scanning technology was used to measure the air gap size distribution between the manikin and different sized protective garments for both pre- and post-exposure conditions. The model consisted of three separate parts to accommodate the unique boundary conditions associated with the garment, air gap and skin. For proper execution of the model, thermo-physical properties of the fabric and size distribution of insulating air gap were used as input to the numerical model. The integrated generalized model was validated using the thermal protective clothing analysis system [36].

This model helped to develop an understanding of the heat transfer process in protective garments and established a systematic method for engineering materials to produce optimum thermal protective performance. Also the research developed a concept to predict skin burn injury resulting from heat transfer through the protective garment. The manikin simulation model reproduced surface heat-flux and skin damage results very well for the single-layer protective garment. Though the burn distribution provided by the experimental analysis and model simulation are good in close fitting areas of the garment, it shows a larger variation in loose fitting areas such as the leg.

11.2.2.3. Monofilament Technical Textiles

Monofilament technical textiles are widely used in different industrial applications like screen printing and filtration. The mechanical properties such as the response of the fabric to contemporary actions of in-plane tensile and local bending are considered to be primary for monofilament technical textiles used in screen printing industry. The experimental characterization in all the cases is not sufficient and must be integrated with analytical or numerical models of the fabrics. The results of experimental characterization are essential for the validation of numerical models. In the last two decades though several numerical models had been developed for technical textile reinforcements, only a few were developed for monofilament technical textiles. These models are based both on analytical energy methods [37-41] or finite element (FE) theory [42-45] and applied mainly to textiles subjected to axial or biaxial loading.

Carvelli et al. [46-47] developed an analytical model to predict the mechanical behavior of dry-monofilament technical textiles mainly used in screen printing industry and presented both the results of the experimental campaign and the numerical modeling. The deformability and strength of the technical textile are essential aspects for the quality of final print.

Therefore, a full understanding of the mechanical properties of the textiles is essential for modeling all the above aspects. The mechanical behavior of the textiles was predicted by a three-dimensional analytical model based on a theory of curved beam. The macroscopic behavior of the monofilament textiles was investigated by taking into consideration a unit cell of the order of the filament diameter (30-70 μm) and assuming the nonlinear constitutive behavior of the fibers obtained by the experiments.

The macroscopic mechanical features of the monofilament textiles were obtained at the meso-scale level with a model of the representative volume [48] assuming the periodicity in the textile geometry. In the model the main assumption was the perfect periodicity of the monofilament textile and the unit cell reproduces the whole fabric structure (Figure 11.6). A three-dimensional FE model was implemented to predict the mechanical behavior of technical textiles. The three-dimensional numerical simulation of a monofilament textile requires the constitutive behavior of the fibers to be as close as possible to the real one.

In this model, the fiber geometry and the contact points of an elementary volume were described by the two circular paths connected by a straight line similar to the first attempt of Peirce [49]. The more recent concept of lenticular shapes which helps to model the flattening of yarns in the contact zones has been used to represent fiber cross section [41]. The fiber adhesion at the crossover points due to the final heat setting was not initially included in the numerical model and an attempt was made to produce a more general model considering the friction between the fibers.

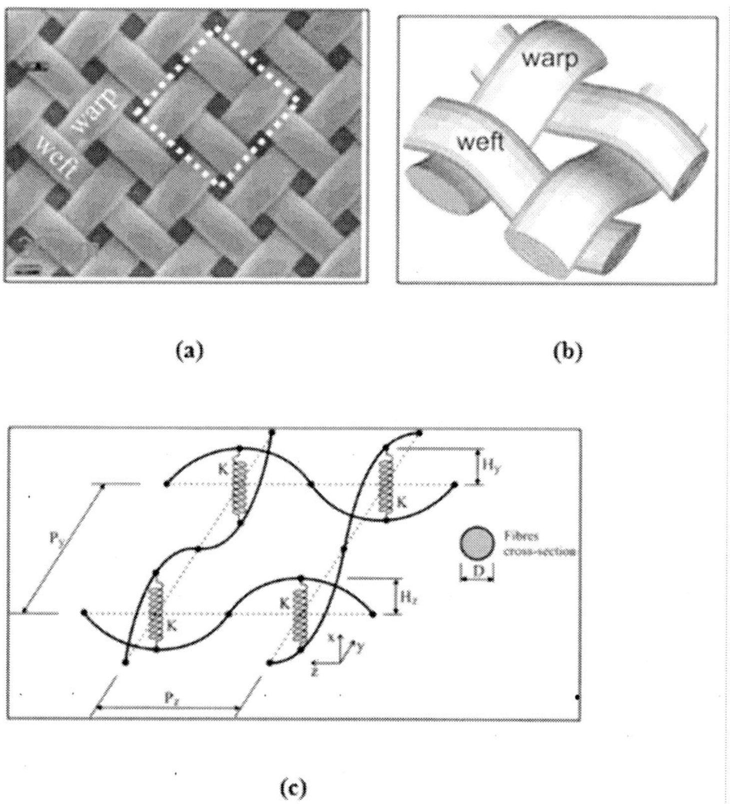

Figure 11.6. (a) Plain woven monofilament technical textile, (b) periodic unit cell and (c) parabolic beam model of the fibers in the unit cell [Source: Reference 46].

The geometry of the fibers in the unit cell was fitted by parabolic segments in the longitudinal direction with circular cross section. The fiber shape was modeled by a set of parabolic beams (four per filament in the unit cell) by adopting the curved beam theory [50]. The mechanical simulation was done imposing displacement boundary conditions and the validation of the proposed model was assessed by comparing the analytical results with the experimental observations on rectangular samples having the size of 250×50 mm for both the textiles in the elastic range and up to failure. A fictitious homogeneous material equivalent was possible to produce to the original textile with the proposed numerical model which can be used in the FE model of textile structures. The proposed numerical model appears to be a toll for the design of various applications of technical textiles where the mechanical aspects are essential. The predicted macroscopic mechanical response showed a satisfactory accuracy. The model seems to be adoptable as a design tool to optimize the fiber dimensions, the geometry and material properties of monofilament technical textiles. Quaglini et al. [51] developed an experimental protocol for mechanical characterization of plain woven technical textiles mainly to access the suitability of the model for technical textiles. The textiles were modeled as orthotropic materials with known directions. The nonlinearity exhibited by the stress-strain behavior was accounted by introducing a "linear-by-step" approximation. The analysis of fabric architecture by scanning electron microscope (SEM) was a preliminary step for assessing the suitability of orthotropic mechanical models and identifying the directions of material symmetry of the fabric. The study was restricted to monofilament polyester technical textiles and uniaxial tests were performed on strip specimens. The fabric was assumed to exhibit a regular yarn pattern with fibers running along two main directions (warp and weft) and intersecting at right angles. Two types of tests were conducted: (1) on-axis tests (specimen was aligned in warp or weft direction), and (2) off-axis tests (specimen was aligned at 45° to warp direction) and uniaxial stress-strain curves were fitted for determination of material coefficients. Both the tests were carried out using a custom biaxial testing machine equipped with two orthogonal loading axes. The stress-strain curves of both the tests were used to identify the full set of elastic parameters of the orthotropic model. The homogeneity of the strain field is a prerequisite for the estimation of reliable macroscopic parameters since edge effects may propagate into the measuring region preventing reliable estimates of stress. The validation of the model was performed by comparing the theoretical output with experimental stress-strain curves. A good agreement was found for the whole curves with a root mean square (rms) error between experiments and calculations less than 0.9 MPa in uniaxial off-axis tests and less than 1.5 MPa in biaxial tests. The performance of the model for the off-axis tests was slightly lower with larger errors at increasing strains than the on-axis tests. The largest deviations between the calculated and measured data occurred within the strain range from 0.04 to 0.06 mm/min, reaching 10% in the worst case and reducing at higher strains. In equi-biaxial stretching conditions the model prediction was more reliable for stresses in the weaker warp directions.

11.2.2.4. Miscellaneous Technical Textiles/Fields

11.2.2.4.1. Production Process

As in industry the long term objective is the optimal design of the production process with respect to the desired product specification, it is necessary to model, simulate and control the production process of technical textiles. Klar et al. [52] presented a hierarchy of mathematical models for the numerical simulation and control of production process of technical textiles and furthermore they predicted and optimized the properties of technical textiles (e.g. homogeneity of mass distribution and directional arrangement). The models ranged from one-dimensional fiber dynamics with stochastic aerodynamic drag and further to efficiently manageable stochastic surrogate models for fiber lay-down to highly complex three-dimensional fluid solid interactions. They were theoretically and numerically analyzed and coupled via asymptotic analysis, similarity estimates and parameter identification. The model hierarchy is applicable to a wide range of industrial relevant production processes and enables the optimization, control and design of technical textiles.

Although technical textiles differ in details, they have three common traits in principle: spinning, entanglement and lay-down. The numerical simulations of the production process involving spinning and turbulent lay-down are performed on the basis of a stochastic generalized string model that is deducted from a Cosserat rod, being capable of large, geometrically nonlinear deformations. The interactions of fiber and turbulent flow are thereby incorporated by a stochastic drag force model derived on top of a k-ε description for the turbulent flow field in [53]. The turbulence effects on the fiber dynamics are modeled by a correlated random Gussian force and its asymptotic limit on a macroscopic fiber scale by Gussian white noise with a flow-dependent amplitude that carries the information of kinetic turbulent energy, dissipation rate and correlation lengths.

Numerical studies for the production process of technical textiles show good agreement to the experimental observations for a nonlinear Taylor drag model [54] and for generalized drag model [55]. As huge amount of physical details are involved, the simulations of the fiber spinning and lay-down usually require an extremely large computational effort and high memory storage. This makes the optimization and control of the production process and the product properties very difficult.

11.2.2.4.2. Permeability of Technical Textiles

The permeability of technical textiles can be determined experimentally. However, this is a time and resource consuming process and no standard procedure is yet available. As in reality textiles are not perfectly periodic structures, the variation on the meso-scale structure has an important influence on the macro-scale permeability. Therefore, it is important to capture the variations in a sufficiently accurate way in order to perform reliable macro-scale simulations. Most of the analytical models give correct result for elementary and structural geometries but not for the stochastic and highly irregular porous technical textiles. Hence, more advanced models are required for the computation of permeability of technical textiles.

Accurate permeability predictions can be obtained by solving the three-dimensional Navier-Stokes or Stokes equations or by solving an equivalent lattice Boltzmann model. Several researchers studied the homogenization of Stokes equations in periodic porous medium [56-57]. Finite Element (FE) and Finite Volume (FV) or Finite Difference (FD) approach can be adopted for a direct numerical solution of the Stokes equations. Simulation tools based on lattice Boltzmann model use a regular grid and avoid the difficult mesh generation. The Monte Carlo method for permeability is based on the idea of generating random paths through the porous sample and the average path length is related to the permeability of the sample [58]. Analytical formulas to determine the permeability of porous medium consisting of a quadratic packing of cylinders for both flow along and transversal direction were derived by Gebart [59]. This method has not been validated yet for the technical porous media and hence unsure of the speed compared to the other known methods.

Verleye et al. [60-61] had investigated the meso-scale permeability i.e. the permeability of a unit cell textile model of multi-scale porous technical textiles for which the macro-scale geometry was taken into account. They presented two methods to predict the permeability, namely the CFD based three-dimensional simulation and the Grid 2D [62] based simulation. The second method takes less time compared to CFD but less accurate. The aim of the work was to construct a network that represents the pore structure in which every pore of the medium is assigned certain conductivity and the overall permeability was computed using the laws of conservation.

Though Washburn model [63] is most commonly applied to describe liquid infiltration into textiles, it has got some limitations such as the model cannot be used to model source driven flows without modification and it cannot be applied to describe spreading in two- or three-dimensions. The other equation which is commonly used to model liquid transport in porous material is Richard's equation [64]. Richard's equation is a semi-empirical formulation based on Darcy's law, where the rate of volumetric liquid flow is proportional to the driving pressure. An essential character of the models based on Richard's equation is that they contain the functional relation between permeability, pressure and moisture.

11.2.2.4.3. Viscoelastic Behavior of Technical Textiles

The finite element description of the nonlinear viscoelastic behavior of technical fabric was presented by Kłosowski et al. [65]. The technical fabric called "Panama" used in this model was made of two polyester thread families woven perpendicularly to each other with the 2/2 weave. The long term uniaxial creep laboratory tests in directions were conducted at five different constant stress levels. The dense net model [66] together with the Schapery one-integral viscoelastic constitutive model [67] was assumed for the fabric behavior characterization and the least square method in the Levenberg-Marquardt variant was used for the parameters identification.

The technical fabric behavior with respect to relations obtained for the warp and weft was described by the dense net model. The dense net model is mainly applicable to the three nodes triangle membrane elements in plane-stress state and also to the four node isoperimetric membrane elements in commercial codes. The advantages of dense net model are:

1. It can properly express shear properties of the fabric by changing the angle between the thread directions and X-axis during deformation process,
2. The thread behavior can be modeled by arbitrary type of uniaxial isotropic constitutive relations, while the anisotropic properties of the technical fabric can be expressed by the FEM, and
3. The model is easily applicable in commercial finite element codes having the three-dimensional membrane element as a user constitutive subroutine.

The nonlinear viscoelastic behavior of the fabric which is assumed to be the most adequate for fabric response analysis under the exploitation loading was adopted for the study. The model described the behavior of technical fabric with respect to relations obtained for the warp and weft. The uniaxial test in both thread directions was satisfactory for physical description of the fabric. A certain number of creep tests applied at different levels were conducted and the results were used for the viscoelastic model parameters' identification. The calculated results confirm that the model of nonlinear viscoelasticity can be applied to describe technical textile's behavior especially the creep response. Due to relative small number of functional parameters which characterize nonlinear viscoelastic materials, this model is convenient to be applied in numerical procedures.

The comparison of calculated and experimental data of the creep curves showed a good correlation. After comparing the calculated results it can be concluded that the viscoelastic behavior of the technical fabric can be described by the one-integral model. In warp and weft directions the numerical curve fitting resulted in a difference of 3.5-9.9% and 0.3-2.6% respectively. Therefore, the Schapery model with the power function characterized more accurate creep behavior in weft direction than warp direction. Also the power function described the strain evaluation better than the exponential function. This research concluded that both the linear and nonlinear viscoelastic identifications based on different material models can be brought together and the results of linear characterization can be applied to the nonlinear description of the material.

11.2.2.4.4. Coated Technical Textiles

The behavior of coated technical textile is complex as it consists of at least two different materials (coating and threads) having radically different physical properties. Threads change their direction during deformation and express physical nonlinearity with the coating. Thus, the numerical applications of the coated technical textiles require special models as general models fail to describe the behavior. Such behavior can be analyzed by a set of difficult laboratory tests such as uniaxial and biaxial tests. Generally, a good description of nonlinear physical properties (mainly threads) is given by the uniaxial tests while a better geometrical nonlinear behavior of a fabric is given by the biaxial tests.

Bahners et al. [68] investigated the concepts of technical fibers for the special coated fibers to be used for textile solar collectors. The technical fiber made of different polymers (polymethyl methacrylate (PMMA) and polyethylene terepthalate (PET)) were characterized with respect to optical properties relevant for the model. The fibers were coated with a matrix of 5% polyvinyl acetate (PVA) in methanol, into which fluorescent dyes were dispersed. The wave guiding behaviors of the technical fiber over greater lengths were quantitatively analyzed by a logarithmic fit. The measured value was found to divert from the ideal fit at larger fiber lengths.

11.2.2.5. Prediction of Properties of Technical Textiles by Different Models

Technical textiles are used in many applications where fluid flow is of great importance such as filtration and wicking. Hence, accurate determination of the permeability of fibrous media is critically important for many industries. Highly permeable nonwovens have been utilized in absorbency products, composites, micro-devices and microbiology applications. Realistic modeling of fluid flow through fibrous materials is essentially important for determining the accuracy of existing theories and developing new predictive tools. The direct numerical treatment of fluid flow in porous media is difficult and time consuming due to the rapid variation of the pore scale. Therefore, thorough understanding of the fluid flow through the fibrous substrates is necessary for the development of new products with specialized applications.

11.2.2.5.1. Computational Fluid Dynamics (CFD)

CFD based on the combination of theory and experiment is a computer based modeling technique used in the analysis and solution of fluid dynamic problems [69]. It has certain advantages over other techniques such as: 1) it can be used for thorough analysis of fluid flow problem at early stages of design cycle, 2) it can be used to solve difficult fluid flow problems which can't be solved by other techniques, and 3) it is faster, cheaper and involves lower risks. It provides the means of testing theoretical advances for conditions unavailable in experimental basis. Recently, CFD has been extensively used in the design of many engineering applications in different areas and it is a preferred means of testing alternative designs before final experiment [70-71]. In the field of textiles, the CFD has been applied to textile drying [72], textile dyeing [73], cotton cleaning and blade coating processes [74].

Wang et al. [69] used a commercial CFD package (FLUENT) to study the unique airflow problem around three-dimensional technical porous media. The study was carried out to optimize the design to improve the web uniformity especially for the finger-shaped nonwoven technical textile intended for hygiene applications. In this study, some basic information on the airflow characteristics was obtained by using an HHF600 air velocity meter. The CFD model was developed for two different web forming systems based on the experimental results. The distribution of air velocity, pressure and fiber in the web forming chamber were simulated. The computational domains for the two web forming systems consisted of the air duct and the mould chamber which were approximated to geometrical models. The fluid flow and related mass transfer phenomena considered in the study were complicated and involved multiple moulds of metal meshes rather than a single mould.

The airflow throughout the computational domain was turbulent based on the calculation of Reynolds number. The re-normalization group (RNG) turbulence model [75], having more universality and improved predictions for turbulent flows was used as high streamline curvature flows and mass transfer were involved in the study. The Euler-Lagrange approach [75], suited for flows where particle streams are injected into a continuous phase flow with a well-defined entrance and exit condition was followed. A fundamental assumption made in this model was that: "the dispersed second phase occupies a low volume fraction (usually less than 10-12%) for the fiber air-flows in both the web forming systems". For both the web forming systems, a three-dimensional 10-node tetrahedron volume element was used to discretize the computational domains.

The characteristics of fluid flow and related mass transfer phenomena in the web forming systems have been simulated and compared, with the emphasis on the uniform distribution of fibers in the duct and mould chamber. The model was evaluated by comparing the actual fabric aerial density distributions of the three-dimensional nonwoven products. It was found that the vertically oriented web forming system significantly improved the uniformity of air velocities and fiber concentration around the moulds compared to the horizontal system. Also the experimental results for the aerial density distributions of the nonwoven products followed similar airflow pattern and fiber distribution obtained by the CFD model simulation.

A three-dimensional imaging technique coupled with CFD simulation has been used for modeling fluid flow through the microstructure of hydroentangled fabric [76]. The hydroentangled fabric was made of nylon fibers. The digital volumetric imaging (DVI) technique from Micro-science Group Inc. was used for obtaining the images of fibrous material. DVI is a block-face fluorescence technique [77] where a fibrous material embedded in a polymeric resin is repeatedly sectioned and imaged. The images obtained by DVI were exported as two-dimensional image stacks in TIFF file format for further analysis. An efficient procedure has been adopted to reconstruct three-dimensional images from two-dimensional images (0.48 to 4.48 μm/pixel) which were subsequently processed for series of numerical simulations on the microstructure of real fibrous media [78].

In order to combine the two-dimensional images, the images needed to be thresholded. Thresholding is a process of splitting the grey levels of the image into two classes in order to separate the objects of interest from the background and is an essential step in image analysis. The important problem involved in thresholding is deciding the optimum threshold value so that the desired classification of the object is achieved. A local contrast enhancement procedure through histogram equalization was employed prior to thresholding to improve the accuracy of thresholding. A fluid domain was constructed by combining the threshold images for the purpose of CFD simulation. There was a direct comparison of the predictions of this work with the predictions of two-dimensional and three-dimensional analytical models. The thickness and permeability of the original samples measured by DVI and standard test methods showed relatively good agreement.

Three-dimensional magnetic resonance imaging (MRI) technique which measures the binary pixel intensities in the three-dimensional Cartesian coordinate system can also be used to study the fluid flow in opaque substrates. Though MRI technique was originally developed for visualization of blood flow for medical purposes, it was being successfully applied in various engineering fields [79]. MRI provides a unique non-destructive methodology to measure the fluid velocities of micro flow at selected positions within a porous fibrous substrate. In many instances the spatial resolution of the MRI technique may not be sufficient to provide data to be used in highly accurate CFD simulations. The fiber surface appears rougher and less even than in the original sample due to the limited resolution achievable by MRI technique which is a limiting factor for the study. MRI has also been used by various researchers to obtain three-dimensional images of porous media [80].

The combination of CFD and MRI has been used for very accurate study of flow phenomenon within textiles in order to have a deeper understanding of the flow of fluids through these fibrous substrates [81]. Any fiber deformation, due to fluid flow is not taken into account by CFD and MRI can detect these spatial structural changes. Hence, results obtained by MRI can provide immediate input for improved CFD models. The CFD calculation requires the knowledge of exact geometry of the flow phenomenon to be studied which is imported into the CFD software using a CAD format. Therefore, a dedicated setup containing a flow cell and an apparatus to control the flow through the flow cell should be used for the detection of flow phenomenon.

11.2.2.5.2. Artificial Neural Network (ANN)

In recent years, artificial neural networks (ANN) have attracted many researchers' attention in different disciplines because of their wide range of applicability, unique ability to handle complex nonlinear problems, enormous processing power, the ability to make sensible selections and the potential to learn by training. An ANN has self-learning ability and can model nonlinear functions [82]. It works similar to a black box with unknown evaluation process and hence making the interpretation difficult. ANNs are the most generic form of artificial intelligence for the emulation of human thinking compared with expert systems and fuzzy logic [83]. The neural networks process information rapidly and can reduce the engineering effort necessary for developing highly computation intensive modeling such as nonlinear Finite Element Analysis (FEA).

Airbags are automotive safety control systems built-into the steering wheel and various other strategic locations of a vehicle. The permeable properties and biaxial stress strain characteristics of airbag fabrics can significantly affect the safety of the occupant in a vehicle as the occupant interacts with the deployed airbag. The properties of engineering fabrics used for manufacturing airbags cannot be modeled easily by the available nonlinear elastic-plastic shell elements. For the airbag simulation studies in the automotive industry a nonlinear membrane element (that incorporates an elaborate tissue material) model has been widely used. This highly computational intensive model does not differentiate between the various physical properties of the fabrics like fiber type, fiber denier and weave pattern.

Keshavaraj et al. [84] demonstrated a simple feed-forward neural network approach to model the complex physical behavior of airbag fabric such as the inflation temperature-pressure drop relationship. They had used two nylon and three polyester fabrics having different calendaring effects for the training and then tested both the fabrics together in the same neural network model.

The proposed model was computationally simple compared to the nonlinear FEA based models and also incorporated the important fabric characteristics not considered by FEA. The model used a network consisting of 3-input node, 4-hidden node and 1-output node abbreviated as 3-4-1 network architecture. The main focus of the model was to predict the changes in the permeability with inflation temperature and internal pressure drop.

The permeability data for fabrics under biaxial strain which were used to train the proposed network architecture were obtained through a blister-inflation technique. The permeability data were divided into two sets namely the training data and the test data. In the blister-inflation technique a blister is created by a pressure drop across the fabric and is maintained while data on permeability and biaxial strain are recorded.

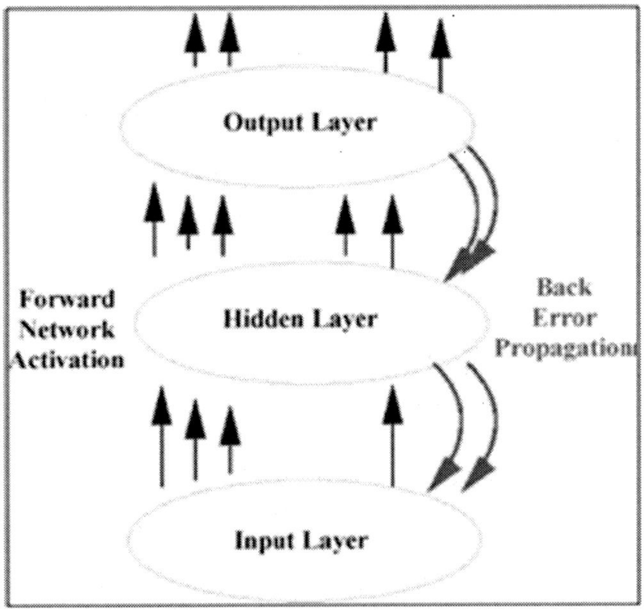

Figure 11.7. Schematic diagram of the back-propagation neural network [Source: Reference 85].

The technique involves the deformation of a flat sheet of test fabric with compressed air into an extended blister and is more sensitive in the lower biaxial strain region (3-5%) that occurs in airbag inflation. The prediction of the proposed neural network model was found to be in excellent agreement with the experimental data. The neural network model performed extremely well for first predictions after training and has the potential to be used for on-line simulation of airbag deployment studies.

Ramaiah et al. [85] developed an ANN model (Figure 11.7) for analyzing the performance of single-layer ballistic fabrics made from Kevlar using the material properties as inputs. The model predicted the dissipation energy and penetration depth during firing and the resistance offered by bulletproof vests. The modeling aspects of the ANNs had been demonstrated by using the data from fragment simulation projectile (FSP) ballistic penetration measurements at 244 m/s. The neural network models adopted for the simulation, design and the optimization of several problems and analysis of non-linear relationships between the complex variables is based on back propagation algorithm inbuilt in MATLAB 7.1 software.

For ballistic protective materials, neural network models are preferred because of their direct applicability, as they do not depend on any theoretical assumptions. Also material investigation results are directly used for building the model and studying the performance. The data collected from the samples was used for comparing the measured and predicted ballistic test parameters. The data base consisting 36 data was divided in to three parts namely training set (25 data), validation set (5 data) and test set (6 data). The results indicated that the neural network model can be successfully used in modeling properties of ballistic fabrics with similarity and better accuracy compared to traditional fiber science methods of study. The optimized networks can be used for the design and development of ballistic protection fabrics without going for further destructive tests. The input and output parameters of the neural network model are shown in Figure 11.8.

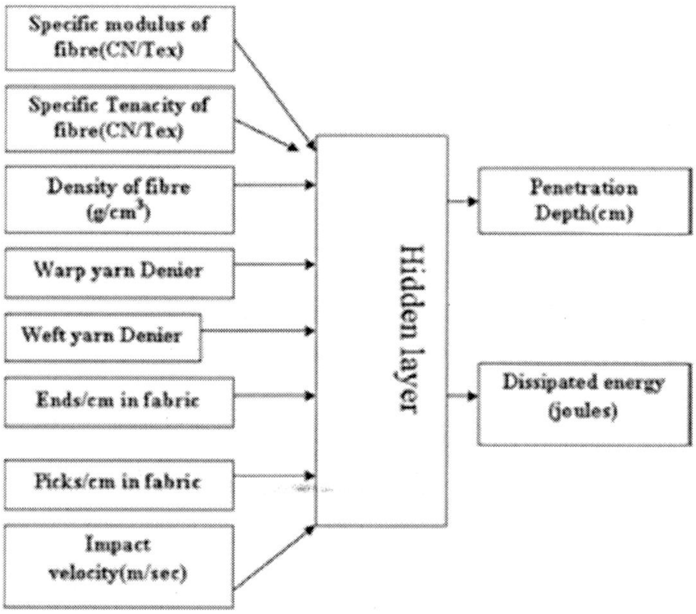

Figure 11.8. Neural network model for Kevlar ballistic fabric [Source: Reference 85].

11.2.2.5.3. Finite Element Analysis (FEA)

A finite element analysis (FEA) has been performed to investigate the effect of composite shell stiffness and the damage development during impact on the dynamic response of a composite motorcycle safety helmet [86]. The composite shell of the safety helmet consisted of an outer layer of woven fabric reinforcement either of carbon, glass or Kevlar and the inner layer of continuous strand glass mat.

The matrix material was chosen to be a polyester resin for both the inner and outer layer. The FEA model consisted of four parts namely the helmet shell, the foam liner, the head-form and the hemispherical test anvil. The foam liner was modeled using a set of 8500 solid elements, the woven ply was modeled as an orthotropic material with damage and the glass mat polyester layer was modeled as an isotropic elastic-plastic material with kinematic hardening.

The helmet drop test simulation used in the analysis was a fairly complicated problem. Three basic dominant failure modes were promoted by the introduction of the composite material model namely fiber failure under tension, matrix failure under tension and matrix failure under compression.

In this model the Chang and Chang [87] failure criterion was used to calculate damage initiation and propagation. Chou and Ko [88] fiber undulation model based on the unidirectional fiber tow properties was used to calculate the woven fabric properties. The dynamic response of the helmet system was almost identical in the case of carbon and glass woven fabric reinforcements.

The delamination and woven ply damage predicted by FEA model showed more pronounced energy absorption mechanisms in Kevlar-fiber shell followed by carbon-fiber shell and glass-fiber shell.

11.3. Future Scopes and Challenges

Although there are some research works on the modeling of smart and technical textiles, still research on modeling is in the infancy stage. The application of modeling in the fields of smart and technical textiles such as photo-sensitive or chromic materials, fiber optics, intelligent coating materials and chemical responsive materials is yet to be explored. The applicability of other networks such as tree network, pulse-propagation network also need more research.

Electronic textiles ostensibly present distributive computing challenges similar to those currently being pursued in adaptive control network and pervasive computing. The specific characteristics and demands of an e-textile info-computing environment add new dimensions to these challenges.

As a newly emerging field, e-textiles impose new challenges not only on modeling and analysis but also on simulation tools used for validation. The issue of concurrency is the fundamental problem in order to efficiently map complex applications onto e-textiles. As e-textiles contain many computational particles distributed on large surfaces, it is of crucial importance to expose the entire concurrency available at application level.

ACKNOWLEDGMENT

The authors acknowledge Sage Publications, Springer Publishing Company, IOP Publishing, IEEE and Elsevier Ltd., for granting permission to reproduce some figures used in this chapter.

REFERENCES

[1] Pierce, F. T. *J. Text. Inst.* 1930, *vol 21*, T377-T416.

[2] Tobushi, H.; Hashimoto, T.; Hayashi, S.; Yamada, E. *J. Intel. Mater. Syst. Str.* 1997, *vol 8*, 711-718.

[3] Hong, S. J.; Yu, W-R.; Youk, J. H.; Cho, Y. R. *Fibers Polym.* 2007, *vol 8*, 377-385.

[4] Tobushi, H.; Okumura, K.; Hayashi, S.; Ito, N. *Mech. Mater.* 2001, *vol 33*, 545-554.

[5] Liu, Y.; Gall, K.; Dunn, M. L.; Greenberg, A. R.; Diani, J. D. *Int. J. Plasticity* 2006, *vol 22*, 279-313.

[6] Gear, J. A.; Lachut, M. J.; Ding, Y. *Anziam J.* 2006, *vol 47*, C137-C151.

[7] Li, Y.; Zhu, Q. Y. *Text. Res. J.* 2004, *vol 74*, 447-457.

[8] Nuckols, M. L. *Ocean Eng.* 1999, *vol 26*, 547-564.

[9] Myers, G. E. *Analytical Methods in Conduction Heat Transfer;* ISBN-13: 9780070442153; McGraw-Hill, New York, 1971.

[10] Hittle, D. C.; Andre, T. L. *ASHRAE Trans.* 2002, *vol 108*, 175-182.

[11] Wang, S.; Li, Y.; Tokura, H.; Hu, J. Y.; Mao, A. Z. Pan (Eds.): *Edutainment 2006*, 2006, 1103-1106.

[12] Li, Y.; Wang, X.; Li, S.; Niu J. L. *Studies Comput. Intell.* 2007, *vol 55*, 255-263.

[13] Li, S.; Li, Y.; Li, F.; Wang, S. *Studies Comput. Intell.* 2007, *vol 55*, 247-254.

[14] Fengzhi, L.; Yi, L. *Modeling Simul. Mater. Sci. Eng.* 2007, *vol 15*, 223-235.

[15] Ying, B. A.; Kwok, Y. L.; Li, Y.; Yeung, C. Y.; Zhu, Q. Y.; Li, F. Z. *Studies Comput. Invest.* 2007, *vol* 55, 235-245.

[16] Martin, T.; Jones, M.; Edmison, J.; Sheikh, T.; Nakad, Z. LCTES, June 2004, Washington DC, USA.

[17] Laerhoven, K. V.; Cakmakci, O. In *Proceedings of the 4th International Symposium on Wearable Computing*, October 2000, USA, 77-83.

[18] CMU Graphics Laboratory. CMU Graphics Lab Motion Capture Database, 2003. Available at: http://mocap.cs.cmu.edu/search.html.

[19] Cottet, D.; Grzyb, J.; Kirsten, T.; Tröster, G. *IEEE Trans. Adv. Pkg.* 2003, *vol 26 (2)*, 182-190.

[20] Gorlick, M. *In Proceedings of the 3rd International Symposium on Wearable Computing*, October 1999, San Francisco, USA, 114-121.

[21] Chedid, M.; Belov, I.; Leisner, P. *Int. J. Clothing Sci. Technol.* 2007, *vol 19*, 59-71.

[22] Henn, A. R. *Interference Technol. Magazine ITEM update*, available at: www.rbitem.com.

[23] Carysse, M.; Puers, R.; Hertleer, C.; Langenhove, L. V.; Egmond, H. V.; Matthys, D. *Sensors Actuators* 2004, *vol 114*, 302-311.

[24] Marculescu, R.; Marculescu, D. *In the of Proc. IEEE/ACM Intl. Symp.* March 2002.

[25] Hiemenz, P. C.; Rajagopalan, R. *Principles of Colloid and Surface Chemistry*, ISBN-10: 0824793978; 3rd Edn, Marcel Dekker, New York, 1997.

[26] Arnold, W. S.; Robb, M. D.; Marshall, I. H. *Composites* 1995, *vol 26*, 739-747.

[27] Peng, X. Q.; Cao, J. *In the Proceedings of the 15th Annual Technical Conference of the American Society for Composites*, 2000.

[28] Landeryou, M.; Eames, I.; Frampton, A.; Cottenden, A. *Int. J. Clothing Sci. Technol.* 2004, *vol 16*, 163-172.

[29] Cheng, S. Y.; Yuen C. W. M.; Kan, C. W.; Cheuk, K. K. L.; Tang, J. C. O. *Text. Res. J.* 2010, *vol 80*, 524-536.

[30] Song, G.; Barker, R. L.; Hamouda, H.; Kuznetsov, A. V.; Chitrphiromsri, P.; Grimes, R. V. *Text. Res. J.* 2004, *vol 72*, 1033-1040.

[31] Barker, R. L.; Lee, Y. M. *Text. Res. J.* 1987, *vol 57*, 331-338.

[32] Shalev, I. *Transient Thermo-physical Properties of Thermally Degrading Fabrics and their Effect on Thermal Protection*; PhD Thesis, North Carolina State University, North Carolina, 1984.

[33] Patankar, S. V. *Numerical Heat Transfer and Fluid Flow;* ISBN-10: 0891165223; Hemisphere publishing, Washington D. C. 1980.

[34] Tannehill, J. C.; Anderson, D. A.; Pletcher, R. H. *Computational Fluid Mechanics and Heat Transfer;* ISBN-13: 9781560320463; 2nd Edn, Taylor and Francis, Washington D. C. 1997.

[35] Beck, J. V. *Parameter Estimation in Engineering and Science;* ISBN-10: 0471061182; John Wiley and Sons, New York, 1977.

[36] Pawar, M.; Barker, R. L. *In the Proceedings of the 5th Scandinavian Symposium on Protective Clothing*, Elsinore, Denmark, 1997.

[37] Kawabata, S.; Niwa, M. M.; Kawai, H. *J. Text. Inst.* 1973, *vol 64*, 21-46.

[38] Kawabata, S.; Niwa, M. M.; Kawai, H. *J. Text. Inst.* 1973, *vol 64*, 47-61.

[39] Lomov, S. V.; Huysmans, G.; Verpoest, I. *Text. Res. J.* 2001, *vol 71*, 534-543.

[40] Lomov, S. V.; Chi, T. T.; Verpoest, I.; Peeters, T.; Roose, D.; Boisse, P.; Gasser, A. *Int. J. Forming Processes* 2003, *vol 6*, 413-442.

[41] Sagar, T. V.; Potluri, P.; Hearle, J. W. S. *Comput. Mater. Sci.* 2003, *vol 28*, 49-62.

[42] Gasser, A.; Boisse, P.; Hanklar, S. *Comput. Mater. Sci.* 2000, *vol 12*, 7-20.

[43] Boisse, P.; Buet, K.; Gasser, A.; Launay, *J. Compos. Sci. Techol.* 2001, *vol 61*, 395-401.

[44] Boisse, P.; Gasser, A.; Hivet, G. *Compos. Part A* 2001, *vol 32*, 1395-1414.

[45] Tarfaoui, M.; Drean, J. Y. *Text. Res. J.* 2001, *vol 71*, 790-795.

[46] Carvelli, V. *Mech. Res. Comm.* 2009, *vol 36*, 573-580.

[47] Carvelli, V.; Corazza, C. ; Poggi, C. *Comput. Mater. Sci.* 2008, *vol 42*, 679-691.

[48] Carvelli, V.; Poggi, C. *Compos. Part A* 2001, *vol 32*, 1425-1432.

[49] Peirce, F. T. *J. Text. Inst.* 1937, *vol 28*, 45-96.

[50] Toniolo, G. La Teoria Della Matrice di Trassmissione, Masson Ed., 1979, Milano, Italy.

[51] Quaglini, V.; Corazza, C.; Poggi, C. *Compos. Part A* 2008, *vol 39*, 1331-1342.

[52] Klar, A.; Marheineke, N.; Wegener, R.; Angew, Z. *Math. Mech.* 2009, *vol 89*, 941-961.

[53] Marheineke, N.; Wegener, R. *J. Appl. Math.* 2006, *vol 66*, 1703-1726.

[54] Marheineke, N.; Wegener, R. *J. Appl. Math.* 2007, *vol 68*, 1-23.

[55] Marheineke, N.; Wegener, R. *J. Fluid Mech.* 2009, *vol 622*, 345-369.

[56] Mikelic, A. *Filt. Porous Med. Ind. Appl.* 2000, *vol 1734*, 127-214.

[57] Sanchez-Palencia, E. *Nonhomogeneous Media and Vibration Theory, Lecture Notes in Physics 127;* ISBN: 3540100008; Springer Verlag : Berlin-New York, 1980.

[58] Hwang, C-O.; Given, J. A.; Mascagni, M. *Phy. Fluids* 2000, *vol 12*, 1699-1709.

[59] Gebart, B. *J. Compos. Mater.* 1992, *vol 26*, 1100-1133.

[60] Verleye, B. *Computation of the Permeability of Multi-Scale Porous Media with Application to Technical Textiles*, PhD Thesis, Katholieke Universiteit Leuven, Leuven, 2008.

[61] Verleye, B.; Lomov, S.; Long, A. C.; Roose, D.; Wong, C. C. *In the Proceedings of the 16th International Conference on Composite Materials*, Japan, 2007.

[62] Wong, C. C.; Long, A. C.; Sherburn, M.; Robitaille, F.; Harrison, P.; Rudd, C.D. *Compos. Part A* 2006, *vol 37*, 847–857.

[63] Washburn, E. W. *Physical Review* 1921, *vol 17*, 273-283.

[64] Richards, L. A. *Physics* 1931, *vol 1*, 318-333.

[65] Kłosowski, P.; Komar, W.; Woźnica, K. *Const. Build. Mater.* 2009, *vol 23*, 1133-1140.

[66] Kłosowski, P.; Zagubień, A.; Woźnica, K. *Arch. Appl. Mech.* 2004, *vol 73*, 661-681.

[67] Papanicolaou, G.C.; Zaoutsos, S. P.; Cardon, A. H. *Compos. Sci. Techol.* 1999, *vol 59*, 1311-1319.

[68] Bahners, T.; Schlossr, U.; Gutmann, R.; Schollmeyer, E. *Solar Ener. Mater. Solar Cells* 2008, *vol 92*, 1661-1667.

[69] Wang, X. Y.; Gong, R. H.; Dong, Z.; Porat, I. *Macromol. Mater. Eng.* 2006, *vol 291*, 210-217.

[70] Gong, R. H.; Fang, C.; Porat, I. *Int. Nonwoven J.* 2000, *vol 9*, 20-24.

[71] Allison, J. L. *Mech. Eng.* 1997, *vol 119*, 82.

[72] Menshutina, N. V.; Kudra, T. *Drying Technol.* 2001, *vol 19*, 1825-1849.

[73] Scharf, S.; Cleve, E.; Bach, E.; Schollmeyer, E.; Naderwitz, P. *Text. Res. J.* 2002, *vol 72*, 783-788.

[74] Iliopoulos, I.; Scriven, L. E. *Phys. Fluids* 2005, *vol 17*, 127-131.

[75] FLUENT 6.0 Documentation, *User's Guide*, 2003.

[76] Jaganathan, S.; Tafreshi, H. V.; Pourdeyhimi, B. *Chem. Eng. Sci.* 2008, *vol 63*, 244-252.

[77] Chinn, D.; Ostendorp, P.; Haugh, M.; Kershmann, R.; Kurfess, T.; Claudet, A.; Tucker, T. *J. Manufact. Sci. Eng.* 2004, *vol 126*, 813-821.

[78] Kerschmann, R. *Filt. Separa.* 2001, *vol 38*, 26-29.

[79] Blümler, P.; Blümich, B.; Botto, R.; Fukushima, E. *Spatially Resolved Magnetic Resonance, Methods, Materials, Medicine, Biology, Rheology, Geology, Ecology, Hardware;* ISBN: 3527296379; Wiley-VCH: Weinheim, Germany, 1998.

[80] Hoferer, J.; Hardy, E. H.; Meyer, J.; Kasper, G. *Filt.* 2007, *vol 7*, 154-158.

[81] NTC Project, Micro-flow in Textiles; F04-GT05.

[82] Wong, A. S. W.; Li, Y.; Yeung, P. K. W. *Text. Res. J.* 2004, *vol 74*, 13-19.

[83] Bose, B. K. *Power Electronics and Variable Frequency Drives;* ISBN-13: 9788186308745; Standard Publishers Distributors: New Delhi, India, 2000.

[84] Keshavaraj, R.; Tock, R. W.; Haycook, D. *J. Appl. Polym. Sci.* 1996, *vol 60*, 2329-2338.

[85] Ramaiah, G. B.; Chennaiah, R. Y.; Satyanarayanarao, G. K. *Mater. Sci. Eng.* 2009, *vol 168*, 100-105.

[86] Kostopoulos, V.; Markopoulos, Y. P.; Giannopoulos, G.; Vlachos, D. E. *Compos. Part B* 2002, *vol 33*, 99-107.
[87] Chang, F. K.; Chang, K. Y. *J. Compos. Mater.* 1987, *vol 21*, 834-855.
[88] Chou, T. W.; Ko, F. K. *Compos. Mater. Series* 3, 1989, Elsevier, Amsterdam.

In: Modeling and Simulation in Fibrous Materials
Editor: Asis Patanaik
ISBN: 978-1-62100-116-4

Chapter 12

APPLICATION OF MODELING AND SIMULATION IN PROTECTIVE AND EXTREME WEATHER CLOTHING

S. A. Chapple **and** ***Asis Patanaik***

CSIR Materials Science and Manufacturing, Polymers and Composites Competence Area, Nonwovens and Composites Group, Port Elizabeth 6000, South Africa

ABSTRACT

The development of protective clothing provides many challenges for researchers. On the one hand, the clothing must provide the desired or regulated level of protection, while on the other, the clothing should not cause the wearer undue discomfort. There are different types of hazards, numerous fibers and fabrics, and various influencing factors to be considered. Because of this, predictive models are used as tools to estimate performance and comfort. This chapter discusses the modeling and simulation of chemical and fire protective clothing, and extreme weather clothing. Modeling and simulation of heat and mass transfer, aerosols, pesticide behavior, bioheat transfer, burn evaluation, and the use of manikins are presented. Some future developments are also discussed.

12.1. INTRODUCTION

In today's world, men and women in the military and emergency services and those in a variety of industries are exposed to a wide range of hazards. These can be in the form of hazardous chemicals, pesticides, radiation, biological hazards, electric arc, fire, extreme cold etc. To combat these hazards, protective clothing is often worn. The clothing can range from everyday work clothes to highly specialized protective suits. A wide range of fibers and fabrics are used. The highest levels of protection are often only achieved at the expense of wearer comfort. There is, therefore, a balance required in the design of protective clothing that will ensure that the garment meets the protective requirements without inhibiting wearer

*E-mail: SChapple@csir.co.za

activity and comfort. Many factors are involved including heat and mass transport, fabric properties such as permeability, porosity, thermal and water vapor resistance, fiber chemical and physical properties, garment fit, drape and design, transport of airborne particles, protection levels, skin responses, other physiological responses etc. (Figure 12.1). Although the performance of protective clothing can be determined experimentally, there is a wide range of hazards to be investigated. This, together with the wide range of fibers and fabrics available, and different methods for determining protective properties, makes it difficult to make appropriate recommendations. There is, therefore, a need for predictive models to estimate protective clothing performance and comfort. Numerous models have been developed which attempt to address this requirement. The models range from simple physics based models to very detailed models based on computational fluid dynamics. In this chapter, some of these models, pertaining specifically to chemical, fire protective and extreme weather clothing are presented.

12.2. Modeling and Simulation in Chemical Protective Clothing

Chemical protective clothing is used to provide an effective barrier between hazardous chemicals and the body. No single material will protect against all chemicals, and this poses challenges when modeling chemical protective fabrics and garments. As with other types of protective clothing, heat and mass transfer are important considerations, for both examination of protective capabilities and wearer comfort. Aerosol particles pose a different challenge; whereas gases and chemical vapors can be captured by activated carbon layers in permeable protective fabrics, aerosol particles can be carried through by airflow. Protection from pesticides has received specific attention not only because of the number of workers that are exposed, but also because protection is often in the form of standard work wear clothing.

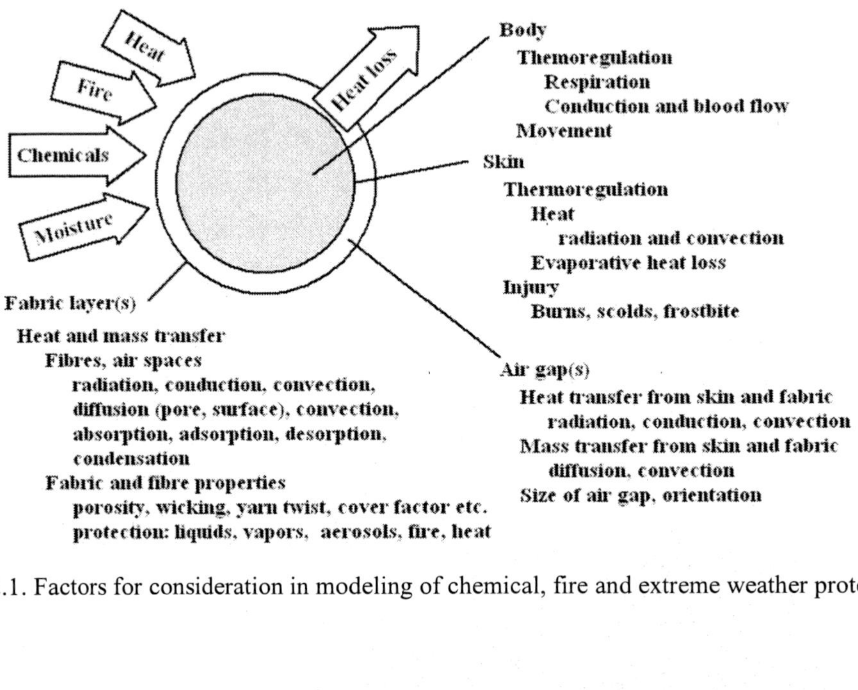

Figure 12.1. Factors for consideration in modeling of chemical, fire and extreme weather protective clothing.

12.2.1. Heat and Mass Transfer .

Gonzalez et al. [1] compared experimental data with physiological outputs determined from model predications in individuals dressed in protective garments. Simulations from two thermal models are tested as predictors of core temperature (T_c). The models evaluated are the United States Army Research Institute of Environmental Medicine (USARIEM) heat strain experimental model and the United Kingdom Loughborough (LUT25) model. For the original USARIEM model construct, abrupt rates of rise of core temperature (T_c) values as a function of time may be observed when the Givoni-Goldman [2] T_{re} (rectal temperature) equation is applied during exercise with different protective clothing systems. Using the Givoni-Goldman equations, the curve prediction of the rectal final steady-state temperature ($T_{re,f}$) is an exponential rise to maximum and the rate constant becomes too elevated. This results in an elevated $T_{re,f}$ at each time point and an over-prediction of model output versus experimental values. To counteract this, the authors implement a time-delay feature in the model that matches each average metabolic heat production:

$$K_{work} = \frac{\left[1 + 3 \cdot \exp\left(-0.3 \cdot \delta\, T_{re,f_{work}}\right)\right]}{225} \tag{12.1}$$

where K_{work} is the exponential rate coefficient for work output (° C/min), and $T_{re,f_{work}}$ is the final T_{re} during work (° C).

The LUT 25 model used is a recoded and adapted version of the Stolwijk-Hardy [3] 25-node model of thermoregulation. Certain adaptations to the model had been made including implementation of the coefficients from the Gagge et al. [4], Pierce model to account for clothing and thermal radiant loads, and intrinsic clothing and water vapor permeability factors [5-6]. Trunk temperature is used to predict the responses of T_{re} from the experimental trials; however, the model calculates T_c increases that become over-predictive in some circumstances, compared with the observed data.

The over-prediction of T_c may be a result of underestimation of evaporative heat transfer coefficients in the model, and may possibly also be affected by the estimations of distribution of blood flow to the core (trunk node) and shell.

The authors show that both models offer reliable predictions of T_c responses (within 1 RMSD) during work in the heat while subjects are wearing protective clothing. Both models simulate equally well the tracking of T_c response during intermittent exercise, however, only the USARIEM model simulated responses adequately for both unacclimated and heat-acclimated phases during continuous work.

Cain and McLellan [7] propose a model to describe the transport of water vapor from the skin of persons working in hot environments while wearing chemical-protective garments. Determination of vapor transport from the skin is more difficult than measuring the transport of heat along the same path. Vapor transport is modeled as a one-dimensional flow through resulting from a series of air and fabric layers between the skin and the environment. To simplify the model, vapor capacitances of the fabrics and inter-clothing air gaps are not considered. The authors use the following to calculate the mean vapor pressure at the skin:

$$\frac{\overline{P_s} - \overline{P_1}}{R_a} = \frac{\overline{P_1} - \overline{P_2}}{R_{12}} \tag{12.2}$$

where $\overline{P_s}$ is the mean vapor pressure at the skin, $\overline{P_1}$ is the vapor pressure adjacent to the skin (below garment), $\overline{P_2}$ is the vapor pressure on the outside of the garment, R_a is the resistance to the vapor flow across an air gap (5 mm) between the skin and first sensor (P_1) and R_{12} is the weighted (2:1) average of water vapor resistance of the upper and lower portions of the body.

The following equation is used to calculate the evaporation rate from the skin:

$$\dot{m}_s = \frac{(\overline{P}_s - P_a)}{R_{sa}} \tag{12.3}$$

where $\dot{m}_s$ is the vapor flow rate per unit area from the skin, P_a is the ambient vapor pressure and R_{sa} is the weighted resistance to water vapor between the skin and the environment which incorporates water vapor resistance of the upper and lower portions of the body, resistances due to air layers between fabrics and the resistance of the protective garment.

The model predicts mean evaporation rates between 52% and 63% of those calculated from measured changes in dress weights. Error analysis did not account for the magnitude of the difference and the authors propose that some of the difference results from a reduction of the resistance of the garment to water vapor due to wicking of liquid sweat through the fabric. Sweat accumulates in areas where the fabric layers are in close contact with the skin and evaporation of this liquid is less efficient for cooling the body.

Sobera et al. [8] used various types of numerical simulation of flow, heat and mass transfer to study the performance of nuclear-biological-chemical (NBC) protective garments. Flow, heat and mass transfer in NBC garments depend on phenomena at different scales; from the micro-scale of the fibers and turbulent flow, to the meso-scale of flow around trunk and limbs, to the macro-scale of a fully clothed subject. The authors use a coupled multi-scale modeling approach to accommodate the large differences in scale encountered. The models are based on large-scale, time-averaged computation fluid dynamics (CFD) descriptions of turbulent flow, using the Reynolds Averaged Navier Stokes (RANS) approach, in which the textile material is modeled as a porous material with particular macroscopic properties. The authors use Fluent 6, a commercial CFD solver, which is based on unstructured finite volume formulation.

The flow around a single clothed limb modeled as a circular cylinder sheathed by a second, larger, porous cylinder is studied using direct numerical simulation (DNS) at the meso-scale. The porous layer (garment) is separated from the inner cylinder by an air gap. The porous layer is treated as a fluid zone, where, according to Darcy's law, the pressure drop is imposed as a sink in the momentum equation. Experimentally obtained resistance values of real protective clothing are used. An infinitely long cylinder is simulated (cylinder length = πD), with periodic boundaries at the end walls). Prior to applying the DNS model, it was first tested in the case of flow around a solid cylinder at sub-critical Reynolds number ($10^3 \leq Re \leq 10^5$, corresponding to air velocities of 0.1 – 10 m/s around a typically sized limb). The model shows that the velocity of the flow in the air gap is much smaller than the outer flow. The inner flow is laminar and periodic with a frequency linked to that of vortex shedding in the wake of the outer flow.

For the micro-scale simulations of flow around textile fibers, Sobera et al. [8] simulated a 3-dimensional flow domain, with a virtual textile plane halfway and parallel to the inflow and outflow boundary of the domain. The inflow boundary conditions (laminarized flow in the vicinity of the textile) are obtained from the meso-scale DNS simulations. Periodic boundary conditions are applied to the edges of the virtual textile. Considering the textile as a 2-dimensional array of cylinders, at a mutual half distance, porosity is given by:

$$\varepsilon = \frac{\delta^2}{(r_f + \delta)^2} \qquad \text{or} \qquad \frac{\partial}{r_f} = \frac{\sqrt{\varepsilon}}{1-\sqrt{\varepsilon}} \tag{12.4}$$

where ε is porosity, δ is the mutual half distance and r_f is the cylinder radius.

Hydraulic permeability is a measure of the fluid conductance in a porous layer, as a function of porosity. It is related to the pressure gradient by Darcy's law:

$$\nabla P = \mu \frac{(v)}{k} \tag{12.5}$$

where ∇P is the pressure gradient, μ is the viscosity, (v) is the velocity in the vicinity of the textile and k is hydraulic permeability.

In DNS at micro-scale, Darcy's law is shown to accurately describe the pressure drop over the textile. For low porosities, where it is more appropriate to use $\frac{(v)}{\varepsilon}$ as the velocity scale rather than (v), the authors propose the following scaling estimate (from Clague et al. [9]) for the hydraulic permeability of an infinite square array of cylinders:

$$k = const \cdot \delta^2 \varepsilon \qquad \text{or} \qquad \frac{k}{r_f^2} = const \cdot \frac{\varepsilon^2}{(1-\sqrt{\varepsilon})^2} \tag{12.6}$$

The findings from the meso-scale and micro-scale DNS are used to validate the simplifications made in a RANS type engineering model. From a comparison made between RANS simulations and meso scale DNS, they conclude that the RANS model had shortcomings in accurately predicting local Nusselt (heat transfer) and Sherwood (mass transfer) numbers, but performed satisfactorily accurate in predicting global heat and mass transfer.

Similar work has been undertaken by Gibson [10]. Gibson uses fabric-covered cylinder models, in conjunction with experimental results, to calculate heat and mass transfer properties of clothing under particular conditions of temperature, relative humidity and environmental wind speed.

Danielsson [11] developed a theoretical heat and mass transfer model for babies placed in a chemical warfare protective bag. The model is based on relations valid for heat and mass transfer rates from a cylinder and flat surface. Thermal strain on a baby was estimated using an electrically heated baby manikin and a water-filled tray.

The combined dry heat transfer is expressed as:

$$(h_c + h_r)_{clothed} = \frac{1}{\left[\frac{1}{(h_c + h_r)_{naked}} + \frac{d}{k}\right]} \tag{12.7}$$

where d is the thickness of the fabric, k is the thermal conductivity of the fabric, h_c is the convective heat transfer coefficient ($Nu_D = \frac{h_c D}{k} = CRa_D^n$, where Nu_D is the Nusselt number, D is the cylinder diameter, k is the heat conductivity of air, Ra_D is the Rayleigh number) and h_r is the radiative heat transfer coefficient (h_r = $A_{eff}\sigma\varepsilon 4T_s^3$, where A_{eff} is 0.57 for a newborn child lying on its back, σ is the Stefan-Boltzmann constant and ε is emissivity of the skin and T_s is skin temperature in degrees Kelvin).

For mass transfer, the evaporative heat transfer coefficient h_e becomes uncertain at very low air speeds, however, an approximation can be estimated from the natural convection at a plane surface h_c:

$$h_e \approx 16.6 h_c \tag{12.8}$$

where h_c is expressed as $Nu_L = \frac{h_c L}{k} = CRa_L^n$ (Nu_L is the Nusselt number, L is the characteristic length of the tray = surface area/circumference, k is the heat conductivity of air and Ra_L is the Rayleigh number). The model for the surface when covered in fabric is:

$$h_{e,clothed} = \frac{1}{\left(\frac{1}{h_{e,naked}} + \frac{d}{\rho}\right)} \tag{12.9}$$

where d is the thickness of the fabric and ρ is the evaporative heat conductivity of the fabric.

The mass transfer coefficients, *hc* and *he*, can be calculated using the iso-vapor pressure temperature T_p:

$$h_c = k_1 \Delta T_p^{0.25} \tag{12.10}$$

$$h_e = k_2 \Delta T_p^{0.25} \tag{12.11}$$

where $k_1/k_2 = 16.6$ is the Lewis relation.

Barry and Hill [12-13] developed models based on computational fluid dynamics (CFD) to predict the performance of protective textile clothing. CFD software solves the governing equations for mass, momentum, and heat transfer in fluids over a 2- and 3-dimensional computational mesh.

However, it did not offer all the models required to address complex multiphase processes in textile fabrics. The authors propose a model in which each fabric layer is a porous membrane of variable porosity. There are four components: solid fibers, gas/vapor mixture, bound liquid and free liquid.

The gas/vapor mixture comprises air, as the main component, and water vapor or vaporized chemical agent. The free and bound liquid can be either water or chemical agent. In the model, bound liquid and free liquid interchange via sorption; bound liquid and vapor phase change via sorption, evaporation or condensation; free water and vapor phase change via evaporation or condensation. Transport of the free liquid occurs via wicking and vapor transport is via diffusion and convection.

The authors use Fluent (version 6.0) CFD software to which they added new models and capabilities. The capabilities include: vapor phase transport, liquid phase transport, vapor/liquid phase change, sorption to fibers and fabric property dependence on moisture content.

In the fabric, volume-averaging techniques [14-15] are used to derive transport equations for mass, momentum, and energy in the gas and liquid phase. Definitions for intrinsic and global phase averages and spatial average for porous media are those given by Whitaker [16-17]. The source term for each transport equation includes quantities that arise due to the variable porosity of the fabric. Permeability is estimated using experimentally determined permeabilities assuming that the flow resistance is proportional to the regain:

$$\frac{1}{K} = \frac{1}{K_{dry}} + \left(\frac{R}{R_{sat}}\right)\left(\frac{1}{K_{sat}} - \frac{1}{K_{dry}}\right) \tag{12.12}$$

where K is the intrinsic permeability, R is fiber regain, $_{dry}$ is dry conditions and $_{sat}$ is saturated conditions. Relative permeability constitutive relationships are based on saturation. The method of Progelhof et al. [18] is used to compute effective thermal conductivity and the Leverett J-function form is used to represent capillary pressure.

The authors use specialized geometry/grid representations of clothed humans to perform 2-dimensional (Figure 12.2) and 3-dimensional simulations of clothing performance. The outer surface of the wearer's skin forms one boundary layer to the clothing model. A simple model of the skin surface imposes the effects of sweating and the rejection of metabolic heat from the body.

Fabric model enhancements include: more accurate treatment of sweat evaporation and skin heat transfer; modifications to the evaporation and condensation model to improve convergence; an evaporation and condensation exchange model between fabric layers and surrounding gas region, and addition of an activated carbon model. Barry and Hill's developments show that CFD modeling is a powerful tool in the development of protective clothing, especially when used in conjunction with experiments and testing.

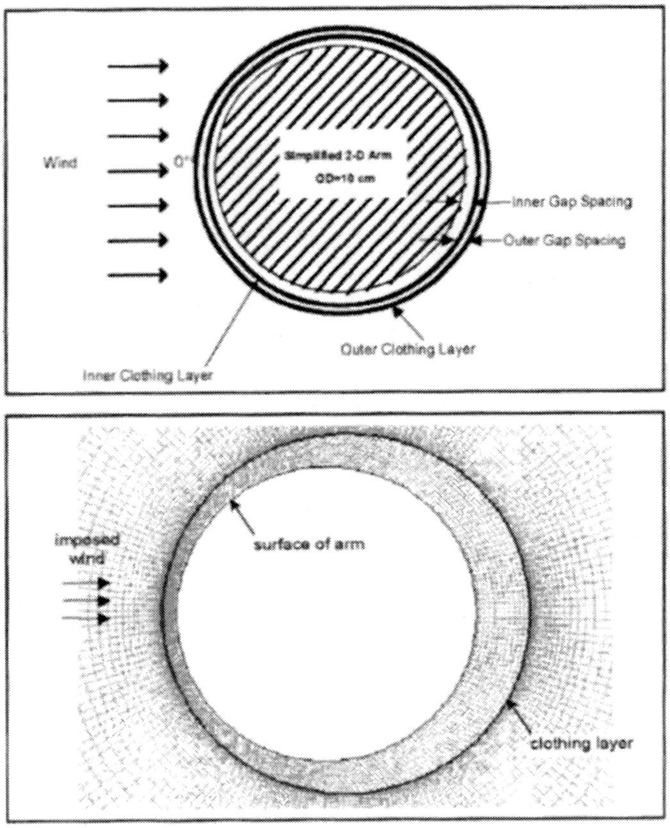

Figure 12.2. Model of a simplified arm [Source: Reference 12]: (A) with uniform gap and two clothing layers; (B) geometry with single clothing layer and nonuniform gap, Reproduced courtesy of International Nonwovens Journal Volume 12, 2003, pg. 27. Publisher, Copyright INDA P.O. Box 1288, Cary, North Carolina 27512-1288, USA. Tel: (919) 233-1210 Fax: (919) 233-1282 Internet: www.inda.org.

12.2.2. Aerosols

Spurny et al. [19] developed a test method for evaluating protective apparel in aerosol hazard conditions. They define a quality factor, Q, for barrier fabric which is the ratio of the efficiency, E, to the pressure drop, ΔP. Better filtration characteristics are indicated by higher values of Q. The effectiveness of the barrier of aerosol protective apparel depends upon its structural properties, such as porosity, thickness, and permeability, and hydrodynamic properties, such as pressure drop.

Fedele et al. [20-21] propose a model, based on fluid dynamic and aerosol transport processes, to evaluate the aerosol protection capacity offered by permeable protective garments. Their model uses geometry to characterize a person's body and the garment system. The person's body is represented by a solid cylinder which is surrounded by an outer circle, representing a permeable fabric layer (garment). An air gap is located between the solid cylinder and outer garment. Air flow's through the permeable fabric layer, driven by external pressure on the garment surface. Aerosol particles contained in the airflow may be deposited in the permeable layer or transported into the air gap. These transported particles may then be deposited on the solid cylinder as airflows round the cylinder. Particles may also be deposited in the permeable fabric layer as the air flows out of the gap when the internal pressure in the gap becomes greater than the external pressure.

Calculations are made for transport of a uniformly mixed aerosol to the garment surface by external air movement; transport through the fabric caused by an external pressure distribution on the surface of the fabric; and deposition on the solid cylinder caused by characteristics of airflow in the gap. The model represents the overall aerosol transport by an equivalent, time-independent process.

Fedele uses a pressure loss coefficient, ζ, defined by Hoerner [22] for flow through a permeable material:

$$\zeta = \frac{(P_1 - P_2)}{\frac{1}{2}\rho_f V} \tag{12.13}$$

where P_1 is the pressure upstream of material, P_2 is the pressure downstream of material, ρ_f is the density of the fabric, and V is the velocity of air through the material.

By substituting for P_1 and P_2 into equation 12.13, Fedele derived the following expression for the ratio of the pressure drop across the fabric and that between the fabric and the body:

$$\frac{\zeta_{fabric}}{\zeta_{gap}} = \frac{\Delta R^3}{3\Gamma\eta R^2} \tag{12.14}$$

where ΔR is the gap between the cylinder and the permeable layer, Γ is permeability, which equals $V/\Delta P$, η is air viscosity, and R is the radius of the solid cylinder, representing the body.

Because the flow through the fabric is greater than that through the gap, the fabric largely determines flow characteristics. Fabric properties include thickness, permeability and filtration efficiency; the latter being a function of aerosol particle size and airflow velocity. Aerosol penetration efficiency, calculated using an expression developed by Lee and Liu [23], is used to calculate the aerosol concentration in the air gap and the deposition in the fabric (garment contamination). Aerosol deposition on the solid cylinder (skin contamination) is calculated from the airflow in the gap. Roberts et al. [24] developed a software program using simple airflow and aerosol fabric penetration and deposition models to determine the skin deposition rates of aerosolized chemical, biological and radiological agents through up to two layers of individual protective equipment. Their calculations for the airflow model for one layer of fabric follow the work of Fedele [21] and Brasser [25] and use simple cylindrical geometries. A Gaussian fit to the pressure distribution for a cylinder in airflow is used [26]. The pressure outside of the cylinder as a function of angle around the body is given as:

$$P_0(\theta) = P_2(-1 + 2e^{-2\theta^2}) \tag{12.15}$$

where $P_0(\theta)$ =pressure on the outside of the cylinder, θ is the angle around the body (given in radians), P2 is hydrostatic pressure = $\frac{1}{2}\rho V^2$, where V is wind velocity and ρ is air density.

A flow model is proposed based on two principles: the volume of air flowing through the garment at a given angle will be equivalent to the change in volume flow in the gap (conservation of flow), and the resistance to flow in the gap will induce a pressure gradient around the body in the gap (Poiseuille's law):

$$\frac{d^2 P_1(\theta)}{d\theta^2} - \beta^2 P_1(\theta) = -\beta^2 P_0(\theta) \qquad (12.16)$$

where $\beta^2 = \frac{12\mu\Gamma R^2}{\Delta R^3}$, and μ is the dynamic viscosity of air, Γ is fabric permeability, *R* is the radius of the body, ΔR is the width of the air gap, $P_0(\theta)$ is the pressure on the outside and $P_1(\theta)$ is the pressure inside the gap.

By solving equation 12.16, using equation 12.15 and boundary conditions that result from flow symmetry and the front and back of the body, Roberts et al. [24] derived the following pressure model for a single layer of clothing:

$$\frac{dP_1(\theta)}{d\theta} = 0; \quad \theta = 0, \pi \qquad (12.17)$$

The single layer model is extended to describe multiple layers of fabric which would be the case in practical situations where a body would be clothed in more than one garment. Roberts et al. [24] propose a simple approach which assumes that the pressure and flow conditions in the first layer are not strongly affected by the presence of the second layer (resistance to flow though the second fabric is much greater than Poiseuille flow resistance through the outer clothing gap).

The pressure in the first layer is then used as the external pressure distribution for solving equation 12.16. Alternatively, the coupling of airflow and pressure between layers is considered (Figure 12.3) and a pair of coupled differential equations are developed and solved to give the pressure distribution in each of the two layers.

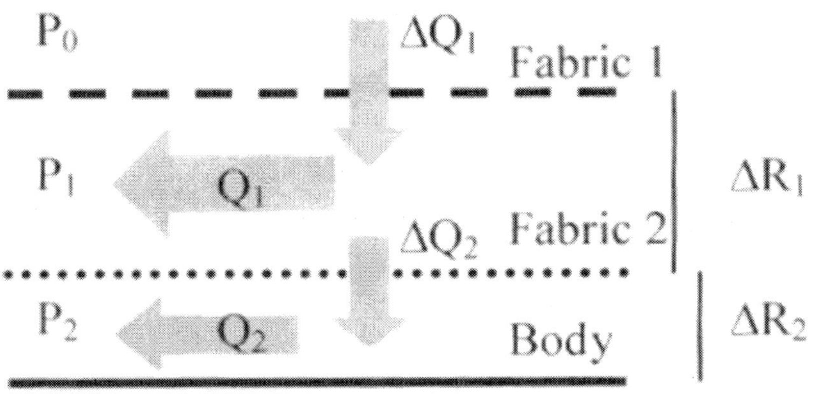

Figure 12.3. Coupled flow model for 2 layers of fabric [Source: Reference 24], Reproduced courtesy of S and T Publications, Defence Science Communications, Defence Science and Technology Organisation, 506 Lorimer St, Fishermans Bend 3207. ph: 03 9626 7842, fax: 03 9626 7133.

The authors propose a fabric filter model based on the single fiber model described by Hinds [27]. This model predicts the filter behavior of fabrics based on fabric thickness, fiber diameter and solidity (solidity = 1 – porosity). Filter behavior in real fabrics, which is influenced in particular by the size distribution and alignment of fibers, differs from that predicted by theory. In order for the single fiber theory to best represent the measured fabric filter behavior, the authors, in their model, treated the parameters of solidity and fiber diameter as "effective" rather than real parameters.

Roberts et al. [24] use the parallel plate model of Hinds [27] for their aerosol deposition model. They assume that the clothing gaps, through which the airflows, are parallel channels. Aerosols are deposited when they come in to contact with the fabric and body surfaces. The rate of deposition is determined by the dimensionless parameter μ_d:

$$\mu_d = \frac{DLH}{Q\Delta R} \tag{12.18}$$

where D is particle diffusivity, L is the length of travel in the gap, H is the height of the body, Q is the volume flow and ΔR is the width of the gap.

The deposition fraction δ depends on the value of $\mu_{d.}$ The actual rate of deposition of the most penetrating aerosols (0.3 μm diameter) is small and the authors conclude that few of the aerosols in the gaps will deposit. This is because in the time that an aerosol spends in a clothing gap (< 1 s) it will not move significantly relative to the airflow that carries it.

Roberts et al. [24] incorporated the physical models into a software program. The threat environment is defined by wind speed and aerosol characteristics (density, size and concentration). The height and width of the body is defined and the program models up to 2 layers of clothing on the body.

The width of the clothing gaps and fabric permeability describe the clothing layers. Filter behavior of the fabrics is based on fabric thickness, fiber diameter and solidity. The model shows that the threat posed by aerosolized chemical, biological and radiological agents is low. This is because the deposition rate of aerosols of the right particle size penetrating the fabrics is low. Two factors influence the low deposition rate. Firstly, the volume of air, and hence the quantity aerosol particles, flowing thought the fabric is relatively small. Secondly, less than a few percent of aerosols that penetrate the fabric will deposit; most exit the clothing with the exiting airflow.

In further work, Roberts and Skvortsov [28] propose the use of a finite solution to the flow problem for multiple clothing layers. The model allows for greater flexibility in describing the clothing and external pressure distribution. The circular body is divided into angular segments. The system is solved by specifying the external pressure, fabric permeabilities and air gap width for each segment. Flow conservation and Poiseuille flow equations are applied to each segment.

Flow is solved using electrical circuit theory, where airflow is analogous to current flow and flow resistance is caused by fabric permeability and Poiseuille flow resistance through the air gaps between fabric layers. Nodal analysis is used to solve sub-elements of the circuit and identifies each junction and Kirchoff's current law is applied to conserve flow at each junction. The model can be modified to represent a complex variation in air gap width and complexities in clothing geometry and structures, such as seams.

12.2.3. Pesticides Behavior

Exposure to pesticides poses a major hazard for agricultural workers. As a subset of chemical protective clothing, pesticide protective clothing has received much research attention. The effectiveness of the protective clothing is dependant on fabric properties, pesticide characteristics and the combination of both.

Goydan et al. [29] developed a prototype model, coded in FORTRAN, for estimating chemical permeation through protective clothing materials. The model uses Fick's law to estimate neat chemical permeation through clothing materials as a function of time. Solubility (S) and the diffusion coefficient (D) are estimated. Solubility estimates are obtained via two approaches: a group contribution approach and an equation of state approach. The author's found that the model is useful for estimating permeation behavior. However, the accuracy of the model is only fair when tested using a range of organic chemicals and clothing materials. The model did not address multicomponent solutions or blends of fibers or laminates. The model is based on the assumption of ideal Fick's law behavior and as such did not deal with intermittent or splash exposure, resistances at the material surface/skin interface, temperature variations and textile properties such as thickness.

Zhang and Raheel [30] developed a predictive statistic model for pesticide penetration in woven fabrics used for protective clothing. Although a model had been developed for nonwoven fabrics [31], workers commonly wear conventional work clothing made from woven fabrics.

The percent penetration was determined by gas chromatographic (GC) and gravimetric analyses. The regression model for the GC analysis is:

$$P1 = -116.5t - 299.4v + 191.7 \tag{12.19}$$

where $P1$ is the percent penetration, t is fabric thickness and v is the solid volume fraction of the fabric. The regression model for the gravimetric analysis is:

$$P2 = -107.7t - 23\gamma + 1.3\gamma^2 - 11.2\eta + 155.4 \tag{12.20}$$

where $P2$ is the percent penetration, γ is the surface tension difference = critical surface tension of a given solid – surface tension of pesticide solution, and η is viscosity of the pesticide solution. In both models, fabric thickness has the most significant influence on pesticide penetration. The two methods are not well correlated; the fabric tends to act as a filter, retaining the active ingredient of the pesticide. This results in an overestimation of penetration in the gravimetric method. A negative relationship between fabric protection performance and air permeability is shown, which is consistent with previous research of Wadsworth et al. [32]. Air permeability is one the properties of fabric comfort, which is a key consideration in the assessment of any garment that is used for protection.

Jain and Raheel [33] extended the work of Zhang and Raheel [30] to include a wider range of woven fabrics, nonwoven fabrics and a range of pesticide variables. The percent penetration (P) models are:

Woven fabrics: $$P = -134.79(t) + 0.47(a) + 329.56(v) - 39.06 \tag{12.21}$$

Nonwoven fabrics: $P = 356.90(t) + 0.66(\gamma_{diff}) - 21.08$ (12.22)

where t is fabric thickness, a is air permeability, v is the solid volume fraction and γ_{diff} is the difference in surface tension. The predictive model for nonwoven fabrics is applicable only to porous nonwoven fabrics. The authors conclude that the physical characteristic of the fabric and pesticide, rather than the pesticide chemistry, play a more significant role in predicting barrier efficacy. Obendorf et al. [34] examined the transfer of pesticides from contaminated work clothing in the dry state to human skin. A synthetic membrane is used to evaluate the kinetic transport of pesticide from contaminated cotton clothing through human skin. The kinetic equation is:

$$y(\%) = 6.38\left(1 - e^{-0.059t}\right) + 0.0395 \qquad (12.23)$$

where y is the pesticide penetration and t is the time (in hours). Fabric weight and fabric finish influence pesticide transport. Obendorf et al. [34] show that the synthetic membrane system is an appropriate model for human skin. Lee and Obendorf [35] used simple fabric and liquid parameters to develop statistical models to estimate the level of protection and thermal comfort of woven fabrics. Fabric thickness is often used as one of the factors in statistical modeling to estimate pesticide penetration. The authors, however, show that the influence of fabric thickness decreases when other fabric parameters are entered into the model. Consequently, fabric thickness is replaced by cover factor and yarn twist factor, which are also better parameters for describing woven fabric geometry than solid volume fraction. The prediction of pesticide penetration through woven fabrics is given as:

$$P = 11 - 97C - 22t_w + 0.7\eta - 5\gamma_s - 4w + 874C^2 + 13t_w^2 - 0.6\gamma_s^2 + 1.5w^2 \qquad (12.24)$$

where C is fabric cover factor, t_w is yarn twist factor, η is the viscosity of the pesticide mixture, γ_s is critical surface tension of fiber, and w is wicking height.

Air permeability is a quality that relates closely to wearer comfort. The authors examine various fabric parameters to develop a statistical model for air permeability. They propose a polynomial model with linear terms of cover factor, fabric thickness, twist factor, and yarn packing factor, and quadratic terms of cover factor, fabric thickness, and packing factor:

$$Air\ permeability = 17 - 127C - 130t - 12t_w - 161\phi + 957C^2 + 229t^2 + 904\phi^2 \qquad (12.25)$$

where t is fabric thickness and ϕ is yarn packing factor.

The authors propose that these models can be used as tools for preparing guidelines for choosing protective clothing in hot and humid environments.

12.3. Modeling and Simulation in Fire Protective Clothing

Fire protective clothing is used primarily to provide an effective barrier against heat and flame and in doing so prevent burn injuries. Depending on the level of protection required, the protective clothing needs to provide protection against the effects of flame, and heat transfer. Heat can be radiative, convective or conductive [36]. Radiant heat levels depend on the temperature difference and the distance between the radiant source and the targeted surface and the reflectivity of the surface. Convection occurs outside of the clothing, within the layers of clothing and in the air gap between the clothing and body. Conduction occurs when the clothing layers are in direct contact with a hot surface and when the clothing layers come into direct contact with the body. Temperatures in fabrics may reach very high levels in seconds. The protective clothing may also have to provide protection against chemicals and moisture transfer and steam burns. The clothing must also be designed with wearer comfort in mind, so physiological issues must be investigated. The modeling of fire protective clothing, therefore, poses many challenges.

12.3.1. Heat and Mass Transfer

A numerical heat transfer model of thin fibrous materials under high heat flux conditions (bench-top burner) was developed by Torvi and Dale [37]. The model is applicable to two common, flame resistant fabrics, Nomex® IIIA and Kevlar®/PBI. A fabric-air gap-test sensor system (Figure 12.4) is used in which heat transfer is assumed to be one-dimensional. The fabric's thermal properties represent the average thermal property values of the fibrous structure. Mass transfer, hot gas flow and fabric structural changes are not considered. The fabric's thermal properties are taken as functions of temperature only. The authors use energy balance equations and models of heat transfer modes to develop a differential equation (equation 12.26), and initial and boundary conditions:

$$C^A(T)\frac{\partial T}{\partial t} = \frac{\partial}{\partial x}\left(k(T)\frac{\partial T}{\partial x}\right) + \gamma q_{rad}(x=0)\exp(-\gamma x) \tag{12.26}$$

for $t = 0$,

$$T(x) = T_i(x) \tag{12.27}$$

for $x = 0$, $t > 0$,

$$-k(T)\frac{\partial T}{\partial x} = q_{conv} = h_{fl}\left(T_g - T_{x=0}\right) \tag{12.28}$$

and for $x = L_b$, $t > 0$

$$T(x = L_b) = T_{amb} \tag{12.29}$$

where C^A is apparent heat capacity, T is temperature, t is time, x is depth, k is thermal conductivity, γ is extinction coefficient, q_{rad} is radiation heat flux, T_i is initial temperature, q_{conv} is convection heat flux, h_{fl} is flame convection heat transfer coefficient, T_g is hot gases temperature, L_b is thickness of the base of air spaces in insulating block and T_{amb} is ambient temperature.

The authors use a suggestion of Stelzer and Welzel [38] to calculate the values of thermal properties and boundary conditions at each step. The average temperature in each element is used to determine the values of the thermal properties of that element. These are then assumed to be constant for the particular time step:

$$C^A(T)\frac{\partial T}{\partial t} = k(T)\frac{\partial^2 T}{\partial x^2} + \gamma q_{rad}(x=0)\exp(-\gamma x) \tag{12.30}$$

A finite element computer program is used to solve the equations. Using the data of Stoll and Chianta [39], the copper test sensor temperatures are used to estimate the time it would take human skin to sustain second-degree burns. The authors show that estimates of the gross thermal response of the materials tested in common bench top tests, and times to exceed the Stoll second-degree burn criterion can be modeled using relatively simple models.

The model has been developed further by Torvi and Threlfall [40]. The model is used to predict fabric temperatures of flame resistant fabrics and skin burn injuries during the cooling phase after burner exposure. Cooling is effected by means of a water-cooled shutter which is positioned between the burner and the front of the test fabric after test exposure. It is assumed that shutter and ambient temperature are at one temperature and that radiant and convection heat transfer only occur between the surface of the fabric and the environment. The apparent heat capacity method is used to account for energies associated with thermochemical reactions and moisture evaporation. The thermal properties of the fabric are taken to be a function of temperature only.

The following differential equations are proposed for the temperature distribution during ($t \le t_{ex}$) and after ($t > t_{ex}$) the exposure:

$$C^A(T)\frac{\partial T}{\partial t} = k(T)\frac{\partial^2 T}{\partial x^2} + \gamma q_{rad}\exp(-\gamma x) \quad (t \le t_{ex}) \tag{12.31}$$

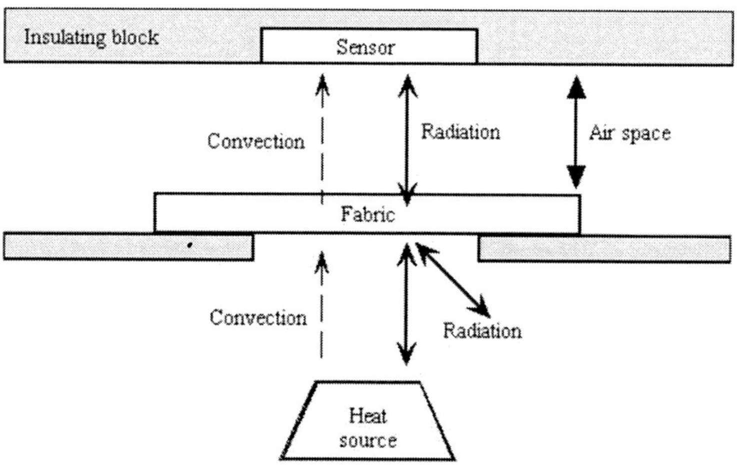

Figure 12.4. Example of bench top fabric-air gap-test sensor set-up for thermal protective fabrics.

$$C^A(T)\frac{\partial T}{\partial t} = k(T)\frac{\partial^2 T}{\partial x^2} \quad (t > t_{ex}) \tag{12.32}$$

The boundary condition on the front of the fabric is given by:

$$-k(T)\frac{\partial T}{\partial x}\bigg| = h_{fl}\left(T_g - T_{x=0}\right) \quad (t \le t_{ex}) \tag{12.33}$$

$$-k(T)\frac{\partial T}{\partial x} = q_{conv} + q_{rad} = h_0\left(T_{amb} - T_{x=0}\right) + \sigma\varepsilon_{fab}\left(T_{amb}^4 - T_{x=0}^4\right) \quad (t > t_{ex}) \tag{12.34}$$

where h_{fl} is the convective heat transfer coefficient for the front of the fabric during exposure and h_o is the convective heat transfer coefficient for the front of the fabric after exposure.

The boundary condition at the air space between the test sensor and insulating block is given by equation 12.29. The initial condition is $T(x) = T_i(x)$, where $T_i(x)$ is the initial temperature distribution in the fabric, air gap and test sensor. Equations 12.31 and 12.32 are solved subject to equations 12.29, 12.33 and 12.34, and the initial condition.

Further improvements to the model have been made by Sawcyn and Torvi [41]. The model calculates heat transfer (radiant and convection) from the test specimen to the test sensor. Flow visualization studies are conducted in the air space (Figure 12.4) to characterize the nature of the convection heat transfer. Heat transfer by convection in the new model is calculated using an empirical correlation. It is assumed that convection only occurs over a portion of the cross-section of the test enclosure. Radiation heat transfer is calculated by dividing the bottom boundary of the enclosure into a series of isothermal regions. The authors show that it is important to consider temperature gradients across the fabric specimen when modeling the heat transfer across the air gap. The authors note that the air gap in bench top tests is horizontal whereas the majority of air gaps in clothing systems are vertical.

Talukdar et al. [42] used CFD modeling for simulation of the air space (Figure 12.4) in bench top tests. Results are compared with the model of Sawcyn and Torvi [41]. Numerical simulations are performed using a CFD source code (FASTEST3D, Institute of Fluid Mechanics, University of Erlangen-Nuremberg, Germany). The CFD model solves the Navier-Stokes equation to calculate the convection heat transfer. A three-dimensional finite volume model is used to calculate radiation heat transfer. It is assumed that the media is transparent. Results give a clear picture of the flow pattern within the air gap. The model shows a higher convective heat flux compared to the calculations done by Sawcyn and Torvi [41]. The authors show that the model is capable of predicting the convection cell which was not possible in previous models. Mell and Lawson [43] developed a heat transfer model for predicting heat flux and temperature in firefighter's protective clothing. The authors model a three-layer protective clothing assembly. This consists of an outer, shell, layer, middle moisture barrier, and inner thermal layer. The layers are separated by air gaps. Heat transfer through the layers is largely due to radiant energy from the external surrounds, and a forward-reverse model is used for radiant heat transfer. A planar geometry and one-dimensional heat transfer is assumed. The fabrics are assumed to be dry and fabric temperatures considered are below the level of thermal degradation, so moisture effects and internal heat generation are not considered in the model. A control volume approach is used to derive the finite difference form of the conservation of energy equation. This method of descretization ensures local energy conservation and requires approximation of only first-order derivatives. The following descretized energy equation, which is solved for each material layer and air gap, is proposed:

$$\overline{c_p \rho \frac{\partial T}{\partial t}}\bigg|^{P} \Delta x dA = k \frac{\partial T}{\partial x}\bigg|_{w}^{e} dA - q_R\big|_{w}^{e} dA \tag{12.35}$$

where the over-bar indicates the quantity is a cell average, c_p is the specific heat, ρ is the mass density, T is temperature, t is time, x is distance measurement into protective clothing, dA is a constant, k is thermal conductivity, q_R is the radiant heat flux, P is the midpoint of central control volume in finite difference scheme, e is interface between control volumes P and E (midpoint of right control volume in finite difference scheme), w is interface between control volumes P and W (midpoint of left control volume in finite difference scheme). Boundary conditions are also defined. The gas/solid boundaries of the shell fabric and thermal layer both involve convective heat transfer which must be accounted for to conserve energy:

$$-q(x_\Gamma) = k_\Gamma (T_S - T_G), \quad \frac{1}{k_\Gamma} = \frac{\delta_e^-}{k_S} + \frac{\delta_e^+}{k_G} + \frac{1}{h_c} \tag{12.36}$$

where q is heat flux, $_\Gamma$ is gas/solid or solid/solid interface, k_Γ is effective conductivity coefficient, $_S$ is solid cell, $_G$ is gas cell, δ_e^- is distance from cell interface e to point P in finite difference scheme, δ_e^+ is distance from cell interface e to point E in finite difference scheme and h_c is surface heat transfer for convection. For the gas/solid boundaries of the middle moisture barrier, for which no thermal boundary exists, continuity of the conductive heat fluxes across the interface is given by:

$$-q(x_\Gamma) = k_\Gamma (T_S - T_G), \quad \frac{1}{k_\Gamma} = \frac{\delta_e^-}{k_S} + \frac{\delta_e^+}{k_G}. \tag{12.37}$$

Model results are compared with only one experimental case. For the inner layers, model prediction compared very well with experimental temperatures (within 5 °C). Temperature predictions on the outer shell were up to 24 °C higher. The authors suggest that an error in the estimates of transmissivity and reflectivity are the most likely source of the modeling error.

Improvements to the model have been made by Lawson et al. [44-46]. The improvements use estimates of thermal conductivity, specific heat and thermo-optical properties (transmittance, reflectance and absorptance) obtained from the thermal data collected from the testing of a variety of fabrics typically used in fire fighters' protective clothing. A detailed mathematic model is developed to study transient heat and moisture transfer through multi-layered fabric assemblies with or without air gaps. First principles are used to derive the governing equations for transient heat and moisture transfer. The equations also account for the effect of moisture on thermodynamic and transport properties. Numerical simulations are used to study heat and mass transfer. A software tool (Protective Clothing Performance

Simulator) is developed which allows users to study transient heat and mass transfer through multi-layered assemblies with or without air gaps. Fabric layer characteristics, air gap thickness, moisture levels, compression and boundary conditions are considered.

Chitrphiromsri and Kuznetsov [47] presented a model that couples heat and moisture transport in firefighters protective clothing during a flash fire exposure. The garment consists of three fabric layers (outer shell, moisture barrier and thermal barrier). The skin also has three layers: epidermis, dermis and subcutaneous. Finite difference method methodology is used to solve the differential equations (equations 12.38-40, 12.47). The equations are: the energy equation for the fabric (based on Gibson's [15] and Torvi's [48] models):

$$\rho c_p \frac{\partial T}{\partial t} + \left(\Delta h_l + \Delta h_{vap}\right)\dot{m}_{sv} = \frac{\partial}{\partial x}\left(k_{eff}\frac{\partial T}{\partial x}\right) + \gamma \cdot q''_{rad} e^{-\gamma x} \tag{12.38}$$

where ρ is the effective density of the fabric, c_p the effective specific heat of the fabric, T is temperature, t is time, Δh_l is the enthalpy of transition from bound water to free liquid water, Δh_{vap} is the enthalpy of evaporation per unit mass, $\dot{m}_{sv}$ is the mass flux of vapor out of the fiber, x is the linear horizontal coordinate, k_{eff} is the effective thermal conductivity of the fabric, γ is the extinction coefficient of the fabric, and q''_{rad} is the incident radiation heat flux from the flame onto the fabric. Free water exists neither on the skin nor in the fabric layer and the gas phase convection contributions due to pressure differences are neglected; the solid phase continuity equation:

$$\rho_w \frac{\partial}{\partial T}\left(\varepsilon_{bw}\right) + \dot{m}_{sv} = 0 \tag{12.39}$$

where ρ_w is the density of the liquid water and ε_{bw} is the volume fraction of water dissolved in the solid phase; the gas phase diffusivity equation:

$$\frac{\partial}{\partial t}\left(\varepsilon_\gamma \rho_v\right) - \dot{m}_{sv} = \frac{\partial}{\partial x}\left(D_{eff}\frac{\partial \rho_v}{\partial_x}\right) \tag{12.40}$$

where ε_γ is the volume fraction of the gas phase ρ_v is the intrinsic density of the water vapor and D_{eff} is the effective diffusivity of the gas phase in the fabric = $\frac{D_a \varepsilon_\gamma}{\tau}$, where D_a is the diffusivity of water vapor in the air and τ is the fabric tortuosity, and the bio-heat transfer equation for the skin (equation 12.47). The Crank-Nicolson scheme [49] is used to discretize the partial differential equations. The Gauss-Seidel point-by-point iterative scheme is then used to solve these equations. An under-relaxation procedure is utilized to avoid divergence of the iteration method. A temperature profile in multiple layers of skin is obtained and the Henriques and Moritz [50] burn integration equation is used to calculate skin burn (See section 12.3.2). Zhu et al. [51] developed a numerical model to investigate heat transfer in a cylinder sheathed by a skin simulant and flame-resistant fabrics when suddenly exposed to convective and radiant heat flux from simulated pre-flashover fire radiation. Wet fabrics and thermal degradation of the fabrics is not considered. A one-dimensional radiant heat transfer mathematical equation is developed based on Torvi and Dale's [37] one-dimensional planar formation and the cylindrical model of Liu [52]:

$$\rho_{fab}(T)c_{fab}(T)\frac{\partial T}{\partial t} = \frac{\partial}{\partial r}\left(\lambda_{fab}(T)\left(\frac{\partial T}{\partial r}\right)\right) + \frac{\lambda_{fab}(T)}{r}\frac{\partial T}{\partial} - \gamma q_{rad}\exp\left(\gamma\left(r - r_{of}\right)\right) \qquad (12.41)$$

where ρ_{fab} is fabric temperature-dependent density, c_{fab} is fabric specific heat, λ_{fab} is fabric thermal conductivity, q_{rad} is the portion of the directly transmitted incident heat flux on the surface of the element due to thermal radiation from the heat source, T is fabric temperature, r_{of} is outer radius of the fabric. The governing equation for the one-dimensional radial heat conduction model for the simulant skin is expressed as:

$$\rho_s c_s \frac{\partial T}{\partial t} = \lambda_s \frac{\partial^2 T}{\partial r^2} + \lambda_s \frac{1}{r}\frac{\partial T}{\partial r} \quad \left(r < r_{os}, t > 0\right) \qquad (12.42)$$

where ρ_s is the skin simulant density, c_s is the skin simulant specific heat, λ_s is the skin simulant thermal conductivity and r_{os} is the outer radius of the skin simulant. The authors found that the numerical model reproduced fabric surface temperatures and skin simulant temperature elevation and required times to the skin burn very well.

Zhu and Zhang [53] developed a novel experimental apparatus to evaluate the heat transmission of protective fabrics subjected to an external heat flux. The fabric has a cold side (maintained at around 36 °C) and a side which is exposed to dry hot air, heated by a hot plate. Thermal conductivity of the fabric is determined by thermal-electric analog principle. The thermal resistance of the sample (R_2) is given by:

$$R_2 = R_1 - R_0 \qquad (12.43)$$

where R_1 is the thermal resistance between the hot plate and the cold plate and R_0 is the thermal resistance of the air between the hot plate and the cold plate. Thermal conductivity is given by:

$$\lambda = \frac{L}{R_2 \cdot A} \qquad (12.44)$$

where L is the thickness of the fabric sample and A is the fabric surface area.

12.3.2. Bioheat Transfer and Evaluation of Burns

Assuming that skin tissue above an isothermal core is maintained at a constant body temperature, Pennes [54] proposed one of the first bioheat models to describe heat transfer in human tissue. The equation incorporates a distinct perfusion term to the standard heat balance equation, which in a simplified form [55] is:

$$\rho c \frac{\partial T}{\partial t} = \nabla \cdot k \nabla T + \rho_{bl} c_{bl} \omega (T_a - T) + q_{met} \tag{12.45}$$

where ρ is density, c is specific heat, T is tissue temperature, k is thermal conductivity, ρ_{bl} is blood density, c_{bl} is specific heat of blood, ω is perfusion rate, T_a is temperature of arterial blood and q_{met} is metabolic heat generation. Because of its simplicity, the model has been adapted by many researchers for analyses of a variety of bioheat transfer phenomena. The model has also come under some criticism regarding the validity of the assumptions underlying the model, for example, assumption of uniform metabolic heating, perfusion rate and thermal conductivity [55-56].

Chem and Holmes [57] developed a model where the heat transfer between small blood vessels and tissue is separated into three modes: a perfusion mode (q_p) (equilibration of blood and tissue temperature), convective mode (q_c) (convective heat transfer from flowing blood against a tissue temperature gradient) and thermal conduction mode (q_{pc}) (conductive heat transfer due to small temperature fluctuations in the blood along the tissue temperature gradient):

$$\rho c \frac{\partial T}{\partial t} = \nabla \cdot k \nabla T + q_p - q_c + q_{pc} + q_{met} \tag{12.46}$$

where ρ is density, c is specific heat, T is tissue temperature, k is thermal conductivity and q_{met} is metabolic heat generation.

In Chitrphiromsri and Kuznetsov's model [47] heat and moisture transport in fire fighters' protective clothing during a flash fire exposure are investigated. The garment consists of three fabric layers (outer shell, moisture barrier and thermal barrier). The skin also has three layers: epidermis, dermis and subcutaneous. The bio-heat transfer equation for the skin, based on the Pennes model [54], is written as:

$$(\rho c_p)_{skin} \frac{\partial T}{\partial t} = \nabla \cdot (k_{skin} \nabla T) + (\rho c_p)_{blood} \omega_b (T_{art} - T) \tag{12.47}$$

where ρ_{skin} is the density of the human skin, $(c_p)_{skin}$ is the specific heat of the human skin k_{skin} is the thermal conductivity of the human skin, ρ_{blood} is the density of the human blood, $(c_p)_{blood}$ is the specific heat of the human blood, ω_b is the blood perfusion and T_{art} is the arterial temperature.

The bioheat (equation 12.47) is used together with the fabric energy, solid phase continuity and gas phase diffusivity equations (equations 12.38-40) to obtain the temperature profile in multiple layers of skin. These are used in Henriques and Moritz [50] burn integral equation to calculate the maximum durations of the flash fire exposure before the human skin can get second and third-degree burns:

$$\Omega = \int_0^t P \exp\left(-\frac{\Delta E}{RT}\right) dt \tag{12.48}$$

where Ω is a quantitative measure of burn damage at the interface or at any depth in the dermis, P is the frequency factor or pre-exponential factor, ΔE is the activation energy for the skin, R is the universal gas constant. T is the temperature at the interface between the epidermis and dermis in the human skin (first and second degree burns) or the temperature at the interface between the dermis and subcutaneous layer (third degree burns). Damage is calculated by integrating the exponential function of tissue temperature history for those temperatures > 44 °C at a specific depth over the total time for which temperatures exceed 44 °C [58]. First degree burns occur when the value of Ω reaches 0.53 at the epidermis / dermis interface; and second degree burns when Ω reaches 1.0 [59]. Third degree burns occur when Ω reaches 1.0 at the dermis/epidermis interface [60].

Zhu et al. [61] developed a cylindrical geometry testing apparatus which incorporates a novel bioheat transfer model to test flame resistant fabric used in fire fighters' protective wear. In the geometry test, heat flows from the radiant source, to the surface of the fabric and then through the fabric. It then flows through in air gap to the surface of a skin simulant. The rise in temperature on the surface of the skin simulant is used to determine the heat flux. This data is then applied to the thermal wave model of bioheat transfer, introduced by Liu et al. [62], the general form of which is:

$$\nabla(k_{skin}\nabla T)+\left[\omega_b\rho_b C_{p,b}(T_b-T)+q_m+q_r+\tau\left(-\omega_b C_{p,b}\frac{\partial T}{\partial t}+\frac{\partial q_m}{\partial t}+\frac{\partial q_r}{\partial t}\right)\right]$$
$$=\rho_{skin}C_{p,skin}\left[\tau\left(\frac{\partial^2 T}{\partial t^2}\right)+\frac{\partial T}{\partial t}\right] \tag{12.49}$$

where ρ ,C , T denote the density, specific heat and temperature, respectively, k is thermal conductivity, ω_b is the perfusion rate of blood, q_m is the metabolic heat generation, and q_r is the heat source of spatial heating, T_b is arterial temperature and regarded constant and τ is the thermal relaxation time of the skin. Equation (12.49) is a hyperbolic equation and is more mathematically complex than the traditional Pennes's bioheat transfer equation. The time to second degree burn is determined using the Henriques and Moritz [50] burn integral equation.

12.3.3. Model Development Involving Manikins

Richards and Fiala [63] compared predicted dynamic physiological responses to wear trials results for fire fighter garments. Data from the Sweating Agile thermal Manikin (SAM) [64] is used in simulation models to predict human responses to transient environmental and personal conditions.

Using SAM, steady state values of dry thermal resistance (I_{cl}) and intrinsic evaporative resistance ($R_{e,cl}$) were determined with and without walking. The intrinsic evaporative resistance is given by:

$$R_{e,cl} = R_{e,t} - \left(R_{e,a} \cdot f_{cl}^{-1}\right) \tag{12.50}$$

where $R_{e,a}$ is the evaporative resistance of the unclothed manikin, f_{cl} is the ratio of the clothed compared to the unclothed body surface and $R_{e,t}$ is the effective evaporative resistance of clothing, defined by:

$$R_{e,t} = \left(P_s - P_a\right) \cdot Q_e^{-1} \tag{12.51}$$

where $(P_s - P_a)$ is the water pressure gradient in P_a across the clothing and Q_e is the evaporative heat flow.

The Institute of Energy and Sustainable Development (IESD) used dynamic model of heat transfer [65-66] for the simulations. The simulations are performed by modeling the boundary conditions measured using SAM. This includes ray-trace-based modeling of the inhomogeneous IR irradiation at the individual surfaces of the exposed surfaces, and the use of measured clothing properties as time-dependent variables for the work and recovery phases of each wear trial. Measured metabolic rates from the wear trials are used in the model. Still and walking conditions are modeled separately. The steady-state approach for multi-layered ensembles proposed by McCullough et al. [67] is used to infer local I_{cl}, $R_{e,cl}$ and f_{cl} values from the overall values. The absorption of radiation at the clothing surfaces is modeled by calculating the required projected area factors for both diffuse and direct IR radiation for each body sector, and considering the geometrical configuration of the wear trials.

The authors found that the predicted physiological responses generally agreed well with the wear trial data. However, local microclimate and heat transfer within the garments need to be considered. To improve the model, they propose that local heat transfer coefficients measured underneath the clothing by each body section of SAM be used. Local I_{cl} and $R_{c,cl}$ values will effect local microclimatic conditions and skin temperatures.

Song et al. [68-69] developed a numerical model to predict skin burn injury resulting from exposure to flash fires. An instrumented manikin (PyroMan®) is dressed in a protective garment. A flash fire is generated and heat transfers by radiation and convection modes to the fabric surface. Heat is transferred through the protective garment by radiation and conduction, and then through the air gap between the garment and manikin body by radiation, convection and conduction. Sensors on the manikin record temperature changes and these are translated into heat flux readings. The estimated total heat transfer coefficient at each sensor is given by:

$$\hat{h}_M = \frac{\hat{q}_M}{T_f(T) - 0.5\left(\hat{T}_{OM} + \hat{T}_{OM-1}\right)} \tag{12.52}$$

where $\hat{h}_M$ is estimated heat transfer coefficient, $\hat{q}_M$ is calculated heat flux, T_f is flame temperature and $\hat{T}_{OM}$ is estimated surface temperature at time t_M.

Predications of skin burn injury are made using Henriques burn integral [50]. A heat transfer equation is applied in conjunction with a Pennes-based [54] skin model to estimate the temperature profile in the dermal layers of the skin. The heat transfer differential equation is given as:

$$\rho_f(T)c_{pf}(T)T\frac{\partial T}{\partial t}=\frac{\partial}{\partial x}\left(k_f(T)\frac{\partial T}{\partial x}\right)+\gamma\cdot q_{rad}e^{-\lambda x} \tag{12.53}$$

where ρ_f, c_{pf} and k_f are the temperature dependent density, specific heat and thermal conductivity of the fabric, respectively, T is the fabric temperature and $\gamma\cdot q_{rad}e^{-\lambda x}$ is the internal heat generated by thermal radiation transferred to the interior of the fabric by the transmissibility of the fabric.

The total amount of radiant and convection energy transferred to the garment was estimated using nude manikin burn test results. The boundary condition at the surface of the garment is expressed as:

$$(q_{conv}+q_{rad})=h_{fl}(T_{g-}T_f) \tag{12.54}$$

where h_{fl} is the total heat transfer coefficient between the flame and the garment.

Conduction and convection heat transfer through the air gap is modeled as a resistance to heat flow between two surfaces:

$$h_{gap}(L_{airgap},T)=Nu\frac{k_{air}(T)}{L_{airgap}} \tag{12.55}$$

where h_{gap} is the heat transfer coefficient for the air gap, L_{airgap} is the size of the air gap, Nu is the Nusselt number and $k_{air}T$ is the thermal conductivity of the air. Three-dimensional modeling was used to measure air gap sizes and distribution. The boundary at the interface between the fabric and skin is expressed as:

$$q_{conv}=h_{gap}(T_f-T_s) \qquad x=L_f,t>0 \tag{12.56}$$

The radiation boundary condition between the fabric and the skin is expressed as:

$$q_{rad}=0.85\sigma\left(T^4_{fabric}(t)-T^4_{skin}(t)\right) \qquad x=L_f,t>0 \tag{12.57}$$

The differential equations describing heat transfer through the fabric, air gap and skin are solved by a finite difference model. Due to the non-linear terms of absorption of incident radiation, Gauss-Seidel point-by-point iterative scheme is used to solve these equations. An under-relaxation procedure is utilized to avoid divergence of the iteration method. The Crank-Nicolson scheme [49] is used to solve the resulting ordinary differential equations in time.

Protective performance is evaluated using the Manikin Thermal Protective Clothing Analysis System. A full test report including contour mapping of burn locations is produced. The authors show that air gaps in protective garments play a vital role in providing thermal insulation when exposed to flash fire conditions. Garment shrinkage during exposure can potentially cause a decrease in the protective performance of the garment.

Gašperin et al. [70] used a manikin system to objectively assess the performance of fire resistant garments. The system acquires temperature readings from the manikin, calculates skin layer temperatures and calculates the Henriques burn integral [50]. Heat transfer in the skin is described by the following partial differential equation [71-72]:

$$(\rho c)_{\xi}\frac{\partial T}{\partial t} = k_{\xi}\frac{\partial^2 T}{\partial x^2} - G(\rho c)_b (T - T_b) \tag{12.58}$$

where T is temperature, k is thermal conductivity, x is the depth, G is the blood perfusion rate, $(pc)_b$ indicates the volumetric heat capacity of the blood, T_b is the blood temperature, ξ indicates the property of a particular skin layer – epidermal, dermal or subcutaneous. The skin's thermal model is solved by means of the Crank-Nicolson scheme [49]. The partial differential equation for heat conduction across sensor is described by:

$$\frac{\partial u}{\partial t} = \alpha\frac{\partial^2 u}{\partial x^2}, \qquad \alpha = \frac{k}{\rho c} \tag{12.59}$$

where α is the diffusion coefficient.

Instead of performing recursive calculations of the heat equation, first for the temperature sensor and then for the skin, an efficient numerical algorithm is used that calculates the temperature at the epidermis/dermis (T_d) and dermis/subcutaneous (T_s) interfaces directly from the temperature recordings by means of simple conversion sums. The calculation is a convolution sum between temperature and impulse response and is calculated off-line which saves on computational load. The temperatures obtained are then used to calculate skin injury.

Gašperin and Juričič [73] extended the work to take into account the variability of the thickness of the skin. Monte Carlo simulations are used to estimate the dispersion of burns over a broad population of humans. Simulations are performed in which the dermal thickness and thermal properties are varied, and in which the epidermal and dermal thicknesses and thermal properties are varied. The authors found that variations in the thickness of the dermis significantly contribute to the final dispersion of the burn areas. The influence of epidermis variability on burn dispersion can be neglected.

12.4. Modeling and Simulation in Extreme Weather Clothing

Extreme cold weather can be hazardous for those not wearing the correct clothing. Extreme weather clothing is used to prevent frostbite, hypothermia, or even death. Often, extreme cold weather clothing must do more than reduce heat loss; it must also keep the wearer dry. Cold weather clothing should be able to provide warmth, wick sweat away from the body and have minimal bulkiness for ease of movement. Often, a series of clothing layers, designed to provide insulation and keep wind, rain and snow away during extreme cold conditions, is worn. As with all protective clothing, a balance must be obtained that maximizes protection while minimizing impact on performance of the wearer. The complex wearer-clothing interactions in extreme weather provide many challenges for modeling.

Because of the overlap with models describing heat and mass transfer in fabrics and clothing, only a few examples relating to extreme weather clothing will be examined.

12.4.1. Heat and Mass Transfer

Farnworth [74] presented a model of the combined conductive and radiative heat flow through fibrous material used for clothing insulation. The amount of heat transferred by convective heat transfer is small compared to that of conductive and radiative heat transport and is neglected. In the case where a sample of fibrous material is held between two plates, one cold and one hot, and in the absence of radiative heat, the heat flow is given by:

$$Q(x) = -k\frac{dT(x)}{dx} \tag{12.60}$$

where T is the temperature, x is the distance from the cold plate to the sample and k is the combined thermal conductivities of air and of the fiber. Equation 12.60 is integrated to give:

$$Q(x) = -\frac{k}{D}(T_2 - T_1) = \text{constant} \tag{12.61}$$

and

$$T(x) = T_1 + (T_2 - T\ )\frac{x}{D} \tag{12.62}$$

where T_2 is the temperature of the hot plate, T_1 is the temperature of the cold plate and D is the thickness of the sample. The temperature within the sample is a linear function of position. Previous calculations have considered this temperature distribution to be imposed on the sample by conduction. The radiant heat transfer is calculated from this temperature distribution and added to the conductive component to give the total heat flow. Radiative heat transfer is usually expressed as a radiative conductivity times the temperature gradient:

$$Q_{RAD} = -k_{RAD}\frac{dT}{dx} \tag{12.63}$$

For insulation thicknesses used in clothing, the radiative and conductive heat flows are about the same size. The contribution of each to the total varies with position within the sample and the temperature distribution can be nonlinear. Consequently, the radiative conductive approach is inappropriate. The authors, therefore, combine equation 12.60 in its differential form with differential equations that describe the flow of thermal radiation:

$$t = \alpha_0 + \alpha_1 x + \alpha_2 e^{-px} + \alpha_3 e^{p(x-D)} \tag{12.64}$$

$$F_T = 2\sigma T_0^4 + 8\sigma T_0^3 \alpha_0 + 8\sigma T_0^3 \alpha_1 x - \beta k \alpha_2 e^{-px} - \beta k \alpha_3 e^{p(x-D)} \tag{12.65}$$

$$F_N = \frac{-8\sigma T_0^3 \alpha_1}{\beta} - kp\alpha_2 e^{-px} + kp\alpha_3 e^{p(x-D)} \tag{12.66}$$

where $$p = \beta \left\{ 1 + \frac{8\sigma T_0^3}{\beta k} \right\}^{1/2} \tag{12.67}$$

and $\alpha_0..\alpha_3$ are constants, F_T is total radiative flux, F_N is net radiative flux and β is the absorption constant = $\frac{f\varepsilon}{R}$, where f is fractional fiber volume, ε is fiber thermal emissivity and R is fiber radius. Using appropriate boundary conditions, the equations are solved simultaneously to determine the heat flows.

Havenith [75] examined some of the differences between detailed and lumped heat exchange models. Heat loss from the body is described as:

$$Heat\ Loss = \frac{gradient \cdot surface\ area}{resistance} \tag{12.68}$$

where the gradient is the temperature or vapor pressure gradient between the skin and the environment, surface area is the body surface area involved and resistance is clothing insulation.

Extreme weather clothing is usually made up of a number of layers with a still air layer on their surfaces and possibly mobile air layers on top of those. Detailed models take into account dry and wet heat transfer. Dry heat transfer tales place by conduction, convection and radiation, the latter two being most important. Wet heat transfer results from sweating. Water moisture on the surface of the skin evaporates remove large amounts of heat from the body. The water vapor molecules are transported to the environment in various ways including diffusion, convection, absorption (adsorption), desorption and condensation. Often, finite element modeling is used for the simulations in which the clothing and air layers are divided into a matrix of elements and heat and mass transport.

Lumped models are often used where the overall influence of the garment on the user is required, for example, ISO 11079, ASTM F2732 [76-77]. For this type of application, complex models are not practical and a calculation of the overall human heat balance is the main basis for the analysis.

Min et al. [78] developed a model to simulate heat and moisture transport from the skin to the environment through fabrics. The model includes radiation heat transfer between surfaces and the surface diffusion along fibers. The horizontal system comprises the skin layer which is separated form the fabric by an air gap (microclimate). The fabric is treated as a porous medium and it is separated from the external environment by a boundary layer. Heat transfer through the microclimate is expressed as:

$$q_{MC} = -k_{MC} \frac{dT}{dx} + \frac{\sigma\left(T_S^4 - T_{MF}^4\right)}{\left(1/e_S + 1/e_{MF} - 1\right)} + j_{MC}\left(\Delta H_{vap}\right) \quad (12.69)$$

where the first term represents thermal conduction through the microclimate ($_{MC}$), the second term represents radiative heat transfer between the skin and fabric and the third term represents heat transfer by moisture diffusion through the microclimate, and where k is thermal conductivity, T is temperature, σ is Stefan-Boltzmann constant, $_{MF}$ is the boundary between microclimate and fabric, e is emissivity, j is diffusion flux and ΔH_{vap} is enthalpy change of vaporization.

The heat and moisture fluxes through the fabric are written as:

$$q_F = -k_{F,eff} \frac{dT}{dx} + j_F \Delta H_{vap} \quad (12.70)$$

where the first term represents thermal conduction through the fabric ($_F$) and the second term represents heat transfer by pore and surface diffusion. The average conductivity is given by:

$$k_{F,eff} = (1-\varepsilon)k_{fabric} + \varepsilon k_{air} \quad (12.71)$$

where the first term is conduction through the solid fibers and the second term is conduction through the air filling the interstices between fibers, and ε is fabric porosity. As the fabric is densely packed and opaque, the radiation effect is included the effective thermal conductivity term.

The heat is transferred from the fabric surface to the environment is written as:

$$q_E = h\left(T_{FE} - T_E\right) + j_E \Delta H_{vap} + \sigma e_F\left(T_{FE}^4 - T_E^4\right) \quad (12.72)$$

where the first term represents convection, the second term represents moisture transport and the third term represents radiation, and h is the heat transfer coefficient, $_E$ is the environment and $_{FE}$ is the boundary layer between the fabric and the environment. At the steady state the heat fluxes through the microclimate, fabric and environment should be balanced. Relative humidity and temperatures at the skin surface and microclimate are assumed to be constant and the temperature and moisture content are continuous at the interfaces. The authors also consider a vertical system where convective motion inside the microclimate is considered. They found that the heat flux was 10% larger in the vertical system than in the horizontal system. Wu and Fan [79] used a three-level two-factor full factorial design to examine the effects of batting constructions on moisture transport through multi-layer garment assemblies under subzero temperature environmental conditions. Perspiration at subzero temperatures may condense to water or freeze with in the insulating layers of protective clothing, reducing thermal resistance and increasing heat loss. A sweating guard plate assembly is used to test the clothing assemblies. Moisture accumulated in the clothing assembly is determined by mass difference. Moisture transmitted through the fabric is determined by subtracting the moisture accumulated in the clothing assembly from the starting mass of the waster supplied in the equipment. The authors found that the type of batting and the position of battings in the construction influenced moisture transport in the clothing assemblies.

12.4.2. Model Development Involving Manikins

Meinander et al. [80] used the IREQ model [81] to choose the insulation levels, and hence clothing systems, required to maintain the body in a state of thermal equilibrium at acceptable levels of body and skin temperature, during four different environmental conditions (0, -10, -25 and -50 °C). Wear trials are compared with the results from sweating manikin tests. The authors found that metabolic rates were higher than predicted, particularly for the clothing systems chosen for -25 and -50 °C. In these systems, strong sweating started at low activity. The authors suggest that uneven clothing weight distribution, garment stiffness, friction between clothing layers and wind could be causes for the higher metabolic rates. These properties were not considered in the predictive model.

Ducharme et al. [82] used computer modeling and manikin data for the selection of military protective gear for 12-h cold survival on land and at sea. Survival time (ST) is estimated using the Cold Exposure Survival Model (CESM), developed at Defense Research and Development Canada. The model uses anthropometric data, environmental conditions (air and water temperature, and wind speed) and thermal resistance values obtained from the manikin tests as variables. Predicted STs are based on torso thermal resistance. This variable was deemed most important during the development of CESM with regard to the onset of lethal hypothermia. The authors showed that manikin data and modeling can be useful in making recommendations for the selection of survival gear.

12.5. Future Trend

Modeling offers a powerful tool that compliments experiments and testing for the development of protective clothing. The objective of modeling and simulation of chemical, fire and extreme weather protective clothing is to realistically describe the clothing and the processes that take place in order to predict their impact on the wearer. Wearer impacts include wearer comfort (thermal stress) and injury. The future will bring the refinement of lumped models and the development of more complex detailed models. These will be driven by advances in the understanding of physiological processes and clothing properties and advances in computational resources. Drivers will be new developments in fibers and fabrics, hybrid materials, the use of semi-permeable membranes and electro-spun membranes, active barriers (carbon, catalysts) multi-functional finishes, and the use of nanotechnology and electronics in protective garments. Modeling activities will examine the effects of these developments on heat and mass transfer, protective capabilities and physiological responses. Of interest will be, the effect of garment layering on protective performance, the effect of friction between layers and heat storage in fabrics on the heat balance, modeling of transport in vertical air gaps, seams and joins, the effect of motion and movement on the various processes, the effects of moisture in under-garments, the condensation of moisture on the skin, accounting for the head, feet and hands response in the total physiological response, the effect of activity levels on fabric and body conditions prior to exposure to the hazard, the effect of external environmental conditions such as temperature, moisture and wind, the effect of uneven levels of exposure to the hazard, for example uneven heating, the effect of force on protective performance, and the effect of fabric degradation from use, laundering and repeated exposure to the hazard on the protective performance of the garments and impact on wearer.

REFERENCES

[1] Gonzalez, R. R.; McClellan, T. M.; Withey, W. R.; Chang, S. K.; Pandolf, K. B. *J. Appl. Physiol.* 1997, *vol 83*, 1017-1032.

[2] Givoni, B.; Goldman, R. F. *J. Appl. Physiol.* 1972, *vol 32*, 812-821.

[3] Stolwijk, J. A. J.; Hardy, J. D. *Handbook of Physiology. Reactions to Environmental Agents*; Am. Physiol. Soc.: Bethesda, MD, 1977, pp 45-67.

[4] Gagge, A. P.; Stolwijk, J. A. J.; Nishi, Y. *ASHRAE Trans.* 1971, *vol 77*, 247-262.

[5] Breckenridge, J. R. In *Clothing Comfort-Interaction of Thermal, Ventilation, Construction, and Assessment Factors*; Hollies, N. R. S.; Goldman, R. F.; Eds.; Ann Arbor Science Publishers: Ann Arbor, MI, 1977.

[6] Woodcock, A. H. *Text. Res. J.* 1962, *vol 32*, 628-632.

[7] Cain, B.; McLellan, T. M. *Int. J. Biometeorol.* 1998, *vol 41*, 183-193.

[8] Sobera, M. P.; Kleijn, C. R.; Brasser, P.; Van den Akker, H. E. A.; In *Proc, ICCS 2004, Poland*; Bubak, M.; Ed.; Springer-Verlag: Berlin, 2004, pp 637-644.

[9] Clague, D.; Kandhai, B.; Zhang, R.; Sloot, P. *Phys. Rev.* 2000, *vol 61*, 616-625.

[10] Gibson, P. *J. Eng. Fibers Fab.* 2009, *vol 4*, 1-8.

[11] Danielsson, U. *Eur. J. Appl. Physiol.* 2004, *vol 92*, 689-693.

[12] Barry, J. J.; Hill, R. W. *Int. Nonwovens J.* 2003, *vol 12*, 25-35.

[13] Hill, R. W.; Barry, J. J. *Int. Nonwovens J.* 2004, *vol 14*, 22-30.

[14] Gibson, P. W. Governing Equations for Multiphase Heat and Mass Transfer in Hygroscopic Porous Media with Applications to Clothing Materials, [Technical Report, Natick/TR-95/004], U.S. Army Natick Research and Development Center, Natick, MA. 1994.

[15] Gibson, P. W. Multiphase Heat and Mass Transfer in Hygroscopic Porous Media with Applications to Clothing Materials, [Technical Report, Natick/TR-97/005], U.S. Army Natick Research and Development Center, Natick, MA, 1996.

[16] Whitaker, S. In *Advances in Heat Transfer*; Irvine, T. F. and Hartnett, J.; Eds.; Academic Press: New York; 1977; vol. 13, pp 119-203.

[17] Whitaker, S. In *Advances in Heat Transfer*; Irvine, T. F. and Hartnett, J.; Eds.; ISBN 0 120200317: Academic Press: New York; 1998; vol. 31, pp 1-102.

[18] Progelhof, R.; Throne, J.; Ruetsch, R. *Polym. Eng. Sci.* 1976, *vol 16*, 615-625.

[19] Spurny, K. R.; Schoermann, J.; Opiela, H. *Am. Ind. Hyg. Assoc. J.* 1990, *vol 51*, 36-43.

[20] Fedele, P. D.; Bergman, W.; McCallen, R.; Sutton, S. *Proc. 1986 Army Science Conf.* 1987.

[21] Fedele, P. D. In *Performance of Protective Clothing: Fourth Volume*; McBriarty, J. P.; Henry, N. W.; Eds.; ASTM Special Technical Publication 1133; American Society for Testing and Materials: Philadelphia, PA, 1992; pp 3-16.

[22] Hoerner, S. F. *Fluid Dynamic Drag*, Hoerner Fluid Dynamic: Bricktown, NJ, 1965, pp 3-23.

[23] Lee, K. W; Lui, B. Y. H. *Aerosol Sci. Tech.* 1982, *vol 1*, 147-161.

[24] Roberts, M.; Jamriska, M.; Skvortsov, A.; McCallum, R. Study on Aerosol Penetration Through Clothing and Individual Protective Equipment, 2009, [Technical report], http://www.dsto.defence.gov.au/corporate/reports/DSTO-TR-2283.pdf

[25] Brasser, P. *Am. Inst. Chem. Eng. J.* 2006, *vol 52*, 3688-3695.

[26] Stuart, I. M.; Denby, E. F. *Text. Res. J.* 1983, *vol 53*, 655-660.

[27] Hinds, W. C. *Aerosol Technology: Properties, Behaviour, and Measurement of Airborne Particles;* ISBN: 9780471194101; John Wiley and Sons: New York, 1999, 2nd Edn, pp 1-504.

[28] Roberts, M. D.; Skvortsov, A. T. *Advance and Appllication Fluid Dynamics.* 2010, Pushma Publishing, pre-print.

[29] Goydan, R.; Carroll, T. R.; Schwope, A. D.; Reid, R. C. Refinement of a Model to Predict the Chemical Permeation of Protective Clothing Materials, 1990, [Technical report, EPA/600/S2-90/028], U.S. Environmental Protective Agency: Cincinnati, OH.

[30] Zhang, X.; Raheel, M. *Bull. Environ. Contam. Toxicol.* 2003, *vol 70*, 652-659.

[31] Lee, S.; Obendorf, S. K. *Text. Res. J.* 2001, *vol 71*, 1000-1009.

[32] Wadsworth, L. C.; Easter, P. E.; Lin, Y. Q. In *Proc. First International Symposium on the Impact of Pesticides, Industrial and Consumer Chemicals on the Near Environment*; Reagan, R.; Johnson, D.; Dusaj, S.; Eds.; 1988.

[33] Jain, R.; Raheel, M. *Bull. Environ. Contam. Toxicol.* 2003, *vol 71*, 437-446.

[34] Obendorf, S. K.; Csiszár, E.; Maneefuangfoo, D.; Borsa, J. *Arch. Environ. Contam. Toxicol.* 2003, *vol 45*, 283-288.

[35] Lee, S.; Obendorf, S. K. *Arch. Environ. Contam. Toxicol.* 2005, *vol 49*, 266-273.

[36] Mäkinen, H. In *Advances in Fire Retardant Materials*; Horrocks, A. R.; Price, D.; Eds.; ISBN-10: 1420079611; Woodhead Publishing, Cambridge; 2008; pp 467-491.

[37] Torvi, D. A.; Dale, J. D. *Fire Technol.* 1999, *vol 35*, 210-231.

[38] Stelzer, J. F.; Welzel R. *Int. J. Numer. Method. Eng.* 1987, *vol 24*, 59-73.

[39] Stoll, A. M.; Chianta, M. A. *Aerospace Med.* 1969, *vol 40*, 1232-1238.

[40] Torvi, D. A.; Threlfall, T. G. *Fire Technol.* 2006, *vol 42*, 27-48.

[41] Sawcyn, C. M. J.; Torvi, D. A. *Text. Res. J.* 2009, *vol 79*, 632-644.

[42] Talukdar, P.; Torvi, D. A.; Simonson, C. J.; Sawcyn, C. M. J. *Int. J. Heat Mass Transfer* 2010, *vol 53*, 526-539.

[43] Mell, W. E.; Lawson, J. R. *Fire Technol.* 2000, *vol 36*, 39-68.

[44] Lawson, J. R.; Walton, W. D.; Bryner, N. P., Amon, F. K. Estimates of Thermal Properties for Fire Fighters Protective Clothing Materials, 2005, NISTIR 7282, National Institute of Standards and Technology: Gaithersburg, MD.

[45] Prasad, K.; Twilley, W.; Lawson, J. R. Thermal Performance of Firefighters' Protective Clothing. 1. Numerical Study of Transient Heat and Water Vapor Transfer, 2002, NISTIR 6681, National Institute of Standards and Technology: Gaithersburg, MD.

[46] Lawson, J. R.; Mell, W. E.; Prasad, K. *Fire Technol.* 2010, *vol 46*, 833-841.

[47] Chitrphiromsri, P.; Kuznetsov, A. V. *Heat Mass Transfer* 2005, *vol 41*, 206-215.

[48] Torvi, D. A. *Heat Transfer in Thin Fibrous Materials under High Heat Flux Conditions;* PhD Thesis, University of Alberta, Edmonton, 1997.

[49] Crank, J.; Nicolson, P. *Proc. Camb. Phil. Soc.* 1947, *vol 43*, 50-67.

[50] Henriques, F. C. Jr.; Moritz, A. R. *Am. J. Pathol.* 1947, *vol 23*, 531-549.
[51] Zhu, F.; Zhang, W.; Song, G. *Fire Safety J.* 2008, *vol 43*, 401-409.
[52] Liu, L. Y. *Mathematical Simulation of Dynamic Heat and Vapor Transmission in Microclimate and the Comfort Model of Clothed Body;* PhD Thesis, China, 2002.
[53] Zhu, F.; Zhang, W. *J. Ind. Text.* 2007, *vol 37*, 175-186.
[54] Pennes, H. H. *J. Appl. Physiol.* 1948, *vol 1*, 93-122.
[55] Valvano, J. W. In *Encyclopedia of Medical Devices and Instrumentation*; ISBN: 0471263583; Webster, J. G.; Ed.; Wiley-Interscience: New York, 2005; vol. 2.
[56] Charny, C. K. In *Bioengineering Heat Transfer*; Cho, Y. I.; Ed.; Academic Press: San Diego, CA, 1992.
[57] Chen, M. M.; Holmes, K. R. *Ann. N. Y. Acad. Sci.*, 1980, *vol 335*, 137-150.
[58] Halin, J.; Mulligan, J.; Hamouda, H.; Barker, R.; Soroka, A. In *Performance of Protective Clothing*; Johnson, J. S.; Mansdorf, S. Z.; Eds.; American Society for Testing Materials: West Conshohocken, PA, 1996.
[59] Weaver, J. A.; Stoll, A. M. *Aerospace Med.* 1969, *vol 40*, 24-30.
[60] Takara, A. N.; Rouse, J.; Stanley, T. 1973, Thermal analysis program, IITRI-J6286, Illinois Institute of Technology, Chicago, IL.
[61] Zhu, F.; Ma, S.; Zhang, W. *Heat Mass Transfer* 2008, *vol 45*, 99-105.
[62] Liu, J.; Xu, C.; Xu, L. X. *IEEE Trans. Biomed. Eng.* 1999, *vol 46*, 420-428.
[63] Richards, M. G. M.; Fiala, D. *Eur. J. Appl. Physiol.* 2004, *vol 92*, 649-653.
[64] Richards, M. G. M.; Mattle, N. G.; Becker, C. In *Proc. 2nd European Conference on Protective Clothing*; Mäkinen, H.; Rossi, R.; Eds.; Montreaux, Switzerland, 2003.
[65] Fiala, D.; Lomas, K. J.; Stohrer, M. *J. Appl. Physiol.* 1999, *vol 87*, 1957-1972.
[66] Fiala, D.; Lomas, K. J.; Stohrer, M. *Int. J. Biometeorol.* 2001, *vol 45*, 143-159.
[67] McCullough, E. A.; Jones, B. W.; Tamura, T. *ASHRAE Trans.* 1989, *vol 91*, 316-328.
[68] Song, G.; Barker, R. *Proc. 83rd World Conference: Quality Textiles for Quality Life.* Shanghai, 2004.
[69] Song, G. *J. Ind. Text.* 2007, *vol 36*, 193-205.
[70] Gašperin, M.; Juričcč, D.; Musizza, B.; Mekjavič, I. *ISA Trans.* 2008, *vol 47*, 198-210.
[71] Crown, E. M.; Dale, J. D. 1992, Evaluation of Flash Fire Protective Clothing using an Instrumental Manikin [Report], University of Alberta.
[72] Walfre, F.; Jie, L.; Guo-Xiang, W.; Nelson, J. S.; Guillermo, A. *Phys. Med. Biol.* 2005, *vol 50*, 387-397.
[73] Gašperin, M.; Juričič, D. *Burns*. 2009, *vol 35*, 970-982.
[74] Farnworth, B. *Text. Res. J.* 1983, *vol 53*, 717-725.
[75] Havenith, G. In *Proc. 11th Int. Conf. Environmental Ergonomics*. Ystad, Sweden, 2005.
[76] ISO 11079. Ergonomics of the Thermal Environment -- Determination and Interpretation of Cold Stress when using required Clothing Insulation (IREQ) and Local Cooling Effects, 2003, International Organization for Standardization, Geneva.
[77] ASTM F2732 – 09. Standard Practice for Determining the Temperature Ratings for Cold Weather Protective Clothing, 2009, ASTM International, West Conshohocken, PA.
[78] Min, K.; Son, Y.; Kim, C.; Lee, Y.; Hong, K. *Int. J. Heat Mass Transfer* 2007, *vol 50*, 5292-5304.
[79] Wu, H.; Fan, J. *Text. Res. J.* 2008, *vol 78*, 988-995.

[80] Meinander, H.; Anttonen, H.; Bartels, V.; Holmér, I.; Reinertsen, R. E.; Soltynski, K.; Varieras, S. *Eur. J. Appl. Physiol.* 2004, *vol 92*, 619-621.
[81] ISO 11079: Evaluation of Cold Environments – Determination of required Insulation (IREQ), 2001, International Organization for Standardization, Geneva.
[82] Ducharme, M. B.; Tikuisis, P.; Potter, P.; *Eur. J. Appl. Physiol.* 2004, *vol 92*, 658-662.

In: Modeling and Simulation in Fibrous Materials
Editor: Asis Patanaik
ISBN: 978-1-62100-116-4

Chapter 13

MODELING RESIN TRANSFER MOULDING PROCESS FOR COMPOSITE PREPARATION

Naveen V. Padaki[1,*] ***and R. Alagirusamy***[2]
[1]Regional Silk Technological Research Station, Central Silk Board, Guwahati, India
[2]Department of Textile Technology, Indian Institute of Technology Delhi, New Delhi, India

ABSTRACT

This chapter deals with the simulation of Resin Transfer Moulding (RTM) process of composite preparation. Determination of permeability of the fibrous material is the most critical aspect for flow front predictions. Two methods of permeability measurements viz., rectilinear and radial flow methods have been briefly described and compared. Permeability properties of the reinforcement are utilized to simulate the resin flow using CFD (Computational Fluid Dynamics) software. This chapter outlines modeling RTM process through simple simulation of resin flow through textile reinforcement during resin transfer moulding method of composite preparation. Simulation steps have been explained with an example of rectangular mould filling multiphase simulation for better understanding.

13.1. INTRODUCTION

Ever increasing demand for faster composite manufacturing has pressed the composite industry to replace hand lay-up technique with alternative fabrication processes such as liquid composite moulding techniques.

Resin transfer moulding (RTM), is a prominent type of liquid injection moulding, defined as a closed moulding process of composite preparation in which catalyzed resin is transferred into an enclosed mold cavity to impregnate a pre-positioned fibrous reinforcement (preform), as presented in the Figure 13.1 [1].

[*]E-mail: naveenvpadaki@gmail.com

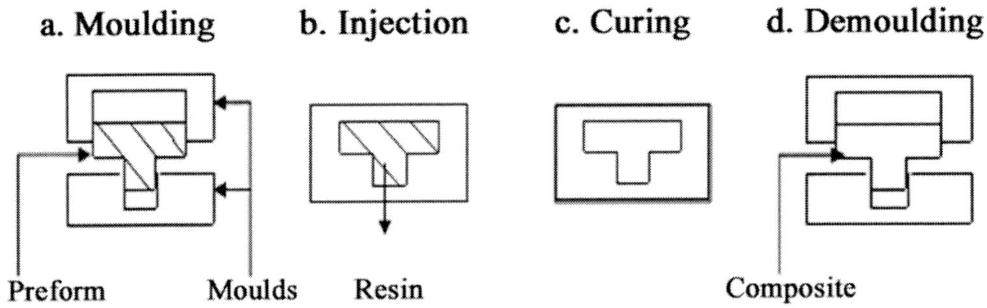

Figure 13.1. RTM Process [Source: Reference 1].

Resin transfer methods represent the fastest growing liquid moulding techniques [1]. Thermoset resins namely epoxy, polyester and vinyl ester resins with lower viscosity are preferred for RTM processes but epoxies have certain advantages like better fatigue life, impact resistance, corrosion resistance and also higher glass transition temperature. Polyester resins are most abundantly used for wide range of applications due to reasonable matrix properties at very affordable cost. The closed mold of RTM offers several advantages over hand lay-up. Parts have tighter tolerances, they are finished on both sides, and higher fiber volume fractions are possible. In addition, the reinforcing fiber layers may be preformed so they can be laid into the mold in one piece, offering considerable time savings.

Even without preforming, RTM offers a timesaving over hand lay-up in the production of complex parts [2]. In addition, the closed mold prevents volatiles from escaping during processing. RTM is a closed mold process for making composite materials to produce better quality parts more cost effectively than the other composite manufacturing methods. However, fluid flow in RTM through the reinforcement tends to be unpredictable for complex shaped composite material.

Hence, CFD (Computational Fluid Dynamics) simulation techniques are adopted as predictive method for resin flow to design the RTM molds and to determine the vent locations in the molds. Accurate prediction of the resin flow through the preform inside the RTM mold is one of the most critical aspects of RTM simulation program. Permeability properties of the textile preform to a great extent influences the resin flow during the RTM process.

13.2. Permeability of Textile Reinforcements

Permeability is a measure of the ease of flow of fluid through a porous medium. A review on permeability properties, experimental evaluation and modeling permeability of textile preforms has been excellently detailed by Wong, 2006. Flow in porous media was first studied experimentally by Henri-Philibert Darcy in 1856 [3]. Later, through adoption of Darcy's law, researchers have introduced this permeability as a parameter for porous media where all the detailed microscopic interactions between the fluid and the porous medium were lumped into the permeability value.

As such, it is a property of the porous medium, which in this case is the fiber reinforcement and is anisotropic. Its value depends on the porosity of the reinforcement, the dimensions of the capillary passages in various directions in the reinforcement and the tortuosity of these passages [4]. Darcy's law for a linear and slow steady state flow through a porous media is given by the equation (13.1) where, K is a flow conductivity proportionality constant or permeability constant of the porous medium with respect to the fluid, ΔP net flow pressure head and L, length of flow in the sample. Permeability constant is a ratio of permeability (k) of the porous media to the fluid viscosity (η) as given by equation (13.2) below:

$$Q = -K\frac{\Delta P}{L} \tag{13.1}$$

$$K = \frac{k}{\mu} \tag{13.2}$$

It is worthy to note that permeability as defined in Darcy's law (equation 13.1) pertains to the steady flow of fluid in a saturated porous medium, whereas the resin transfer process involves unsteady flow of resin into an unsaturated preform. The above holds true when analyzing the macroscopic flow through a fiber preform, i.e. flow in production of a composite part. However, one must be aware that the flow behavior in a fiber preform is markedly different when analyzed at the mesoscopic length scale, which concerns the interaction between the fluid and the intricate structure of the fiber bundles.

Experimental measurements remain the most reliable way to determine permeability. The two most widely used techniques for measurement of the in-plane permeability are the rectilinear flow method and the radial flow method. In a rectilinear flow experiment the reinforcement is usually placed in a rectangular mould and resin is injected from a side gate and allowed to permeate the reinforcement in only one direction.

In contrast, in a radial flow experiment resin is injected through a central circular gate into a square or circular mould containing the reinforcement and proceeds to follow an elliptical flow front. In both experimental methods, a transparent cover is usually used to permit observation of the flow front progression and by measuring the pressure drop and flow rate, one can then use Darcy's law to find the permeability in the flow direction. It is generally agreed that both experimental methods give comparable permeability values. The procedures for both the rectilinear and radial flow experiments for fabrics have been described in detail by Adams, Miller, Rebenfeld and co-researchers [5-8] Gebart and Lidstrom [9] and Parnas and Salem [10] amongst others.

Heardman et al. [11] described a radial flow method to measure permeability under transient flow, where neither the pressure nor the flow-rate is considered constant. They obtained good agreement with experiments using the normal constant pressure setup for assemblies of woven fabrics. They also showed that, using error analysis on the equation used to calculate permeability, the estimated error due to the parameters used to determine permeability (e.g. pressure, viscosity and time) is in the range of 8-16%. However, they did not quantify the range of measured permeability values for their reinforcements and hence did not compare it to the estimated error.

Initial studies showed that directional permeability is independent of driving pressure, fluid velocity, fluid viscosity and fiber surface wetting properties [12-14] performed a series of constant flow rate experiments using a different flow rate for each set. They reported that while Darcy's law was obeyed in each set of constant flow rate experiments, the permeability increased with increasing injection flow rate. Chan et al. [13] explained this observation by the preferential channeling of the liquid through the pore spaces as the pressure gradient goes up.

13.3. RTM Flow Modeling Approaches

Fluid flow in RTM moulds is very complex. One complication involves the treatment of multiphase fluid flow. Prior to injection there is air, or void space if a vacuum is drawn, and solid fibers. During injection there are multiple flow areas with resin and fibers in one region and fibers and air or void space in another. Often times a transition region exists between the two where air and resin coexist in varying concentrations. Another complication involves dealing with the multiple length scales present in the fiber preform. A model that includes all the geometric complexities of the porous medium inside the mould would be overwhelming for a reasonable sized mould, let alone take into account the multi-phase flow regions. Thus it is common to make several assumptions when evaluating flow through RTM moulds.

There are two common approaches. The first approach is to combine porous media theory with more generalized flow models such as Stokes flow to take into account micro-geometric effects. This approach is referred to as micro-modeling. Another approach is to use porous media theory and apply it to the entire mould, macro-flow. Micro-flow models can take into account complex geometry regions of a mould, while macro-flow models can consider larger, more complex moulds [15].

A number of standard numerical techniques are available to solve the resultant partial differential equation derived from Darcy's law, including the finite difference method (FDM), the boundary element method (BEM) and the finite element method (FEM) [16]. Since mould filling is a moving boundary problem, these numerical techniques can be broadly divided into two approaches into moving grid or fixed grid [17-18]:

i) Moving grid: The computational domain covers only the resin saturated region in the mould and needs to be remeshed at each time step as the flow front progresses. This scheme provides an accurate representation of the flow front but is very computationally intensive and may encounter problems with multiple flow fronts and injection gates.
ii) Fixed grid: The computational domain is discretized only once for each simulation. An additional algorithm is needed to simulate the flow front progression, giving a less accurate representation of the flow front compared to a moving grid scheme.

CFD provides the most rigorous approach to analyze fluid dynamics problems computationally, solving the full set of equations of motion [16]. It means that it will require a longer time to obtain the solution. Details of CFD in general are covered in many textbooks, such as Anderson [19] and Ferziger and Peric [20].

Among many available CFD software packages, Fluent® marketed by Fluent Inc. is one of the most widely used software. It includes a pre-processor, GAMBIT™, and the CFD solver with the company name which also has advanced post-processing capabilities. Fluent® solves the Navier-Stokes equations of motion using the finite volume method.

It has an extensive range of physical modeling and multiphysics capabilities, including adaptive meshing, turbulence models, heat transfer, and multiphase models. This chapter will discuss the flow experiment and a simple three-dimensional CFD simulation model, and will compare the results obtained for the RTM resin flow through the textile reinforcement. The CFD calculations were performed using Fluent® high-speed flow CFD software. Fluent solves the integral form of the time-dependent Reynolds-averaged Navier-Stokes equations in three dimensions using a structured grid design. For all flows, Fluent solves conservation equations for mass and momentum.

13.4. Flow Model Considerations and Theoretical Aspects

The flow model generally has two main input variables, namely vacuum pressure at the inlet/outlet vents and fabric particulars. The fluid pressure accuracy is determined by the mesh resolution in the direction of the pressure gradients. Pressure gradients will typically be highest around sharp geometry transitions such as inlet and outlet locations in the case of RTM.

The resolution of the flow front location will be determined by the distance from one node to the next. Since the multiphase model with volume of fluid (VOF) method does not locate the flow front exactly, the uncertainty in the flow front locations is the length of the element. The typical assumptions made in the formulation of the model include [21]:

1. The preform is a homogeneous, porous and isotropic medium. The fabric properties required for the flow model are permeability and porosity. Fabric porosity is not influenced by its composition hence the fabric is assumed to be homogeneous. Also the permeability of the fabric is experimentally evaluated using flow imaging technique which provides an average value considering composition and directional effects in the fabric. In case of anisotropic performs, permeability values in different directions can also be defined in the flow model at material property section.
2. The flow is quasi-steady state.
3. Capillary and inertial effects are neglected (low Reynolds number flow).
4. The fluid is assumed to be Newtonian (its viscosity is independent of shear rate), and incompressible.
5. The fluid does not leak from the mold cavity.

The continuity equation for the fluid is given by equation 13.3 as given below:

$$\frac{\partial v_i}{\partial x_i} = 0 \qquad (13.3.)$$

where v_i is the interstitial velocity vector. As the fluid flows through the pores of the preform, the interstitial velocity of the resin is defined by equation 13.4 as below:

$$v_i = \frac{q_i}{\emptyset} \tag{13.4}$$

where q_i is the superficial velocity vector and Ø is the porosity of the preform. The porosity was assumed to be constant in this simulation. Using the assumptions that the preform is a porous medium and that the flow is quasi-steady state, the momentum equation can be replaced by Darcy's law (equation 13.5):

$$q_i = -\frac{k}{\mu}\frac{\partial P}{\partial x_j} \tag{13.5}$$

where μ is the fluid viscosity, k is the permeability tensor of the preform, and P is the fluid pressure. As the resin is incompressible and substituting (13.5) into (13.2) gives the governing differential equation of the flow:

$$\frac{\partial}{\partial x_i}\left(\frac{k}{\mu}\frac{\partial P}{\partial x_j}\right) = 0 \tag{13.6}$$

This second order partial differential equation (13.6) can be solved when the boundary conditions are prescribed. The boundary condition for the outlet to the mold is a prescribed pressure condition:

$$P_{\,outlet} = P_{\,outlet}(\mathrm{t}) \tag{13.7}$$

The tracking of the flow front or the interface between the phases is accomplished by the solution of a continuity equation for the volume fraction of one (or more) of the phases in VOF formulation. For the 'q'th phase, equation (13.8) is given by:

$$\frac{\partial \alpha_q}{\partial t} + \vec{v}.\nabla\alpha_q = \frac{S_{\alpha q}}{\rho_q} \tag{13.8}$$

where, α_q is the volume fraction of 'q'th phase and $S_{\alpha q}$ is the mass of 'q'th phase [22]. Since the resin cannot through the mold wall, the final boundary condition necessary to solve equation (13.6) is that the velocity normal to the wall at the boundary of the mold must be zero:

$$\bar{v}\,\bar{n} = 0 \tag{13.9}$$

where $\bar{n}$ is the vector normal to the mold wall.

13.5. Single and Multi Phase Models

Two different 3D models are possible to simulate the resin flow through textile preforms during RTM operation. First model is a simple single-phase model where unsteady flow of only resin through the textile preform can be modeled to arrive at the saturated inlet and outlet flow rate values. Based on this flow rate (*f*), mold filling time (t_{fill}) can be calculated using the relation (13.10) [23]:

$$t_{fill} = \frac{\emptyset\ \rho\ (h\ W\ L)}{f} \tag{13.10}$$

where, ρ is resin density, $\emptyset$ fabric porosity and (hWL) volume available in the mold. The limitations of this model are non-transient behavior simulation and also lack of video streaming possibilities to generate simulated flow pattern. Prior to resin injection inside the mold, through the preform there is air, or void space if a vacuum is drawn. During injection there are multiple flow areas with resin and fibers in one region and fibers and air or void space in another.

Many times transition region exists between the two, where air and resin coexist in varying concentrations. Multiphase flow model can be developed for the RTM flow simulation to incorporate the effects of both the phases. This model not only provides visuals of resin flow (flow front) in the mold but also depicts the transient behavior of flow of two phases (air and resin) where the initial air phase is replaced by the resin during mold fill process.

Interphase of such multiphase is modeled using VOF - multiphase options in the simulation software. This option evaluates the flow pattern considering unit volume of the flow area in mold as fraction of resin. Value 1 means flow volume saturated by resin and 0.5 means half volume occupied by resin and other half by air.

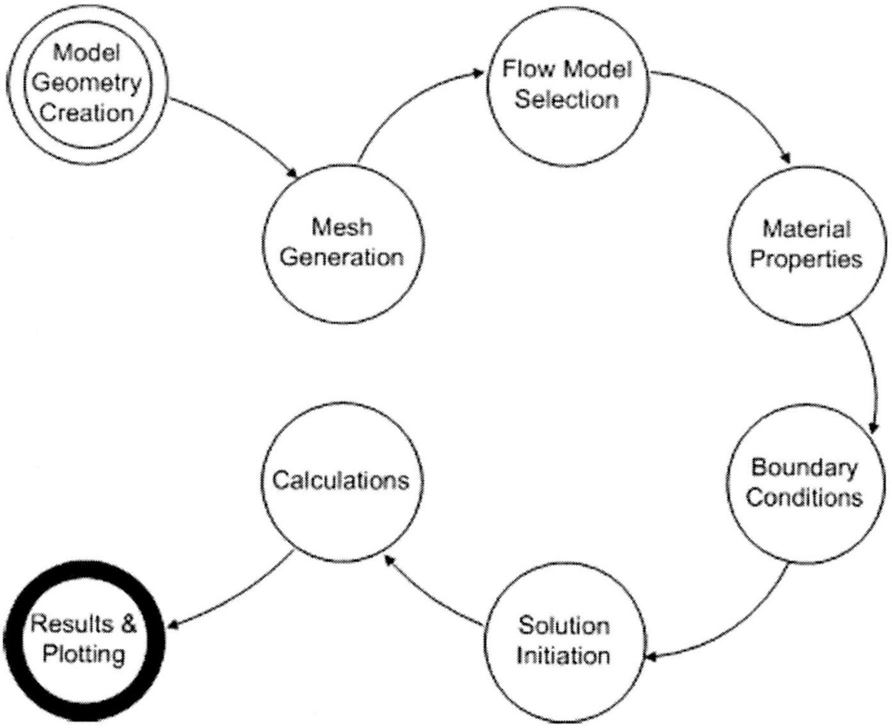

Figure 13.2. Steps of RTM Flow Simulation [Source: Reference 1].

13.6. Simulation of RTM Flow

Simulation of resin flow through RTM mould assembly can be accomplished by use of software's such as Fluent. Typical steps involved in such simulation are as provided in the flow chart (Figure 13.2) [1]. The details involved in each step have been explained below along with providing an example for multiphase flow through a fabric inside the RTM assembly. A mould assembly of cavity 20 cm x 20 cm x 1.5 mm for fabric along with inlet and outlet vents of 4.5 mm diameter has been used as example.

a. Creation of model geometry

Model geometry comprising of fabric inside the RTM mould assembly, inlet vent and outlet vent on one surface of the fabric has to be created using softwares such as GAMBIT. An illustration of the model geometry comprising of fabric of size 200 mm x 200 mm x 1.5 mm, inlet and outlet on one surface of the fabric having 4.5 mm diameter each at opposite sides of the fabric, 10 mm away from the fabric edge has been is as illustrated by Figure 13.3. [1].

b. Mesh generation

Although many types of automated meshing softwares are available, a simple inbuilt scheme such as Cooper scheme can be used to automatically generate an unstructured, hexahedral/tetrahedral mesh for the generated model. Mesh needs to be examined, which can be accomplished using the inbuilt tools of software to assess the quality of the resultant mesh, as properties of the mesh such as skewness can greatly affect the accuracy and robustness of the CFD solution. For the illustrated model, fabric volume mesh was created for thickness of 1 mm, with 80646 nodes and 39922 elements where both inlet and outlet face meshes consisted of 33 elements and 25 nodes each. Mesh was examined using the inbuilt tool of GAMBIT to assess the quality of the resultant mesh. The range of skewness of the 3D element (brick) was examined for Equiangle skew and the results are tabulated in Table 13.1 [1]. In quality assessment of the mesh for skewness, smaller range values are more desirable which are observed in the meshed geometry with more than 95% elements in the range 0-0.2. The meshed structure of fabric model geometry of 1 mm thickness is shown in Figure 13.4 [1]. The file was saved for Fluent 6 solver with fabric volume as fluid type material.

Table 13.1. Mesh examination values [Source: Reference 1]

Skewness Range (Equiangle skew)	Active elements (%)	
0 – 0.1	88.34	89.52
0.1 – 0.2	7.12	7.14
0.2 – 03	2.21	1.92
0.3 – 0.4	1.83	1.06
0.4 – 0.5	0.47	0.36

c. Grid import and Multiphase Model selection in Fluent

The mesh geometry file is imported into Fluent 3D solver; the imported grid is checked and scaled to actual units of measurements. Segregated solver (default) was selected for the incompressible resin flow through fabric during RTM process (low velocities of the fluid-low Reynolds number). 1st order implicit, physical velocity porous formulation for 3D unsteady flow was opted in the model-solver options. Viscous laminar model was selected for physical model (laminar flow).

d. Material properties and Boundary conditions

New material by name 'resin' was created with specified density (1140 kg/m^3) and viscosity (0.60 kg/m.s) along with the existing 'air' (density 1.225 kg/cc, viscosity (1.7894e-5 kg/m.s). Multiphase VOF option was selected under the model option, air was defined as the primary phase and resin was set as secondary phase. Gravity (-9.81 m/s^2) was activated in the operating conditions panel in the 'z' direction, density of air (1.225 kg/cc) was specified under variable density parameter for better convergence of solution. Mixed mode (both for air and resin) boundary conditions for the inlet (pressure inlet, 0 pascal) and outlet (pressure outlet, -97325 pascals in 'z' direction) were set. Mixed mode fabric permeability (viscous resistance, $1/m^2$) and fluid porosity (1-fabric porosity) were defined for the fabric. No slip boundary conditions were set (default) for the walls for both the phases. For resin phase, 1 was set under the volume fraction for inlet and 0 was set for back-flow volume fraction for outlet boundary conditions.

e. Solution initiation and monitoring

Solution controls were defined for the flow calculation and the relaxation factors for pressure (0.3), density (1), body forces (1), momentum (0.1) and volume fraction (0.3) were defined. Discretization schemes for pressure (standard), pressure-velocity coupling (PISO), momentum (first order upwind) and volume fraction (first order upwind) were selected for the solution control. Solution initialization was set from outlet, relative to cell reference frame. A converged solution was defined for residual of magnitude (0.001) for continuity and velocities (x, y and z). Residuals and inlet/outlet mass flow rate were monitored during computation. Animation monitoring and frame savings were set for resin volume fraction images for the fabric (interior), inlet and outlet at defined time steps, so that video can be made of the flow front through streaming the frames.

f. Calculations and plotting of results

Computation of solution is commenced by iterating at 0.01 second time step size for 4000 time steps (for 1mm fabric thickness) and 2000 time steps (for 1.5 mm fabric thickness) with 3 iterations per each time step. The solution converged to the defined tolerances around 500-550 iterations for all multiphase simulation of RTM resin flow in textile fabrics. A representative residual convergence plot is as shown by Figure 13.5 [1]. Animated frame of resin volume fraction for the illustrated example at 1 second and 5 second are as shown in Figure 13.6 [1]. Figures 13.7 and 13.8 [1] provide the resin mass flow rate plots with respect to flow time at inlet and outlet, respectively.

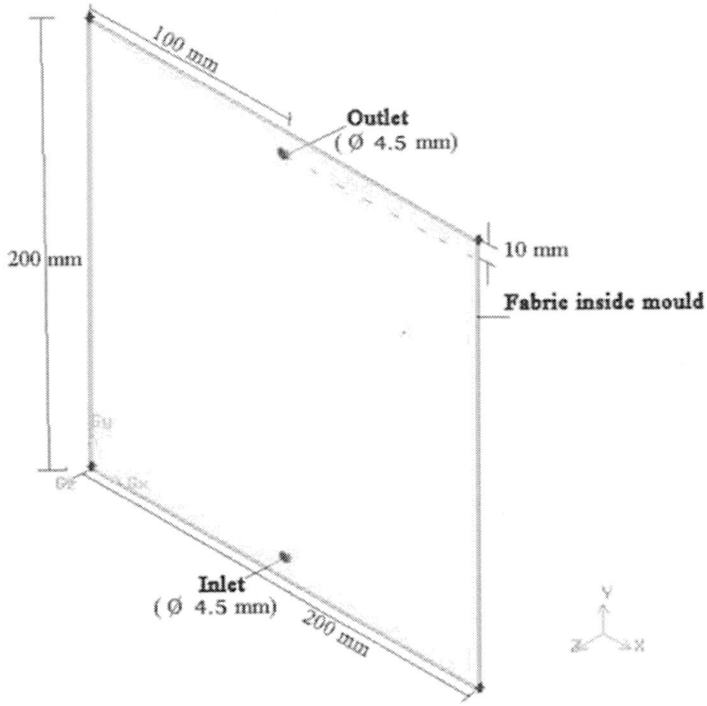

Figure 13.3. Creation of model geometry of the fabric with inlet and outlet [Source: Reference 1].

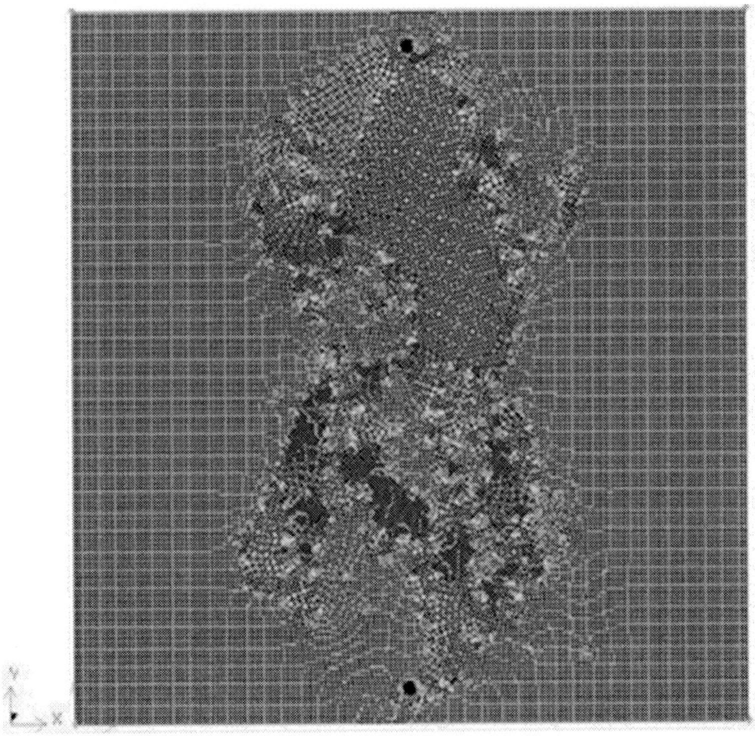

Figure 13.4. Meshed fabric geometry [Source: Reference 1].

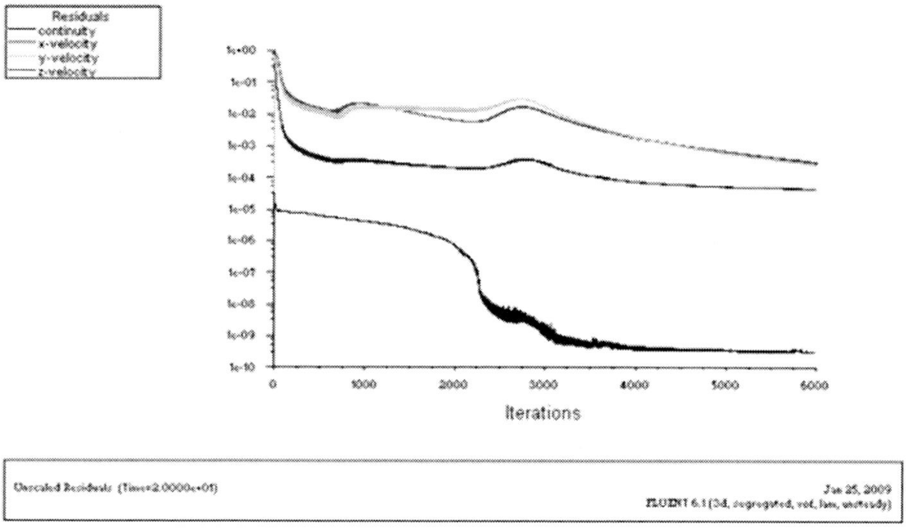

Figure 13.5. Residual plot of Multiphase model [Source: Reference 1].

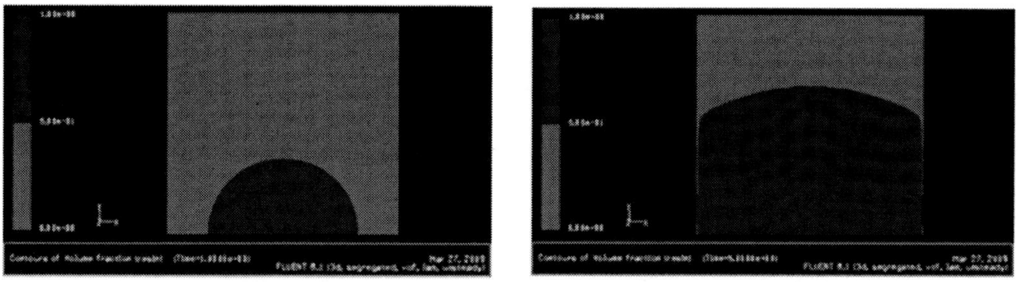

1 Second **5 Second**

Figure 13.6. Multiphase model-contour of resin flow during RTM process [Source: Reference 1].

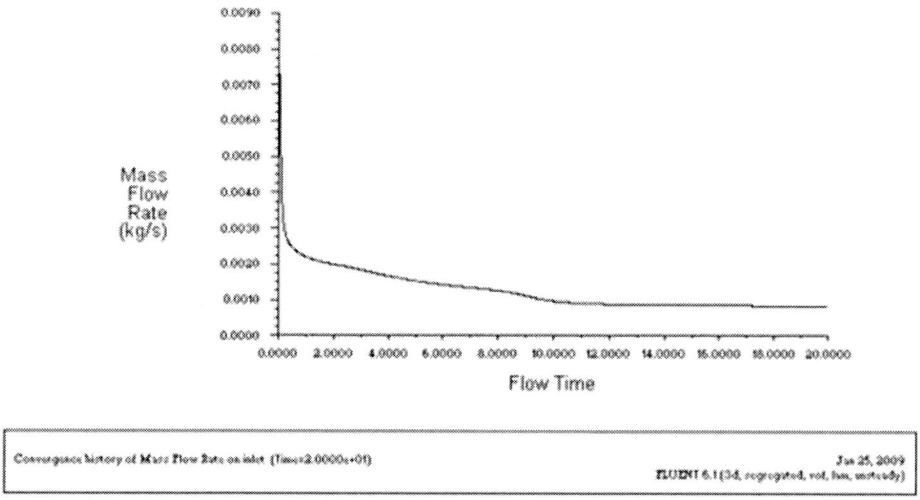

Figure 13.7. Multiphase model – Mass flow rate at inlet [Source: Reference 1].

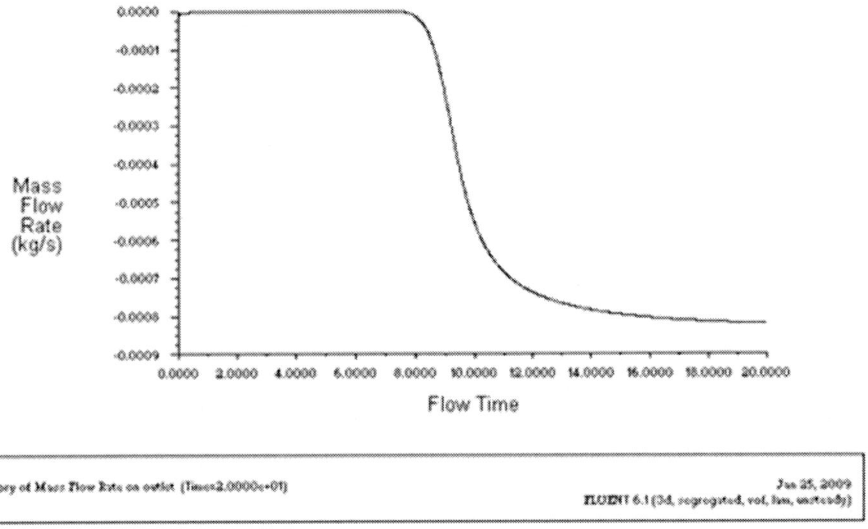

Figure 13.8. Multiphase model – Mass flow rate at outlet [Source: Reference 1].

13.7. Applications of RTM Modeling

a. Mould design and vent location optimization

The principal drawback to RTM process of composite preparation is the start-up cost. RTM requires expensive injection equipment, and RTM molds are expensive to produce. Most mold design and particularly the number and locations of injection and vent ports, is done by trial and error, and currently it is more of an art than a science. Accurate predictions of resin flow are necessary in order to properly locate injection and vent ports. Fabricating RTM molds are much more expensive to make than ones for hand lay-up, thus hindering their development. Multiple injection ports may be necessary for large and complex parts to keep injection times under the gel time of the resin. This complicates flow patterns and may result in a mould that traps air, creating dry spots, or ones that fill incompletely. Currently, mould design is more of an art than a science; the odds that the initial layout of injection ports to have problems appear to be high. Since moulds are very costly, accurate prediction of resin flow in the mould is desirable. Hence there has been a great deal of research in flow modeling for RTM [15]. RTM modeling not only helps mould designing for complex parts but also aids optimizing the vent locations in the mould through feasibility simulation studies.

b. Disturbances in mould filling operations

During an RTM system design, filling the mold as quickly as possible is an important goal as it will reduce the cycle time and increase the profitability of the process. Also for quick-curing resins, it lowers the possibility of resin gelling and hardening before the mold is completely filled. Many factors influence the mold filling time during RTM process [24] such as resin viscosity, defects in the preform, preform compaction during mould closing, interaction of textile layers during mould closing and accidental flow paths in mould cavity. RTM simulation provides excellent option to simulate and analyze the influence of such factors before hand to optimize the RTM system fabrication for efficient utilization.

13.8. FUTURE TREND

RTM modeling has already gained importance in mould design for complex shapes and vent locations for composite manufacture without defects. The simulation and modeling techniques would strive to achieve RTM process perfection, improvement in manufactured composite quality and reduction of manufacture cost.

Most likely, simulation models would zero the gap between the predicted and actual manufacturing process though incorporation of various process and material variables. RTM simulation would also target to contribute towards the overall objective of composite commercialization through cost reduction in nonrecurring costs, recurring costs, and direct operating costs such as durability, maintainability, reliability, and repairability.

The next stage of RTM modeling would attempt the prediction of composite material properties through integration of constituent component properties and RTM processing methodologies. Similar to the concept of closed mould composite manufacturing like RTM, simulation and modeling are here to stay and would play active role in composite manufacture processes.

REFERENCES

[1] Padaki N. V. Influence of Multilayer Interlocked Woven Structure on Impregnation Behavior of Performs and Impact Properties of the Composites; PhD Thesis, Indian Institute of Technology, New Delhi, 2010; pp 1-158.

[2] Kruckenberg, T. M.; Paton, R. Resin Transfer Moulding for Aerospace Structures; ISBN-10: 0412731509; Kluwer Academic Publishers: Dordrecht, 1998, pp 1-543.

[3] Scheidegger, A. E. The Physics of Flow through Porous Media; ISBN-10: 0802018491; University of Toronto Press: Toronto, 1974, 3rd Edn, pp 1-353.

[4] Parnas, R. S. Preform Permeability in Resin Transfer Moulding for Aerospace Structures; Kruckenberg, T. M., Paton, R.; Eds.; ISBN-10: 0412731509; Kluwer Academic Publishers: Dordrecht, 1998, pp 1-543.

[5] Adams, K. L.; Miller, B.; Rebenfeld, L. *Polym. Eng. Sci.* 1986, vol 26(20), 1434-1441.

[6] Adams, K. L.; Rebenfeld, L. *Text. Res. J.* 1987, vol 57(11), 647-653.

[7] Hirt, D. E.; Adams, K. L.; Prud'homme, R. K.; Rebenfeld, L. *J. Therm. Insul.* 1987, vol 10, 153-157.

[8] Adams, K. L.; Russel, W. B.; Rebenfeld, L. *Int. J. Multiphase Flow.* 1988, vol 14(2), 203-210.

[9] Gebart, B. R.; Lidstrom, *P. Polym. Compos.* 1996, vol 17(1), 43-51.

[10] Parnas, R. S.; Salem, *A. J. Polym. Compos.* 1993, vol 14(5), 383-394.

[11] Heardman, E.; Lekakou, C.; Bader, *M. G. Compos. Sci. Techol.* 2004, vol 64, 1239-1249.

[12] Adams, D. F.; Miller, A. K. *Mater. Sci. Eng.* 1975, vol 19, 245-260.

[13] Chan, A. W.; Larive, D. E.; Morgan, R. J. J. Compos. Mater. 1993, vol 27(10), 996-1004.

[14] Lekakou, C.; Edwards, S.; Bell, G.; Amico, S. C. Paper Presented in FPCM-7 Conference, Newark, USA, 7th – 9th February, 2004.

[15] Cairns, D. S. Fluid Flow Modeling of Resin Transfer Moulding for Composite Material Wind Turbine Blade Structures; Technical report no. SAND2004-0076, Sandia National Laboratories, New Mexico, USA, 2004, pp 1-178.
[16] Wong, C.C. Modeling the Effects of Textile Preform Architecture on Permeability; PhD Thesis, University of Nottingham, UK, 2006, pp 1-209.
[17] Shojaei, A.; Ghaffarian, S. R.; Karimian, S. M. H. Polym. Compos. 2003, vol 24(4), 525-544.
[18] Shojaei, A.; Ghaffarian, S. R.; Karimian, S. M. H. *Compos. Sci. Techol.* 2003, vol 63(7), 1931-1948.
[19] Anderson J. D. Computational Fluid Dynamics: The Basics with Applications; ISBN10: 0070016852; McGraw Hill Inc.: Singapore, 1995, pp 1-574.
[20] Ferziger, J. H.; Peric, M. Computational Methods for Fluid Dynamics; ISBN-10: 3540420746; Springer-Verlag: Berlin, 2001, pp 1-423.
[21] Verleye, B. Computation of the Permeability of Multi-Scale Porous Media with Application to Technical Textiles; PhD Thesis, Katholieke Universiteit Leuven, Leuven, 2008, pp 1-172.
[22] Fluent 6.1 Tutorial Guide, vol 1 and 2, Fluent Inc., Lebanon, 2003.
[23] Pillai, K. M. *J. Compos. Mater*. 2004, vol 38(23), 2097-2118.
[24] Gokce, A.; Advani, S. G. Design and Manufacture of Textile Composites; ISBN-10: 0849325935; Long, A. C.; Ed.; Woodhead publishing Ltd: Cambridge, 2005, pp 1-476.

In: Modeling and Simulation in Fibrous Materials
Editor: Asis Patanaik
ISBN: 978-1-62100-116-4

Chapter 14

APPLICATION OF MODELING AND SIMULATION IN PREDICTING FIRE BEHAVIOR OF FIBER-REINFORCED COMPOSITES

E. D. McCarthy [*] ***and B. K. Kandola***
Centre for Materials Research and Innovation, University of Bolton, UK

ABSTRACT

The mathematical modeling of combustion in structural laminates is a complex task requiring consideration of the heat transfer, chemistry and kinetics of polymer decomposition and combustion. It also requires adequate characterization of the mass transport of volatiles through melts and chars as well as the effect of temperature and resin mass loss on thermal expansion, ply delamination and retention of the mechanical properties of the composite immediately before, during and after combustion.

A variety of workers have tackled this challenge in numerous works, and while there are many established macroscopic models of 'fire-in-a-room' scenarios, these tend to concentrate on large-scale engineering and safety situations rather than processes and mechanisms occurring at the microscopic scale within resins and residual chars during combustion. This chapter discusses the motivation for computer-modeling of structural composite behavior in fire at the microscopic level of resin and fiber chemistry. The current capabilities of composite fire models and the extent of their deployment in composite design with respect to fire resistance are discussed. Lastly, the remaining deficiencies in the predictive capability of these models with respect to actual composite behavior under fire load are identified.

There are a number of improvements required for a fully representative model for combustion of structural composites and its effect on residual mechanical strength which require developments in the following areas; a) a more accurate determination of ignition temperature under a known fire load, b) a full, quantitative profile of the volatiles released during combustion enabling the calculation of more accurate heat of combustion, and c) better determinations of surface radiative emissivity and convective heat transfer coefficient.

[*]E-mail: eddie.mccarthy@gmail.com

Thermo-mechanical models are hybrids of heat transfer and finite element models of residual strength. Through-thickness temperature profiles and resin mass-loss data from the thermal components of these models have been used respectively to predict residual modulus and increase in pressure due to accumulated volatiles within the burning composite material. It remains for such models to a) adequately represent mixed modes of mechanical failure b) accurately model ply layer delamination, and c) correctly model the directionality of thermally-induced strains as a function of ply orientation.

The increasing amount of published work in this area indicates that a hybrid thermo-mechanical composite failure model may soon be achieved at the microscopic level of resin chemistry. The achievement of such a model is expected to enhance the ability of polymer chemists to formulate improved resins, as well as the capacity of engineers to design improved fire-resistant structures.

14.1. Introduction

The fire-resistance and mechanical resilience of epoxy-based laminate structures to thermal radiation is of crucial importance to their use and specification in key aerospace and marine applications. In particular, the ability of these fiber-reinforced polymers to retain structural integrity and mechanical strength for the longest possible time after ignition is a key objective. In cases where it is not possible to prevent ignition completely or even delay it significantly using fire retardant additives, it is then necessary to assess the time at which a composite will lose its structural integrity, the rate with which this will occur, and most importantly, the severity of this strength loss. Normally, to do this, it is necessary to perform experiments which reproduce as closely as possible the worst hazards to which the composite may be exposed and thereby derive data which allows material designers and engineers to specify materials and designs with maximum confidence. However, it is becoming increasingly desirable to model the combustion process analytically using mathematical simulation so as to screen new resins and additives early in the design process, well in advance of experiments and industrial scale-up and assess their inherent resistance to heat exposure and fire damage as accurately as possible. This is desirable to reduce both development costs and time-to-market for such materials and provide an effective initial screening for candidate composite materials before scale-up tests commence. Such theoretical models should also help engineers to better understand and control the thermal and chemical processes underlying heat damage and combustion, providing an early opportunity to reformulate where inherent weaknesses are identified. In the longer term, highly accurate predictive computer models of the combustion process, which have been closely validated by experimental data, offer the potential to partially or even wholly replace experimental testing as the principal means of validating the fire resistance of structural composites. The mathematical modeling of combustion in structural laminates is a complex task requiring consideration of heat transfer, the chemistry and kinetics of pyrolysis and combustion, the mass transport of gas through melts and chars as well as the effect of temperature on retention of mechanical properties of the laminate immediately before, during and after ignition and also accompanying loss in material. A number of distinguished workers have tackled this challenge in numerous researches reported so far. Henderson, [1], provided the first most fundamental formulation of the problem with a solution strategy which has remained the starting point for most workers. More recently workers such as Gibson and Mouritz, [2],

Drysdale, [3-4], Lyon, [5-6], Staggs, [7-12], and Galgano et al. [13], have provided refinements of the Henderson model which examine different aspects of the combustion problem using different modeling approaches. This chapter discusses the motivation and drivers for computer-modeling of structural composite behavior in fire. Here we will discuss the current capabilities of composite fire models, the extent of their application in composite design with respect to fire resistance, and locate the key gaps in their predictive capability with respect to actual composite behavior under fire load i.e. why are microscopic fire models for structural materials are required, where do they fit into the design process, what are the current deficiencies in their construction, and how may these be corrected?

The principal goals of modeling the fire behavior of structural composites are:

- An improved understanding of the causes and mechanisms of mechanical failure of structural composites as influenced by resin decomposition and the change in thermal and mechanical properties of the components with combustion time.
- An exact determination of the onset and extent of ignition / combustion of a structural composite with fully understood composition and physico-chemical properties, under a known external heat flux, with the ability to predict the release rates of volatiles and heat.
- An improved ability to predict the moment of structural failure under a known heat load for a composite with known resin / fiber composition, using a database of known virgin physical properties and appropriate mathematical models developed from first principles for heat transfer, decomposition and the chemistry of solid and gas phase combustion.
- An enhanced capacity to determine the effect of fire retardants added to the resin employing knowledge of the decomposition characteristics of the resin/additive system within the composite during heat exposure and combustion. A quantitative, comprehensive comparison of the relative performances of fire-retardant resins against untreated control resins is desired.

Generally a model should satisfy all four requirements listed above to be comprehensive and capable of closely predicting the response of a composite material to a heat load. In the next section we examine the current state of predictive composite fire modeling, the types of models available, and some of the specific techniques and software resources used. In this way it is hoped to illustrate how far fire modeling of composite materials has progressed so far, and how much further it must go before it becomes an accepted everyday tool of fire engineers worldwide.

14.2. Current Fire Modeling: Macroscopic versus Microscopic Models

There are as many different types of fire models as there are design and operational objectives for engineers, ranging from macroscopic "fire-in-a-room" scenarios to the microscopic study of the behavior of structural composite material at the level of resin chemistry. Different models are required depending on the specific aspect of the fire phenomenon which is of interest whether this be the residual mechanical performance of the composite, the concentration of toxic volatiles, the extent of heat release during combustion or another parameter. However, in general it is possible to distinguish two principal categories of fire model which are distinct in their nature and objectives; those which are *microscopic* and those which are *macroscopic*. As the names suggest, these refer to small and large-scale fire scenarios respectively and will be discussed here presently.

The most important distinction between the two types of model is in their application. Microscopic models are of most relevance and application to the formulation of composite ***resins***, since they are specifically concerned with the physico-chemical processes which occur in the resin during decomposition and combustion. Conversely, macroscopic fire models are of more relevance to the design and orientation of composite *components* within a larger construction such as building compartments, as they examine variables such as large-scale heat transfer, flame length, and fire spread.

Comparing the two approaches, it is not surprising that macroscopic models are considerably more accepted and further advanced in engineering practice, because of the immediate relevance of their output to the practical design of engineering structures at large scale. However, microscopic modeling, which is the principal concern of the present chapter, is less well established, most likely because its application is one step away from application to front-line design. However this perhaps conceals the fundamental advantages it can offer in understanding and modifying the fire response of composite materials including fire retardants, both established and novel. Thus, microscopic modeling has the potential to become a crucial tool in the appropriate formulation and optimization of resin and retardant chemistry, if it can be developed to a sufficient level of predictive accuracy. Recent developments in this field indicate that the drive towards more representative and sophisticated microscopic modeling is rapidly accelerating towards this goal.

14.2.1. Macroscopic Models

Macroscopic fire models are designed to represent large-scale fire scenarios and have an established pedigree in fire safety engineering, being employed not only to describe real-life fires at compartment-scale, but also being featured within established fire standards and building codes. Perhaps the most elementary and recognizable output of such models are the time-temperature curves which have become standard design references within fire engineering. Different hypothetical fire scenarios are distinguished, each of which is allocated its own standard time-temperature curve, and some of these fire scenarios have become classified within established codes and standards e.g. BSEN1991-1-2: (2002) [14]. According to this standard, fire categories within a compartment include, but are not limited to, *standard, external, hydrocarbon,* and *smouldering.* In reality of course a compartment fire can be considered as a hybrid of some or all of these idealized models. The associated standard time-temperature curves, presented as equations, are summarized in Table 14.1, and describe the response of the gas temperature, T within the compartment over time, t during any given fire event. Macroscopic fire events are also characterized geometrically, typically employing a selection of standard room scenarios, with different locations and types of flame relative to wall panels, windows and other features. Some of these geometrical scenarios have been characterized mathematically to a high degree of sophistication. One example of such a model is reproduced in Figure 14.1. [15]. In this figure, three design variations are illustrated for the case of a flame being produced from a compartment through an open window under non-forced draught conditions. Geometrical parameters as well as parameters describing the extent and shape of the flame are defined for each case. Exact equations result for each of these parameters. More comprehensive and complex expressions can be generated for such situations and are comprehensively discussed by Bailey [15] and elsewhere in the literature. Many of these expressions are also incorporated into design protocols and standards e.g. European design standard EN 1991-1-2, Annex B which include expressions for flame length, compartment temperature and heat release rate with respect to the design scenario in Figure 14.1. Such expressions are then incorporated into a larger section of design protocols and standards with respect to building design and fire proofing across EU member states.

Clearly, the extent to which macroscopic modeling outputs have become incorporated into recognized building codes reflects not only the extent to which these models have been accepted by the engineering community, but also their practical utility in calculating fire loads and other engineering indicators which contribute towards the final design of a building or structure. It is not the intention of the present chapter to describe macroscopic modeling principles in greater detail, except to illustrate how successful modeling output can ultimately become an established, and frequently essential, component of the engineering design process. This standard must be achieved by microscopic modeling techniques if they are to become similarly accepted within the composite manufacturing and component design industries. In the next section, we examine the current state of advanced microscopic models of composite materials, and analyze how they must develop further in order to become truly useful at the front line of resin formulation and composite design.

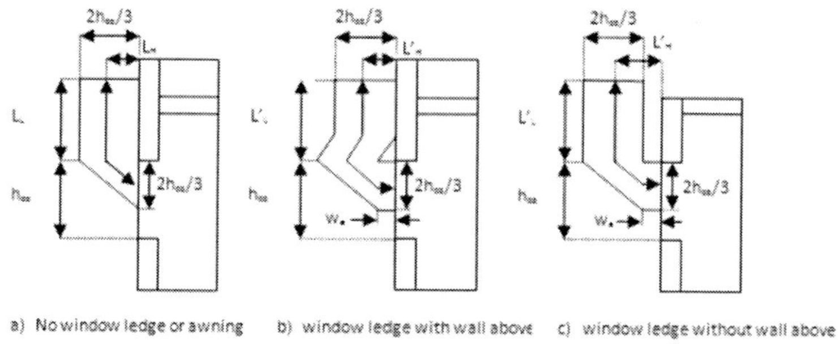

Flame dimension	Wall above, $h_{eq} < 1.25\ w_t$	No wall above or $h_{eq} > 1.25w_t$
Flame length, L_f, (axis of flame)	Calculated as in case (a)	
Flame width	Equals window width, w_t	
Flame depth	= 2/3 window height or $2h_{eq}/3$	
Flame height, L_L'	$= L_L - (1+\sqrt{2})w_a$	$=L_L-w_a$
Horizontal projection of flame, L_H'	$L'_H=L_H$	$L'_H=L_H+w_a$

Figure 14.1. Three geometrical scenarios in a macroscopic fire model for a non-forced draught window fire [Source: Reference 15].

Table 14.1. Various recognized fire scenarios and their associated time-temperature curves [Source: Reference 15]

Fire Scenario	Time-temperature curve
Standard	$T = 20 + 345\log_{10}(8t+1)$
External	$T = 660(1 - 0.687\exp(-0.32t) - 0.313\exp(-3.8t)) + 20$
Hydrocarbon	$T = 1080(1 - 0.325\exp(-0.167t) - 0.675\exp(-2.5t)) + 20$
Smouldering	$T = \begin{cases} 154t^{0.25} + 20\,for(0 \le t \le 21\,\text{min}) \\ 345\log_{10}[8(t-20)+1] + 20\,for(t > 20\,\text{min}) \end{cases}$

14.2.2. Microscopic Models

Microscopic models may properly be defined as those which examine the small-scale behavior of composite *materials* rather than *components* such as wall-panels within the context of a macroscopic "fire-in-a-room" scenario. As such they are predominantly concerned with the physical and chemical properties of a given material as they pertain to its heat transfer behavior and flammability. The geometrical aspects of the problem are of less importance than for a typical macroscopic model, and most microscopic models are defined for simple planar geometries, such as laminate or sandwich panel, where there may be one or more layers of material under consideration, and the area and thickness of the laminate are the only geometrical parameters required. An example of a microscopic scenario is provided in Figure 14.2, (by these authors), for an elementary slab with insulated bottom and side surfaces, irradiated by a heat source on its top surface. The key processes under examination in a microscopic model are conduction of heat through the solid bulk of the material, radiation and convection of heat to and from the exposed surface, the decomposition of the resin with increasing temperature, and most importantly, both solid and gaseous phase ignition and combustion processes. Another process of importance is heat or *enthalpy* loss from the solid by mass transfer of volatiles to the headspace adjacent to the solid.

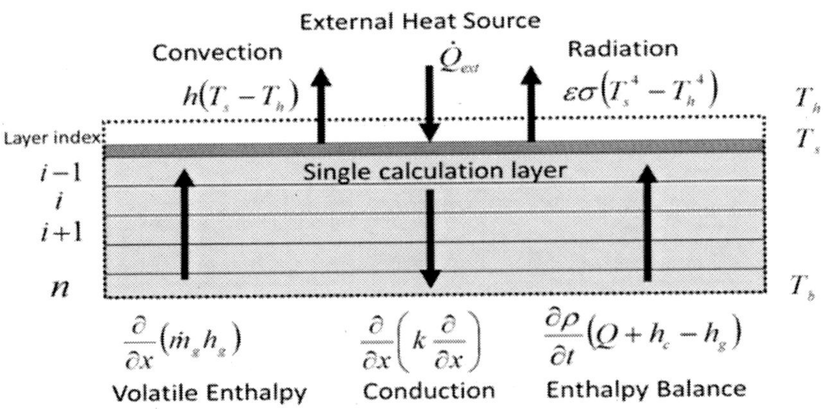

Figure 14.2. Schematic representation of a structural laminate exposed to a steady-state external heat flux, and the movements of heat associated with the system.

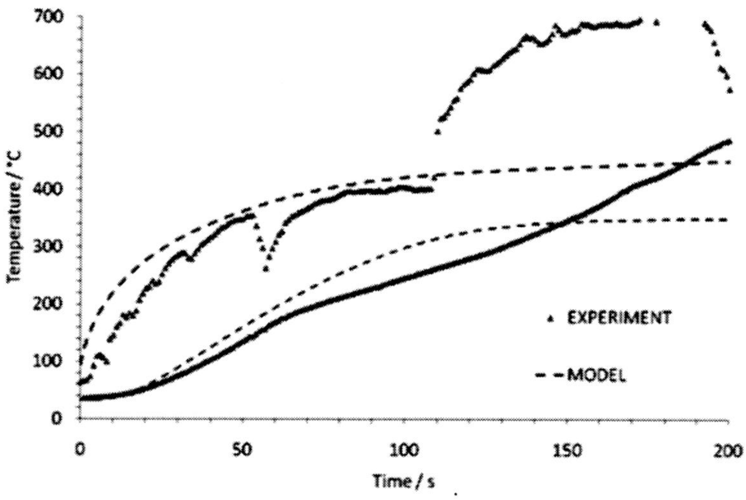

Figure 14.3. Typical comparison of conventional Henderson-type temperature prediction vs. real temperature data obtained with embedded thermocouples in an epoxy-E-glass plaque.

Thus for a microscopic model to be truly representative of the actual combustion process, it requires that two principal conditions are satisfied:

a) The mass and energy balance of the solid and its surroundings must be correctly defined by an appropriate heat transfer expression which incorporates conduction, convection, enthalpy loss and the heats of decomposition and combustion of the resin.
b) The correct physical, chemical and mass transfer properties of the system must be determined experimentally, which is usually done separately for the resin and the reinforcing medium, respectively, and used to derive accurate overall properties for the composite material.

Both these conditions must be adequately addressed in order to create a truly representative model of the combustion process within a material. We will discuss each of these conditions later in sections 14.3.1.2 and 14.3.1.3, outlining some of the essential components required by a properly-functioning microscopic model.

14.2.3. Current Modeling Techniques and Software

This section examines the various generic modeling techniques and current state-of-the-art computer packages and commercial software which are available to model real-life scenarios of both heat transfer and change in structural-mechanical property during fire events.

14.2.3.1. Current Macroscopic Fire Models

There are a variety of macroscopic fire modeling packages available to service the requirements of engineers and material experts across the building, aircraft and marine construction sectors. The demand for these models is based not only on the need for design information, but frequently on the need for numerical output compulsory to satisfy fire standards and legislation as well as building codes, (e.g. the European standard EN 1991-1-2 which requires time-temperature curves under different geometrical configurations of a compartment). Thus macroscopic fire modeling has become an established and increasingly important sector within the construction industry, which is reflected in the considerable number and variety of commercial fire models available on the market today. Some specimen models are listed in Table 14.3, and are described in considerably more detail by Olenick and Carpenter [16]. In general, macroscopic models can be divided into five of basic categories:

1. Zone models: These are applied to fire situations in partially enclosed volumes i.e. allowance is made for the presence of doors, windows, and vents in the problem definition. These models can be equally well applied to both single and multi-room configurations. A particular distinction is made between a lower fire plume and an upper smoke layer within the room (s).
2. Field models: This type of model treats the room as a collection of thousands of control volumes, (finite volumes), and solves the mass and energy balance equations inside each control volume. This allows for a more detailed solution compared to zone models, where, as we have seen, only two zones, plume and smoke, are distinguished. This multi-cell nature of a field model allows it to handle considerably more complex geometries than a typical zone model.
3. Egress models: These are used to predict the time available to occupants to evacuate a live fire scene. The outputs of such models are often quoted to demonstrate design code compliance and for determining where likely congestion areas will develop during an emergency evacuation. An increasing trend both in number and sophistication of these models was identified by Olenick and Carpenter [16], and attributed to a progressive shift of fire modeling towards a *performance-based* design approach from the more *prescriptive* practices prevalent in the past.
4. Detector response models: These models are used to predict the response time of a detector device to a fire event, so as to predict how quickly an active intervention will be made by a fire reaction system, such as sprinklers or foam sprayers. Such models use calculated fire heat-release to determine the trigger point of a thermal indicator for example, by calculating the time taken for the local air temperature to reach a trigger value. As such they have very specific inputs and outputs and limited objectives specific to certain devices in contrast to more ambitious categories of models, such as egress or fire endurance models.
5. Fire endurance models: These models focus specifically on the mechanical endurance of structures during fire, and so are generally finite element models which calculate changes in stress in response to temperature change within particular structural components of different materials, namely steels, masonries or polymer composites. They can be written in isolation or combined with either zone or field models to produce a comprehensive description of the fire scenario in terms of heat and mass flows as well as component stresses. Thus, such models can be used to predict the moment of structural failure in a member with variable accuracy. One important specific result of such models is a temperature-time profile through the thickness of a structural component such as an I-beam for example.

Table 14.2. Some specimen commercial macroscopic fire modeling packages available, (August 2010). Three examples are provided in each category, from a range of international sources, [Source: Reference 16]

Package	Author(s)	Application	Reference
Zone Models			
ARGOS, (Denmark)	Danish Institute of Fire Technology	Multi-compartment zone model	[17]
HEMFAST, (US)	NIST	Furniture in a room	[18]
POGAR, (Russia)	Higher Engineering Fire-Technical School, Moscow	Single compartment zone model	[19]
Field Models			
CFX, (UK)	AEA Technology, Harwell, UK	CFD software, applicable to fire and explosions	[20]
FIRE, (Australia)	Novozhilov, V.; Harvie, D. J. E.; Green, A. R.; Kent, J. H.	CFD model with water sprays	[21]
KAMELEON FireEx, (Norway)	SINTEF Energy Research.	CFD model for fire *linked to* a finite element code for thermal response of structures	[22]
Egress Models			
EGRESS, (UK)	Ketchell, N.; Cole, S.; Webber, D. M.; Marriott, C. A.; Stephens, P. J.; Brearley, I. R.; Fraser, J.; Doheny, J.; Smart, J.	Cellular automatic evacuation of multiple people through complex geometries.	[23]
EVACS, (Japan)	Takahashi, K.; Tanaka, T.; Kose, S.	Evacuation model for determining optimal design	[24]
WAYOUT, (Australia)	Shestopal, V. O.; Grubits, S. J.	Egress part of the FireWind suite of programs	[25]
Detector Response Models			
DETACT-QS, (US)	Evans, D. D.; Stroup, D. W.	Calculates thermal detector activation time under unconfined ceilings, arbitrary fire	[26]
PALDET, (Finland)	Technical Research Centre of Finland, Bjorkman, J.; Huttunen, O.; Kokkala, M.	Response of sprinklers and fire detectors under an unconfined ceiling	[27]
G-JET, (Norway)	Jensen, G.; Lonvik, T.; Heskestad, A. W.	Smoke detection model	[28]
Fire Endurance Models			
CIRCON, (Canada)	Lie, T. T.; Celikkol, B.	Fire resistance of loaded, reinforced concrete columns with a circular cross section	[29]
TASEF, (Sweden)	Sterner, E.; Wickstrom, U.	Finite element method for temperature analysis of structures exposed to fire	[30]
FIRES-T3, (US)	University of California, NIST	Finite element heat transfer for 1-, 2-, or 3-D conduction	[31]

It is this type of macroscopic model which is closest in structure and intent to the microscopic models introduced in this chapter, to the extent that fire behavior within solid components, as opposed to gaseous atmospheres is concerned. The principal difference between current fire endurance models as practiced and that to be discussed presently is one of scale, i.e. microscopic models feature much smaller sample masses. In a microscopic model, the size of the modeled cells would be far smaller, and hence more accurate and representative of actual fire response at the scale of materials, to the extent that meaningful distinctions could be made between different constituent resins. This is perhaps not something which can be best achieved with existing macroscopic models due to more immediate, large-scale, structural design requirements and priorities. There are in each of the above categories a selection of miscellaneous software packages which incorporate some or all of the above modeling capabilities to varying extents, which might be regarded as "master" models, each of which reflect different market priorities depending on the sector in which they are used (Table 14.2), [16]. The collective success of all of these macroscopic models has inspired a similar drive towards the development and incorporation of microscopic models into the composite design process, providing at once the theoretical basis, programming experience and industrial credibility to allow microscopic modeling of composite materials, both thermal and structural to become integral to the design process.

Table 14.3. Key microscopic fire models developed from the Henderson model to the present [Source: Reference 39]

Author(s)	Features	Additions / Limitations	Reference
Henderson, (1985)	Full and original treatment of physical heat transfer within composite. The basis of all current *composite* models.	A physical heat transfer model, accounting for decomposition & volatile mass transfer, but not ignition, combustion, thermal expansion or the formation of voids or cracks.	[1], [39-44]
Sullivan, (1990-93)	Finite element model for thermal decomposition of composites	Added a consideration of pressure effects and thermal expansion to the Henderson model. Lacked thermo-mechanical treatment of crack formation.	[45- 48]
Pering, (1980)	Response of tensile and shear properties of composites to heat and/or fire load.	Post-fire tensile properties related to mass loss; post-fire shear properties were related to char thickness as predicted by the thermal model	[49]
McManus, (1992)	Thermo-mechanical model for carbon-phenolic/ carbon-carbon composites	Added modeling of ablation of the polymer surface. Most comprehensive model at that time.	[50-51]
Dimitrienko, (1997)	Thermo-mechanical behavior of composite structures at high temperature.	Absence of model for ablation, (n/a for thermosets) or treatment of the propagation of cracks / voids with increase in temperature.	[52]
Milke and Vizzini, (1991)	3D thermal response model for anisotropic composite laminate	Lack of treatment of ignition or combustion reactions	[53]
Gibson, (1995)	Thermal performance of thick composite laminates in hydrocarbon fires	Lack of treatment of ignition or combustion reactions	[54]

14.2.3.2. Current Microscopic Fire Models

The earliest studies in thermal modeling of materials were those of burning wood which were made from the 1940s onwards by Bamford, Munson, Kung, Murty, Kansa and Fredlund, [33-38]. The equivalent track record in structural polymer composite modeling can be traced from Henderson's paper, [1], on the thermal modeling of polymer composites which applied their methodologies to the latter material. An excellent recent review of both the historical development and current state of microscopic fire modeling of composites is provided by Mouritz et al. [39]. These experienced authors provide a survey of the historical development of microscopic composite fire models, noting the key incremental developments and the subtle differences in the treatment of fire processes between contemporary models which exist up to the present day. A summary of the key models presented in this genre, [40-54], is given in Table 14.3, [39].

All of these models share an almost identical problem description, and focus mainly on the thermal effect of the fire on the physico-chemical condition of the composite material as it decomposes. However in most of them there lacks a complete description of the actual ignition or combustion processes relevant to the fire itself in the gaseous phase which is normally represented within these models as a simplified 'overhead' heat flux.

Thus there is little provision within most models to predict the ignition point of a combustion event or specifically describe the effect of a solid- or gas- phase combustion reaction on the residual mechanical strength of the composite material, although key papers by Dimitrienko, [55-56], have developed the mathematical basis for determining the opposite effect of thermo-mechanical erosion of materials on their combustion in the presence of high-speed gas flows tangential to the material surface. However these models are limited in that they apply specifically to a scenario of tangential surface flow of gas over a composite surface as opposed to the typical situation shown in Figure 14.2, where there is, at least initially, a static mass of gas directly over the laminate surface, e.g. a standard cone calorimetry test.

Secondly, Dimitrienko studies the effect of mechanical erosion on combustion rates rather than the opposite effect of combustion on mechanical properties, which is at issue here. Nevertheless, many of the mathematical structures and methodologies developed by Dimitrienko could be adapted to a 'static' combustion problem, and represent a good basis for further progress.

Recent work by Galgano et al. [13], on the thermal response of sandwich panels, does include a simplified two-step decomposition and combustion reaction sequence as part of its model, but does not include a prediction of ignition temperature or time from thermodynamic first principles, in the manner suggested by Kanury, [61]. It is also limited by simplifying assumptions such as invariance of volume and constant internal gas pore pressure. Nevertheless, it represents one of the more advanced examples of predictive modeling in the available literature at present, together with work by Stoliarov et al. [58-59], and Latenberger et al. [61-62], which will be discussed later in this chapter.

However, to date, most models are mainly 'thermal' being limited to a consideration of conduction, convection, physical expansion, volatile mass transport and the chemical decomposition of the resin.

Nevertheless, these models can often be highly accurate in predicting mass loss and temperature profiles in a composite section, and latterly strength loss, to the extent that they are sufficient for most practical applications, e.g. Kandare et al. [57].

However there are examples of temperature data for burning composites, (e.g. see a set of temperature curves recently obtained by these authors at University of Bolton, (Figure 14.3), where existing models are still insufficient to produce a representative result for either true resin mass loss or temperature distribution within a component, and it is under these circumstances that the true limitations of non-combustion models are exposed. This limitation underlines the present necessity for sufficient models of the ignition and combustion events to be incorporated into these models in order that they become fully capable of capturing all aspects of composite burning behavior. Recent developments in the specific area of ignition and combustion modeling by various workers will be discussed in section 14.3.1.3

14.3. Modeling Heat Transfer and Decomposition: Finite Difference Analysis

This section will examine the modeling of heat transfer, resin decomposition and both solid and gas phase combustion using finite difference programming, as it has evolved from the physical Henderson model to more complex models which can discriminate the point of ignition and accurately predict mass loss, temperature profiles, and the flux rates of volatile and heat release.

14.3.1. One-Dimensional, Sectional Composite Models

14.3.1.1. Descriptive Equation for Microscopic Model

The classic description of heat transfer in a combusting composite laminate or plane slab has been provided by Henderson, [1], and this remains the foundation for practically all variations of mass and energy balances attempted since then, (see equation 14.1). The terms of this equation are included in Figure 14.2. to illustrate the locations within the slab where they are relevant.

$$\rho c_p \frac{\partial T}{\partial t} = \frac{\partial}{\partial x}\left(k\frac{\partial T}{\partial x}\right) - \frac{\partial}{\partial x}\left(\dot{m}_{gas}h_{gas}\right) - \frac{\partial \rho}{\partial t}\left(Q_{pyr} + h_{char} - h_{gas}\right) \tag{14.1}$$

Here, the left term represents the flux of the heat storage within the laminate, which is balanced on the right hand side by terms for conduction within the laminate, loss of gas enthalpy through volatile escape, and a term balancing the heat of pyrolysis with the changing balance between gas and char enthalpy which accompanies the reaction. However the equation in this form does not explicitly provide for a heat balance between the laminate and the external surroundings such as a radiant heat source. Therefore it is necessary to include some extra terms in the model to provide for this.

To do this we begin by recognizing that there are three fundamental mechanisms of heat transfer; conduction, convection and radiation, each of which play a part in the thermal response of a laminate specimen exposed to an external heat flux. This allows us to write an equation 14.2, for the heat balance at the surface line of the laminate described in Figure 14.1.

$$\dot{Q}_{net} = \dot{Q}_{ext} - \varepsilon\sigma(T_s - T_h)^4 - h_{conv}(T_s - T_h) - k(T_s - T_{i+1}) \tag{14.2}$$

If we draw a boundary around the surface and headspace in a way which isolates the surface layer, $i = 1$, from the rest of the laminate bulk we can eliminate the conduction term from the expression, which simplifies the treatment of the problem to that of the exchange of heat between the surface alone and the headspace i.e. equation 14.3:

$$\dot{Q}_{net} = \dot{Q}_{ext} - \varepsilon\sigma(T_s - T_h)^4 - h_{conv}(T_s - T_h) \tag{14.3}$$

It is assumed that, at least initially, the transmission of heat through the laminate from layer 1, (which includes the surface, *s*), to layer *n*, (the insulated base of the laminate), is by conduction, since it is assumed that the laminate, (prior to pyrolysis and the internal evolution of gaseous volatiles), is principally solid with minimal porosity. In this circumstance it is possible to represent a temperature increment between two successive layers within the laminate *i* and *i+1*, respectively, as a function of the instantaneous heat flow of conduction, and the specific heat capacity and density of the laminate, (14.4):

$$\frac{dQ_i}{dt} = \frac{d(x_i \rho c_{p,c} T_i)}{dt} = \left(\frac{kdT}{dx_{i+1}} - \frac{kdT}{dx_i}\right) \tag{14.4}$$

This may be re-written in discretised form for use in finite-difference analysis,

$$\frac{\Delta Q_i}{\Delta t} = \frac{x_i \rho c_{p,c} \Delta T_i}{\Delta t} = \left(\frac{k(T_{i-1} - T_i)}{(x_i + x_{i-1})/2} - \frac{k(T_i - T_{i+1})}{(x_i + x_{i+1})/2}\right) \tag{14.5}$$

Furthermore, two variations of equation 14.5 are written for transmission near the surface, ($i = 1$ or s), and base, ($i = n$), of the laminate; (14.6) and (14.7) respectively:

$$\frac{\Delta Q_1}{\Delta t} = \frac{x_1 \rho c_{p,c} \Delta T_1}{\Delta t} = \left(Q_{net} - \frac{k(T_1 - T_2)}{(x_1 + x_2)/2}\right) \tag{14.6}$$

$$\frac{\Delta Q_n}{\Delta t} = \frac{x_n \rho c_{p,c} \Delta T_n}{\Delta t} = \left(\frac{k(T_{n-1} - T_n)}{(x_n + x_{n-1})/2}\right) \tag{14.7}$$

Equations 14.3, 14.5, 14.6 and 14.7 may be used to construct a finite difference solution for a laminate exposed to an external heat flux, *in the absence of any pyrolysis of the laminate or combustion of the pyrolysis products*. As such, this form of the model is only sufficient to represent the physical movement of heat into and from the laminate alone. However, where pyrolysis and combustion processes are active within the slab, i.e. (for the case that it is a combustible polymer composite material), it is necessary to include extra terms for the heats of pyrolysis and combustion within the finite difference equations 14.5, 14.6 and 14.7 to produce a modified set of equations given by 14.8, 14.9, and 14.10. These extra terms are $Q_{pyr,i}$ and $Q_{comb,I}$ representing the heats of resin decomposition and volatile combustion respectively.

$$\frac{\Delta Q_i}{\Delta t} = \frac{x_i \rho c_{p,c} \Delta T_i}{\Delta t} = \left(\frac{k(T_{i-1} - T_i)}{(x_i + x_{i-1})/2} - \frac{k(T_i - T_{i+1})}{(x_i + x_{i+1})/2} + Q_{pyr,i} + Q_{comb,i} \right) \tag{14.8}$$

$$\frac{\Delta Q_1}{\Delta t} = \frac{x_1 \rho c_{p,c} \Delta T_1}{\Delta t} = \left(Q_{net} - \frac{k(T_1 - T_2)}{(x_1 + x_2)/2} + Q_{pyr,1} + Q_{comb,1} \right) \tag{14.9}$$

$$\frac{\Delta Q_n}{\Delta t} = \frac{x_n \rho c_{p,c} \Delta T_n}{\Delta t} = \left(\frac{k(T_{n-1} - T_n)}{(x_n + x_{n-1})/2} + Q_{pyr,n} + Q_{comb,n} \right) \tag{14.10}$$

20

We now have a set of equations which allow us to solve for the temperature profile through a laminate using known values of heat capacity, density, thermal conductivity and the heats of decomposition and combustion of the resin and pyrolysis products respectively. However it remains to establish a proper model for the movement of gas formed within the laminate to the surface and into the headspace above the laminate, a problem which is concerned with the hydrodynamics of the molten laminate. To do this we use an expression of the form of equation 14.11, proposed by Staggs et al. [8]:

$$\dot{m}_{v,i}(x,t) = -\beta_T \rho_i (1 - f_{char,i}) \left[\frac{T_{i-1} - T_i}{(x_{i-1} + x_i)/2} \right] \tag{14.11}$$

Here, the instantaneous mass flux of gas at time, t and position x within the laminate, is the product of a thermal gas diffusivity coefficient, β_T, [m^2/ s/K], the density of the polymer melt, ρ, the proportion of gas formed by pyrolysis within the laminate, (1-f_{char}), and the temperature gradient through the laminate. Physically, the basis of the equation is that the velocity of gas increases in the direction of falling viscosity gradient within the laminate. The latter, in turn, is inversely related to the temperature gradient, hence the negative sign on the right hand side of the equation. The result is that when this model is employed in a combustion model, an upward flow of gas through the laminate is simulated, consistent with the positive temperature gradient calculated in this direction from the heat balance.

The model described is now sufficient to enable an initial model of the combustion process, including as it does terms for heat conduction through the bulk, convection and radiation exchanges at the surface, degradation and combustion occurring both at the surface and within the slab bulk, and finally, volatile mass and enthalpy transport from within the bulk to the surface of the laminate and into the headspace. However, to use this model effectively, two crucial elements must be adequately addressed:

1. The supply of accurate values for physical parameters derived from appropriate experimental data. Such values should ideally be temperature sensitive; if possible they should be expressible as continuous analytical functions of temperature. In the absence of an analytical function, an empirical relationship between each of the properties and temperature may suffice where adequately fitted by empirical data. The sourcing of this data is discussed in the next section.
2. An adequate mechanism within the model to distinguish the point of ignition in the composite material; such a mechanism is normally designed to calculate the ignition temperature, or alternatively, the ignition time, based on ignition criteria which are calculated from the first principle of thermodynamics. The next section details two recent models which present some ideas towards implementing such a predictive mechanism for ignition.

Table 14.4. Variables and parameters essential for the operation of a microscopic combustion model

Variable/Unit	Symbol	Source	How used in model
Heat Transfer Parameters			
Densities of resin, matrix, and volatile gases	ρ	Gravimetric determination, (solids); thermodynamic tables, (gases)	To calculate correct mass balance of species during decomposition and combustion processes
Thermal conductivity	k	Lee's Disc method	To calculate heat conduction through bulk of laminate
Convective heat transfer coefficient	h_{conv}	Calculated from surface and headspace temperature data at a known heat flux	To calculate heat released from surface to headspace by convection
Emissivity	ε		To calculate heat released from surface to headspace by radiation
Kinetic Parameters			
Arrhenius Pre-exponent	A	Thermo gravimetric analysis, modeling	To calculate the rates of decomposition of the resin and combustion of the volatiles released by this decomposition
Activation Energy	E_a		
Enthalpies			
Heat of decomposition	ΔH_{decomp}	Differential scanning calorimetry, bomb calorimeters etc.	To calculate heat absorbed by resin during decomposition
Heat of combustion	ΔH_{comb}		To calculate heat released during both solid and gaseous combustion processes
Mass transfer coefficient	β	Permeation measurements in composite material	To calculate the rate of mass and enthalpy transfer from the laminate bulk to the headspace

14.3.1.2. Determination of Relevant Properties for Microscopic Model

There are many parameters and variables required to enable the model outlined in the previous section to be utilized. These comprise heat transfer variables such as thermal conductivity, kinetic parameters such as the activation energies for the combustion of involved species, and mass transfer coefficients for volatiles evolving from the laminate during resin decomposition. A comprehensive list of the required variables, where they are sourced, and where in the model they are used is provided by Table 14.4.

Most of the variables listed may be determined readily by experiment or else studied in the thermodynamic or chemistry literature. However some of the variables such as surface radiative emissivity and convective heat transfer coefficient can be more difficult to ascertain. More difficult still is a definitive value for the mass transfer coefficient of volatiles, because of the complexity of measuring gas mass transfer through a burning thermoset composite laminate. Incorrect values for these parameters taken together can contribute to a substantial error in the predictive capacity of a model, even where the balance of the other properties have been correctly evaluated and deployed.

It could be said that whereas the theoretical statement of the combustion problem is at an advanced stage of development, the correct determination of experimental parameters is at best incomplete, if not always inaccurate. Thus, much of the remaining work required to create representative microscopic models requires the need for new or improved experimental methods to determine emissivity and surface convection coefficient, in addition to an effective measurement of volatile mass transfer through thermoset laminates. The final component to be added is a comprehensive species profile of the volatile gas mixture released, to enable a correct calculation of the actual enthalpy of combustion of the gas released during the decomposition of the laminate resin.

14.3.1.3. Predictive Modeling of Ignition Temperature

This section discusses two particular examples of one-dimensional combustion-predictive heat transfer models recently developed, those of Stoliarov and Lyon, [55], and McCarthy et al. [56], which aim to capture the point of ignition with associated mass loss, thermal profile, and release rates of volatiles and heat accompanying the decomposition and combustion of structural composites.

The model by Stoliarov and Lyon, [58-59], is the basis of the *Thermokin*® computer package which presents a one dimensional finite-element statement of the composite combustion problem similar to that defined in Figure 14.2.

It calculates solid heat conduction, surface convection and radiation in a manner similar to the models related in Table 14.3. It also presents a theoretical treatment of the mass transfer of volatiles out of the composite, and the additional gas-phase heat supplied to the composite during gas-phase combustion.

The expressions for volatile mass transfer used are given by equations 14.12a and 14.12b, for the two cases of constant thermal expansion coefficient and constant pressure respectively

$$\gamma = 0, \alpha = const \quad J_g = \frac{\lambda \rho_g A \alpha \partial P_g}{P \partial x}$$
$$\gamma = 1, P = const \quad J_g = \frac{\lambda \rho_g A P \partial \alpha_g}{P \partial x} \qquad (14.12a/b)$$

Secondly, the *Thermokin*® package simulates the effect of gaseous phase combustion in the headspace of the composite and is modeled theoretically by a supplementary heat flux, additional to the imposed external heat source, called the flame *feedback.* The feedback heat flux is only activated within the model when the condition of equation 14.13 is obeyed; i.e. when the mass flux of volatiles leaving the surface layer of the composite exceeds a critical value at which ignition occurs. The definition of this critical gas mass flux and the corresponding critical heat flux and temperature of ignition from which it is calculated are given by equation s 14.14 – 14.16, given by Lyon and Quintiere, [5].

$$\sum_{comps} \frac{\dot{m}_g}{\dot{m}_{g,crit}} > 1 \quad Q_{rad}(t) = Q_{rad}(t) + Q_{flm} \tag{14.13}$$

$$\dot{m}_{g,crit} = \left(\frac{h_{conv}\dot{Q}_{crit}}{\rho_g c_{p,0} h_{comb}} \right) \cdot \left(1 - \frac{(T_{ign} - T_a)}{(T_{flame} - T_a)} \right) \tag{14.14}$$

$$\dot{Q}_{crit} = \rho_g c_{p,0} \cdot \left(\frac{T_s - T_a}{1 - \eta_{eff}} \right) \tag{14.15}$$

$$T_{ign} \approx \left[\frac{T_0 h_g}{c_p} \right]^{0.5} \tag{14.16}$$

Equation 14.14, the critical mass flux is a function not only of the critical heat flux at the surface, but also of the convective heat transfer coefficient at the surface, the density and specific heat capacity of the combusting gas, the enthalpy of combustion of this gas and the ignition, flame and ambient temperatures which persist. Stoliarov and Lyon, [58-59] reported good predictivity of their model with respect to cone calorimetric mass and heat flux data

The model by McCarthy et al. [60], is based on the theoretical slab geometry represented in Figure 14.2. and employs a thermal calculation algorithm based closely on the conventional Henderson model. However, it also includes calculation routines which attempt to account for volatile mass transfer through a composite laminate, and most importantly, incorporates a tested literature correlation for the prediction of ignition temperature. This correlation is used as a switch criterion within the code of the model to enable the heats of solid and gas-phase combustion to be calculated. These values are then included in the finite difference algorithm and used to calculate the through-thickness temperature profile of the laminate.

These extra heat terms are intended to allow the programme to predict temperature surges at various points of the slab which may be directly and specifically related to the combustion phenomena occurring immediately after the ignition temperature. The approximation used to predict ignition temperature is described by Lyon and Quintiere, [5], given by equation 14.16, where T_{ign} is the predicted ignition temperature of the laminate material, [K], h_g is its heat of gasification / decomposition, [J/kg], and c_p is its specific heat capacity, [J/kg/ K].

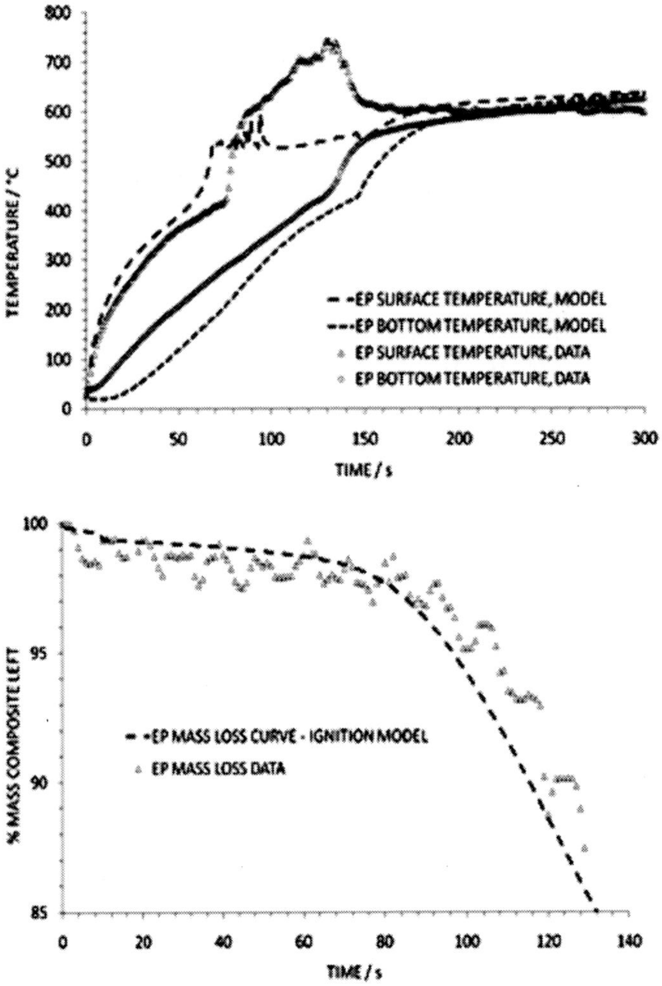

Figure 14.4. Experimental and Predicted Temperature and Mass Loss Profiles for an Epoxy/E-glass composite plaque at 50 kW/m^2, (Henderson model with prediction of ignition/ combustion) [Source: Reference 60].

The use of this simple correlation as the basis of a novel predictive combustion module, linked to the Henderson model, has been presented recently by McCarthy et al. [58], within a novel finite difference code which achieves good predictivity of both the plaque temperature and mass loss of an epoxy/E-glass composite under an imposed asymmetric heat load of 50 kW/m^2 in a standard cone calorimeter experiment (Figure 14.4). This is despite the fact that the gas-phase combustion has been simplified as that of pure methane, when in reality a more complex gas mixture is released by most epoxy resin formulations, before accounting for any volatile emissions from included fire-retardants.

These initial encouraging results of integrating the classical Henderson model and the correlation of equation 14.16 to create an improved microscopic composite fire model demonstrate the value of combining established techniques and data from multiple sources in novel ways so as to improve the ability of formulation chemists and fire-protection engineers to not only predict the particular behavior of a new resin under a fire load, but also to diagnose the underlying physico-chemical causes underlying its response. If a proven version of a predictive model such as this can accurately predict the response of a particular resin or fire-retardant system, much experimental time and resource can be saved in its evaluation at an early stage of product development.

Conversely, if a well-validated model is incapable of adequately predicting the behavior of a novel resin, despite the use of accurately measured and validated physical properties and decomposition parameters, it can serve as an early indicator of anomalous or unusual chemistry within the new formulation, which may require further diagnosis and/or correction as appropriate to the proposed application of that resin within a composite structure. Alternatively, the failure of a trusted model could indeed highlight flaws or errors in the experimental methodologies used to determine key physical properties or kinetic parameters of decomposition and/or combustion.

However, at time of writing, there is no validated microscopic composite combustion model of such unquestioned predictive power that it can be used as a gold-standard indicator of anomalies in the measurement of resin physical properties, still less is there one of such a pedigree that experimental results will be questioned before the structures, assumptions and content of the model itself. Before such a model could be established the following principal requirements would need to be addressed:

1. *An exact analytical determination of the ignition temperature*: The correlation used by McCarthy et al. [60] employed an established correlation, (equation 14.16), [5], for ignition temperature which, though qualitatively effective within their model, remains a simplified approximation of the true value. A more accurate value would require an analytical expression which employed weighted values for both heat of gasification / decomposition and specific heat capacity of the resin. Alternative approaches to the prediction of ignition could be pursued using correlations for ignition delay time as outlined by Kanury, [59], which is based more rigorously on thermodynamic principles, and from which perhaps a more accurate value for ignition temperature could be calculated.
2. *A fully representative quantitative profile of the volatiles released:* Obtaining such a gas-profile requires quantitative gas analysis using *in-situ* gas chromatography. At present there exists the capacity to analyze such gases within anaerobic gas testing units used to determine heat of gasification during pyrolysis, but the altered composition of such gases in the presence of a simultaneous combustion reaction is not as easily determined because of the logistical difficulties of obtaining stable gas samples in the presence of fully-developed gas-phase combustion, where significant quantities of airborne soot and dust particles are being generated, and sampling equipment must withstand high gas-phase combustion temperatures for sustained sampling periods.
3. *More accurate gas diffusivity/permeability parameters:* There is a need for a dynamic resin-char-matrix gas mass transfer coefficient which can accurately measure transfer of gas between successive plies of a polymer composite during the decomposition and combustion process. The acquisition of such a coefficient would ideally require permeation testing of the composite material during combustion under a pressure gradient through the laminate, a process for which, at present, no satisfactory methodology exists because of the difficult working environment which exists in the highly turbulent atmosphere of fully-developed gas-phase combustion

and the unpredictable structure and intrinsic instability of a resin-fiber-char network during and after combustion.

4. *More accurate determinations of in-situ surface heat transfer coefficient:* At present it is conventional to employ a single constant value of convective heat transfer coefficient to model the re-convection of heat from an exposed surface of a polymer composite material. However it is clear that this assumption cannot be satisfactory in a situation where the temperature of both the surface and the gas in immediate proximity are changing significantly and a sudden gas-phase plume develops in the gas headspace immediately over the surface of the laminate at ignition. Thus, it is necessary to develop expressions which reflect not only the temperature sensitivity of the convective coefficient, but also the hydrodynamic environment of the solid-gas interface at the surface during the three phases of pre-heating, ignition, combustion and thermal equilibration after flame-out, which also affect the calculation of this value. The effect of changing surface roughness on the coefficient would also be a factor in such a study.
5. *More accurate determination of reflective radiative surface emissivity*: The radiative emissivity of the composite surface, similarly to the heat transfer coefficient (4), is a function of the resin-char composition of the surface at any point during the combustion process, and is also affected by the surface profile and roughness. Hence it is not sufficient to use one value for this parameter in the model for the duration of the exposure event. However, characterizing the *in-situ* dynamic behavior of this parameter in a turbulent combustive atmosphere in the presence of soot, flame and airborne particles is a particularly challenging task, so that modelers may well be forced to rely on binary before/after shifts in the emissivity based on measurements made on the virgin resin-fiber composite and char-fiber residue respectively.

Of the five measures mentioned above, perhaps the most important are a refinement of ignition temperature, (1), and a proper quantitative analysis of volatiles produced during resin decomposition, (2), both of which can be expected to enhance the predictive model significantly; however the remaining refinements might only become significant for very specific, well-defined problem definitions where, for example, discrepancies of five to ten degrees Kelvin on the surface of the calculation slab might have a very significant effect on the result of the model. It is anticipated that implementation of some or all of these measures would result in a model of true quantitative fidelity to cone calorimeter data, and could ultimately be used as a partial, or even complete, substitute for experiment in the early stages of composite formulation and fire testing.

14.3.2. Three-Dimensional Models

In section 14.2.3.1 we saw that macroscopic fire models have been developed to treat the problem of heat conduction in three dimensions, (see [29]), so that the methodology and software for creating such models is already well-established. Computational fluid dynamics, CFD, and finite element/volume, (FE/FV), techniques are specifically designed to deal with problems of heat transfer in solids and bulk fluids, and are, at present, almost standard tools within the process industries as aids to chemical reactor visualization and design. However, the relative turbulence and chaos of the combustion process in comparison to the more elementary task of modeling the temperature response of a heated metal bar or a standard bulk liquid mixing problem can present considerable challenges to even the most advanced of these packages. In section 14.3.1 we have examined the current state of one-dimensional, (1D), modeling of composite combustion and discovered the added complexities of modeling ignition point and heat feedback in addition to the standard thermal Henderson model. While the 1D problem has almost been solved in a qualitative sense, it is clear that for a number of reasons, the modeling community is far from a conclusive solution, partly because of inadequate data for the physical properties of a composite during the burning phase, and partly because of the absence of a conclusive analytical solution for the ignition temperature based on thermodynamic principles. Minor deficiencies in values for volatile mass transfer coefficient, surface radiative emissivity and heat transfer coefficient contribute even further to an incomplete model even in one dimension, creating both temporal and spatial errors in the 1D temperature and mass-loss predictions, which would be greatly magnified if such a solution, were generalized in three dimensions. There is, in addition, the question of the almost infinite varieties of geometry which could be faced by a CFD composite combustion code, with all the permutations of volatile gas flow, heat reflection and char formation which would accompany problem descriptions far more complex than the familiar asymmetric plaque configuration described in Figure 14.2. In this situation the twin problems of random char-network formation and inadequate volatile mass flow models would become considerably more pronounced than in the current one-dimensional model. Key challenges would be an adequate geometrical model to represent the process of char-network formation, and a reliable empirically-validated model for gas flow based on a properly calculated gas mass transfer coefficient. It will be in the development of a three-dimensional model of the composite combustion problem that these specific challenges, which can sometimes be over-looked within a one-dimensional 'layer' or finite element 'slice' model, will assume greater importance and urgency among fire modeling engineers.

14.3.3. Online Open-Source Models: GPYRO

The advent of the internet has enabled a new generation of 'open-source' computer programs to be developed which are readily downloadable and can be adapted to the particular modeling needs of users across a range of technical disciplines. One prominent example of such an open-source model in the area of fire modeling is the GPYRO model developed by Lautenberger et al. University of California, Berkeley in collaboration with workers at the National Institute of Standards and Technology, MD, [62- 64].

GPYRO is a code which has been designed to work with the Fire Dynamics Simulator program developed by workers at NIST. It can also be used in isolation to perform simulations under a wide range of modeling conditions specified by the user. Specific phenomena modeled by the GPYRO code include heat and fluid flows as well as gas pressure distribution under both reactive and non-reactive conditions in both porous and non-porous media. Published results for GPYRO have focused on the pyrolysis of a selection of standard fuels under both oxidative and non-oxidative conditions, [62], the oxidative pyrolysis of wood, [63], and the propagation of smouldering in polyurethane foam, [64].

14.4. Modeling of Mechanical Properties: Finite Element Analysis

This section examines the application of conventional finite element analysis to the determination of the residual mechanical strength of structural composites during a fire event. Here we examine how a finite element code can distinguish material strength loss as a result of heat generated by combustion, what the physical models underlying this recognition are, and finally, how they are both explained theoretically and validated experimentally.

In section 14.2 we encountered numerous macroscopic models already in use to model the effect of fire damage on the structural integrity of polymer composite components in a variety of common applications. Such models belong to the fire-endurance category and normally consist of finite element computer programmes, which are primarily designed to deal with mechanical properties, and are normally decoupled from the thermal modeling we have discussed in section 14.3.

However, there are an increasing number of these macroscopic models which are integrating the thermal modeling of such components with the corresponding mechanical behavior, models which can be termed 'thermo-mechanical'. The creation of such models necessarily requires the programmer to distinguish the specific effect of temperature on the strength of a material, something which is difficult enough to achieve for a bulk component in one homogeneous material.

However, when a material is a structural composite of resin and continuous reinforcement such as carbon fiber or woven E-glass, the task of representation by finite element methods is considerably more complicated. Regardless of this difficulty, there are a considerable number of workers who specialize in these types of thermo-mechanical models, [64-88], who have published modeling work specifically in this area for many years. The key concept used in such models is the idea of temperature-dependent material modulus, and how this can be calculated in tandem with through-thickness solid-temperature models generated in the Henderson tradition.

The fundamental statement of the temperature-modulus relationship is given by equation 14.17 [57]:

$$E_{f(R,T)} = \left(\frac{E_{f(0)} + E_{f(R,550)}}{2} - \frac{E_{f(0)} - E_{f(R,550)}}{2} \tanh(\phi(T - T_k)) \right) (R_{c,T})^n \tag{14.17}$$

where, $E_{f(R,T)}$, is the residual flexural modulus of a polymer composite specimen after heat exposure at a given temperature, T, and $E_{f(0)}$ is the room temperature flexural modulus of the undamaged specimen. $E_{f(R,550)}$ is the residual flexural modulus of the specimen after exposure determined at 550°C, φ is a material fitting parameter, T_k is the exposure temperature by which specimens will have lost 50% of their room temperature flexural modulus, $R_{c,T}$, is the remaining mass fraction of combustible matter and n is a dependency index relating flexural modulus retention to mass loss.

In principle, a full temperature field for a polymer composite produced by an advanced thermal model can now be used to generate a predictive field for modulus throughout the corresponding solid, whether the model is one or three dimensional. Such a direct translation of finite difference/volume values for temperature into values for corresponding modulus, do not necessarily require any finite element programming, but the potential exists that these predicted cellular outputs of residual modulus could be adopted as appropriate inputs to a finite difference code.

In one particular example, recent work by Kandare et al. [57], has demonstrated good correlation between the thermal modulus retention predicted by equation 14.17 using a Henderson-type temperature model, and actual flexural modulus data for a selection of burnt polymer composite samples in various states of final decomposition.

Further embellishments to thermo-mechanical models include improved expressions relating solid resin mass loss to pressure change in the context of gas pore formation within the composite during decomposition and volatile formation, one example being a theoretical pressure-decay model by Dimitrienko, [52], (equation 14.18), where p represents the pressure within the laminate at an arbitrary instant during decomposition, and p_0 is the initial pressure within the laminate prior to the beginning of decomposition. The parameters A and E are the respective Arrhenius pre-exponent and activation energy for the decomposition reaction of resin treated as a first order decay. Thus, the Dimitrienko model represents an attempt to directly link the known decomposition rate of the resin to the evolution of pressure within the voids evacuated by the volatilized material, and indirectly provides the potential to link resin decomposition rate directly to other phenomena such as general thermal expansion of the laminate, and ply delamination.

$$\frac{\partial m}{\partial t} = A\left[1 - \frac{p}{p_0}\exp\left(-2E/RT\right)\right]^{0.5} \exp\left(-E/RT\right) \tag{14.18}$$

Specific estimates have been proposed for overpressures developed by gases generated within composite voids and between plies, most notably by Florio and co-workers [43-44]. Pressures as high as 15 atm were initially indicated in calculations by the latter, but lower subsequent figures, (~9 atm), were more readily validated experimentally.

Nevertheless, there remain considerable deviations between the composite pressure decay indicated by experimental data and that predicted according to their models. Moreover, neither of these contributions proposes a direct relationship between pressure gradient and a mass transfer coefficient of the volatile gas mixture out of the laminate and towards the heat / (ignition) source (s), i.e. there is no model for the physical mechanism of gas transport of a sort proposed by Staggs, [8]. Equation 14.18 is rather primarily focused on directly relating solid resin mass loss, and hence material failure, to overpressure, based on the assumption of an accumulating volatile phase, which neglects volatile escape from the structure.

Other related phenomena which have received attention within some thermo-mechanical models are those of thermal expansion and ply delamination within polymer composites during thermal degradation leading to mechanical failure. For example, two equations to describe expansion effects have been proposed by Florio et al. [87], (equations 14.19 and 14.20).

The first represents expansion in the composite prior to the onset of polymer decomposition into char and volatile species i.e. in the initial heating phase of the composite. Equation 14.20 represents the considerably more complicated scenario which exists once polymer decomposition is initiated:

$$\varepsilon(t) = \varepsilon_T - \alpha_v (T - T_\infty) \tag{14.19}$$

$$\frac{\partial \varepsilon(T)}{\partial t} = \alpha_v F \frac{\partial T}{\partial t} + \alpha_c (1 - F) \frac{\partial T}{\partial t} + \frac{\partial}{\partial t}\left(\eta T + \zeta \frac{m}{m_0} \right) \tag{14.20}$$

Here ε_T is the strain of the total composite, and α_v and α_c are the respective thermal expansion coefficients of the virgin resin and char, [1/K], F is the void fraction in the composite, η is the viscosity of the resin, [Pa s], T is the composite temperature, [K], and ζ is a mass-loss related strain coefficient.

McManus and Springer, [88], developed a strain model with similar intent to that of Flory, but constructed differently, (equation 14.21).

$$\varepsilon(t) = S\sigma - \Lambda_{ij} \Delta p + \alpha_{ij} \Delta T + \beta_{ij} \Delta(MC) + \chi_{ij} \Delta v \tag{14.21}$$

Here, the first term represented strain applied on the laminate from an external source, the second the contribution from the increasing pressure due to gas generation and expansion within the laminate pores, while the third and fourth terms respectively modeled thermal expansion due both to the material itself and water of vaporization. The last term modeled the specific strain exerted by the formation of char. The McManus model requires experimental values for each of the parameters used, and has been satisfactorily validated using real observations of intra-ply delamination failure within polymer composites, [88]. It may also be noted that the pressurization within the laminate is not explicitly related to resin mass loss as it is indicated by Dimitrienko, [52], (equation 14.18), so that the potential to modify the second term of equation 14.21 to incorporate a function of resin mass loss could be explored.

Overall, despite the doubtless advances made by authors in the development of representative thermo-mechanical models, there still remain some substantial challenges before a fully predictive and integrated thermo-mechanical solution to the problem of composite failure during combustion can be finalized. Some of the remaining requirements are as follows:

- *Adequately representing mixed modes of failure* i.e. predicting real composite failure which in reality is a hybrid of idealized failure modes such as tension, compression, shear, torsion etc. So far, uni-mode failure models have predominated so that it is particularly desired to develop an integrated model which will encompass simultaneous failure in multiple modes. Such integrated models will ultimately require carefully designed experiments which simulate multi-mode failure in such a way that realistic input data can be generated, and meaningful validation can be achieved. So far, thermo-mechanical models tend to be validated in one particular failure mode, whether flexural, tensile or buckling, which results in tightly fitted empirical temperature-modulus models in that particular failure mode, which will not necessarily apply when another mode operates in parallel.
- *Accurately modeling ply layer delamination,* an objective which will require more truly mechanistic descriptions of de-bonding at the resin-fiber interface as a continuous function of temperature in the presence of fully developed combustion

and charring. At present, models such as those of McManus and Florio are constructed mechanistically using additive terms for various contributing phenomena, which represent each of these phenomena as being equally important; however the relative contributions of each of these phenomena to composite strains has not been determined experimentally, something which would aid the refinement of the model in correctly weighting each of the additive terms in the model. The practical result of fitting an equation such as equation 14.21 to real strain data would be to devise individual weighting coefficients for each of the terms by a type of multi-factorial regression, to produce a hybrid empirical-mechanistic model.

- *Correctly modeling the directionality of strain as a function of ply orientation:* The strains calculated by expansion models are often linear, (Florio), and act in one particular direction only. A fully satisfactory mechanical model, similarly to the thermal model, should have the ability to represent strain acting in all directions through the composite, and, in addition, to recognize the relative contributions of strains associated with different failure mechanisms i.e. those caused by tension, compression, torsion and flexure. Achieving and validating this type of model would require very careful permutation design of experiments to study each particular configuration of failure modes so as to perfect a mixed mode thermo-mechanical failure model taking account of strain vectors in all directions, as functions of a known simultaneous temperature field fluctuating in time.

Thus it can be seen that while there are a range of substantial challenges which must be addressed in order to advance thermo-mechanical modeling, both the diversity of ideas generated and the rate of development in this area are perhaps sufficient to indicate that a fully validated, hybrid thermo-mechanical model of composite thermal response, decomposition, combustion and mechanical failure may soon be achieved.

14.5. Future Challenges in Composite Fire Modeling

Throughout this chapter we have summarized the current state of fire modeling, both at the macroscopic level of 'fire-in-a-room' scenarios, and at the microscopic level where the physico-chemical decomposition, ignition and combustion of the resins within structural polymer composites are examined. It has been seen that many of the methodologies and approaches established within macroscopic modeling are transferrable to predictive microscopic modeling systems. We have also discussed the current availability of academic and commercial fire models in both categories. We have outlined in particular how existing microscopic models can be better utilized and adjusted, with reference to recent published work. Lastly we have examined how microscopic composite fire models can be better promoted and adapted for everyday use by design, safety and fire engineers.

From a survey of existing composite heat transfer models it is clear that there are recognized deficiencies in existing microscopic models which are required to be addressed, namely; a) the absence of a sufficiently exact analytical determination of composite ignition temperature under known fire load, and b) a fully representative quantitative profile of the volatiles released during combustion, (so as to more accurately calculate the aggregate heat of combustion of the volatile mixture). Moreover, there lack sufficient determinations of both the *in-situ* surface convective heat transfer coefficient and reflective radiative surface emissivity for exposed composite surfaces.

In addition to the refinement of purely thermal, heat-transfer modeling of polymer composite degradation, there is also a family of integrated models which combine treatments of both heat transfer and mechanical properties, usually known as *thermo-mechanical* models. Work on such models is focused on developing and validating more adequate relationships between the outputs of heat transfer programs such as through-thickness solid component temperature gradients, and the corresponding results of finite element-type programs designed to calculate residual strengths and modulus within the same calculation space for any specified composite component.

Current work on such models is focused on achieving three principal improvements; a) adequately representing mixed modes of failure i.e. failure under tension, compression, shear, and torsion, b) accurately modeling ply layer delamination, an objective which will require a proper description of de-bonding at the resin-fiber interface and c) correctly modeling the directionality of strain as a function of ply orientation. The strains calculated by expansion models are often linear, (Flory), and act in one particular direction only. A fully satisfactory mechanical model, similarly to the thermal model, should have the ability to represent strain acting in all directions through the composite, as heating, decomposition, and combustion processes progress. Lastly, there exists the potential to link resin mass loss, volatile pressure accumulation and composite strain by means of combining two specific approaches of Florio et al. [87] and McManus and Springer [88].

The integrated modeling of heat transfer, ignition, combustion and mechanical failure of structural composites is an area of active development across many science and engineering disciplines internationally. It combines the efforts of mathematicians, polymer scientists, experimentalists and engineers, and requires genuine ingenuity in predicting the complex and somewhat chaotic natural phenomenon of combustion with accuracy. The particular challenges of predicting the point of ignition and establishing a useful relationship between temperature and residual strength within a structural composite are the greatest challenges involved. However, recent work in this area indicates that a hybrid thermo-mechanical composite failure model may soon be achieved. The achievement of such a model is expected to enhance the ability of polymer chemists to formulate improved resins, and engineers to design improved fire-resistant structures.

REFERENCES

[1] Henderson, J. B.; Wiebelt, J. A.; Tant, M. R. *J. Compos. Mater.* 1985, *vol 19*, 579-595.

[2] Gibson, A. G.; Mouritz, A. P. *In Solid Mechanics and Its Applications*; ISBN: 1402049455143; Gladwell G. M. L.; Ed.; Springer: 2006, 140-168.

[3] Grant, G.; Drysdale, D. *Fire Safety Journal* 1995, *vol 24*, 247-278.

[4] Kuang-Chung, T.; Drysdale, D. *Fire Safety Journal* 2002, *vol 37*, 697-706.

[5] Lyon, R. E.; Quintiere, J. G. *Combust. Flame* 2007, *vol 151*, 551-559.

[6] Lyon, R. E. *Polym. Degrad. Stab.* 1998, *vol 61*, 201-210.

[7] Staggs, J. E. J.; Whitelely, R. H. *Fire Mater.* 1999, *vol 23*, 63-69.

[8] Staggs, J. E. J. *Polym. Degrad. Stab.* 2003, *vol 82*, 297-307.
[9] Staggs, J. E. J. *Polym. Degrad. Stab.* 2001, *vol 74*, 433-439.
[10] Watt, S. D.; Staggs, J. E. J.; McIntosh, A. C.; Brindley, J. *Fire Safety Journal* 2001, *vol 26*, 421-436.
[11] Staggs, J. E. J. *Fire Safety Journal,* 2000, *vol 34*, 69-80.
[12] Staggs, J. E. J. *Fire Safety Journal*, 1999, *vol 32*, 17-34.
[13] Galgano, A.; Di Blasi, C.; Branca, C.; Milella, E. *Polym. Degrad. Stab.* 2009, *vol 94*, 1267-1280.
[14] BS EN 1991-1-2:2002, *Eurocode 1. Actions on Structures Exposed to Fire*, ISBN: 0580408310; 2002.
[15] Bailey C. Website: One-stop Shop in Structural Fire Safety Engineering', University of Manchester, 2010
[16] http://www.mace.manchester.ac.uk/project/research/structures/strucfire/default.htm.
[17] Olenick, S. M.; Carpenter, D. J. *SFPE J. Fire Protect. Eng.* 2003, *vol 13(2),* 87-110.
[18] Danish Institute of Fire Technology, ARGOS User's Guide, Datovej 48, DK 3460, Birkeroed, 1991.
[19] Dietenberger, M. A. Validated Furniture Fire Model with FAST (HEMFAST), NISTGCR-89-564, National Institute of Standards and Technology, June 1989.
[20] Astapenko, V.; Koshmarov, Y.; Molchadski, I.; Shevlyakov, A. Moscow, Stroiizdat, 1988 (from Higher Engineering Fire-Technical School, Moscow).
[21] CFX-5 User Manual, AEA Technology, Harwell, UK, 2000.
[22] Novozhilov, V.; Harvie, D. J. E.; Green, A. R.; Kent, J. H. *Combust. Sci. Technol.* 1997, *vol 123(1–6),* 227–245.
[23] Kameleon FireEx 99 User Manual, SINTEF Energy Research Report TRF5119, Trondheim, Norway, 1999.
[24] Ketchell, N.; Cole, S.; Webber, D. M.; Marriott, C. A.; Stephens, P. J.; Brearley, I. R.; Fraser, J.; Doheny, J.; Smart, J. Institution of Civil Engineers, London, 1993.
[25] Takahashi, K.; Tanaka, T.; Kose, S. *In the Proceedings of the 2nd International Symposium on Fire Safety Science*; International Association for Fire Safety Science, 1988.
[26] Shestopal, V. O.; Grubits, S. J. *In the Proceedings of the 4th International Symposium on Fire Safety Science,* IAFSS, 1994, 625–632.
[27] Evans, D. D.; Stroup, D. W. *Fire Technol.* 1986, *vol 22*, 54–65.
[28] Bjorkman, J.; Huttunen, O.; Kokkala, M. Research Notes 1036, Technical Research Centre of Finland.
[29] Jensen, G.; Lonvik, T.; Heskestad, A. W. *In the Proceedings of the 11th International Conference on Automatic Fire Detection AUBE 1999*, University of Duisburg, Germany, 1999.
[30] Lie, T. T.; Celikkol, B. *ACI Mater. J.* 1991, *vol 88(1),* 84-91.
[31] Dodd, A. B.; Lautenberger, C.; Fernandez-Pello, A. C. *In the Proceedings of the Combustion Institute,* 2009, *vol 32,* 2497.
[32] Sterner, E.; Wickstrom, U. TASEF –Swedish National Testing and Research Institute (SP), SP Report 1990:05, Boras, 1990.
[33] Bresler, B.; Iding, R.; Nizamuddin, Z. FIRES-T3: A Computer Program for the Fire Response of Structure-Thermal (Three-Dimensional Version), UCB FRG 77-15,

University of California, Berkeley, NIST GCR 95-682, National Institute of Standards and Technology, 1996.

[34] Bamford, C. H.; Malan, D. H. *Proc. Camb. Philos. Soc.* 1946, *vol 42*, 166–182.

[35] Munson, T. R.; Spindler, R. J. RAD-TR-61-10, AVCO Corporation; May 1961.

[36] Kung, H. C. *Combust. Flame* 1972, *vol 18*, 185–195.

[37] Murty, K. A. *Combust. Flame* 1972, *vol 18,* 75–83.

[38] Kansa, E. J.; Perlee, H. E.; Chaiken, R. F. *Combust. Flame* 1977, *vol 29*, 311–324.

[39] Fredlund, B. *Fire Safety Journal,* 1993, *vol 20,* 39–69.

[40] Mouritz, A. P.; Feih, S.; Kandare, E.; Mathys, Z. Gibson, A. G.; Des Jardin, P. E.; Case, S. W.; Lattimer, B. Y. *Compos. Part A*, 2009, *vol 40,* 1800–1814.

[41] Tant, M. R.; Henderson, J. B.; Boyer, C. T. *Composites* 1985, *vol 16*, 121–126.

[42] Henderson, J. B.; Doherty, M. P. *High Temp–High Press* 1987, *vol 19*, 95–102.

[43] Henderson, J. B.; Wiecek, T. E. *J. Compos. Mater.* 1987, *vol 21*, 373–393.

[44] Florio, J.; Henderson, J. B.; Test, F. L. *High Temp–High Press* 1989, *vol 21*, 157–165.

[45] Florio, J. Henderson, J. B.; Test, F. L.; Hariharan, R. *Int. J. Heat Mass Transfer* 1991, *vol 34*, 135–147.

[46] Sullivan, R. M.; Salamon, N. J. *Int. J. Eng. Sci.* 1992, *vol 30,* 431–441.

[47] Sullivan, R. M.; Salamon, N. J. *Int. J. Eng. Sci.* 1992, *vol 30,* 939–951.

[48] Sullivan, R. M. *A Finite Element Method for Thermochemically Decomposing Polymers*; PhD Thesis, The Pennsylvania State University, Pennsylvania, 1990.

[49] Sullivan, R. M. *J. Compos. Mater.* 1993, *vol 27*, 408–434.

[50] Pering, G. A.; Farrell, P. V.; Springer, G. S. *J. Compos. Mater.* 1980, *vol 14*, 54–66.

[51] McManus, H. L.; Springer, G. S. *J. Compos. Mater.* 1992, *vol 26*, 206–229.

[52] McManus, H. L.; Springer, G. S. *J. Compos. Mater.* 1992, *vol 26,* 230–255.

[53] Dimitrienko, Y. I. *Composites* 1997, *vol 28A*, 453–461.

[54] Milke, J, Vizzini, A. J. *J. Compos. Technol. Res.* 1991, *vol 13,* 145–151.

[55] Gibson, A. G.; Wu, Y. –S.; Chandler, H. W.; Wilcox, J. A. D.; Bettess, P. *Rev. L'Inst Franc. Petrol* 1995, *vol 50,* 69–74.

[56] Dimitrienko, Y. I. *Combust. Flame* 2000, *vol 122,* 211– 226.

[57] Dimitrienko, Y. I. *Combust. Flame* 1997, *vol 111,* 161– 174.

[58] Kandare, E.; Kandola, B. K.; McCarthy, E. D.; Myler, P. Edwards, G.; Jifeng, Y.; Wang, Y. C. *J. Compos. Mater.* doi: 10.1177/0021998310385024.

[59] Stoliarov, S. I.; Lyon, R. Thermo-kinetic Modeling of Pyrolysis, (Thermokin)', Online at: *www.fire.tc.faa.gov/.../files/Fire.../StoliarovModelPres.pdf*

[60] Stoliarov, S. I.; Crowley, S.; Walters, R. N.; Lyon, R. E. *Combust. Flame* 2010, doi.10.1016/j.combustflame.2010.03.011.

[61] McCarthy, E. D.; Kandola, B. K.; Myler, P. *Paper Presented in the 12th International Conference on Fire Retardant Polymer Materials*, Poznan, 31st August – 3rd September 2009.

[62] Kanury, A. M. *In Combustion Science and Technology Book Series*; ISBN: 067702690; Glassman, I.; Ed.; Gordon and Breach: London, 1975.

[63] Lautenberger C. *A Generalised Pyrolysis Model for Combustible Solids*, PhD Thesis, University of California, Berkeley, Fall 2007.

[64] Lautenberger, C.; Fernandez-Pello A. C. *Combust. Flame* 2008, *vol 156(8),* 1503.

[65] Dao, M.; Asaro, R. J. *Composites* 1999, *vol 30A*, 123–131.

[66] Krysl, P.; Ramroth, W.; Asaro, R. J. *In the Proceedings of the SAMPE Technical Conference and Exhibition;* Long Beach, CA, 16–20th May 2004.

[67] Krysl, P.; Ramroth, W. T.; Steward, L. K.; Asaro, R. J. *Int. J. Numer. Methods Eng.* 2004, *vol 61*, 49–68.

[68] Gu, P, Asaro, R. J. *Compos. Struct.* 2009, *vol 88*, 461–467.

[69] Asaro, R. J.; Lattimer, B.; Ramroth, W. *Compos. Struct.* 2009, *vol 87*, 382–393.

[70] Lua, J.; O'Brien, J. *In the Proceeding of Composites in Fire 3*, Newcastle-upon-Tyne, UK, 9–10th September 2003.

[71] Lua, J.; Key, C. T.; O'Brien, J.; Lopez-Anido, R. A.; El-Chiti, F. W.; Dagher, H, et al. *In the Proceedings of the Joint American Society for Composites/American Society for Testing and Materials Committee D30*, Atlanta, GA, 17–20th October 2004.

[72] Key, C. T.; Lau, J. *Composites* 2006, *vol 37A*, 1005–1014.

[73] Lua, J.; O'Brien, J.; Key, C. T.; Wu, Y.; Lattimer, B. Y. *Composites* 2006, *vol 37A*, 1024–1039.

[74] Lua, J. *Composites* 2007, *vol 38A*, 1019–1037.

[75] Bausano, J. V.; Boyd, S. E.; Lesko, J. J.; Case, S. W. *In the Composites 2004 Convention and Trade Show American Composites Manufacturers Association*; Tampa, Florida, 2004.

[76] Boyd, S. E.; Lesko, J. J.; Case, S. W.; Bausano, J. V. *In the Proceedings of the Composites in Fire – 4*, University of Newcastle upon Tyne, CompositeLink, 2005, 15–16th September 2005.

[77] Boyd, S. E.; Bausano, J. V.; Lesko, J. J.; Case, S. W. *Modeling of Naval Composite Structures in Fire*, In Couchman, L.; Mouritz, A. P. Ed.; Acclaim Printing: Melbourne, 2006.

[78] Bausano, J. V.; Lesko, J. J.; Case, S. W. *Composites* 2006, *vol 37A*, 1092–1100.

[79] Boyd, S. E.; Lesko, J. J.; Case, S. W. *Compos. Sci. Technol.* 2007, *vol 67*, 3187–3195.

[80] Boyd, S. E.; Case, S. W.; Lesko, J. J. *Composites* 2007, *vol 38*, 1462–1472.

[81] Zhang, Z.; Case, S. W.; Lua, J. *In the Proceedings of the 17th International Conference on Composite Materials*, Edinburgh, UK, 27– 31 July 2009.

[82] Summers, P. T.; Lattimer, B. Y.; Feih, S. *In the Proceedings of the 17th International Conference on Composite Materials*, Edinburgh, UK, 27–31 July 2009.

[83] Lui, L.; Holmes, J. W.; Kardomateas, G. A.; Birman, V. *In the Proceedings of Composites in Fire – 4*, University of Newcastle upon Tyne, CompositeLink; 2005, 15–16th September 2005.

[84] Liu, L.; Kardomateas, G. A.; Birman, V.; Holmes, J. W.; Simitses, G. J. *Composites* 2006, *vol 37A*, 972–980.

[85] Gibson, A. G.; Wu, Y.-S.; Evans, J. T.; Mouritz, A. P. *J. Compos. Mater.* 2006, *vol 40*, 639–658.

[86] Feih, S.; Mathys, Z.; Gibson, A. G.; Mouritz, A. P. *Compos. Sci. Technol.* 2007, *vol 67*, 551–564.

[87] Feih, S.; Mathys, Z.; Gibson, A. G.; Mouritz, A. P. *Composites* 2007, *vol 38A,* 2354–2365.

[88] lorio, J.; Henderson, J. B.; Test, F. L. *High Temp–High Press* 1989, *vol 21*, 157–165.

[89] McManus, H. L.; Springer, G. S. *J. Compos. Mater.* 1992, *vol 26*, 206–229.

In: Modeling and Simulation in Fibrous Materials
Editor: Asis Patanaik
ISBN: 978-1-62100-116-4

Chapter 15

APPLICATIONS OF MODELING IN ELECTROSPINNING NANOFIBERS

Valencia Jacobs*

CSIR Materials Science and Manufacturing, Polymers and Composites Competence Area, Port Elizabeth 6000, South Africa
Department of Textile Science, Faculty of Science, Nelson Mandela Metropolitan University, Port Elizabeth, South Africa

ABSTRACT

This chapter reports the applications of statistical modeling in determining optimum parameters during synthesis of nanofibers. Nanofibers of poly ethylene oxide (PEO) and chitosan were obtained by electrospinning following a factorial design which was then used to study the effect of solution and processing parameters, namely, tip-to-collector distance, applied voltage, polyelectrolyte, electric field strength, concentration of chitosan, ratio of solvents - TFA/DCM on fiber diameter. It also encompasses the influence of electrospinning parameters on the structural morphology and diameter of electrospun nanofibers from previous reports.

15.1. INTRODUCTION

With the appearance of nanotechnology, researchers are becoming more interested in understanding the unique properties of nanomaterials. This interest has been elucidated by the promise that these nanosize materials hold in diverse applications. The technique of electrostatic spinning or electrospinning may be traced back over 400 years when Gilbert [1] showed that when a piece of rubbed amber is placed near a water droplet on a smooth surface, a cone can be formed. In 1745, Bose [2] was the first to describe the process of electro-hydrodynamic spraying of fluids and later in 1882, Rayleigh [3] expanded on the field by studying thin liquid jets when placed in electric fields and their stability condition. Early, in

* E-mail: VJacobs@csir.co.za

the past century, the first published work in this field was by Zeleny [4] in 1914. His experiments focused on the causes leading to the eventual break-up of the jet once it has left the capillary in a high electric field. In the succeeding years, more experimental work was performed on the deformation and breaking of water-drops by strong electric fields but no theoretical explanations were given at the time [5-6]. Further research showing that a monodispersity of droplet sizes could be obtained within the range of 1.0–0.1 micrometer was performed by Vonnegut and Neubauer [7]. Their work showed that electro-hydrodynamic spraying was capable of being manipulated within an electric field and as a result, numerous authors ventured into the field [8]. It was not, however, until 1934 to 1944, when Formhals published a series of patents [9–13], describing an experimental setup for the production of polymer filaments using an electrostatic force. In 1969, Taylor [14] derived the condition for the critical electric potential needed to transform the droplet of liquid into a cone (commonly referred to as the Taylor Cone) and to exist in equilibrium under the presence of both electric and surface tension forces as:

$$V^2{}_c = 4\frac{H^2}{L^2}\left(\ln\frac{2L}{R} - 1.5\right)(0.117\pi R\gamma),$$

where Vc is the critical voltage, H is distance between the capillary exit and the ground, L is length of the capillary with radius R, and γ is the surface tension of the liquid. Taylor's further investigations on various viscous solutions lead to establishment that in order to balance the surface tension of the polymer with the electrostatic force, an angle of 49.3 degrees is required.

In the 1980s, the electrospinning process has reclaimed popularity doubtlessly due to a growing interest in nanotechnology, as ultrathin fibrous structures derived from various polymers with diameters down to nanometers can be easily fabricated with this process. During this time, nearly every polymer that was soluble was a potential candidate for the electrospinning process. Many of the early experiments were focused on materials that were soluble in water including PEO due to their low cost and the overall availability of water [15].

Electrospinning is known as a simple and relatively inexpensive process in which electrostatic fields are introduced into a polymer solution to produce nanofibers [16-40]. Due to their large surface to mass ratio and high pore volume, such fibers find a wide application in a variety of areas including wound dressing, filtration, drug delivery, composites, sensors, and others [21-26]. A typical electrospinning setup consists of a high voltage power supply, a spinneret and a metallic collector screen. Polymer solution is loaded into the spinneret that is positioned such that the polymer solution forms a suspended droplet at the nozzle of the spinneret. Electric fields are introduced into the polymer solution and transform the droplet into a conical shape, due electrostatic repulsion. When the electrostatic force overcomes the surface tension of the solution droplet, a jet is formed. The jet travels in straight-line until disturbed by external forces that cause the jet to destabilize and speedily whip in a chaotic motion. The strong thinning of the jet as well as the strong elongated deformation taking place during electrospinning has been attributed to this mode of instability. In the process, the solvent evaporates in air as a result dry nanofibers are deposited on the collector screen resulting in a nonwoven mesh of nanofibers. The process can be adjusted by varying a number of parameters that governs the electrospinning technique. These parameters include solution properties (polymer concentration, molecular weight, viscosity, surface tension, dielectric constant and conductivity), processing parameters (applied voltage, electric field strength, tip-to-collector distance, flow rate) and ambient conditions (temperature and humidity). Thus, the traditional way of studying the influence of one factor at a time is not a correct method in such a highly variable process. This, however, leads to contradictory results as reported in the published literatures [15, 19-26, 41-43] and overlooking the influence of other parameters associated with the process. To deal with such a problem, response surface methodology, which incorporates design of experiments, is a very useful tool [44-45]. The advantage of the above is that not only the individual effect of the parameters can be known qualitatively and quantitatively, the interaction effect of the variables can also be studied.

15.2. Factors Affecting Electrospinning

Though easily realizable in the laboratory, electrospinning is a complex phenomenon to analyze because of the coupling between the electric field and the deformation of the fluid which in turn is determined by the rheology of the material. This section deals with the influence of many parameters like solution properties, processing and ambient conditions, on the transformation of polymer solution into electrospun nanofibers.

15.2.1. Concentration of Polymer Solution

Although the range of concentrations that produce fibers will obviously vary depending on the polymer-solvent system used, the forces of viscosity and surface tension will determine the upper and lower boundaries of processing window, if all other variables are held constant. At low viscosities ($\eta <$ 1 poise), surface tension is the dominant influence on fiber morphology and below a certain concentration, drops will form instead of fibers. At high concentrations ($\eta >$ 20 poise), processing will be hindered by inability to control and maintain the flow of a polymer solution at the tip of the spinneret and by the cohesive nature of the high viscosity [46]. Previous research demonstrates how the nanofiber structural morphology and diameter is influenced by the concentration of polymer solution [26, 46-48]. Demir et al. [26] demonstrated the dependence of the average fiber diameter on solution concentration using a power law relationship for Polyurethane fibers as:

$$d \alpha C^3$$

where d is the diameter of electrospun nanofibers, C is the solution concentration and 3 is the scaling exponent which differs greatly between different polymers, and for the same polymers with different molecules or the same molecules with different properties.

In 2001, Dietzel et al. [46] described the discrepancy in the fiber diameter and morphology to the shape of the jet-originating surface. They discovered that at lower concentrations the fibers took long to dry resulting in varying morphologies on the structure of electrospun nanofibers. At low polymer concentration, defects in the form of beads and droplets have been observed [26, 47-48]. Increasing the solution viscosity by increasing the polymer concentration yields uniform fibers along with the formation of a few beads [46, 49]. Increasing the concentration of a polymer solution can also affect its surface tension [46]. In some other cases (Figure 15.1), a combination of these morphologies may result in beaded fibers due to the influence of lower polymer concentration while smoother nanofibers are formed at higher concentrations.

Other studies revealed that the formation of bead defects in electrospun fibers has been initiated by varying solution surface tension and solution charge density, and by charge neutralization [42]. As attractive characteristics of electrospinning, these structures may also exhibit wide variations in their shapes and surface morphologies [50-51].

15.2.2. Molecular Weight

The length of the polymer chain is determined by the molecular weight of the polymer, which in turn have an effect on the concentration. However, one of the required conditions for electrospinning to take place in order to form fibers is that the solution must contain a polymer of adequate molecular weight and sufficient viscosity [19, 39]. Normally, when a polymer of higher molecular weight is dissolved in a solvent, its viscosity will be higher than the same polymer of lower molecular weight. Jacobs et al. [19] investigated the influence molecular weight on the 5% and 6% aqueous solution of PEO (M_w= 3 x 10^5 g/mol) and PEO (M_w= 9 x 10^5 g/mol).

They observed that the nanofibers obtained from 6% PEO (M_w= 9 x 10^5 g/mol) shows beadless and smooth fibers while those nanofibers from 6% PEO (M_w= 3 x 10^5 g/mol) indicated fibers with beads (Figure 15.2).

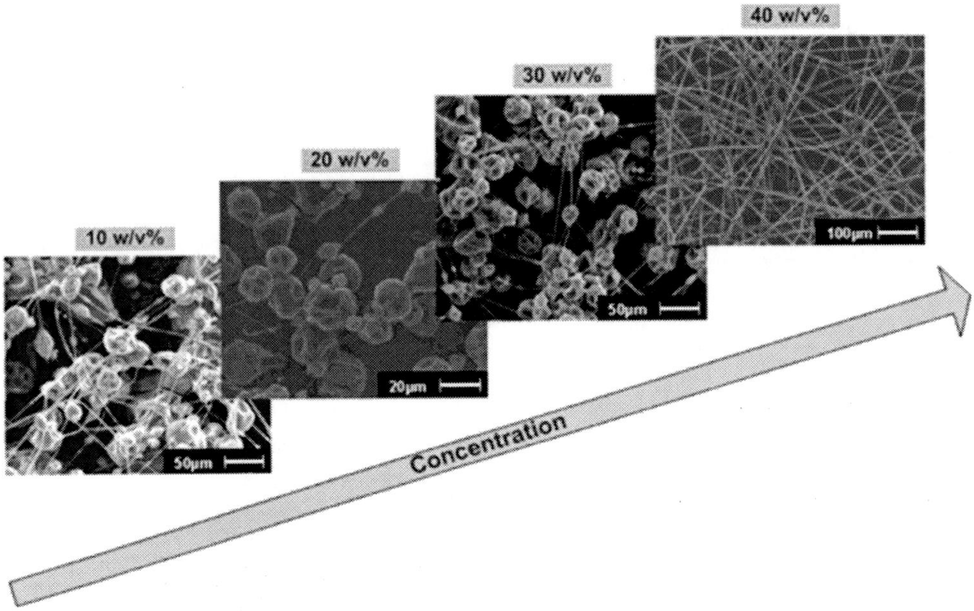

Figure 15.1. SEM micrographs showing the effect of concentration on the electrospun Poly (vinyl phenol) nanofibers from 10, 20, 30 and 40 w/v% at 15cm and applied voltage of 15 kV.

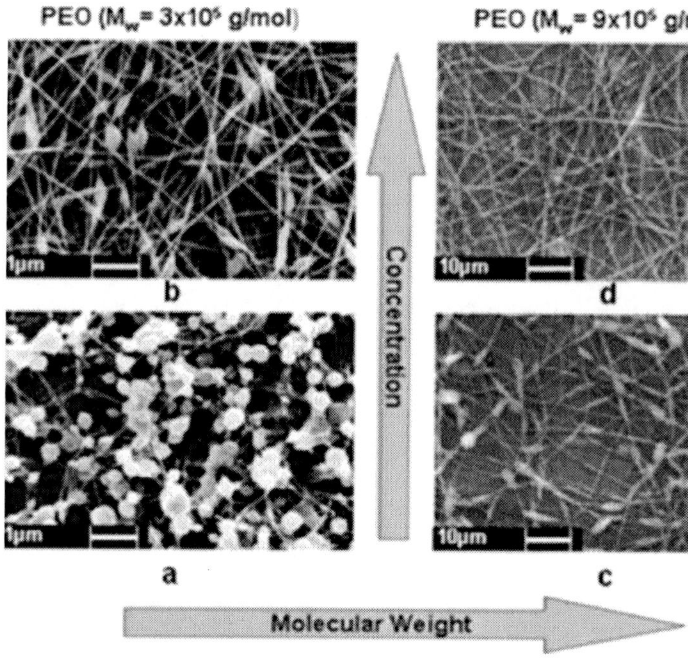

Figure 15.2. SEM micrographs showing the effect of molecular weight on the microstructures of electrospun nanofibers from (a) 5%, (b) 6% PEO (Mw, 3 x 10^5 g/mol) aqueous solution; (c) 5%, (d) 6% PEO (Mw, 9 x 10^5 g/mol) aqueous solution at 15 cm and applied voltage of 10 kV (0.6 kV/cm). All other parameters were kept constant.

To ascertain the effect of polymer molecular weight on the morphology of the nanofibers, a critical chain overlap concentration was determined by Gupta et al. [52] and found that decreased with increasing molecular weight [52]. Researchers have prepared oligomer-sized phospholipids from lecithin solutions into nonwoven membranes through electrospinning. The results revealed that high molecular weights are not always essential for the electrospinning process if sufficient intermolecular interactions can provide a substitute for the inter-chain connectivity obtained through chain entanglements [53-54].

15.2.3. Applied Voltage

Electrospinning is known to be caused by the charges on the jet. These charges can be influenced by electric fields which in turn can affect the jet pathway. Hence many attempts to control the electrospinning process through tempering with electric field, is one of the vital factors. Applied electric field is one of the most important parameters in the electrospinning process due to its direct influence on the dynamics of the fluid flow [39]. In the electrospinning process, a high voltage is introduced into a polymer solution such that charges are induced within the fluid. However, the changes in the applied voltage will be reflected on the shape of the suspending droplet at the nozzle of the spinneret, its surface charge, dripping rate, velocity of the flowing fluid and hence on the structural morphology of electrospun fibers. At introduction of the electrostatic force (+5kV), there would be a deformation of the droplet into a conical shape where the surface tension is dominant (Figure 15.3).

When the applied voltage surpasses the critical value, the electrostatic force overcomes the surface tension of the polymer solution and a jet would be ejected (+10kV). An increase in applied voltage caused a change in the formation of the jet initiating point, and hence the structure and morphology of fibers [46]. They concluded that the change in the spinning current is related to the change in the instability mode. Megelski et al. [51] investigated the voltage dependence on the fiber diameter using polystyrene. They discovered that the fiber size diminished from about 20 μm to 10 μm with an increase in voltage from 5kV to 12kV. Buchko et al. [55] studied electrospinning of silk-like polymer fibers with fibronectin functionality and concluded that there was a decrease in the fiber diameter with an increase in the applied voltage. Further voltage-diameter relationship has been found in the electrospinning of polyurethane fibers [26], polyamide- 6 fibers [48] and PEO fibers [46].

15.2.4. Polyelectrolyte and Solution Conductivity

Since polymeric nanofibers are widely used in many industries such as textiles, composites, filtration, wound dressing, tissue engineering and electronics, the surface properties of these fibers are of utmost importance. The surface properties of such fibers can be modified utilizing various techniques or treatments to suit particular application.

When placed in water or in any other ionizing solvent, a polyelectrolyte has a tendency of dissociating into highly charged polymeric molecule. Since the ionic charges have a direct relation with the electric conductivity of the solution, the addition of a polyelectrolyte increases the charge density on the surface of the ejected jet, leading to an increase in the electric charges carried by the jet [19]. As a result, more stretching and elongation of the jet takes place during electrospinning, and thereby causing reduction in the fiber diameter [56] and less bead formation. Son et al. [57] discovered that the critical voltage for electrospinning to occur is reduced by the addition of ions that increases the conductivity of the solution. Another factor determined by Choi and coworkers [58] is the increased instability of the jet due to inclusion of charged ions in the electrospun polymer solution, thus, leading to increased jet path and finer fibers. Fong et al. [42] concluded that the net charge density carried by the electrospinning jet is another important factor which largely influences the morphology of electrospun products besides the viscosity and the surface tension of the solution. Their experiments showed that beads became smaller and spindle-like with an increase of the net charge density.

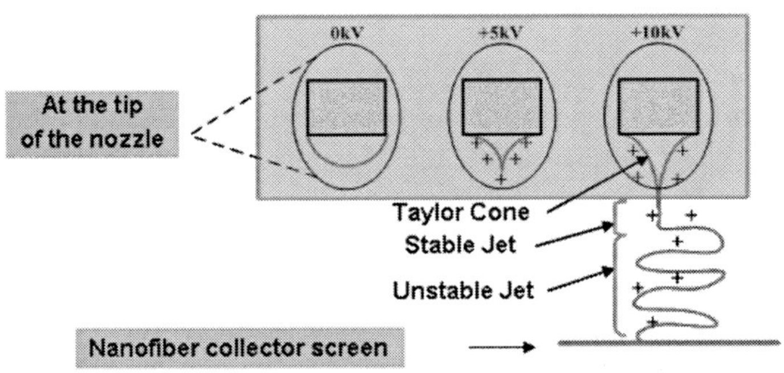

Figure 15.3. Schematic illustration of a Taylor cone formation with increase in applied voltages.

15.2.5. Tip-to-Collector Distance

The distance between the nozzle and the collector has a direct effect on both the flight time of the jet as well as the electric field strength. During electrospinning process, it is important to adjust the tip-to-collector distance in such a way that the jet has enough flight time for a solvent to evaporate so as to prevent the production of wet fiber and possible bead formation. From previous studies on electrospun polyacrylonitrile-iron acetylacetone, Du and Zhang [47] discovered that when the tip-to-collector distance is too small the jet may not have enough time to dry leading to varying structural morphology with spindles. Likewise, when the distance is too long the jet has enough time to dry but with reduced electric field strength non-uniform fibers can be produced. The effect of tip and the collector distance on fiber morphology is not as significant as other parameters and this has been observed with electrospinning of PVA [59], gelatine [60], chitosan [61] and poly (vinylidene fluoride) [62].

15.2.6. Solvent Effect

The type of the solvent used during electrospinning has significant effect on the resultant nanofibers. This is due to some of the parameters associated with the polymer solution such as dielectric effect that governs the formation of electrospun nanofibers. Generally, a solvent with a higher dielectric effect reduces the bead formation and encourages the formation of thinner fibers [39, 57]. Solvents such as dichloromethane [63], chloroform [64], acetic acid [65] may be added to a solution to increase dielectric constant and improve on the structural morphology of electrospun nanofibers.

15.2.7. Feed Rate

Modifying the feed rate of the polymer solution from the nozzle is another way of controlling the structural morphology and diameter of electrospun nanofibers, provided all other parameters are held constant. When the feed rate is increased, the there is corresponding increase in fiber diameter and bead size due to a greater volume that is drawn from nozzle [39, 47, 52]. Few studies have systematically investigated on the relationship between solution feed or flow rate on fiber morphology and size [51, 66]. Previous research indicates that high flow rates result in beaded fibers due to unavailability of proper drying time prior to reaching the collector [65, 67, 68-69].

15.3. Role of Fiber Diameter and Structural Morphology

This section describes the function of two important characteristics of electrospun nanofibers, namely, fiber diameter and structural morphology, on a variety of applications. However, the development of useful applications of the electrospun nanofibers requires a thorough understanding of the electrospinning parameters as the structural morphology and diameter of the electrospun nanofiber will have an influence on the final product. Owing to their unique properties such as large surface area, small pore size and high pore volume, such fibers with substances of different compositions [43, 63] have been widely used for applications ranging from biomedical, filtration, composites, protective clothing, sensors and others.

15.3.1. Biomedical Applications

The usage of electrospun nanofibrous candidates for biomedical applications has attracted much research attention in the past several years. Such material with flexible structural morphology have been demonstrated as suitable substrates for tissue engineering, affinity membranes, wound dressing and controlled drug delivery.

For tissue engineering, the electrospun scaffold should physically resemble the nanofibrous features of extracellular matrix (ECM) with appropriate physicochemical and biological properties [15, 70]. Electrospun nanofibrous nonwoven membranes utilised in wound dressing have pore sizes ranging from 500nm to 1μm which are small enough to prevent exogenous bacterial invasion and for exuding fluid from the wound [15, 71-73]. Owing to their high specific surface area, nanofibrous drug system has a higher overall release rate than the bulk film [66, 74-76] while affinity membrane have a great potential to separate the targeted bio-molecules via ligands attached on the surface of the nanofibrous membrane [39, 77-80].

15.3.2. Filtration Applications

For high filtration efficiency, a fibrous material should have adequate characteristics needed to fulfill influential filter parameters. Such parameters include small diameter and high porosity which makes electrospun nanofibers attractive to the application [80, 81-83]. Further studies revealed that filtration efficiency of air filters; molecular filters and dust filter bags were improved using various polymers [84-85].

15.3.3 Composites Applications

The inclusion of electrospun nanofibrous mats in the composite structures can provide the high modulus and strength to weight ratios, which generally cannot be achieved by other engineered material alone.

Due to large specific surface area and irregular pore structure, nanofibers have the potential to increase the interactions between the fibers and the polymer matrix [86-87]. Dzenis and Reneker [88] investigated unidirectional laminated composites using polybenzimidazole (PBI) nanofibers. They discovered that energy release rate of the material was improved.

Further studies were conducted on the bending performance of nylon 6 electrospun nanofibers reinforced dental resin composites and it was revealed that the addition of 5% weight fraction improved the adhesion between nylon fibers and the matrix polymer [89].

15.3.4. Protective Clothing Applications

Ideal protective clothing requires a lightweight and breathable fabric which is permeable to both air and water vapour, insoluble in all solvents and highly reactive with gases and other deadly chemical agents. With their large specific surface area, high pore density and small pore sizes, electrospun nanofibers are capable of neutralising chemical agents without hindering air and water vapour permeability to the clothing [89-90].

15.3.5. Other Applications

Up until now, electrospun nanofibrous membranes have been studied as chemical sensors, gas sensors, biosensors and optical sensors [91-94]. Ding et al. [95] used quartz crystal microbalances (QCM) sensors coated with poly (acrylic acid)/poly (vinyl alcohol), PAA/PVA electrospun nanofibers and discovered an increase in the sensitivity of QCM sensors. Similar studies were conducted on MoO_3, WO3, NO_2 and TiO_2 nanofibers [96-99]. These nanofibers exhibited an increase in sensitivity when compared to traditional sensors. In some instances, nanofibers are used as batteries [100-103], catalyst [104-110] and enzyme [111-120] carriers, due to their interconnected porous nature, large surface area, and high permeability to reactants.

15.4. Modeling for Predicting Electrospun Nanofiber Diameter

This section deals with the optimization of electrospinning process and solution parameters using statistical modeling to obtain uniform fibers. Two polymers namely, PEO and chitosan, have been used to investigate the interaction effects on electrospinning parameters, respectively.

15.4.1. Optimization of Electrospinning Process

Electrospinning is a highly erratic and much interactive process due to simultaneous influence of multiple variables or parameters. In such a process, studying the influence of one factor at a time is not recommended as this may result in outcomes similar to those reported earlier. Also, other parameters associated with the process might be disregarded. To overcome such a challenge, response surface methodology which takes into cognizance the interaction effect of the variables as well as the individual effect of the parameters can be utilized. Also, the prediction of the optimization region involving various parameters can also be shown. Furthermore, the number of experiment trials required is less compared to studying individual parameter, and therefore cost-effective. In this regard, Box and Behnken factorial design is a very useful tool in dealing with multiple variables acting simultaneously [44]. In the subsequent sections, process and parameter optimization of PEO and chitosan nanofibers are further explored.

The structural morphology and diameter of the nanofibers were determined by ESEM FEI quanta 200 without gold coating and 100 readings were taken for measuring average fiber diameters. Systat 10 was used for the process optimization and the response surface methodology.

15.4.2. Process and Parameter Optimization of PEO Nanofibers

Based on the experimental observations and electro-hydrodynamic theories, mathematical models have been developed by several groups to investigate the electrospinning process. All these studies provide a better understanding of the mechanism responsible for the process of electrostatic spinning. More-so, they may assist the researchers enormously in the design of new setups that may provide a better control over the structural morphology and diameter of electrospun nanofibers. The electrospinning process can be adjusted by varying a number of parameters that governs the technique. These parameters include solution properties (polymer concentration, molecular weight, viscosity, surface tension, dielectric constant and conductivity), processing parameters (applied voltage, electric field strength, tip-to-collector distance, flow rate) and ambient conditions (temperature and humidity). Thus, the traditional way of studying the influence of one factor at a time is not a suitable method in such a highly variable process. This, however, leads to contradictory results as reported in the published literatures [122-123]. Also, the influence of other parameters associated with the process is overlooked in the process. Orthogonal experimental design was utilized in the parameter study of nylon 6 electrospinning. The viscosity of the solution, salt content, and the strength of the electric field were the main parameters determining the diameter of the fibers. The influence of process parameters on the catalytic carbon nanofibers formation using a 2^3 factorial design was reported and it was concluded that the effects of the main parameters along with the effects due to interactions between them are significant and important in the growth of carbon nanofibers.

PEO nanofibers were electrospun by varying distance between the nozzle and collector screen, applied voltage as well as the concentration of the poly allylamine hydrochloride (PAH, a polyelectrolyte). The variation in PAH concentration was considered as the solution parameter due to its influence on the charge density of the polymer solution [17]. Box and Behnken factorial design with three variables (i.e. distance, applied voltage and PAH %) varied at three levels (-1, 0 and +1) was selected for the experimental purpose (Figure 15.4). The interaction effect between concentration of PAH and the distance played the most significant role, followed by the interaction between distance and applied voltage, and last by the applied voltage alone in obtaining uniform nanofibers.

i. Morphology of PEO Nanofibers and Response Surface Function

In all the 15 cases of the experiments, uniform and bead free nanofibers with fiber diameters ranging from 134-192nm were obtained and a less variation in diameters is shown as indicated by the CV%, in brackets (Table 15.1). It shows that accurate selection of the parameters played a crucial role in obtaining uniform nanofibers. Some of the randomly selected images from the SEM for the experimental combination numbers of 3, 7, 8 and 15 are shown in Figure 15.5 which also confirms this aspect.

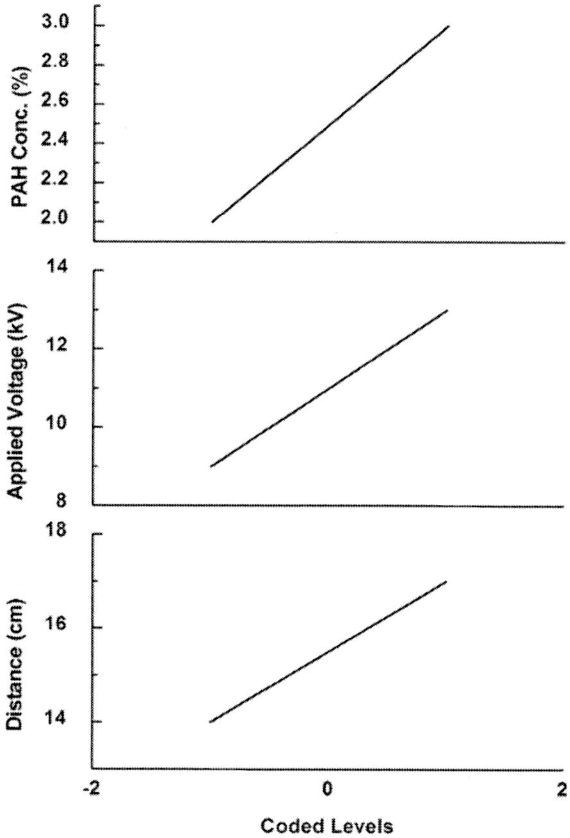

Figure 15.4. Coded levels of variables for the factorial design of experiment.

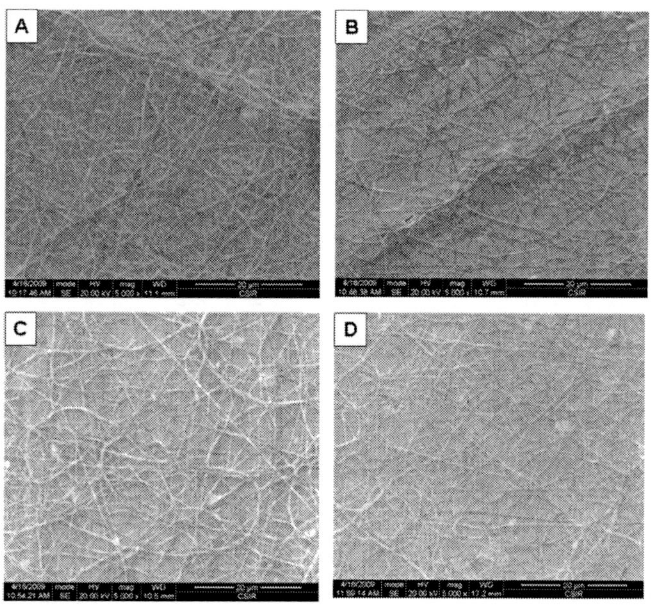

Figure 15.5. SEM images showing the morphology of nanofibers for experimental combination numbers: (a) T3, (b) T7, (c) T8 and (d) T15.

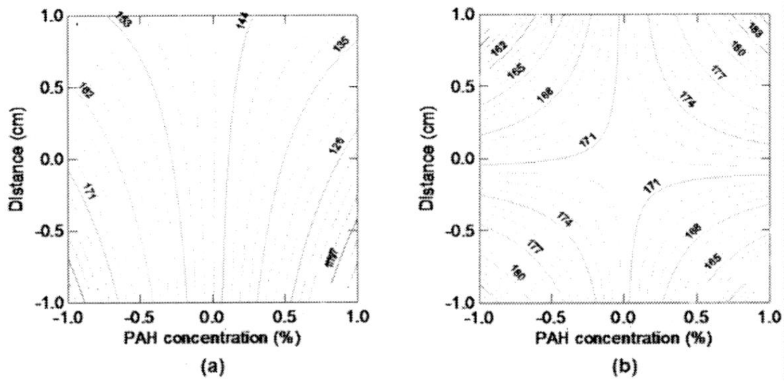

Figure 15.6. Contour plots of interaction effect of levels of PAH concentration and distance on nanofiber diameters for variable voltages: (a) 9 kV and (b) 13 kV.

The coefficient of the parameters (distance, applied voltage, PAH concentration), significance probability (*P*-value) and correlation coefficient were obtained (Table 15.2). The significant test was tested at 95% confidence interval. *P*-value in Table 15.2 is a measure of statistical significance, if it is less than 0.05, then the variable has significant effect on average fiber diameter and vice versa. The interaction between concentration of PAH and the distance played the most dominant role, as shown by its coefficient among all the selected parameters. The coefficient values of the interaction between distance and applied voltage and applied voltage alone were more or less similar. In such a scenario, response surface plots may determine which coefficients play the dominant role, which is discussed in the following sections.

Table 15.1. Coded levels and actual values of variables for different experimental combinations along with average fiber diameter

Experimental combination number	Coded level of variables x_1-----x_2-----x_3			Actual values x_1-----x_2-----x_3			Average fiber diameter (nm)
T1	-1	-1	0	14	9	2.5	178 (11.21)
T2	+1	-1	0	17	9	2.5	134 (14.98)
T3	-1	+1	0	14	13	2.5	159 (11.92)
T4	+1	+1	0	17	13	2.5	164 (14.03)
T5	-1	0	-1	14	11	2	190 (10.66)
T6	+1	0	-1	17	11	2	136 (11.75)
T7	-1	0	+1	14	11	3	157 (12.60)
T8	+1	0	+1	17	11	3	161 (11.30)
T9	0	-1	-1	15.5	9	2	139 (14.45)
T10	0	+1	-1	15.5	13	2	192 (9.73)
T11	0	-1	+1	15.5	9	3	138 (12.22)
T12	0	+1	+1	15.5	13	3	174 (14.78)
T13	0	0	0	15.5	11	2.5	158 (11.30)
T14	0	0	0	15.5	11	2.5	154 (11.77)
T15	0	0	0	15.5	11	2.5	151 (13.20)

*Values in the parenthesis indicate CV% in the measurement of nanofiber diameters.

Table 15.2. Analysis of variance for three variables (distance, applied voltage, PAH concentration), significance probability (*P*-value) and correlation coefficient

Term	Coefficient	*P*-value*
Constant	$C_0 = 159.000$	0.000
Distance: x_1	$C_1 = -11.125$	0.013
Applied voltage: x_2	$C_2 = 12.500$	0.007
Distance × Applied voltage: $x_1.x_2$	$C_{12} = 12.250$	0.040
PAH × Distance: $x_3.x_1$	$C_{31} = 14.500$	0.019

Denotes $R^2 = 0.774$

$$y = C_0 + C_1x_1 + C_2x_2 + C_{12}x_1x_2 + C_{31}x_3x_1.$$

Using response surface methodology, the dependence of mean fiber diameter on distance, applied voltage and PAH concentration was determined as follows:

$$y = 159 - 11.125x_1 + 12.5x_2 + 12.250x_1x_2 + 14.5x_3x_1 \quad (15.1)$$

where y is the average fiber diameter, x_1, x_2 and x_3 are the coded values of distance, applied voltage and PAH concentration, respectively. The square of correlation coefficients between the experimental and calculated values obtained from the response surface equation (R^2) is 0.774. This value indicates a correlation between the selected process and solution parameters in obtaining uniform fiber diameter.

ii. Response Surfaces of PEO Fiber Diameter as a Function of PAH Concentrations and Distances for Different Values of Applied Voltage

The response surface of fiber diameter as a function of different levels of selected parameters and the predicted optimized regions is discussed subsequently along with the changes in the morphology of electrospun nanofibers. Three different cases of parameters have been considered as per their statistical significance and dominance. Consider the interaction effect of levels of PAH concentration and distance on fiber diameters for different values of applied voltage (Figure 15.6). The applied voltage was kept at -1 level (9 kV), 0 level (11 kV) and +1 level (13 kV), respectively.

The response surface equation for average fiber diameter for -1 level of distance is given by substituting $x_2 = -1$ in equation 15.1. Similarly, other two equations were obtained by substituting $x_2 = 0$ and $x_2 = 1$ in equation 15.1. The equations are:

$$\text{For } x_2 = -1,\ y = 146.5 - 23.375 * x_1 + 14.5 * x_3 * x_1 \quad (15.2)$$

$$\text{For } x_2 = 0,\ y = 159 - 11.125 * x_1 + 14.5 * x_3 * x_1 \quad (15.3)$$

$$\text{For } x_2 = +1,\ y = 171.5 + 1.125 * x_1 + 14.5 * x_3 * x_1 \quad (15.4)$$

The contour plots of effect of levels of PAH concentration and distance on fiber diameters for an applied voltage of 9 kV is shown in Figure 15.6.a. The values of average fiber diameter at -1 level, 0 level and +1 level of PAH are 185 nm, 146 nm and 112 nm, respectively. With the increase in levels of PAH concentration from -1 to 0 levels, there is a 21% decrease in fiber diameter.

Further increase in levels of PAH concentration from 0 to +1 level, there is a 23% decrease in fiber diameter. Overall, with the increase in levels of PAH concentration from -1 to +1 level, there is a 39% decrease in fiber diameter, resulting in much thinner fibers. This can be attributed to the increasing charge density carried by the jet with the increasing PAH concentration. Thus, leading to an increase in stretching and thinning of the jet and ultimately making the fiber thinner [12, 17, 124]. Also, Coulombic repulsion force increases with an increase in PAH concentration, further elongating the jet making the fiber diameter smaller.

iii. Response Surfaces of PEO Fiber Diameter as a Function of Distances and Applied Voltages for Different Values of PAH Concentrations

Consider the second case of the interaction effect of levels of distance and applied voltage on fiber diameters for different values of PAH concentration (Figure 15.7). The interaction effect of levels of distance and applied voltage on fiber diameters for a PAH concentrations of 2% with the values of fiber diameter at -1, 0 and +1 levels of distance are 183 nm, 146 nm and 115 nm, respectively.

With the increase in levels of distance from -1 to 0 levels, there is a 20% decrease in fiber diameter and from 0 to +1 levels, there is a 21% decrease in fiber diameter.

The combined influence of distance, applied voltage and PAH concentration on fiber diameter shows that +1 level of distance (17 cm), -1 level of applied voltage (9 kV) and -1 level of PAH concentration (2%) give uniform nanofiber with the lowest diameter of 115 nm. The optimized region for the above selected parameters lies just below this region.

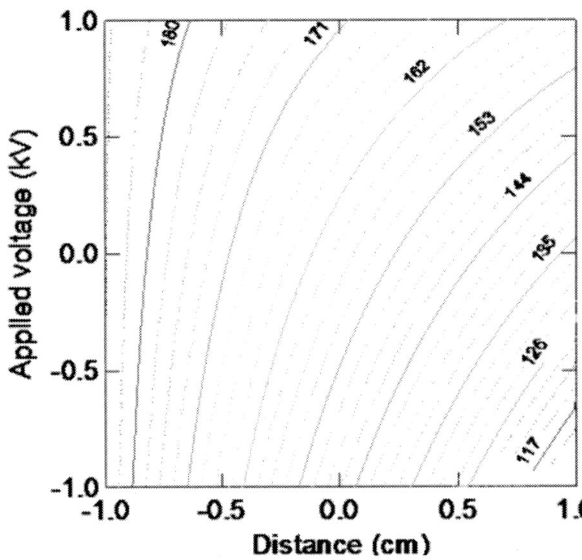

Figure 15.7. Contour plots of interaction effect of levels of distance and applied voltage on nanofiber diameters for 2 % PAH concentrations.

15.4.3. Process and Parameter Optimization of Chitosan Nanofibers

The electrospinning of chitosan is a complex process due to the behaviour of the solution of this polymer such as polycationic specification, high molecular weight, and wide molecular weight distribution. Though no interaction effect of variables reported, there are however, very few studies done on the influence of the electrospinning control parameters that significantly affect the spinnability and the quality of the electrospun chitosan nanofibers. In these studies, only individual effect of parameters is reported. Homayoni et al. [125] investigated the effect of molecular weight, moisture uptake and solvent on the structure and diameter of electrospun chitosan nanofibers. They discovered that the diameter of the optimum product was 140 nm and presented 13.95% moisture uptake, which was 74% higher than that of the polymer powder. Chitosan hydrolyzed for 48 h was spinnable in lower concentrations of acetic acid and the solution of this polymer at 7 and 7.5 wt% in acetic acid 80% and 70% lead to nanofibers with mean diameters of 250 and 284 nm, respectively. Desai and co-workers [126] studied the influence of process and solution parameters (molecular weight, blend ratios, polymer concentration, choice of solvent, and degree of deacetylation of chitosan) and concluded that each of the above-mentioned factors governs the fiber shape and size of electrospun nanofibers. Chitosan from crab shells with 85% deacetylation was prepared from various concentrations and electrospun under different electrospinning conditions. The coded levels of the experimental parameters (Table 15.3), the experimental designs and average fiber diameters at three levels (Table 15.4) were obtained for all 15 samples. The above-mentioned parameters were varied at three levels from -1, 0 and +1 during electrospinning of chitosan nanofibers as per the Box and Behnken factorial experimental design: electric field strength; 0.65, 0.70, 0.75 kV/cm, concentration of chitosan; 5, 5.5, 6 w/v%, ratio of solvents - TFA/DCM; 70/30, 75/25, 80/20 v%. The response surface methodology was applied to predict fiber diameters from the processing parameters.

i. Morphology of Chitosan Nanofibers and Response Surface Function

The electrospun chitosan nanofibers showed both beaded and smooth structural morphology and the fiber diameters range from 169-233nm (Table 15.5).

Response surface morphologies were utilized to analyze the experimental data (Table 15.5) and a second order polynomial equation was fitted by multiple regression analysis [44-45]. The quality of fit for the model was estimated by the coefficients of determination (R^2) and the analysis of variances. After the coefficients were examined and the model was fine-tuned, all the insignificant coefficients were omitted. As a result, the quadratic response model was fitted to the following equation:

$$Y=\beta_0+\sum_{i=1}^{3}\beta_i x_i+\sum_{i=1}^{3}\beta_{ii}x_i^{2}+\sum_{i=1}^{2}\sum_{j=i+1}^{3}\beta_{ij}x_i x_j \qquad (15.5)$$

where Y is the predicted response value, x_i is the ith independent factor. β_0, β_i, β_{ii}, and β_{ij} are regression coefficients. β_0 is the constant term, β_i linear effect term, β_{ii} is squared effect term, and β_{ij} is interaction effect term. By multiple regression analysis, coefficient of the parameters (electric field strength, ratio of solvents - TFA/DCM and concentration of chitosan), significance probability (*P*-value) and R^2 were obtained (Table 15.5). The significance test was conducted at 95% confidence interval. *P*-values as a measure of statistical significance are shown in Table 15.5, if it is less than 0.05, then the variable has significant effect on average fiber diameter and vice versa. The interactions between electric field strength and concentration of chitosan as well as interactions between ratio of solvents - TFA/DCM and electric field strength showed the most significant effect, followed by concentration of chitosan, and lastly by electric field strength as shown by calculated coefficients and P-values. The response surface equation for an average fiber diameter is given by:

$$y = 198.8 - 9.750x_1 + 13.625x_2 + 14x_1x_2 + 14x_3x_1 \tag{15.6}$$

where, y is the average fiber diameter, x_1, x_2 and x_3 are the coded values of electric field strength, concentration of chitosan and ratio of solvents - TFA/DCM, respectively. The square of correlation coefficient (R^2) between the experimental and calculated values obtained from the response surface equation is 0.769. This value indicates a good correlation between the experimental and the predicted values determined form the model.

ii. Response Surfaces of Chitosan Fiber Diameter Function of Electric Field Strength and Ratio of solvents - TFA/DCM for Different Concentrations of Chitosan

The response surface of fiber diameter as a function of stipulated variables are discussed subsequently along with the changes in the structural morphology of nanofibers. However, three different cases have been considered as per their statistical significance in obtaining uniform fiber diameters and dominance starting with the one with the highest significance.

Consider the first case of the effect of levels of electric field strength and ratio of solvents - TFA/DCM on fiber diameters for different levels of chitosan concentration (Figure 15.8). The concentration of chitosan was kept at -1 level (5%), 0 level (5.5%), and +1 level (6%), respectively. The response surface equation for average fiber diameter for -1, 0 and +1 levels of chitosan concentration is given by substituting x_2 = -1 in equation 15.6. Likewise, other two equations were obtained by substituting x_2 = 0 and x_2 = 1 in equation 15.6 as follows:

$$\text{where } x_2 = -1, y = 185.175 - 23.75x_1 + 14x_3x_1 \tag{15.7}$$

$$\text{where } x_2 = 0, y = 198.8 - 9.75x_1 + 14x_3x_1 \tag{15.8}$$

$$\text{where } x_2 = 1, y = 212.425 + 4.25x_1 + 14x_3x_1 \tag{15.9}$$

The contour plots of interaction effect of electric field strength and ratio of solvents - TFA/DCM on fiber diameters at 5% concentration of chitosan is shown in Figures 15.8a and

15.8b. The values of fiber diameter at -1, 0 and +1 levels of electric field strength are 223 nm, 209 nm and 195 nm, respectively. Also, the fiber diameters at -1, 0 and +1 levels of ratio of solvents - TFA/DCM were 223 nm, 186 nm and 150 nm, respectively. However, this plot suggests that the fiber diameter changes are more sensitive to the ratio of solvents - TFA/DCM than the electric field strength. With the increase in electric field strength from -1 to 0 level, there is 6% decrease in fiber diameter and from 0 to +1 level, about 7% decrease in fiber diameter is observed. Overall, with the increase in electric field strength from -1 to +1 level, 13% decrease in fiber diameter results. This is attributed to the increased stretching of the jet at higher charge density as a result of increased electric field strength leading to formation of finer fibers [19, 26, 40].

With the increase in levels of ratio of solvents - TFA/DCM from -1 to 0 level, there is 17% decrease in fiber diameter and from 0 to +1 level, there is further 19% decrease in fiber diameter. Overall, with the increase in levels of ratio of solvents - TFA/DCM from -1 to +1 level, there is about 33% decrease in fiber diameter, thus yielding finer fibers. This may be due to the mixture of solvent (TFA/DCM, 70/30) reported as the best to electrospin chitosan. Owing to its lower dielectric constant and boiling point much lesser than that of water, the presence of DCM in the system increased the rate of evaporation of the solvent, which reduced the excessively strong charge density originated by TFA, thus, resulting in ultrafine fibers. Similarly, the contour plots of interaction effect of levels of electric field strength and ratio of solvents - TFA/DCM on fiber diameters at 6% concentration of chitosan is shown in Figure 15.8.b. With the increase in electric field strength from -1 to +1 level, there is 12% decrease in fiber diameter. This is following a similar trend observed in Figure 15.8.a, where a decrease in fiber diameter as a result of increased electric field strength is elucidated by increased electrostatic force that is encouraging the elongation of the jet, yielding thinner fibers. Also, an increase in ratio of solvents - TFA/DCM from -1 to +1 level indicates a 9% decrease in the diameter of fibers. This might be due to the reasons mentioned above, where a ratio of solvent (TFA/DCM, 70/30) is sufficient enough to balance the evaporation rate by DCM and reducing excessive charge density created by TFA, leading to finer fibers.

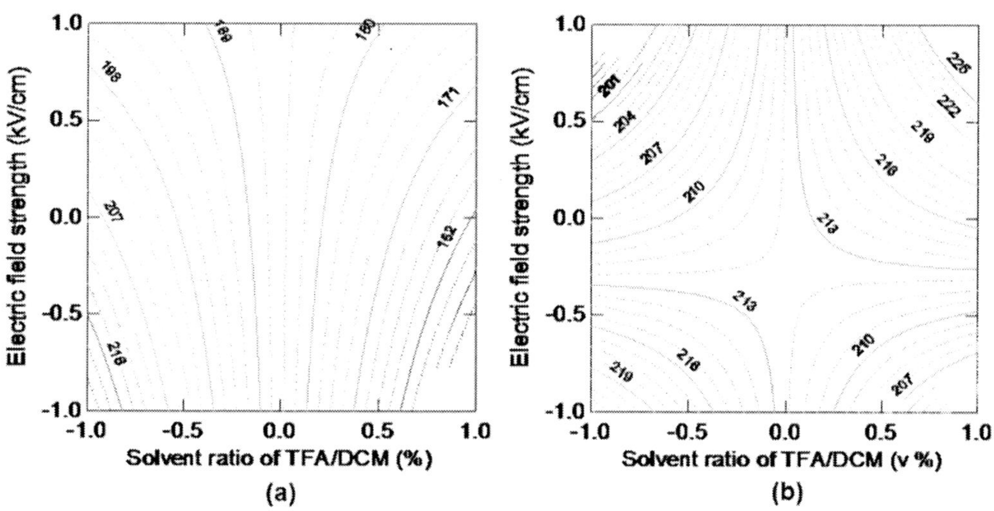

Figure 15.8. Contour plots of interaction effect of levels of (a) and (b), Electric field strength and ratio of solvents - TFA/DCM on nanofiber diameters for variable concentration of chitosan.

Table 15.3. Levels of variables for design of experiment

Variables (or parameters)	Coded levels		
	-1 – 0 - +1		
Electric field strength, x_1 (kV/cm)	0.65	0.70	0.75
Concentration of chitosan, x_2, (%)	5	5.5	6
Ratio of solvents - TFA/DCM x_3, (%)	70/30	75/25	80/20

Table 15.4. Coded levels and actual values of variables for different experimental combinations along with average fiber diameter

Experimental combination number	Coded level of variables x_1 - x_2 - x_3			Actual values x_1 - x_2 - x_3			Average fiber diameter (nm)
P1	-1	-1	0	14.95/23	5	75/25	211
P2	+1	-1	0	20.25/27	5	75/25	169
P3	-1	+1	0	14.95/23	6	75/25	193
P4	+1	+1	0	20.25/27	6	75/25	207
P5	-1	0	-1	14.95/23	5.5	70/30	229
P6	+1	0	-1	20.25/27	5.5	70/30	176
P7	-1	0	+1	14.95/23	5.5	80/20	199
P8	+1	0	+1	20.25/27	5.5	80/20	202
P9	0	-1	-1	17.50/25	5	70/30	189
P10	0	+1	-1	17.50/25	6	70/30	233
P11	0	-1	+1	17.50/25	5	80/20	174
P12	0	+1	+1	17.50/25	6	80/20	219
P13	0	0	0	17.50/25	5.5	75/25	191
P14	0	0	0	17.50/25	5.5	75/25	194
P15	0	0	0	17.50/25	5.5	75/25	196

Table 15.5. Analysis of variance for three variables (electric field strength, concentration of chitosan, ratio of solvents - TFA/DCM), significance probability (*P*-value) and correlation coefficient

Term	Coefficient		*P*-value
Constant	C_0	198.000	0.000
Electric field strength: x_1	C_1	-9.750	0.028
Concentration of chitosan: x_2	C_2	13.625	0.005
Electric field strength × Concentration of chitosan: $x_1.x_2$	C_{12}	14.000	0.026
Ratio of solvents - TFA/DCM × Electric field strength: $x_3.x_1$	C_{31}	14.000	0.026
			R^2
			0.769

iii. Response Surfaces of Chitosan Fiber Diameter Function of Function of Electric Field Strength and Concentration of Chitosan for Different Values of Ratio of solvents - TFA/DCM

Consider the second case of the interaction effect of levels of electric field strength and concentration of chitosan for different values of ratio of solvents - TFA/DCM (Figure 15.9). The ratio of solvents - TFA/DCM was kept at -1 level (70/30), 0 level (75/25) and +1 level (80/20), respectively. The response surface equation of fiber diameter for -1, 0 and +1 levels of solvents - TFA/DCM is derived from substituting x_3= -1, x_3= -1 and x_3= -1 in equation 15.6 as follows:

$$x_3 = -1, y = 198.8 - 23.75x_1 + 13.625x_2 + 14x_1x_2 \quad (15.10)$$

$$x_3 = 0, y = 198.8 - 9.75x_1 + 13.625x_2 + 14x_1x_2 \quad (15.11)$$

$$x_3 = 1, y = 198.8 + 4.75x_1 + 13.625x_2 + 14x_1x_2 \quad (15.12)$$

The contour plots of interaction effect of levels of electric field strength and concentration of chitosan for ratio of solvents - TFA/DCM, 70/30 and 75/25 are displayed in Figure 15.9. The average fiber diameter at -1 level, 0 level and +1 level of electric field strength are 220 nm, 184 nm and 151 nm, respectively. With an increase in levels of electric field strength from -1 to 0 levels, there is a 16% decrease in fiber diameter. Further increase in levels of electric field strength from 0 to +1 levels, there is a 18% decrease in fiber diameter. Overall, with the increase in levels of electric field strength from -1 to +1 level, there is a 31% decrease in fiber diameter, which indicates fibers are getting finer.

This might be due to the reasons mentioned above, that the increasing charge density carried by the jet as a result of increasing electric field strength, encourages the stretching and thinning of the jet and ultimately making the fibers thinner. With an increase in concentration of chitosan from -1 to +1 level, there is no change in fiber diameter (Figure 15.9.a). The cause might be the influence by other two parameters namely, electric field strength and ratio of solvents - TFA/DCM.

Similar trend is observed for Figure 15.9.b, with the interaction effect of levels of electric field strength and concentration of chitosan for different values of ratio of solvents - TFA/DCM at -1 level (70/30), 0 level (75/25) and +1 level (80/20), respectively. With an increase in levels of electric field strength from -1 to 0 levels, there is a 12% decrease in fiber diameter. Further increase in levels of electric field strength from 0 to +1 levels, there is a 12% decrease in fiber diameter.

Overall, with the increase in levels of electric field strength from -1 to +1 level, there is a 22% decrease in fiber diameter. The results are concurrent with the results in Figure 15.8.a and Figure 15.8.b as discussed above. This is attributed to the coulombic repulsive forces in the jet, stretching the viscoelastic solution. Increase in electric field strength leads to an increase in charge density, causing the jet to accelerate faster encouraging more stretching and production of thinner fibers. Again, the increase in concentration of chitosan from -1 to

+1 level, indicates no change in fiber diameter (Figure 15.9.b). This might be due electric field strength and ratio of solvents - TFA/DCM.

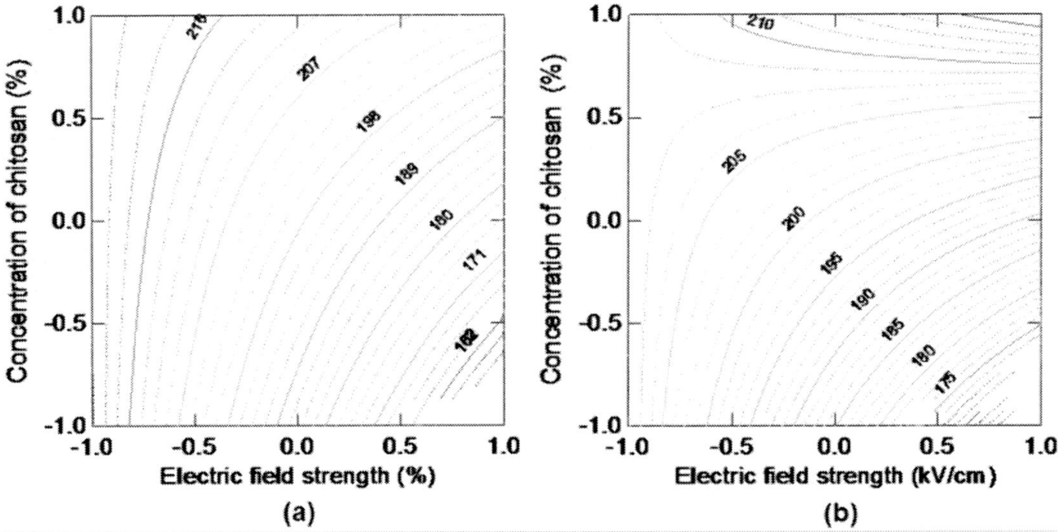

Figure 15.9. Contour plots of interaction effect of levels of (a) and (b), Electric field strength and concentration of chitosan on nanofiber diameters for variable ratio of solvents - TFA/DCM.

15.5. CONCLUSION

In this chapter, the fundamentals of electrospinning process, the formation of electrospun nanofibers and their potential applications have been discussed. As is evident from the aforementioned discussions on the applications of nanofibers, the potential of nanofibrous materials in advanced applications is unlimited. Researchers are making constant efforts to exploit the high surface area and porosity properties of electrospun nanofibers to develop value added and sophisticated high-tech materials. The influence of electrospinning processing and parameters on the structural morphology and diameter of electrospun nanofibers, have been explored. Further optimization of these parameters has been undertaken using response surface methodology to obtain uniform smooth fibers. From the statistical modeling approach, a processing window which can be used as a basis for the optimization of process and solution parameters in electrospinning PEO and chitosan has been identified. For PEO, the combined influences of PAH concentration, distance and applied voltage on fiber diameter gave uniform nanofiber with the lowest diameter of 112 nm. Similarly, for chitosan, the combined influence of electric field strength, ratio of solvents - TFA/DCM and concentration of chitosan on fiber diameter uniform nanofibers with the lowest diameter of 176 nm. This study showed that the interaction between the different variables played a significant role, rather than one particular parameter in obtaining uniform nanofibers. It opens up a new direction in optimizing the electrospinning process to precisely control the fiber diameter from the selected parameters.

15.6. Future Trend and Challenges

Although electrospun nanofibrous membranes have been shown to be commercially viable as air filtration, continuous research in electrospinning for other applications may lead to more products in the near future. With the advancement in the electrospinning technique and the increasing performance of the electrospun nanofibers, this technology may find niche market in various applications. To fulfil the requirements of these applications; the electrospinning parameters and conditions should be regulated properly to deter the nanofiber morphology from being damaged. In some other instances, some polymers do not posses any specific functional groups and therefore their surfaces have to be specifically functionalised for successful applications. These processes however, may pose challenges as the costs of the electrospinning technology may be elevated. To overcome such hurdles, researchers in the nanofiber-based research may have to collaborate, to bring the cost down. Also researchers need to use modeling and simulation tools for a better understanding of the process, which in turn will be helpful in predicting the desired properties.

References

[1] Gilbert, W. *de Magnete*. 1600.

[2] Bose, G. M. *Recherches sur la cause et sur la veritable theorie del'electricite*, Wittenberg, 1745.

[3] Raleigh, X. L. *Philos. Mag.* 1884, 184.

[4] Zeleny, J. *Phys. Rev.* 1914, *vol 3,* 69-91.

[5] Macky, W. A. *Proc. Roy. Soc. A.* 1931, *vol 133,* 565-587.

[6] Nolan, J. *J. Proc. Roy. Irish Acad.* 1926, *vol 37A*, 28-39.

[7] Vonnegut, B.; Neubauer, R. L. *J. Colloid Sci.* 1952, *vol 7*, 616-622.

[8] Lyons, J.; Ko, F. *Polymer News* 2005, *vol 30*, 170-178.

[9] Formhals, A. *US Patent*, 1,975,504, 1934.

[10] Formhals, A. *US Patent*, 2,160,962, 1939.

[11] Formhals, A. *US Patent*, 2,187,306, 1940.

[12] Formhals, A. *US Patent*, 2,323,025, 1943.

[13] Formhals, A. *US Patent*, 2,349,950, 1944.

[14] Taylor, G. I. *Proc. Roy. Soc. London*, 1969, *vol A313,* 453-475.

[15] Huang, Z. M.; Zhang, Y. Z.; Kotaki, M.; Ramakrishna, S. *Compos. Sci. Technol.* 2003, vol *63,* 2223-2253.

[16] Patanaik, A.; Anandjiwala, R. D.; Rengasamy, R. S.; Pal, H.; Ghosh, A. *Text. Prog.* 2007, *vol 39*, 67-120.

[17] Patanaik, A.; Jacobs, V.; Anandjiwala, R. D. *Paper Presented in 86th Textile Institute World Conference*, Hong Kong, 18-22nd November, 2008.

[18] Theron, S. A.; Yarin, A. L.; Zussman, E.; Kroll, E. *Polymer* 2005, *vol 46,* 2889–2899.

[19] Jacobs, V.; Anandjiwala, R. D.; Maaza, M. *J. Appl. Polym. Sci.* 2010, *vol 115,* 3130-3136.

[20] Patanaik, A.; Jacobs, V.; Anandjiwala, R. D. *J. Membr. Sci.* 2010, *vol 352,* 136-142.

[21] Subbiah, T.; Bhat, G. S.; Tock, R. W.; Parameswaran, S.; Ramkumar, S. S. *J. Appl. Polym. Sci.* 2005, *vol 96,* 557-569.
[22] Agarwal, S.; Wendorff, J. H.; Greiner, A. *Adv. Mater.* 2009, *vol 21,* 3343-3351.
[23] Agarwal, S.; Wendorff, J. H.; Greiner, A. *Polymer* 2008, *vol 49,* 5603-5621.
[24] Ding, B.; Wang, M.; Yu, J.; Sun, G. *Sensors* 2009, *vol 9,* 1609-1624.
[25] Li, X.; Ding, B.; Lin, J.; Yu, J.; Sun, G. *J. Phys. Chem. C.* 2009, *vol 113,* 20452-20457.
[26] Demir, M. M.; Yilgor, I.; Yilgor, E.; Erman, B. *Polymer* 2002, *vol 43,* 3303-3309.
[27] MacDiarmid, A. G.; Jones, W. E.; Norris, I. D.; Gao, J. A.; Johnson, T.; Pinto, N. J.; Hone, J.; Han, B.; Ko, F. K.; Okuzaki, H.; Laguno, M. *Synth. Met.* 2001, *vol 199,* 27-30.
[28] Yun, M. H.; Myung, N. V.; Vasquez, R. P.; Lee, C. S.; Menke, E.; Penner, R. M. *Nano Lett.* 2004, *vol 4,* 419-422.
[29] Li, M.; Guo,Y.; Wei, Y.; MacDiarmid, A. G.; Lelkes, P. Y. *Biomaterials* 2006, *vol 27,* 2705-2715.
[30] Wnek, G. E.; Carr, M. E.; Simpson, D. G.; Bowlin, G. I. *Nano Lett.* 2003, *vol 3,* 213-216.
[31] Schreuder-Gibson, H.; Gibson, P.; Senecal, K.; Sennett, M.; Walker, J.; Yeomans, W.; Ziegler, D.; Tsai, P. P. *J. Adv. Mater.* 2002, *vol 34,* 44-45.
[32] Gibson, P.; Gibson, H. S.; Rivin, D. *Colloids Surf. A* 2001, *vol 469,* 187–188,
[33] Kim, C.; Yang, K. S. *Appl. Phys Lett.* 2003, *vol 83,* 1216-1218.
[34] Choi, S. W.; Jo, S. M.; Lee, W. S.; Kim, Y. R. *Adv. Mater.* 2003, *vol 15,* 2027-2032.
[35] Liu, H.; Kameoka, J.; Czaplewski, D. A.; Craighead, H. G. *Nano Lett.* 2004, *vol 4,* 671-675.
[36] Aussawasathien, D.; Dong, J. H.; Dai, L. *Synth. Met.* 2005, *vol 154,* 37-40.
[37] Manesh, K. M.; Gopalan, A. I.; Lee, K. P.; Santhosh, P.; Song, K. D.; Lee, D. D. *IEEE Trans. Nanotechnol.* 2007, *vol 6,* 513-518.
[38] Dong, H.; Prasad, S.; Nyame, V.; Jones, W. E. Jr. *Chem. Mater.* 2004, *vol 16,* 371-373.
[39] Ramakrishina, S.; Fujihara, K.; Teo, W. E.; Lim, T. C.; Ma, Z. *An Introduction to Electrospinning of Nanofibers*; ISBN: 9812564152; World Scientific: Singapore, 2005.
[40] Jacobs V.; Patanaik A.; Anandjiwala R. D. *Paper Presented in 1st International Biomaterials-Africa*, Pretoria, South Africa, 20-22nd September, 2009.
[41] Reneker, D. H.; Chun, I. *Nanotechnology* 1996, *vol 7,* 216-223.
[42] Fong, H.; Chun, I.; Reneker, D. H. *Polymer* 1999, *vol 40,* 4585–4592.
[43] Lin, T.; Wang, H.; Wang, H.; Wang, X. *Nanotechnology* 2004, *vol 15,* 1375-1381.
[44] Box, G. E. P.; Behnken, D. W. *Technometrics* 1960, *vol 2,* 455-475.
[45] Montgomery, D. C. *Design and Analysis of Experiments*; ISBN- 10:0471316490; John Wiley & Sons: New York, 2002, 5th Edn, pp 451-454.
[46] Deitzel, J. M.; Kleinmeyer, J.; Harris, D.; Beck Tan, N. C. *Polymer* 2001, *vol 42,* 261–272.
[47] Du, J.; Zhang, X. *J. Appl. Polym. Sci.* 2008, *vol 109,* 2935–2941.
[48] Mit-Uppatham, C.; Nithitanakul, M.; Supaphol, P. *Macromol. Chem. Phys.* 2004, *vol 205,* 2327–2338.
[49] Jarusuwannapoom, T.; Hongroijanawiwat, W.; Jitjaicham, S.; Wannatong, L.; Nithitanakul, M.; Pattamaprom, C.; Koombhongse, P.; Rangkupan, R.; Supaphol, P. *Eur. Polym. J.* 2005, *vol 41,* 409–421.

[50] Koombhongse, S.; Liu, W.; Reneker, D. H. *J. Polym. Sci. Part B: Polym. Phys.* 2001, *vol 39,* 2598-2606.
[51] Megelski, S.; Stephens, J. S.; Rabolt, J. F.; Chase, D. B. *Macromolecules* 2002, *vol 35,* 8456-8466.
[52] Gupta, P.; Elkins, C.; Long, T. E.; Wilkes, G. L. *Polymer* 2005, *vol 46,* 4799–4810.
[53] Burger, C.; Hsiao, B. S.; Chu, B. *Annu. Rev. Mater. Res.* 2006, *vol 36,* 333–368.
[54] Mckee, M. G.; Wilkes, G. L.; Colby, R. H.; Long, T. E. *Macromolecules* 2004, *vol 37,* 1760–1767.
[55] Buchko, C. J.; Chen, L. C.; Yu, S.; Martin, D. C. *Polymer* 1999, *vol 40,* 7397-7407.
[56] Zong, X.; Ran, S.; Fang, D.; Hsiao, B. S.; Chu, B. *Polymer* 2002, *vol 43,* 4403-4412.
[57] Son, W. K.; Lee, J. H.; Youk, S. C.; Park, W. H. *J. Polym. Sci. B-Polym. Phys.* 2004, *vol 42,* 5-11.
[58] Choi, J. S.; Lee, S. W.; Jeong, L.; Bae, S. H.; Min, B. C.; Youk, J. H.; Park W. H. *Macromolecules* 2004, *vol 34,* 249-256.
[59] Zhang, C.; Yuan, X.; Wu, L.; Han, Y.; Sheng, J. *Eur. Polym. J.* 2005, *vol 41,* 423–32.
[60] Ki, C. S.; Baek, D. H.; Gang. K. D.; Lee, K. H.; Um, I. C.; Park, Y. H. *Polymer* 2005, *vol 46,* 5094–5102.
[61] Geng, X.; Kwon, O. H.; Jang, J. *Biomaterials* 2005, *vol 26,* 5427–32.
[62] Zhao, Z. Z.; Li, J. Q.; Yuan, X. Y.; Li, X.; Zhang, Y. Y.; Sheng, J. *J. Appl. Polym. Sci.* 2005, *vol 97,* 466–474.
[63] Yang, Q. B.; Li, Z. Y.; Hong, Y. L.; Zhao, Y.Y.; Qiu, S. L.; Wang, C.; Wei, Y. *J. Polym. Sci. Pol. Phys.* 2004, *vol 42,* 3721-3726.
[64] Berkland, C.; Pack, D. W.; Kim, K. *Biomaterial* 2004, *vol 25,* 5649-5658.
[65] Wannatong, L.; Sirivat, A.; Supaphol, P. *Polym. Int.* 2004, *vol 53,* 1851-1859.
[66] Zong, X.; Ran, S.; Fang, D.; Hsiao, B. S.; Chu, B. *Polymer* 2002, *vol 43*, 4403-4412.
[67] Yuan, X. Y.; Zhang, Y. Y.; Dong, C. H.; Sheng, J. *Polym. Int.* 2004, *vol 53,* 1704–1710.
[68] Kim, K. H.; Jeong, L.; Park, H. N.; Shin, S. Y.; Park, W. H.; Lee, S. C. *J. Biotechnol.* 2005, *vol 120,* 327–339.
[69] Zuo, W. W.; Zhu, M. F.; Yang, W.; Yu, H.; Chen, Y. M.; Zhang, Y. *Polym. Eng. Sci.* 2005, *vol 45*, 704–709.
[70] Nerem, R. M.; Saltzman, A. *Tissue Eng.* 1995, *vol 1,* 3-13.
[71] Hong, K. H. *Polym. Eng. Sci.* 2007, *vol 47,* 43-49.
[72] Jia, J.; Duan, Y. Y.; Wang, S. H. *J. US-China Med. Sci.* 2007, *vol 4,* 52-54.
[73] Ignatova, M.; Manolova, N.; Rashkov, I. *Eur. Polym. J.* 2007, *vol 43,* 1609-1623.
[74] Zeng, J.; Xu, X.; Chen, X.; Liang, Q.; Bian, X.; Yang, L.; Jing, X. *J. Control. Release* 2003, *vol 92,* 227-231.
[75] Verreck, G.; Chun, I.; Rosenblatt, J.; Peeters, J.; Van Dijck, A.; Mensch, J.; Noppe, M.; Brewster, M. E. *J. Control. Release* 2003, *vol 92,* 349-360.
[76] Zussman, E.; Yarin, A. L.; Weihs, D. A. *Expt. Fluid.* 2002, *vol 33,* 315-320.
[77] Roper, D. K.; Lightfoot, E. N. *J. Chromatogra. A*, 1995, *vol 702,* 3-26.
[78] Ma, Z.; Kotaki, M.; Inai, R. *Tissue Eng.* 2005, *vol 11,* 101-109.
[79] Stankus. J. J.; Guan, J.; Wagner, W. R. *J. Biomed. Mater. Res. Part A*, 2004, *vol 70A,* 603-614.
[80] Gopal, R.; Kaur, S.; Ma, Z.; Chan, C.; Ramakrishna, S.; Matsuura, T. *J. Membr. Sci.* 2006, *vol 281,*581-586.

[81] Schreuder-Gibson, H.; Gibson, P.; Wadsworth, L.; Hemphill, S.; Vontorcik, *J. Adv. Filtr. Sep. Technol.* 2002, *vol 15,* 525-537.

[82] Groitzsch, D.; Fahrbach, E. *US Patent* 4,618,524, 1986.

[83] Wang, X.; Chen, X.; Yoon, K.; Fang, D.; Hsiao, B. S.; Chu, B. *Environ. Sci. Technol.* 2005, *vol 39,* 7684-7691.

[84] Maus, R.; Goppelsroder, A.; Umhauer, H. *Atmos. Environ.* 1997, *vol 31,* 2305-2310.

[85] Tsaia, P. P.; Schreuder-Gibson, H.; Gibson, P. *J. Electrostatics* 2002, *vol 54,* 333–341.

[86] Bergshoef, M. M.; Vansco, G. *J. Adv. Mater.* 1999, *vol 11,* 1362-1365.

[87] Fang, J.; Niu H.; Lin, T.; Wang, X. *Chinese Sci. Bull.* 2008, *vol 53,* 2265-2286.

[88] Dzenis, Y. A.; Reneker, D. H. *US Patent*, 6,265,33, 2001.

[89] Smith, D.; Reneker, D. H.; Schreuder, G. H.; Mello, C.; Sennett, M.; Gibson, P. *US Patent Application*, *PTC/US00/27776*, 2001.

[90] Graham, K.; Gogins, M.; Schreuder-Gibson, H. *Int. Nonwovens J.* 2004, *vol 13,* 21-27.

[91] Wang, X.; Drew, C.; Lee, S. H. *Nano Lett.* 2002, *vol 2,* 1273-1275.

[92] Lee, S. H.; Ku, B. C.; Wang, X.; Samuelson, L. A.; Kumar, J. *Mater. Res. Soc. Sym. Proc.* 2002. *vol 708,* 403-408.

[93] Kwoun, S. J.; Leo, R. M.; Han, B.; Ko, F. K. *Int. Freq. Control Sym. Exhb.* 2000, 52-57.

[94] Celin, S. M.; Pandit, M.; Kapoor, J. C.; Sharma, R. K. *Surf. Sci. Rep.* 2003, *vol 24,* 3-54.

[95] Ding, B.; Kim, J.; Miyazaki, Y.; Shiratori, S. *Sensor Actuat B-Chem.* 2004, *vol 101,* 373-380.

[96] Gouma, P. I. *Rev. Adv. Mater. Sci.* 2003, *vol 5,* 147-154.

[97] Wang, G.; Ji, Y.; Huang, X. *J Phys Chem B* 2006, *vol 110,* 23777-23778.

[98] Sawicka, K. M.; Prasad, A. K.; Gouma, P. I. *Sens. Lett.* 2005, *vol 3,* 31-35.

[99] Kim, I. D.; Rothschild, A.; Lee, B. H. *Nano Lett.* 2006, *vol 6,* 2009-2013.

[100] Kim, J. R.; Choi, S. W.; Jo, S. M. *J. Electrochem. Soc.* 2005, *vol 152,* A295- A300.

[101] Li, X.; Cheruvally, G.; Kim, J. K. *J. Power Sources* 2007, *vol 167,* 491—498.

[102] Kim, J. K.; Cheruvally, G.; Choi, J. W. *J. Electrochem. Soc.* 2007, *vol 154,* A839-A843.

[103] Choi, S. W.; Kim, J. R.; Jo, S. M. *J. Electrochem. Soc.* 2005, *vol. 152,* A989-A995.

[104] Demir, M. M.; Gulgun, M. A.; Menceloglu, Y. Z. *Macromolecules* 2004, *vol 37,* 1787-1792.

[105] Yu, J.; Liu, T. *Acta Polym. Sin.* 2007, *vol 6,* 514-518.

[106] Patel, A. C.; Li, S.; Wang, C. *Chem. Mater.* 2007, *vol 19,* 1231-1238.

[107] Stasiak, M.; Studer, A.; Greiner, A. *Chem. A. Eur. J.* 2007, *vol 13,* 6150-6156.

[108] Chen, L.; Bromberg, L.; Hatton, T. A. *Polymer* 2007, *vol 48,* 4675-4682.

[109] Zhan, S.; Chen, D.; Jiao, X. *J. Phys. Chem. B* 2006, *vol 110,* 11199-11204.

[110] Jin, M.; Zhang, X.; Emeline, A. V. *Chem. Commun.* 2006, *vol 43,* 4483-4485.

[111] Matatov-Meytal, Y.; Sheintuch, M. *Appl. Catal. A*, 2002, *vol 231,* 1-16.

[112] Li, S. F.; Chen, J. P.; Wu, W. T. *J. Mol. Catal. B: Enzym.* 2007, *vol 47,* 117-124.

[113] Huang, X. J.; Ge, D.; Xu, Z. K. *Eur. Polym J.* 2007, *vol 43,* 3710-3718.

[114] Ye, P.; Xu, Z. K.; Wu, J.; Innocent, C.; Seta, P. *Macromolecules* 2006, *vol 39,* 1041-1045.

[115] Lee, K. H.; Ki, C. S.; Baek, D. H. *Fibers Polym.* 2005, *vol 6,* 181-185.

[116] Jia, H.; Zhu, G.; Vugrinovich, B. *Biotech. Prog.* 2002, *vol 18,* 1027-1032.

[117] Kim, T. G.; Park, T. G. *Biotech Prog*, 2006, *vol 22,* 1108-1113.
[118] Kim, B. C.; Nair, S.; Kim, J. *Nanotechnology* 2005, *vol 16,* 382-388.
[119] Wang, Z. G.; Xu, Z. K.; Wan, L. S. *Macromol. Rapid. Commun.* 2006, *vol 27,* 516-521.
[120] Wang, Z. G.; Ke, B. B.; Xu, Z. K. *Biotech. Bioeng.* 2007, *vol 97,* 708-720.
[121] Tan. S. H.; Inai, R.; Kotaki, S.; Ramakrishna, S. *Polymer* 2005, *vol 46,* 6128-6134.
[122] Gu, S. Y.; Ren, J.; Vancso, G. J. *Eur. Polym. J.* 2005, *vol 41*, 2559-2568.
[123] Biber, E.; Gündüz, G.; Mavis, B.; Colak, U. *Appl. Phys. A,* 2010, *vol 99,* 477-487.
[124] Supaphol, P.; Mit-Uppatham, C.; Nithitanakul, M. *J. Polym. Sci. B Polym. Phys.* 2005, *vol 43,* 3699-3712.
[125] Homayoni, H.; Ravandi, S. A. H.; Valizadeh, M. *Carbohy. Polym.* 2009, *vol 77,* 656-661.
[126] Desai, K.; Kit, K.; Li, J.; Zivanovic, S. *Biomacromolecules* 2008, *vol 9,* 1000-1006.
[127] Li, L.; Hsieh, Y. L. *Polymer* 2005, *vol 46,* 5133-5139.

INDEX

A

B

C

D

E

F

G

H

I

J

K

L

M

N

O

P

Q

R

S

T

U

V

W

Y

Z